MINING GRAPH DATA

BICENTENNIAL

1807

WILEY

2007

BICENTENNIAL

THE WILEY BICENTENNIAL–KNOWLEDGE FOR GENERATIONS

$\mathcal{E}$ach generation has its unique needs and aspirations. When Charles Wiley first opened his small printing shop in lower Manhattan in 1807, it was a generation of boundless potential searching for an identity. And we were there, helping to define a new American literary tradition. Over half a century later, in the midst of the Second Industrial Revolution, it was a generation focused on building the future. Once again, we were there, supplying the critical scientific, technical, and engineering knowledge that helped frame the world. Throughout the 20th Century, and into the new millennium, nations began to reach out beyond their own borders and a new international community was born. Wiley was there, expanding its operations around the world to enable a global exchange of ideas, opinions, and know-how.

For 200 years, Wiley has been an integral part of each generation's journey, enabling the flow of information and understanding necessary to meet their needs and fulfill their aspirations. Today, bold new technologies are changing the way we live and learn. Wiley will be there, providing you the must-have knowledge you need to imagine new worlds, new possibilities, and new opportunities.

Generations come and go, but you can always count on Wiley to provide you the knowledge you need, when and where you need it!

WILLIAM J. PESCE
PRESIDENT AND CHIEF EXECUTIVE OFFICER

PETER BOOTH WILEY
CHAIRMAN OF THE BOARD

MINING GRAPH DATA

EDITED BY

Diane J. Cook
School of Electrical Engineering and Computer Science
Washington State University
Pullman, Washington

Lawrence B. Holder
School of Electrical Engineering and Computer Science
Washington State University
Pullman, Washington

WILEY-INTERSCIENCE

A JOHN WILEY & SONS, INC., PUBLICATION

Library of Congress Cataloging-in-Publication Data

Mining graph data / edited by Diane J. Cook, Lawrence B. Holder.
    p. cm.
    Includes index.
    ISBN-13 978-0-471-73190-0
    ISBN-10 0-471-73190-0 (cloth)
  1. Data mining. 2. Data structures (Computer science) 3. Graphic methods.
I. Cook, Diane J., 1963- II. Holder, Lawrence B., 1964-
    QA76.9.D343M52   2006
    005.74—dc22

                                                            2006012632

10 9 8 7 6 5 4 3

To Abby and Ryan, with our love.

CONTENTS

PREFACE

Data mining, or **knowledge discovery in databases**, is a large area of study and is populated with numerous theoretical and practical textbooks. In this book, we take a focused and comprehensive look at one topic within this field: *mining data that is represented as a graph*. We attempt to cover the full breadth of the topic, including graph manipulation, visualization, and representation, mining techniques for graph data, and application of these ideas to problems of current interest.

The book is divided into three parts. Part I, Graphs, offers an introduction to basic graph terminology and techniques. In Part II, Mining Techniques, we take a detailed look at computational techniques for extracting patterns from graph data. These techniques provide an overview of the state of the art in frequent substructure mining, link analysis, graph kernels, and graph grammars. Part III, Applications, describes application of mining techniques to four graph-based application domains: chemical graphs, bioinformatics data, Web graphs, and social networks.

The book is targeted toward graduate students, faculty, and researchers from industry and academia who have some familiarity with basic computer science and data mining concepts. The book is designed so that individuals with no background in analyzing graph data can learn how to represent the data as graphs, extract patterns or concepts from the data, and see how researchers apply the methodologies to real datasets.

For those readers who would like to experiment with the techniques found in this book or test their own ideas on graph data, we have set up a Web page for the book at http://www.eecs.wsu.edu/MGD. This site contains additional information on current techniques for mining graph data. Links are also given to implementations of the techniques described in this book, as well as graph datasets that can be used for testing new or existing algorithms.

With the advent of and continued prospect for large databases containing relational and graphical information, the discovery of knowledge in such data is an important challenge to the scientific and industrial communities. Fielded applications for mining graph data from real-world domains has the potential to make significant contributions of new knowledge. We hope that this book accelerates progress toward meeting this challenge.

ACKNOWLEDGMENTS

We would like to acknowledge and thank the many people who contributed to this book. All of the authors were very willing to help and contributed excellent material to the book. The creation of this book also initiated collaborations that will continue to further the state of the art in mining graph data. We would also like to thank Whitney Lesch and Paul Petralia at Wiley for their assistance in assembling the book and to thank the faculty and staff at the University of Texas at Arlington and at Washington State University for their continued encouragement and support of our work. Finally, we would like to thank our children, Abby and Ryan, for the joy they bring to our lives and for forcing us to talk about topics other than graphs at home.

CONTRIBUTORS

Indrajit Bhattacharya University of Maryland
College Park, Maryland

Horst Bunke Institute of Computer Science and Applied Mathematics
University of Bern
Bern, Switzerland

Deepayan Chakrabarti Yahoo! Research
Sunnyvale, California

Diane J. Cook School of Electrical Engineering and Computer Science
Washington State University
Pullman, Washington

Walter Didimo Dipartimento di Ingegneria Elettronica e dell'Informazione
Università degli Studi di Perugia
Perugia, Italy

Christos Faloutsos School of Computer Science
Carnegie Mellon University
Pittsburgh, Pennsylvania

Thomas Gärtner Fraunhofer AIS
Schloß Birlinghoven
Sankt Augustin, Germany

Lise Getoor University of Maryland
College Park, Maryland

David Gibson IBM Almaden Research Center
San Jose, California

Seth A. Grennblatt 21st Century Technologies, Inc.
Austin, Texas

Jiawei Han University of Illinois at Urbana-Champaign
Urbana-Champaign, Illinois

Lawrence B. Holder School of Electrical Engineering and Computer Science
Washington State University
Pullman, Washington

Tamás Horváth Fraunhofer AIS
Schloß Birlinghoven
Sankt Augustin, Germany

Takahiko Ito NARA Institute of Science and Technology
Ikoma, Nara, Japan

Istvan Jonyer Department of Computer Science
Oklahoma State University
Stillwater, Oklahoma

George Karypis Department of Computer Science & Engineering
University of Minnesota
Minneapolis, Minnesota

Nikhil Ketkar School of Electrical Engineering and Computer Science
Washington State University
Pullman, Washington

Ravi Kumar Yahoo! Research, Inc.
Santa Clara, California

Michihiro Kuramochi Department of Computer Science & Engineering
University of Minnesota
Minneapolis, Minnesota

Quoc V. Le Statistical Machine Learning Program
NICTA and ANU Canberra
Canberra, Australia

Giuseppe Liotta Dipartimento di Ingegneria Elettronica e dell'Informazione
Università degli Studi di Perugia
Perugia, Italy

Michel Liquière LIRMM
Montpellier, France

Sherry E. Marcus 21st Century Technologies, Inc.
Austin, Texas

Kevin S. McCurley Google, Inc.
Mountain View, California

Akira Mogi Institute of Scientific and Industrial Research
Osaka University
Osaka, Japan

Hiroshi Motoda Institute of Scientific and Industrial Research
Osaka University
Osaka, Japan

Melanie Moy 21st Century Technologies, Inc.
Austin, Texas

Michel Neuhaus Institute of Computer Science and Applied Mathematics
University of Bern
Bern, Switzerland

Phu Chien Nguyen Institute of Scientific and Industrial Research
Osaka University
Osaka, Japan

Kouzou Ohara Institute of Scientific and Industrial Research
Osaka University
Osaka, Japan

Takashi Okada Department of Informatics
School of Science & Engineering
Kwansei Gakuin University
Sanda, Japan

Masashi Shimbo NARA Institute of Science and Technology
Ikoma, Nara, Japan

Alex J. Smola Statistical Machine Learning Program
NICTA and ANU Canberra
Canberra, Australia

Andrew Tomkins Google, Inc.
Santa Clara, California

Takashi Washio Institute of Scientific and Industrial Research
Osaka University
Osaka, Japan

Stefan Wrobel Fraunhofer AIS
Schloß Birlinghoven
Sankt Augustin, Germany
and
Department of Computer Science III
University of Bonn,
Bonn Germany

Xifeng Yan Department of Computer Science
University of Illinois at Urbana-Champaign
Urbana-Champaign, Illinois

Mohammed Zaki Department of Computer Science
Rensselaer Polytechnic Institute
Troy, New York

1

INTRODUCTION

LAWRENCE B. HOLDER AND DIANE J. COOK

School of Electrical Engineering and Computer Science
Washington State University, Pullman, Washington

The ability to mine data to extract useful knowledge has become one of the most important challenges in government, industry, and scientific communities. Much success has been achieved when the data to be mined represents a set of independent entities and their attributes, for example, customer transactions. However, in most domains, there is interesting knowledge to be mined from the relationships between entities. This relational knowledge may take many forms from periodic patterns of transactions to complicated structural patterns of interrelated transactions. Extracting such knowledge requires the data to be represented in a form that not only captures the relational information but supports efficient and effective mining of this data and comprehensibility of the resulting knowledge. Relational databases and first-order logic are two popular representations for relational data, but neither has sufficiently supported the data mining process.

The graph representation, that is, a collection of nodes and links between nodes, does support all aspects of the relational data mining process. As one of the most general forms of data representation, the graph easily represents entities, their attributes, and their relationships to other entities. Section 1.2 describes several diverse domains and how graphs can be used to represent the domain. Because one entity can be arbitrarily related to other entities, relational databases and logic have difficulty organizing the data to support efficient traversal of the relational links.

Graph representations typically store each entity's relations with the entity. Finally, relational database and logic representations do not support direct visualization of data and knowledge. In fact, relational information stored in this way is typically converted to a graph form for visualization. Using a graph for representing the data and the mined knowledge supports direct visualization and increased comprehensibility of the knowledge. Therefore, mining graph data is one of the most promising approaches to extracting knowledge from relational data.

These factors have not gone unnoticed in the data mining research community. Over the past few years research on mining graph data has steadily increased. A brief survey of the major data mining conferences, such as the Conference on Knowledge Discovery and Data Mining (KDD), the SIAM Conference on Data Mining, and the IEEE Conference on Data Mining, has shown that the number of papers related to mining graph data has grown from 0 in the late 1990s to 40 in 2005. In addition, several annual workshops have been organized around this theme, including the KDD workshop on Link Analysis and Group Detection, the KDD workshop on Multi-Relational Data Mining, and the European Workshop on Mining Graphs, Trees and Sequences. This increasing focus has clearly indicated the importance of research on mining graph data.

Given the importance of the problem and the increased research activity in the field, a collection of representative work on mining graph data was needed to provide a single reference to this work and some organization and cross fertilization to the various topics within the field. In the remainder of this introduction we first provide some terminology from the field of mining graph data. We then discuss some of the representational issues by looking at actual representations in several important domains. Finally, we provide an overview of the remaining chapters in the book.

1.1 TERMINOLOGY

Data mining is the extraction of novel and useful knowledge from data. A *graph* is a set of nodes and links (or vertices and edges), where the nodes and/or links can have arbitrary labels, and the links can be directed or undirected (implying an ordered or unordered relation). Therefore, *mining graph data*, sometimes called *graph-based data mining*, is the extraction of novel and useful knowledge from a graph representation of data. In general, the data can take many forms from a single, time-varying real number to a complex interconnection of entities and relationships. While graphs can represent this entire spectrum of data, they are typically used only when relationships are crucial to the domain. The most natural form of knowledge that can be extracted from graphs is also a graph. Therefore, the *knowledge*, sometimes referred to as *patterns*, mined from the data are typically expressed as graphs, which may be subgraphs of the graphical data, or more abstract expressions of the trends reflected in the data. Chapter 2 provides more precise definitions of graphs and the typical operations performed by graph-based data mining algorithms.

While data mining has become somewhat synonymous with finding frequent patterns in transactional data, the more general term of *knowledge discovery* encompasses this and other tasks as well. *Discovery* or *unsupervised learning* includes not only the task of finding patterns in a set of transactions but also the task of finding possibly overlapping patterns in one large graph. Discovery also encompasses the task of *clustering*, which attempts to describe all the data by identifying categories or clusters sharing common patterns of attributes and relationships. Clustering can also extract relationships between clusters, resulting in a hierarchical or taxonomic organization over the clusters found in the data. In contrast, *supervised learning* is the task of extracting patterns that distinguish one set of graphs from another. These sets are typically called the positive examples and negative examples. These sets of examples can contain several graph transactions or one large graph. The objective is to find a graphical pattern that appears often in the positive examples but not in the negative examples. Such a pattern can be used to predict the class (positive or negative) of new examples. The last graph mining task is the visualization of the discovered knowledge. *Graph visualization* is the rendering of the nodes, links, and labels of a graph in a way that promotes easier understanding by humans of the concepts represented by the graph.

All of the above graph mining tasks are described within the chapters of this book, and we provide an overview of the chapters in Section 1.3. However, an additional motivation for the work in this book is the important application domains and how their data is represented as a graph to support mining. In the next section we describe three domains whose data is naturally represented as a graph and in which graph mining has been successful.

1.2 GRAPH DATABASES

Three domains that epitomize the tasks of mining graph data are the Internet Movie Database, the Mutagenesis dataset, and the World Wide Web. We describe several graph representations for the data in these domains and survey work on mining graph data in these domains. These databases may also serve as a benchmark set of problems for comparing and contrasting different graph-based data mining methods.

1.2.1 The Internet Movie Database

The Internet Movie Database (IMDb) [41] maintains a large database of movie and television information. The information is freely available through online queries, and the database can also be downloaded for in-depth analysis. This database emerged from newsgroups in the early 1990s, such as rec.arts.movies, and has now become a commercial entity that serves approximately 65 million accesses each month.

Currently, the IMDb has information on 468,305 titles and 1,868,610 people in the business. The database includes filmographies for actors, directors, writers, composers, producers, and editors as well as movie information such as titles, release dates, production companies and countries, plot summaries, reviews and ratings, alternative names, genres, and awards.

Given such filmography information, a number of mining tasks can be performed. Some of these mining tasks exploit the unstructured components of the data. For example, Chaovalit and Zhou [9] use text-based reviews to distinguish well-accepted from poorly accepted movies. Additional information can be used to provide recommendations to individuals of movies they will likely enjoy. Melville et al. [33] combine IMDb movie information (title, director, cast, genre, plot summary, keywords, user comments, reviews, awards) with movie ratings from EachMovie [14] to predict items that will be of interest to individuals. Vozalis and Margaritis [42] combine movie information, ratings from the GroupLens dataset [37], and demographic information to perform a similar recommendation task. In both of these cases, movie and user information is treated as a set of independent, unstructured attributes.

By representing movie information as a graph, relationships between movies, people, and attributes can be captured and included in the analysis. Figure 1.1(a) shows one possible representation of information related to a single movie. This hub topology represents each movie as a vertex, with links to attributes describing the movie. Similar graphs could be constructed for each person as well. With this representation, one task we can perform is to answer the following question:

What commonalities can we find among movies in the database?

Using a frequent subgraph discovery algorithm, subgraphs that appear in a large fraction of the movie graphs can be reported. These algorithms may report discoveries such as movies receiving awards often come from the same small set of studios [as shown in Fig. 1.1(b)] or certain director/composer pairs work together frequently [as shown in Fig. 1.1(c)].

By connecting people, movies, and other objects that have relationships to each other, a single connected graph can be constructed. For example, Figure 1.2 shows how different movies may have actors, directors, and studios in common. Similarly,

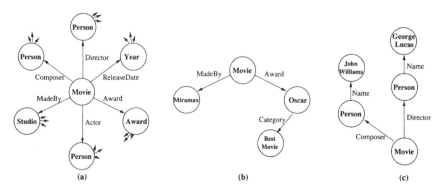

Figure 1.1. (a) Possible graph representation for information related to a single movie. (b) One possible frequent subgraph. (c) Another possible frequent subgraph.

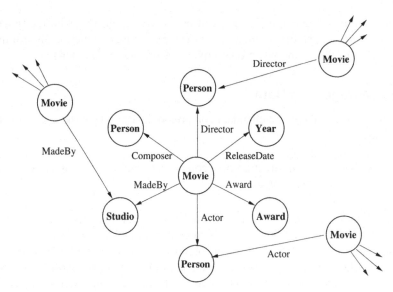

Figure 1.2. Second graph representation in which relationships between data points are represented using labeled edges.

different actors may appear in the same movie, forming a relationship between these people. Analysis of this connected graph may answer questions such as:

What common relationships can we find between objects in the database?

For the movie graph, a discovery algorithm may find a recurring pattern that movies made by the same studio frequently also have the same producer. Jensen and Neville [21] mention another type of discovery that can be made from a connected graph. In this case, an emerging film star may be characterized in the graph by a sequence of successful movies in which he or she stars and by winning one or more awards.

Other analyses can be made regarding the topology of such graphs. For example, Ravasz and Barabasi [36] analyzed a graph constructed by linking actors appearing in the same movie and found that the graph has a distinct hierarchical topology. Movie graphs can also be used to perform classification. As an example, Jensen and Neville [21] use information in a movie graph as shown in Figure 1.2 to predict whether a movie will make more than $2 million in its opening weekend. In a separate study, they use structure around nominated and nonnominated movies to predict which new movies will be nominated for awards [32].

These examples show that patterns can be learned from structural information that is explicitly provided. However, missing structure can also be inferred from this data. Getoor et al.'s [16] approach learns a graph linking actors and movies using IMDb information together with demographic information based on actor ZIP codes. Mining algorithms can be used to infer missing links in the movie graph. For example, given information about a collection of people who starred together in a movie, link completion [17, 28] can be used to determine who the remaining individuals are who starred in the same movie. Such link completion algorithms can also be used to determine when one movie is a remake of another [21].

1.2.2 Mutagenesis Data

The Mutagenesis dataset is a chemical compound dataset where the task is explicitly defined as:

> Given the molecular structure of a compound, identify the compound as mutagenic or not mutagenic.

The Mutagenesis dataset was collected to identify mutagenic activity in chemical data [10]. Mutation is a structural alteration in DNA (deoxyribonucleic acid). Occasionally, a mutation improves an organism's chance of surviving or has no observable effect. In most cases, however, such DNA changes harm human health. A high correlation is also observed between mutagenicity and carcinogenicity. Some chemical compounds are known to cause frequent mutations. Mutagenicity cannot be practically determined for every compound using biological experiments, so accurate evaluation of mutagenic activity from chemical structure is very desirable.

Structure–activity relationships (SARs) relate biological activity with molecular structure. The importance of SARs to drug design is well established. The Mutagenesis problem focuses on obtaining SARs that describe the mutagenicity of nitroaromatic compounds or organic compounds composed of NO or NO_2 groups attached to rings of carbon atoms. Analyzing relationships between mutagenic activity and molecular structure is of great interest because highly mutagenic nitroaromatic compounds are carcinogenic.

The Mutagenesis dataset collected by Debnath et al. [10] consists of the molecular structure of 230 compounds, such as the one shown in Figure 1.3. Of these compounds, 138 are labeled as mutagenic and 92 are labeled nonmutagenic. Each compound is described by its constituent atoms, bonds, atom and bond types, and partial charges on atoms. In addition, the hydrophobicity of the compound ($\log P$), the energy level of the compound's lowest unoccupied molecular orbital (LUMO), a Boolean attribute identifying compounds with three or more benzyl rings (I1), and a

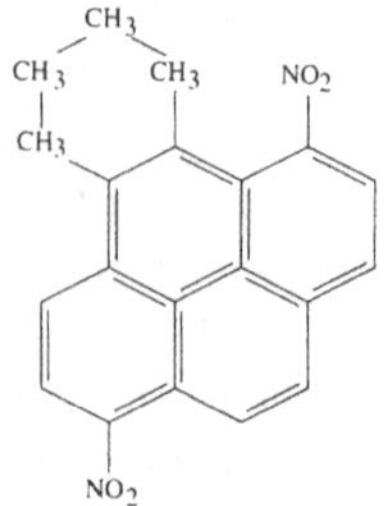

Figure 1.3. 1,6,-Dinitro-9,10,11,12-tetrahydrobenzo[*e*]pyrene.

Boolean attribute identifying compounds that are acenthryles (Ia). The mutagenicity of the compounds has been determined using the Ames test [1]. While alternative datasets are being considered by the community as challenges for structural data mining [29], the Mutagenesis dataset provides both a representative case for graph representations of chemical data and an ongoing challenge for researchers in the data mining community.

Some work has focused on analyzing these chemical compounds using global, nonstructural descriptors such as molecular weight, ionization potential, and various physiocochemical properties [2, 19]. More recently, researchers have used inductive logic programming (ILP) techniques to encode additional relational information about the compounds and to infuse the discovery process with background knowledge and high-level chemical concepts such as the definitions of methyl groups and nitro groups [23, 40]. In fact, Srinivasan and King in a separate study [38] show that traditional classification approaches such as linear regression improve dramatically in classification accuracy when enhanced with structural descriptors identified by ILP techniques.

Inductive logic programming methods face some limitations because of the explicit encoding of structural information and the prohibitive size of the search space [27]. Graphs provide a natural representation for the structural information contained in chemical compounds. A common mining task for the Mutagenesis data, therefore, is to represent each compound as a separate graph and look for frequent substructures in these graphs. Analysis of these graphs may answer the following question:

> What commonalities exist in mutagenic or non-mutagenic compounds that will help us to understand the data?

This question has been addressed by researchers with notable success [5, 20]. A related question has been addressed as well [11, 22]:

> What commonalities exist in mutagenic or nonmutagenic compounds that will help us to learn concepts to distinguish the two classes?

An interesting twist on this task has been offered by Deshpande et al. [12], who do not use the substructure discovery algorithm to perform classification but instead use frequency of discovered subgraphs in the compounds to form feature vectors that are then fed to a Support Vector Machine classifier.

Many of the graph templates used for the Mutagenesis and other chemical structure datasets employ a similar representation. Vertices correspond to atoms and edges represent bonds. The vertex label is the atom type and the edge label is the bond type. Alternatively, separate vertices can be used to represent attributes of the atoms and the bonds, as shown in Figure 1.4. In this case information about the atom's chemical element, charge, and type (whether it is part of an aromatic ring) is given along with attributes of the bond such as type (single, double, triple) and relative three-dimensional (3D) orientation. Compound attributes including log

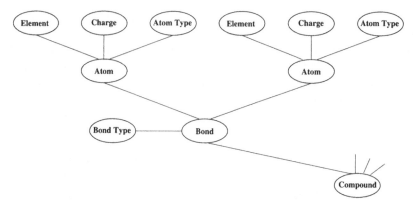

Figure 1.4. Graph representation for a chemical compound.

P, LUMO, I1, and Ia can be attached to the vertex representing the entire chemical compound.

When performing a more in-depth analysis of the data, researchers often augment the graph representation with additional features. The types of features that are added are reflective of the type of discoveries that are desired. Ketkar et al. [22], for example, add inequality relationships between atom charge values with the goal of identifying value ranges in the concept description. In Chapter 14 of this book, Okada provides many more descriptive features that can be considered.

1.2.3 Web Data

The World Wide Web is a valuable information resource that is complex, dynamically evolving, and rich in structure. Mining the Web is a research area that is almost as old as the Web itself. Although Etzioni coined the term "Web mining" [15] to refer to extracting information from Web documents and services, the types of information that can be extracted are so varied that this has been refined to three classes of mining tasks: Web content mining, Web structure mining, and Web usage mining [26].

Web content mining algorithms attempt to answer the following question:

What patterns can I find in the content of Web pages?

The most common approach to answering this question is to perform mining of the content that is found within each page on the Web. This content typically consists of text occasionally supplemented with HTML tags [8, 43]. Using text mining techniques, the discovered patterns facilitate classification of Web pages and Web querying [4, 34, 44].

When structure is added to Web data in the form of hyperlinks, analysts can then perform Web structure mining. In a Web graph, vertices represent Web pages and edges represent links between the Web pages. The vertices can optionally be labeled by the domain name (as Kuramochi and Karypis describe in Chapter 6), and edges are typically unlabeled or labeled with a uniform tag. Additional vertices can be attached to the Web page nodes that are labeled with keywords or other textual information found in the Web page content. Figure 1.5 shows a sample graph of this type for a collection of three Web pages. With the inclusion of this hypertext information, Web page classification can be performed based on structure alone (Gartner and co-workers describe this in Chapter 11 of this book) or together with Web content information [18]. Algorithms that analyze Web pages based on more than textual content can also potentially glean more complex patterns, such as "there is a prevalence of data mining web pages that have links to job pages and links to publication pages."

Other researchers focus on insights that can be drawn using structural information alone. Chakrabarti and Faloutsos. (Chapter 4) and others [7, 24] have studied the unique attributes of graphs created from Web hyperlink information. Such hyperlink graphs can also be used to answer the following question:

What patterns can I find in the Web structure?

In Chapter 6, Kuramochi and Karypis discover frequent subgraphs in these topology graphs. In Chapter 16, Tomkins and Kumar show how new or emerging communities of Web pages can be identified from such a graph. Analysis of this graph leads to identification of topic hubs and authorities [25]. Authorities in this case are highly ranked pages on a given topic, and hubs represent overview sites with links to strong authority pages. The PageRank program [6] precomputes page ranks based on the number of links to the page from other sites together with the probability that a Web surfer will visit the page directly, without going through intermediary sites. In Chapter 12, Shimbo and Ito also demonstrate how the relatedness of Web pages can be determined from link structure information. Finally,

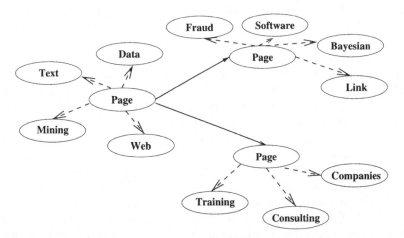

Figure 1.5. Graph representation for Web text and structure data. Solid arrows represent edges labeled "hyperlink" and dashed arrows represent edges labeled "keyword."

Desikan and Srivastava [13] have been investigating methods of finding patterns in dynamically evolving graphs, which can provide insights on trends as well as potential intrusions.

The third type of question that is commonly addressed in mining the web is:

What commonalities can we find in Web navigation patterns?

Answering this question is the problem of Web usage mining. Although mining clickstream data on the client side has been investigated [30], data is most easily collected and mined from Web servers [39]. Didimo and Liotta (Chapter 3) provide some graph representations and visualizations of navigation patterns. As Berendt points out [3], a graph representation of navigation allows the individual's website roadmap to be constructed. From the graph one can determine which pages act as starting points for the site, which collection of pages are typically navigated sequentially, and how easily (or often) are pages within the site accessed. Navigation graphs can be used to categorize Web surfers and can ultimately assist in organizing websites and ranking Web pages [3, 31, 35, 45].

1.3   BOOK OVERVIEW

The intention of this book is to provide an overview of the state of the art in mining graph data. To this end, we have gathered writings from colleagues that contribute to varied aspects of the problem. The aspects we focus on here are basic graph tools, techniques for mining graph data, and noteworthy graph mining applications.

Chapter 2 kicks off the *graph tools* part of the book by providing working definitions of key terms including graphs, subgraphs, and the operation that underlies much of graph mining, graph isomorphism. Here, Bunke and Neuhaus examine graph isomorphism techniques in detail and evaluate their merits based on the type of data that is available and the task that must be performed. Didimo and Liotta provide a thorough overview of graph visualization techniques in Chapter 3. They show how graph drawing algorithms assist and enhance the mining process and show how many of the techniques are customized for particular mining and other graph-based tasks. In Chapter 4, Chakrabarti and Faloutsos describe how the R-MAT algorithm can be used to generate graphs that exhibit properties found in real-world graphs. The ability to generate such graphs is useful for developing new mining algorithms, for testing existing algorithms, and for performing mining tasks such as anomaly detection.

In Part II, we highlight some of the most popular *mining techniques* that are currently developed for graph data. Chapters 5 through 7 focus on methods of discovering subgraph patterns from graph data. Yan and Han (Chapter 5) and Kuramochi and Karypis (Chapter 6) investigate efficient methods for extracting frequent substructures from graph data. Cook, Holder, and Ketkar (Chapter 7) evaluate subgraph patterns based on their ability to compress the input graph. Discovered subgraphs can be used to generate a graph grammar that is descriptive of the data, as shown by

Jonyer in Chapter 8. In contrast, Ohara et al. (Chapter 9) allow discovered subgraphs to represent features in a supervised learning problem. These subgraphs represent the attributes in a decision tree that can be learned from graph data. In Chapter 10, Liquière presents an alternative method for inducing concepts from graph data. By defining a partial order over the graph-based examples, the classification space can be viewed as a lattice and classical algorithms can be used to construct the concept definition from this lattice. In Chapter 11, Gärtner et al. define kernels on structural data that can be represented as a graph. The result can be applied to graph classification, making this problem tractable.

The next two chapters focus on properties of portions of the graph (individual edges, nodes, or neighborhoods around nodes), rather than on the graph as a whole, to perform the mining task. Shimbo and Ito, in Chapter 12, define an inner product of nodes in a graph. The resulting kernel can be used to analyze Web pages based on a combination of two factors: importance of the page and the degree to which two pages are related. Bhattacharya and Getoor use this location information in Chapter 13 to perform graph-based entity resolution. Edge attributes and constructed clusters of nodes can be used to identify the unique (nonduplicated) set of entities in the graph and to induce the corresponding entity graph.

The final part of the book features a collection of graph mining *applications*. These applications cover a diverse set of fields that are challenging and relevant, and for which data can be naturally represented as a graph. In Chapter 14, Okada provides an overview of chemical structure mining, including graph representations of the data and graph-based algorithms for analyzing the data. Zaki uses tree mining techniques to analyze bioinformatics data in Chapter 15. Specifically, the Sleuth algorithm is used to mine subtrees and can be applied to bioinformatics data such as RNA (ribonucleic acid) structures and phylogenetic subtrees. Tomkins and Kumar apply graph algorithms to Web data in Chapter 16 in which dense subgraphs are extracted that may represented communities of websites. Finally, Greenblatt and co-workers introduce a variety of graph mining tools in Chapter 17 that are effective for analyzing social network graphs. These applications are by no means comprehensive but illustrate the types of fields for which graph mining techniques are needed, and define the challenges that continue to drive this growing field of study.

REFERENCES

1. B. N. Ames, J. Mccann, and E. Yamasaki. Methods for detecting carcinogens and mutagens with the salmonella/mammalian-microsome mutagenicity test. *Mutation Research*, 31(6):347–364, 1975.
2. J. M. Barnard, G. M. Downsa, and P. Willet. Descriptor-based similarity measures for screening chemical databases. In H. J. Bohm and G. Schneider, eds. *Virtual Screening for Bioactive Molecules*, Wiley, New York, 2000.
3. B. Berendt. The semantics of frequent subgraphs: Mining and navigation pattern analysis. In Proceedings of WebKDD, Chicago, Illinois, 2005.
4. T. Berners-Lee, J. Hendler, and O. Lassila. The semantic Web. *Scientific American* 279(5):34–43, 2001.

5. C. Borgelt and M. R. Berthold. Mining molecular fragments: Finding relevant substructures of molecules. In Proceedings of the IEEE International Conference on Data Mining, Maebashi City, Japan, pp. 51–58 2002.

6. S. Brin and L. Page. The anatomy of a large-scale hypertextual (web) search engine. *Computer Network and ISDN Systems* 30:107–117, 1998.

7. A. Broder, R. Kumar, F. Maghoul, P. Raghavan, S. Rajagopalan, R. Stat, and A. Tomkins. Graph Structure in the Web: Experiments and models. In Proceedings of the World Wide Web Conference, Amsterdam, The Netherlands, 2000.

8. S. Chakrabarti. Data mining for hypertext: A tutorial survey. *ACM SIGKDD Explorations* 1(2):1–11, 2000.

9. P. Chaovalit and L. Zhou. Movie Review Mining: A comparison between supervised and unsupervised classification approaches. In Proceedings of the Thirty-Eigh.h Annual Hawaii International Conference on System Sciences, Waikoloa, Hawaii, 2005.

10. A. K. Debnath, R. L. Lopez de Compadre, G. Debnath, A. J. Shusterman, and C. Hansch. Structure-activity relationship of mutagenic aromatic and heteroaromatic nitro compounds: Correlation with molecular orbital energies and hydrophobicity. *Journal of Medicinal Chemistry* 34(2):786–797, 1991.

11. M. Deshpande, M. Kuramochi, and G. Karypis. Automated approaches for classifying structures. In Proceedings of the Workshop on Data Mining in Bioinformatics, Edmonton, Alberta, Canada, 2002.

12. M. Deshpande, M. Kuramochi, N. Wale, and G. Karypis. Frequent substructure-based approaches for classifying chemical compounds. *IEEE Transactions on Knowledge and Data Engineering* 17(18):1036–1050, 2005.

13. P. Desikan and J. Srivastava. Mining Temporally Evolving Graphs. In Proceedings of WebKDD, Seattle, Washington, 2004.

14. EachMovie. http://research.compaq.com/SRC/eachmovie.

15. O. Etzioni. The World wide web: Quagmire or gold mine? *Communications of the ACM* 39(11):65–68, 1996.

16. L. Getoor, N. Friedman, D. Koller, and B. Taskar. Learning probabilistic models of relational structure. In Proceedings of the International Conference on Machine Learning, Williamstown, Massachusetts, 2001.

17. A. Goldenberg and A. Moore. Tractable learning of large bayes net structures from sparse data. In Proceedings of the International Conference on Machine Learning, 2004.

18. J. Gonzalez, L. B. Holder, and D. J. Cook. Graph-based relational concept learning. In Proceedings of the International Machine Learning Conference, 2002.

19. C. Hansch, R. M. Muir, T. Fujita, C. F. Maloney, and M. Streich. The correlation of biological activity of plant growth-regulators and chloromycetin derivatives with hammett constants and partition coefficients. *Journal of the American Chemical Society* 85:2817–2824, 1963.

20. A. Inokuchi, T. Washio, T. Okada, and H. Motoda. Applying the apriori-based graph mining method to mutagenesis data analysis. *Journal of Computer Aided Chemistry* 2:87–92, 2001.

21. D. Jensen and J. Neville. Data mining in social networks. In Workshop on Dynamic Social Network Modeling and Analysis, Washington, DC, 2002.

22. N. Ketkar, L. B. Holder, and D. J. Cook. Qualitative comparison of graph-based and logic-based multi-relational data mining: a case study. In Proceedings of the KDD Workshop on Multi-Relational Data Mining, 2005.

23. R. D. King, S. H. Muggleton, A. Srinivasan, and M. J. E. Sternberg. Structure-activity relationships derived by machine learning: The use of atoms and their bond connectivities

to predict mutagenicity by inductive logic programming. In Proceedings of the National Academy of Sciences, Vol. 93, pp. 438–442, National Academy of Sciences, Washington, DC, 1996.

24. J. Kleinberg and S. Lawrence. The structure of the web. *Science* 294, 2001.

25. J. M. Kleinberg. Authoritative sources in a hyperlinked environment. *Journal of the ACM* 46:604–632, 1999.

26. P. Kolari and A. Joshi. Web mining: Research and practice. *IEEE Computing in Science and Engineering* 6(4):49–53, 2004.

27. S. Kramer, B. Pfahringer, and C. Helma. Mining for causes of cancer: Machine learning experiments at various levels of details. In Proceedings of the Conference on Knowledge Discovery and Data Mining, pp. 233–226, Newport Beach, California, 1997.

28. J. Kubica, A. Goldenberg, P. Komarek, and A. Moore. A comparison of statistical and machine learning algorithms on the task of link completion. In Proceedings of the KDD Workshop on Link Analysis for Detecting Complex Behavior, 2003.

29. H. Lodhi and S. H. Muggleton. Is Mutagenesis Still Challenging? In Proceedings of the International Conference on Inductive Logic Programming, pp. 35–40, 2005.

30. A. Maniam. Graph-based click-stream mining for categorizing browsing activity in the world wide web. Master's thesis, University of Texas at Arlington, 2004.

31. J. E. McEneaney. Graphic and numerical methods to assess navigation in hypertext. *International Journal of Human-Computer Studies* 55:761–786, 2001.

32. A. McGovern and D. Jensen. Identifying predictive structures in relational data using multiple instance learning. In Proceedings of the International Conference on Machine Learning, 2003.

33. P. Melville, R. Mooney, and R. Nagarajan. Content-boosted collaborative filtering for improved recommendations. In Proceedings of the National Conference on Artificial Intelligence, pp. 187–192, 2002.

34. A. Mendelzon, G. Michaila, and T. Milo. Querying the world wide web. In Proceedings of the International Conference on Parallel and Distributed Information Systems, pp. 80–91, 1996.

35. R. Meo, P. L. Lanzi, M. Matera, and R. Esposito. Integrating web conceptual modeling and web usage mining. In Proceedings of WebKDD, 2004.

36. E. Ravasz and A.-L. Barabasi. Hierarchical organization in complex networks. *Physical Review E*, 67, 2003.

37. P. Resnick, N. Iacovou, M. Sushak, P. Bergstrom, and J. Riedl. Grouplens: An open architecture for collaborative filtering of netnews. In Proceedings of the ACM Conference on Computed Supported Cooperative Work, pp. 175–186, 1994.

38. A. Srinivasan and R. D. King. Feature construction with inductive logic programming: A study of quantitative predictions of biological activity aided by structural attributes. *Data Mining and Knowledge Discovery* 3(1):37–57, 1999.

39. J. Srivastava, R. Cooley, M. Deshpande, and P.-N. Tan. Web usage mining: Discovery and applications of usage patterns from web data. *SIGKDD Explorations* 1(2):1–12, 2000.

40. M. J. E. Sternberg and S. H. Muggleton. Structure activity relationships (SAR) and pharmacophore discovery using inductive logic programming (ILP). *QSAR and Combinatorial Science* 22, 2003.

41. The Internet Movie Database. http://www.imdb.com.

42. E. Vozalis and K. Margaritis. Recommender systems: An experimental comparison of two filtering algorithms. In Proceedings of the Ninth Panhellenic Conference in Informatics, 2003.

43. R. Weiss, B. Velez, and M. Sheldon. HyPursuit: A hierarchical network search engine that exploits context-link hypertext clustering. In Proceedings of the Conference on Hypertext and Hypermedia, pp. 180–193, 1996.

44. O. R. Zaiane and J. Han. Resource and knowledge discovery in global information systems: A preliminary design and experiment. In Proceedings of the International Conference on Knowledge Discovery and Data Mining, pp. 331–336, 1995.

45. M. J. Zaki. Efficiently mining frequent trees in a forest. In Proceedings of the International Conference on Knowledge Discovery and Data Mining, 2002.

Part I

GRAPHS

2

GRAPH MATCHING—EXACT AND ERROR-TOLERANT METHODS AND THE AUTOMATIC LEARNING OF EDIT COSTS

HORST BUNKE AND MICHEL NEUHAUS

Institute of Computer Science and Applied Mathematics
University of Bern, Bern, Switzerland

2.1 INTRODUCTION

In recent years, the use of graph representations has gained popularity in pattern recognition and machine learning [1–3]. The main advantage of graphs is that they offer a powerful way to represent structured data. Among other applications, attributed graphs have been used to address the problem of graphical symbol recognition [4], character recognition [5, 6], shape analysis [7], biometric person authentication by means of facial images [8] and fingerprints [9], computer network monitoring [10], Web document analysis [11], and data mining [12].

The process of evaluating the structural similarity of graphs is commonly referred to as graph matching. A large variety of methods addressing specific problems of structural matching have been proposed [13]. Graph matching systems can roughly be divided into systems matching structure in an exact manner and systems matching structure in an error-tolerant way. Although exact graph matching offers a rigorous way to describe the graph matching problem in mathematical terms, it is generally only applicable to a restricted set of real-world problems. Error-tolerant graph matching, on the other hand, is able to cope with strong inner-class distortion, which is often present in real-world problems, but is generally computationally less efficient.

Mining Graph Data, Edited by Diane J. Cook and Lawrence B. Holder
Copyright © 2007 John Wiley & Sons, Inc.

Edit distance [14, 15] is one of the most intuitive dissimilarity measures on graphs. However, the applicability of error-tolerant graph matching using edit distance strongly depends on the definition of suitable edit costs. In the past, edit cost functions have mostly been designed in a manual fashion, due to the lack of methods to automatically learn edit costs.

In this chapter we provide an introduction to, and overview of, basic graph matching methods. We describe exact matching paradigms, such as graph isomorphism, subgraph isomorphism, and maximum common subgraph, as well as error-tolerant matching approaches, such as graph edit distance, and present two methods for automatically learning edit costs from a sample set of graphs.

2.2   DEFINITIONS AND GRAPH MATCHING METHODS

Attributed graphs with an unrestricted label alphabet is one of the most general ways to define graphs. It turns out that the definition given below is sufficiently flexible for a large variety of pattern recognition applications.

Definition 2.1 (Graph)   Given a node label alphabet L_V and an edge label alphabet L_E, we define a (directed attributed) graph g by the four-tuple $g = (V, E, \mu, \nu)$, where

- V denotes a finite set of nodes.
- $E \subseteq V \times V$ denotes a set of edges.
- $\mu : V \rightarrow L_V$ denotes a node labeling function.
- $\nu : E \rightarrow L_E$ denotes an edge labeling function.

The set V can be regarded as a set of node identifiers and is often chosen to be equal to $V = \{1, \ldots, |V|\}$. While V defines the nodes, the set of edges E represents the structure of the graph. That is, a node $u \in V$ is connected to a node $v \in V$ by an edge $e = (u, v)$ if $(u, v) \in E$. The labeling functions can be used to integrate information about nodes and edges into graphs by assigning attributes from L_V and L_E to nodes and edges, respectively. Usually, there are no constraints imposed on the label alphabets. In practical applications, however, label alphabets L are often defined as vector spaces of a fixed dimension, $L = R^k$, or discrete sets of symbols, $L = \{s_1, \ldots, s_k\}$. Theoretically, nodes and edges may also have more complex labels such as strings or graphs themselves.

The graph definition introduced above includes a number of special cases. To define undirected graphs, for instance, we require that $(v, u) \in E$ for every edge $(u, v) \in E$ such that $\nu(u, v) = \nu(v, u)$. In the case of nonattributed graphs, the label alphabets are defined by $L_V = L_E = \{\phi\}$, so that every node and every edge gets assigned the null label $\emptyset$. The empty graph is defined by $g_\varepsilon = (\emptyset, \emptyset, \mu_\varepsilon, \nu_\varepsilon)$.

For some applications, it is important to detect if a smaller graph is present in a larger graph—for instance, if the larger graph represents an aggregation of objects and the smaller graph a specific object in the larger context. This intuitively leads to the formal definition of a subgraph.

Definition 2.2 (Subgraph) Let $g_1 = (V_1, E_1, \mu_1, \nu_1)$ and $g_2 = (V_2, E_2, \mu_2, \nu_2)$ be graphs. Graph g_1 is a subgraph of g_2, written $g_1 \subseteq g_2$, if

- $V_1 \subseteq V_2$.
- $E_1 = E_2 \cap (V_1 \times V_1)$.
- $\mu_1(u) = \mu_2(u)$ for all $u \in V_1$.
- $\nu_1(u, v) = \nu_2(u, v)$ for all $(u, v) \in E_1$.

Conversely, graph g_2 is called a supergraph of g_1 if g_1 is a subgraph of g_2. Sometimes the second condition of this definition is replaced by $E_1 \subseteq E_2$, and a subgraph fulfilling the more stringent condition given above is called an induced subgraph. The notion of subgraph can be used to approach more complex problems such as the largest common part of several graphs, which will be discussed in greater detail below. In the following section, a number of standard techniques to perform graph matching are described.

2.2.1 Exact Graph Matching

In exact graph matching, the objective is to determine whether or not the labels and structure, or part of the structure, of two graphs are identical. While it is easy to determine equality of patterns in case of feature vectors or strings, the same computation is much more complex for graphs. Because the nodes and edges of a graph cannot be ordered in general, unlike the components of a feature vector or the symbols of a string, the problem of graph equality, known as graph isomorphism, is computationally very demanding.

Definition 2.3 (Graph Isomorphism) Let $g_1 = (V_1, E_1, \mu_1, \nu_1)$ and $g_2 = (V_2, E_2, \mu_2, \nu_2)$ be graphs. A graph isomorphism between g_1 and g_2 is a bijective function $f : V_1 \rightarrow V_2$ satisfying

- $\mu_1(u) = \mu_2(f(u))$ for all nodes $u \in V_1$.
- For every edge $e_1 = (u, v) \in E_1$, there exists an edge $e_2 = (f(u), f(v)) \in E_2$ such that $\nu_1(e_1) = \nu_2(e_2)$.
- For every edge $e_2 = (u, v) \in E_2$, there exists an edge $e_1 = (f^{-1}(u), f^{-1}(v)) \in E_1$ such that $\nu_1(e_1) = \nu_2(e_2)$.

Two graphs g_1 and g_2 are called isomorphic if there exists a graph isomorphism between them.

From the last definition we conclude that isomorphic graphs are identical in terms of structure and labels. To establish an isomorphism, one has to map each

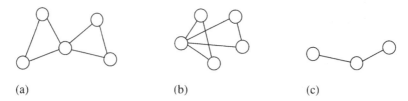

Figure 2.1.   Three graphs: Graph (b) is isomorphic to (a) and (c) is isomorphic to a subgraph of (a).

node from the first graph to a node of the second graph such that the edge structure is preserved and the node and edge labels are consistent. An illustration of two isomorphic graphs without labels is shown in Figure 2.1(a) and 2.1(b).

The most straightforward approach to checking the isomorphism of two graphs is to traverse a search tree considering all possible node-to-node correspondences [16]. The expansion of tree branches is continued until the edge structure implied by the node mapping does not correspond in both graphs. If nodes and edges are additionally endowed with labels, matching nodes and edges must also be consistent in terms of their labels. Reaching a leaf node of the search tree is equivalent to successfully mapping all nodes without violating the structure and label constraints and is therefore equivalent to having found a graph isomorphism. In general, the computational complexity of this procedure is exponential in the number of nodes of both graphs.

Closely related to graph isomorphism is the problem to detect if a smaller graph is present in a larger graph. If graph isomorphism is regarded as a formal notion of graph equality, subgraph isomorphism can be seen as subgraph equality.

Definition 2.4 (Subgraph Isomorphism)   Let $g_1 = (V_1, E_1, \mu_1, \nu_1)$ and $g_2 = (V_2, E_2, \mu_2, \nu_2)$ be graphs. An injective function $f : V_1 \to V_2$ is called a subgraph isomorphism from g_1 to g_2 if there exists a subgraph $g \subseteq g_2$ such that f is a graph isomorphism between g_1 and g.

A subgraph isomorphism exists from g_1 to g_2 if the larger graph g_2 can be turned into a graph that is isomorphic to the smaller graph g_1 by removing some nodes and edges. For an illustration of two graphs with an existing subgraph isomorphism between them, see Figures 2.1(a) and 2.1(c). Subgraph isomorphism can also be determined with the procedure outlined above for graph isomorphism [16]. It is known that subgraph isomorphism belongs to the class of NP-complete problems.

The definition of subgraph isomorphism leads us to the formal definition of the largest common part of two graphs.

Definition 2.5 (Maximum Common Subgraph, MCS)   Let $g_1 = (V_1, E_1, \mu_1, \nu_1)$ and $g_2 = (V_2, E_2, \mu_2, \nu_2)$ be graphs. A graph $g = (V, E, \mu, \nu)$ is called a common subgraph of g_1 and g_2 if there exist subgraph isomorphisms from g to g_1 and from g to g_2. A common subgraph of g_1 and g_2 is called maximum common

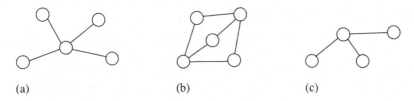

(a)   (b)   (c)

Figure 2.2.   Two graphs: (a) and (b) and a maximum common subgraph (c).

subgraph (MCS) if there exists no other common subgraph of g_1 and g_2 that has more nodes than g.

A maximum common subgraph of two graphs represents the maximal part of both graphs that is identical in terms of structure and labels. Note that, in general, the maximum common subgraph is not uniquely defined, that is, there may be more than one common subgraph with a maximal number of nodes. A standard approach to computing maximum common subgraphs is based on solving the maximum clique problem in an association graph [17]. The association graph of two graphs represents the whole set of possible node-to-node mappings that preserve the edge structure of both graphs. Finding a maximum clique in the association graph, that is, a fully con-nected maximal subgraph, is equivalent to finding a maximum common subgraph. Two graphs and a maximum common subgraph are shown in Figure 2.2.

Object similarity, or dissimilarity, is a key issue in many pattern recognition tasks. A simple graph dissimilarity measure can be derived from the maximum com-mon subgraph of two graphs. Intuitively speaking, the larger a maximum common subgraph of two graphs is, the more similar are the two graphs. This observation leads to a graph dissimilarity measure that is able to cope with structural errors. The dissimilarity of two graphs is defined by relating the size of the maximum common subgraph to the size of the larger of the two graphs [18].

Definition 2.6 (MCS Distance)   Let $g_1 = (V_1, E_1, \mu_1, \nu_1)$ and $g_2 = (V_2, E_2, \mu_2, \nu_2)$ be graphs and $g_{MCS} = (V_{MCS}, E_{MCS}, \mu_{MCS}, \nu_{MCS})$ be their maximum common subgraph. The MCS distance, d_{MCS}, of g_1 and g_2 is then defined by

$$d_{MCS}(g_1, g_2) = 1 - \frac{|V_{MCS}|}{\max\{|V_1|, |V_2|\}} \qquad (2.1)$$

Note that whereas the maximum common subgraph of two graphs is not uniquely defined, the MCS distance is. If two graphs are isomorphic, their MCS distance is 0; if two graphs have no part in common, their MCS distance is 1. The MCS distance accounts for a certain amount of tolerance toward errors, as two graphs need not be completely identical for a successful match. However, a small MCS distance, and hence a high graph similarity, can only be obtained if large portions of both graphs are isomorphic. Two other graph distances based on the MCS of a pair of graphs have been reported in [19, 20].

2.2.2 Error-Tolerant Graph Matching

In graph representations based on real-world patterns, it is often the case that graphs from the same class differ in terms of structure and labels. Hence the graph matching systems need to take structural errors into account. Graph edit distance [14, 15] offers an intuitive way to integrate error tolerance into the graph matching process and is applicable to virtually all types of graphs. Originally, edit distance has been developed for string matching [21], and a considerable amount of variants and extensions to edit distance have been proposed for strings and graphs. The key idea is to model structural variation by edit operations reflecting modifications in structure, such as the removal of a single node or the modification of an attribute attached to an edge. A standard set of edit operations consists of a node insertion, node deletion, node substitution, edge insertion, edge deletion, and edge substitution operation. Applying a node deletion operation, for instance, is equivalent to removing a node from a graph, and applying an edge substitution operation is equivalent to changing the value of an edge label.

Definition 2.7 (Edit Path)   Let $g_1 = (V_1, E_1, \mu_1, \nu_1)$ and $g_2 = (V_2, E_2, \mu_2, \nu_2)$ be graphs. Any bijective function $f : \hat{V}_1 \rightarrow \hat{V}_2$, where $\hat{V}_1 \subseteq V_1$ and $\hat{V}_2 \subseteq V_2$, is called an edit path from g_1 to g_2.

Edit paths are used to establish correspondences between some nodes of the first graph and some nodes of the second graph. An edit path $f : \hat{V}_1 \rightarrow \hat{V}_2$ assigns to each node $u \in \hat{V}_1$ a matching node $f(u) \in \hat{V}_2$, which is equivalent to substituting node u by node $f(u)$. Accordingly, the remaining nodes of the first graph, $V_1 \setminus \hat{V}_1$, and the remaining nodes of the second graph, $V_2 \setminus \hat{V}_2$, are not assigned to a node of the other graph. Hence, this is equivalent to deleting all nodes that belong to $V_1 \setminus \hat{V}_1$ from g_1, and inserting all nodes from $V_2 \setminus \hat{V}_2$ into g_2. In a similar manner, we obtain edge edit operations from an edit path. If a node is removed, for example, all edges connected to that node are removed as well. Thus, every edit path as defined above is equivalent to a sequence of edit operations transforming the first graph into the second one. In some cases it is more convenient to specify the edit path in terms of edit operations, for example $f = \{u_1 \rightarrow v_3, u_2 \rightarrow \varepsilon, \ldots, \varepsilon \rightarrow v_6\}$, instead of an equivalent function f. Using the edit operation notation, $u_1 \rightarrow v_3$ denotes the substitution of node $u_1 \in V_1$ by node $v_3 \in V_2$, $u_2 \rightarrow \varepsilon$ denotes the deletion of $u_2 \in V_1$, and $\varepsilon \rightarrow v_6$ denotes the insertion of $v_6 \in V_2$.

With substitutions, deletions, and insertions for both nodes and edges at our disposal, any graph can be transformed into any other graph by iteratively applying edit operations. Consequently, the concept of graph editing can be used to define a dissimilarity measure on graphs. To quantify how strongly an edit operation modifies the structure of a graph, it is common to use an edit cost function that assigns a cost value to each edit operation. An edit operation associated with a low cost is assumed to only slightly alter the graph under consideration, while an edit operation with a high cost is assumed to strongly modify the graph. To obtain a cost function on edit paths, we simply accumulate individual edit operation costs of the edit path. The problem of measuring the dissimilarity of two graphs is then equivalent to the problem of finding the edit path that models the structural difference of two graphs in the least costly way. Consequently, the graph edit distance of two graphs is defined by the minimum cost edit path from the first to the second graph.

Definition 2.8 (Graph Edit Distance)   Let $g_1 = (V_1, E_1, \mu_1, \nu_1)$ and $g_2 = (V_2, E_2, \mu_2, \nu_2)$ be graphs, and let $c(f)$ denote the cost of edit path f. If $\{f_1, \ldots, f_m\}$ is the set of edit paths from g_1 to g_2, the edit distance of g_1 and g_2 can be defined by

$$d(g_1, g_2) = \min_{i \in \{1, \ldots, m\}} c(f_i) \qquad (2.2)$$

If the two graphs under consideration are very similar in terms of structure and labels, it can be assumed that only minor edit operations are required to transform the first into the second graph, which results in a low-cost edit path. In this case, the resulting edit distance will be small. Conversely, if the two graphs differ significantly, every edit path will necessarily include strong modifications and hence result in high costs.

The edit cost functions can be used to tailor edit distance to specific applications and datasets. The insertion, deletion, and substitution costs of nodes, for instance, determine whether it is cheaper to delete a node u and subsequently insert a node v instead of substituting node u with v, which means that a deletion and insertion is preferred over a substitution in an optimal edit path. Usually, edit operation costs are limited to nonnegative values. If, in addition, a cost function satisfies the conditions of positive definiteness and symmetry as well as the triangle inequality at the level of single edit operations, the resulting edit distance is known to be a metric [22], which legitimates the use of the term distance in edit distance. Also note that the MCS distance defined in Eq. (2.1) can be regarded as an instance of edit distance under certain cost functions [23].

To compute the edit distance of graphs, one usually resorts to a tree search procedure evaluating all possible edit paths [15]. In some cases, the running time and memory requirements can be reduced by applying heuristics to the tree search. Yet, the overall computational complexity is rather high. For unconstrained graphs, the time and space complexity of an edit distance computation is exponential in the number of nodes involved. Thus, graph edit distance is in general only feasible for small graphs. For larger graphs, an approximate edit distance algorithm has been proposed [24] that computes an upper bound of the exact edit distance.

A generic procedure for the computation of graph edit distance is outlined in Algorithm 2.1. The basic idea is to keep a set of partial edit paths to be processed (OPEN) and terminate as soon as a complete edit path with minimal costs is available. In step 5, the best minimum-cost candidate is retrieved from the OPEN list. If the candidate path is a complete edit path, that is, if all nodes from g_1 and from g_2 are involved, the algorithm returns the optimal edit path. Otherwise, in steps 9–15, new edit operations are added to the candidate path, resulting in the addition of a number of new partial edit paths to OPEN. If a node u_{k+1} from g_1 has not yet been processed, all substitutions from u_{k+1} to nodes in g_2 that are consistent with

Algorithm **2.1** *Computation of Graph Edit Distance*

Input: Nonempty graphs $g_1 = (V_1, E_1, \mu_1, \nu_1)$ and $g_2 = (V_2, E_2, \mu_2, \nu_2)$, where $V_1 = \{u_1, \ldots, u_{|V_1|}\}$ and $V_2 = \{v_1, \ldots, v_{|V_2|}\}$

Output: A minimum-cost edit path from g_1 to g_2, e.g., $f = \{u_1 \rightarrow v_3, u_2 \rightarrow \varepsilon, \ldots, \varepsilon \rightarrow v_6\}$

1: Initialize OPEN to the empty set
2: For each node $w \in V_2$, insert the substitution $\{u_1 \rightarrow w\}$ into OPEN
3: Insert the deletion $\{u_1 \rightarrow \varepsilon\}$ into OPEN
4: **loop**
5:    Retrieve minimum-cost partial edit path f_{min} from OPEN
6:    **if** f_{min} is a complete edit path **then**
7:       Return f_{min} as the solution
8:    **else**
9:       Let $f_{min} = \{u_1 \rightarrow v_{i_1}, \ldots, u_k \rightarrow v_{i_k}\}$
10:       **if** $k < |V_1|$ **then**
11:          For each $w \in V_2 \setminus \{v_{i_1}, \ldots, v_{i_k}\}$, insert $f_{min} \cup \{u_{k+1} \rightarrow w\}$ into OPEN
12:          Insert $f_{min} \cup \{u_{k+1} \rightarrow \varepsilon\}$ into OPEN
13:       **else**
14:          Insert $f_{min} \cup \bigcup_{w \in V_2 \setminus \{v_{i_1}, \ldots, v_{i_k}\}} \{\varepsilon \rightarrow w\}$ into OPEN
15:       **end if**
16:    **end if**
17: **end loop**

the present candidate path are considered (step 11), as well as the deletion of u_{k+1} (step 12). If all nodes from g_1 have been processed, the remaining unmatchable nodes from g_2 are eventually inserted (step 14). It is straightforward to see that this algorithm returns a minimum cost edit path from g_1 to g_2.

For specific pattern classification tasks, the edit distance of graphs is often evaluated with classifiers based on the nearest-neighbor paradigm. In this case, we simply assume that patterns that are similar in terms of edit distance belong to the same class, and we therefore classify an unknown graph according to the most similar known graph. More elaborate nearest-neighbor classifiers consider more neighbors than only the closest one and also take the distribution of the neighbors into account.

2.3 LEARNING EDIT COSTS

One of the major difficulties in the application of edit distance is the definition of adequate edit costs. The edit costs essentially govern how the structural matching is performed. For some graph representations, it may be crucial whether a node is missing or not, while for other representations, the connecting edges are more important than the nodes. The question of how to define edit costs can therefore only be addressed in the context of an application-specific graph representation.

A simple edit cost model that has been used often is based on the distance of labels [25]. Here it is assumed that labels are elements of the n-dimensional space of real numbers. The idea is to assign edit costs to substitutions that are proportional to the Euclidean distance of the two labels. Substituting an edge by another edge with the same label therefore does not involve any costs. For nonidentical labels, the further the two labels differ from each other, the higher will be the corresponding substitution cost. Insertions and deletions are often assigned constant costs in this model. The advantage of this simple model is that only a few parameters are involved, and the edit costs are defined in a very intuitive way. However, it turns out that for some applications this model is not sufficiently flexible. For instance, it does not take into account that some label components may be more relevant than others. Also, the absolute values of the labels are not evaluated, but only the distance of labels, which means that all regions of the label space are equally weighted in terms of edit costs.

2.3.1 Learning Probabilistic Edit Costs

To address the problems discussed above, a probabilistic model for edit costs was proposed [26]. The edit costs are defined such that they can be learned from a labeled sample set of graphs. The idea is to model the distribution of edit operations in the label space. The probability distribution is then adapted so as to assign lower costs to those edit paths that correspond to pairs of graphs from the same class, while higher costs are assigned if a pair of graphs belong to different classes. To this end, we adopt a stochastic view on the process of editing a graph into another one, motivated by a cost learning method for string edit distance [27].

In the first step, we define a probability distribution on edit operations. While the general framework can be implemented for various distribution models, we propose a system of Gaussian mixtures to approximate the unknown distribution. Assuming that edit operations are statistically independent, we obtain a probability measure on edit paths. If $G(\ .\ |\mu, \Sigma)$ denotes a multivariate Gaussian density with mean μ and covariance matrix Σ, the probability of an edit path $f = \{e_1, \ldots, e_k\}$, for instance, $f = \{u_1 \rightarrow v_3, u_2 \rightarrow \varepsilon, \ldots, \varepsilon \rightarrow v_6\}$, is given by

$$p(f) = p(e_1, \ldots, e_k) = \prod_{j=1}^{k} \beta_{t_j} \sum_{i=1}^{m_{t_j}} \alpha_{t_j}^i \, G(e_j | \mu_{t_j}^i, \Sigma_{t_j}^i) \qquad (2.3)$$

where t_j denotes the type of edit operation e_j. Every edit operation type is additionally provided with a weight β_{t_j}, a number of mixture components m_{t_j}, and for each component $i \in \{1, \ldots, m_{t_j}\}$ a mixture weight $\alpha_{t_j}^i$. During training, pairs of graphs that are required to be similar are fed into the cost model. The objective of the learning algorithm is to derive the most likely edit paths between the training graphs and adapt the model so as to further increase the probability of the optimal edit paths. If we consider the probability of all edit paths between two graphs, we obtain a probability distribution on pairs of graphs. Formally, given two graphs g and g' and the set of edit paths $\{f_1, \ldots, f_m\}$ from g to g', the probability of g and g' is

$$p_{\text{graph}}(g, g') = \int_{i \in \{1, \ldots, m\}} dp(f_i | \Phi) \qquad (2.4)$$

where Φ collectively denotes the parameters of the edit path distribution. The resulting learning procedure leads to a distribution that assigns high probabilities to graphs from the same class. If graphs from different classes are sufficiently distinct in terms of structure and labels, we assume that while increasing the probability of intraclass pairs, the probability of interclass pairs will not be increased significantly. Hence the relative similarity of graphs from the same class will be increased.

To train the probability model, we resort to an Expectation Maximization (EM) algorithm. EM can be used to perform a maximum-likelihood estimation in case of missing or hidden information [28, 29]. In the context of our model, this means that the learning procedure actually estimates maximum-likelihood model parameters with respect to the training sample, which corresponds to the intuitive description of the training process given above. The EM learning algorithm iteratively adapts the hidden model parameters to better reflect the observed edit operations. The hidden data in our model are the underlying parameters of the Gaussian mixtures, namely mean vectors, covariance matrices, and weighting factors β_{t_j} and $\alpha^i_{t_j}$, whereas the observed data correspond to edit operations occurring on edit paths between graphs required to be similar. In the case of Gaussian mixtures, the EM training algorithm is guaranteed to converge to a stationary point of the likelihood surface—in most cases it even converges to a local maximum of the likelihood. To avoid having to define a constant number of mixture components, we employ a greedy initialization technique for Gaussian mixtures [30]. The idea is to initialize the mixture with a single component, train the system until convergence, and iteratively insert new components until the model cannot be improved any further. For more details about the training algorithm, the reader is referred to [26].

After training we derive edit costs from the probabilistic model. Consistent with the maximum-likelihood training criterion, we assume that the probability of an edit operation reflects its importance in matching graphs of the same class. Hence, an edit operation with a high probability is frequently involved in matching graphs of the same class and should therefore be assigned a small cost. A low probability edit operation should conversely be blocked from optimal edit paths and be assigned a high cost. Thus, for an edit operation e, we suggest to derive costs from probabilities by

$$c(e) = -\log p(e) \qquad (2.5)$$

The resulting edit cost model has the advantage of being able to discriminate between more and less relevant regions of the attribute space.

2.3.2 Self-Organizing Edit Costs

The probabilistic edit cost model introduced in Section 2.3.1 is potentially limited
in that a certain number of graph samples are required for the accurate estimation of
the distribution parameters. For small samples, we therefore propose an alternative
way to derive edit costs from samples by means of self-organizing maps.

Self-organizing maps (SOMs) [31] are two-layer neural networks, consisting of
an input layer and a competitive layer. The role of the input layer is to forward
input patterns to the competitive layer. The competitive neurons are arranged in a
homogeneous grid structure such that every neuron is connected to its neighbors.
The idea is that the competitive layer reflects the space of input elements, that is,
every competitive neuron corresponds to an element of the input space. For pattern
representations in terms of feature vectors, this is usually accomplished by assigning
weight vectors of the same dimension as the input space to neurons. Upon feeding
an input pattern into the network, neurons of the competitive layer compete for the
position of the input element. To this end, the weight of the closest competitive
neurons and their neighbors are adapted so as to shift them toward the position of
the input element. The longer this procedure is carried out, the more the competitive
neurons tend to migrate to areas where many input elements are present. That is,
the competitive layer can be regarded as a model of the pattern space, where the
neuronal density reflects the pattern density. This unsupervised learning procedure is
called self-organization as it does not rely on predefined classes of input elements. A
basic self-organization procedure is given in Algorithm 2.2. The similarity value α
of two neurons in line 7 is usually computed by means of an activation function with
respect to the grid distance of the two neurons. In line 8, the weight of neurons close

---

Algorithm 2.2 *Self-Organization Procedure*

```
Input:      Initial map and set of input vectors
Output:     Map after self-organization
```

1: Initialize competitive neuron weights with random or
   predefined values;
2: **while** terminating condition does not hold **do**
3:   Randomly choose a vector v_{in} from the set of all
     input vectors;
4:   Determine neuron n_{win} with the most similar weight
     to v_{in};
5:   **for all** competitive neurons n with weight v **do**
6:     **if** neuron n is close to n_{win} in the grid **then**
7:       Let $\alpha \in [0, 1]$ define the similarity of n and
         n_{win};
8:       $v = v + \alpha (v_{in} - v)$;
9:     **end if**
10: **end for**
11: **end while**

to the input pattern are shifted toward the input vector. During learning it is common to iteratively reduce the neighborhood radius and the learning intensity. That is, the similarity value α decreases in successive iterations resulting in a convergence of the self-organization process.

Similarly to the probablistic model described above, SOMs can be used to model the distribution of edit operations. The idea is to use a SOM to represent the label space. A sample set of edit operations is derived from pairs of graphs from the same class in a manner that is equivalent to the probabilistic model. The self-organization process turns the initially regular grid of the competitive layer into a deformed grid. From the deformed grid, we obtain a distance measure for substitution costs and a density estimation for insertion and deletion costs. In the case of substitutions, the SOM is trained with pairs of labels, where one label belongs to the source node (or edge) and one to the target node (or edge). The competitive layer of the SOM is then adapted so as to draw the two regions corresponding to the source and the target label closer to each other. The edit cost of node and edge substitutions is then defined proportional to the distance in the trained SOM. That is, instead of measuring label distance by the Eucliden distance, we measure the deformed distance in the corresponding SOM. In the case of insertions and deletions, the SOM is adapted so as to draw neurons closer to the inserted or deleted label. The edit cost of insertions and deletions is then defined according to the competitive neural density at the respective position. That is, the more neurons at a certain position in the competitive layer are, the lower is the respective insertion or deletion cost. The self-organizing training procedure for substitutions, insertions, and deletions hence results in lower costs for those pairs of graphs that are in the training set and belong to the same class. For further details, see [32].

Unlike the probabilistic model, the self-organizing model can also be used to specifically train pairs of graphs required to be dissimilar. Hence, not only the similarity of graphs can be modeled explicitly but also the dissimilarity. In addition, the self-organizing model is also applicable to small training sets. On the other hand, the probabilistic model is capable of learning general edit cost, while the self-organizing model is restricted to Euclidean label distances.

2.4 EXPERIMENTAL EVALUATION

In this section we present an experimental evaluation of the two edit cost models introduced in Section 2.3. For this purpose we use a database of graphs representing line drawings of capital letters. First, 15 prototypes are manually created, one for each capital letter consisting of straight lines only (letters A, E, F, H, I, K, L, M, N, T, V, W, X, Y, Z). To generate a graph sample, we apply a distortion process to each prototype drawing that randomly displaces, removes, and inserts lines. The strength of the distortions can be controlled by a distortion parameter. By repeating this procedure we obtain a database of line drawings. These line drawings are then converted into graphs by representing endpoints of lines by nodes, endowed with a position label, and lines by edges without label. A line drawing of letter A and the corresponding attributed graph are shown in Figure 2.3.

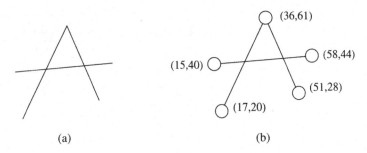

Figure 2.3.   (a) A letter line drawing and (b) the corresponding attributed graph.

First, we conduct an evaluation of the cost learning process. For a database of three letter classes containing 11 graphs each, we train a probabilistic cost model and compare the initial costs before learning and the final cost after learning. To obtain a visualization of the graph distribution according to edit costs, we derive from the edit distance matrix a Euclidean embedding by means of multidimensional scaling [33]. An illustration of the distribution before and after learning is shown in Figure 2.4. It can clearly be observed that the training process leads to compact and well-separated clusters of graphs.

In our next experiment, we train a self-organizing model for a database consisting of all 15 letter classes. At every training iteration, we evaluate the resulting graph clusters by means of a cluster validation index. That is, at every training step we compute an index value reflecting the clustering quality in quantitative terms. The validation index we use is the C-index [34] (modified such that it assigns high index values to compact and well-separated clusters). The progress of the self-organizing training is illustrated in Figure 2.5(a) on a weakly distorted database of

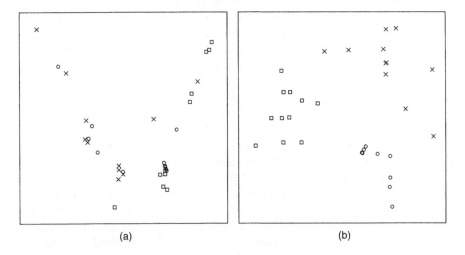

Figure 2.4.   Distribution of three graph classes: (a) before and (b) after probabilistic learning.

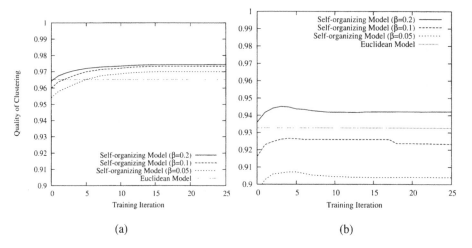

Figure 2.5. SOM learning on (a) weakly distorted letter graphs and (b) strongly distorted letter graphs.

letter graphs and in Figure 2.5(b) on a strongly distorted database. The straight line indicates the clustering quality of the simple Euclidean model without any learning, and parameter β of the self-organizing model denotes a parameter weighting the preference of substitutions over insertions and deletions. The improvement of the clustering quality during self-organization can clearly be seen. The fact that the strongly distorted database is more difficult to cope with than the weakly distorted database is also visible in the diagrams.

Finally, we turn to a performance evaluation of the cost models on datasets of various distortions. By varying the distortion parameter in the generation process of the graph database, we obtain datasets of several degrees of distortions, that is, of varying recognition difficulty. On each dataset we then proceed by training a proba-bilistic model and a self-organizing model. The quality of clustering resulting from the learned edit costs is illustrated in Figure 2.6(a). We find that the probabilistic model outperforms the self-organizing model particularly on difficult datasets. The validation index of the self-organizing model clearly degrades with an increasing degree of distortion, while the probabilistic model is able to maintain a higher level of clustering quality. As the self-organizing model is superior only for the least strongly distorted dataset, we conclude that the probabilistic model is best applied to strongly distorted data in the presence of large training sets, whereas the self-organizing model is rather suitable for weakly distorted data or small training sets.

This conclusion is also confirmed for a different kind of distortion, where a graph dataset is additionally deformed by applying a shearing transformation to the line drawings, or graphs. The stronger the shearing operation, the more difficult the letter recognition is supposed to be. An illustration of the classification accuracy of the two learning models and the simple Euclidean model described at the beginning of this section is provided in Figure 2.6(b). As expected, the simple Euclidean model is unable to cope with strong distortions. For small distortions the self-organizing

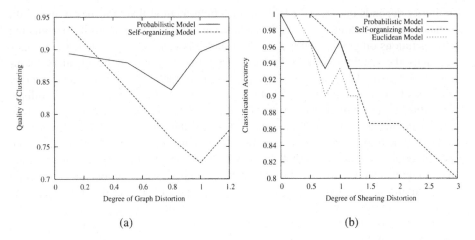

Figure 2.6. Performance in relation to degree of distortion for (a) graph distortion operator and (b) shearing operator.

model is best, while for strong distortions the probabilistic model clearly outperforms the other two.

2.5 DISCUSSION AND CONCLUSIONS

Graph matching has successfully been applied to various problems in the field of pattern recognition and machine learning. In the case of exact graph matching, the graph extraction process is assumed to be structurally flawless, that is, the conversion of patterns from a single class into graphs always results in identical structures or substructures. Otherwise graph isomorphism or subgraph isomorphism detection are rather unsuitable, which seriously restricts the applicability of graph isomorphism algorithms. The main advantages of isomorphism algorithms are their mathematically stringent formulation and the existence of well-known procedures to derive optimal solutions.

While the maximum common subgraph concept belongs to the category of exact graph matching methods, too, it provides us with a less restricted notion of graph matching than methods based on isomorphism. If two graphs have a large part in common, they can be positively matched, even if the remaining parts differ completely from each other. While the theoretical computational complexity of graph isomorphism and maximum common subgraph algorithms is exponential in the number of nodes, exact graph matching can be quite efficient in practice, particularly in the presence of node and edge labels.

Error-tolerant methods, sometimes also referred to as inexact or error-correcting methods, are characterized by their ability to cope with errors, or noncorresponding parts, in structure and labels of graphs. Hence, for two graphs to be positively matched, they need not be identical at all but only similar. The notion of graph similarity depends on the error-tolerant matching method that is to be applied. Edit distance offers a definition of graph similarity in an intuitive manner by introducing edit operations on graphs. Although edit distance is generally applicable to all kinds of graphs, the suitability of edit distance is largely dependent on the adequate definition of edit costs. For this reason, we propose two methods to automatically derive edit costs from a sample set of graphs. The first method is based on a probabilistic estimation of the edit operation distribution, and the second method is based on self-organizing maps in the label space. Experimental results show that the learning methods can be used to obtain cost functions that outperform manually designed cost functions. We also find that the two methods are particularly well-performing in complementary situations. The probabilistic method, which requires a sufficiently large training set, is applicable to strongly distorted data, whereas the self-organizing method is superior on weakly distorted data or on small training sets. The automatic learning of cost functions is an important step toward making graph edit distance more widely and easily applicable.

ACKNOWLEDGMENT

This research was supported by the Swiss National Science Foundation NCCR program "Interactive Multimodal Information Management (IM)2" in the Individual Project "Multimedia Information Access and Content Protection."

REFERENCES

1. Special Section on Graph Algorithms and Computer Vision. *IEEE Transactions on Pattern Analysis and Machine Intelligence*, 23(10):1040–1151, 2001.
2. Special Issue on Graph Based Representations. *Pattern Recognition Letters*, 24(8): 1033–1122, 2003.
3. Special Issue on Graph Matching in Pattern Recognition and Computer Vision. *International Journal of Pattern Recognition and Artificial Intelligence* 18(3):261–517, 2004.
4. J. Lladós and G. Sánchez. Graph matching versus graph parsing in graphics recognition. *International Journal of Pattern Recognition and Artificial Intelligence*, 18(3):455–475, 2004.
5. P. N. Suganthan and H. Yan. Recognition of handprinted Chinese characters by constrained graph matching. *Image and Vision Computing*, 16(3):191–201, 1998.
6. J. Rocha and T. Pavlidis. A shape analysis model with applications to a character recognition system. *IEEE Transactions on Pattern Analysis and Machine Intelligence*, 16(4):393–404, 1994.
7. A. Shokoufandeh and S. Dickinson. A unified framework for indexing and matching hierarchical shape structures. In *Proceedings of 4th International Workshop on Visual Form*, Vol. 2059 of LNCS, pp. 67–84. Springer, Berlin, 2001.
8. A. Tefas, C. Kotropoulos, and I. Pitas. Using support vector machines to enhance the performance of elastic graph matching for frontal face authentification. *IEEE Transactions on Pattern Analysis and Machine Intelligence* 23(7):735–746, 2001.

9. D. Maio and D. Maltoni. A structural approach to fingerprint classification. In *Proceedings of 13th International Conference on Pattern Recognition*, Vienna, Austria, pp. 578–585, 1996.
10. H. Bunke, M. Kraetzl, P. Shoubridge, and W. D. Wallis. Detection of abnormal change in time series of graphs. *Journal of Interconnection Networks*, 3(1–2):85–101, 2002.
11. A. Schenker, M. Last, H. Bunke, and A. Kandel. Classification of web documents using graph matching. *International Journal of Pattern Recognition and Artificial Intelligence* 18(3):475–496, 2004.
12. L. B. Holder and D. J. Cook. Graph-based relational learning: Current and future directions. *SIGKDD Explorations, Special Issue on Multirelational Data Mining*, 5(1):90–93, 2003.
13. D. Conte, P. Foggia, C. Sansone, and M. Vento. Thirty years of graph matching in pattern recognition. *International Journal of Pattern Recognition and Artificial Intelligence*, 18(3):265–298, 2004.
14. A. Sanfeliu and K. S. Fu. A distance measure between attributed relational graphs for pattern recognition. *IEEE Transactions on Systems, Man, and Cybernetics (Part B)*, 13(3):353–363, 1983.
15. B. T. Messmer and H. Bunke. A new algorithm for error-tolerant subgraph isomorphism detection. *IEEE Transactions on Pattern Analysis and Machine Intelligence* 20(5):493–504, 1998.
16. J. R. Ullman. An algorithm for subgraph isomorphism. *Journal of the Association for Computer Machinery*, 23(1):31–42, 1976.
17. G. Levi. A note on the derivation of maximal common subgraphs of two directed or undirected graphs. *Calcolo*, 9:341–354, 1972.
18. H. Bunke and K. Shearer. A graph distance metric based on the maximal common subgraph. *Pattern Recognition Letters*, 19(3):255–259, 1998.
19. M.-L. Fernandez and G. Valiente. A graph distance metric combining maximum common subgraph and minimum common supergraph. *Pattern Recognition Letters*, 22(6–7):753–758, 2001.
20. W. D. Wallis, P. Shoubridge, M. Kraetzl, and D. Ray. Graph distances using graph union. *Pattern Recognition Letters*, 22(6):701–704, 2001.
21. R. A. Wagner and M. J. Fischer. The string-to-string correction problem. *Journal of the Association for Computer Machinery*, 21(1):168–173, 1974.
22. H. Bunke and G. Allermann. Inexact Graph Matching for Structural Pattern Recognition. *Pattern Recognition Letters*, 1:245–253, 1983.
23. H. Bunke. On a relation between graph edit distance and maximum common subgraph. *Pattern Recognition Letters*, 18:689–694, 1997.
24. M. Neuhaus and H. Bunke. An error-tolerant approximate matching algorithm for attributed planar graphs and its application to fingerprint classification. In Proceedings of 10th International Workshop on Structural and Syntactic Pattern Recognition, LNCS 3138, Springer, Berlin, pp. 180–189, 2004.
25. B. Le Saux and H. Bunke. Feature selection for graph-based image classifiers. In Proceedings of 2nd Iberian Conference on Pattern Recognition and Image Analysis LNCS 3523, Springer, Berlin, 2005.
26. M. Neuhaus and H. Bunke. A probabilistic approach to learning costs for graph edit distance. In J. Kittler, M. Petrou, and M. Nixon, eds. Proceedings 17th International Conference on Pattern Recognition, Cambridge, United Kingdom, Vol. 3, pp. 389–393, 2004.

27. E. Ristad and P. Yianilos. Learning string edit distance. *IEEE Transactions on Pattern Analysis and Machine Intelligence* 20(5):522–532, 1998.
28. A. P. Dempster, N. M. Laird, and D. B. Rubin. Maximum likelihood from incomplete data via the EM algorithm. *Journal of the Royal Statistical Society*, 39(1):1–38, 1977.
29. R. A. Redner and H. F. Walker. Mixture densities, maximum likelihood and the EM algorithm. *Society for Industrial and Applied Mathematics (SIAM) Review* **26**, 195–239 (1984).
30. N. Vlassis and A. Likas. A greedy EM algorithm for Gaussian mixture learning. *Neural Processing Letters*, 15(1):77–87, February 2002.
31. T. Kohonen. *Self-Organizing Maps*. Springer, Berlin, 1995.
32. M. Neuhaus and H. Bunke. Self-organizing maps for learning the edit costs in graph matching. *IEEE Transactions on Systems, Man, and Cybernetics (Part B)* 35(3):503–514, 2005.
33. T. Cox and M. Cox. *Multidimensional Scaling*. Chapman and Hall, London, 1994.
34. L. Hubert and J. Schultz. Quadratic assignment as a general data analysis strategy. *British Journal of Mathematical and Statistical Psychology*, 29:190–241, 1976.

3

GRAPH VISUALIZATION AND DATA MINING

WALTER DIDIMO AND GIUSEPPE LIOTTA

*Dipartimento di Ingegneria Elettronica e dell'Informazione,
Università degli Studi di Perugia, Perugia, Italy*

3.1 INTRODUCTION

Graph visualization, also known as graph drawing, is an emerging discipline that addresses the problem of conveying the structure of data space and of producing new knowledge of it by using diagrams. In this introductory section we briefly glance at the research area of graph visualization and then a discussion on typical applications of graph drawing that can be of interest in the context of data mining.

3.1.1 Graph Drawing at a Glance

A large body of graph drawing literature has been published in the last two decades. Surveys (see, e.g., [40, 50, 72, 102, 113]), book chapters (see, e.g., [84, 115]), an annual symposium [10, 35, 39, 67, 80, 83, 86, 92, 96, 97, 116, 118], a number of journal special issues (see, e.g., [30, 32, 45–47, 71, 79, 85, 93]), and various books (see, e.g., [41, 75, 78, 95, 98, 107]) have been devoted to this research area.

A graph drawing algorithm receives as input a combinatorial description of a graph and returns as output a drawing of the graph. Various drawing conventions have been proposed for the representation of graphs in the plane. Usually, each

Mining Graph Data, Edited by Diane J. Cook and Lawrence B. Holder
Copyright © 2007 John Wiley & Sons, Inc.

vertex is mapped to a distinct point and each edge (u, v) is mapped to a simple Jordan curve between the points representing u and v. In particular, the edges are represented as polygonal chains in a *polyline drawing*, as chains of alternating horizontal and vertical segments in an *orthogonal drawing*, and as segments in a *straight-line drawing*.

Within a given drawing convention, a graph has infinitely many drawings; however, one of the fundamental properties of a drawing is its readability, that is, the capability of conveying the information associated with the graph in a clear way. The *readability* of a drawing is expressed by means of aesthetic criteria, which can be formulated as optimization goals for the drawing algorithms. These aesthetic criteria have been selected through an extensive analysis of human-drawn diagrams from various contexts (see, e.g., [114]) and have been validated through experimental studies on human understanding of diagrams (see, e.g., [100, 101]). Some aesthetic criteria are general, others depend on the drawing convention adopted and on the particular class of graphs considered (trees, planar graphs, hierarchical graphs, etc.). They have been frequently used as quality measures in the evaluation and comparison of graph drawing algorithms. For a survey on experimental studies of graph drawing algorithms see, for example, [117].

3.1.2   Graph Drawing Applications for Data Mining

Graphs and their visualizations are essential in data exploration and understanding, particularly for those applications that need to manage, process, and analyze huge quantities of data. The benefits of graph drawing techniques for data mining purposes include analysis through visual exploration, discovery of patterns and correlations, and abstraction and summarization. Among the different areas that benefit from technological solutions imported from graph drawing, we list the following.

Internet Computing. Visualization of interconnections among routers are relevant for Internet service providers (ISPs) that need to manage large and complex networks. ISPs are also interested in representing the structure of the Internet at a higher level (connection among autonomous systems) in order to understand the position of their partners and competitors in the Internet. The visualization of the graph of the autonomous systems and their connection policies can be used, for example, to analyze and capture recurrent patterns in the Internet traffic and to detect routing instabilities (see, e.g., [23, 29, 44, 51]).

Social Sciences. Social network analysts use graph-theoretic methodologies to identify important actors, crucial links, subgroups, roles, and network characteristics in such graphs. In this context, graph visualization is essential to make complicated types of analysis and data handling, such as identification of recurrent patterns and of relationships, more transparent, intuitive, and readily accessible. For example, the centrality role of an element in a collaboration network can be more easily detected by visual inspection of a radial drawing of the network itself (see, e.g., [17]), that is, a drawing where vertices lie on concentric circles. Examples and references of visual analysis of different types of social networks, including citation networks and co-authorship networks, can be found, for example, in [2, 16, 19].

Software Engineering. The display of schematic diagrams, like Unified Modeling Language (UML) diagrams, is a valuable tool in reverse engineering to understand the architecture of object-oriented software systems and to detect the roles, the connections, and the dependencies of their classes. The need of visual representations of software diagrams becomes fundamental when a system consists of a large number of classes and relations among them. Different automatic graph drawing plug-ins for existing Computer Aided Software Engineering (CASE) tools and several ad hoc algorithms and systems have been developed in this context (see, e.g., [5, 57, 68, 69]).

Information Systems. To design, maintain, update, and query databases, users and administrators cope with the complexity of the database schemas describing the structure of the data. A graphical representation of such schemas greatly improves the friendliness of a database application. Also, a critical task in the engineering of information systems is the detection of the data types, that is, entities and relationships, that are involved in the system. Inspecting large databases with the aim of identifying the key data is not feasible in practice without using diagrammatic interfaces. Examples of automatic graph drawing systems and algorithmic techniques for the visual representation of databases can be found in [3, 4, 6, 36, 111].

Homeland Security. One of the most crucial aspects concerning homeland security is the detection of organized crimes, like, for example, terrorism and narcotics trafficking. Organized crimes often rely on the cooperation and the aggregation of individuals and groups who may play different roles. Intelligence and law enforcement agencies typically have large volumes of raw data collected from multiple sources, and often incomplete, that need a careful analysis to possibly reconstruct criminal organizations. Recent approaches to criminal network analysis rely on graph- drawing-based tools that produce graphical representation of criminal networks and in some cases also provide analytical functionalities to detect patterns of interactions, subgroups, and implications (see, e.g., [24, 25, 119]). For example, Figure 3.1 shows a layout of a criminal activity network (CAN) [24].

Web Searching. Developing effective technologies for mining data from the World Wide Web is a very fertile research field. The effectiveness and the efficiency of most common search engines rely on the structural properties of the Web graph (Web resources and their hyperlinks). Valuable insight into the link structure and the page rank of the Web graph can be provided by automatic visualization systems, as it is shown, for example, in a paper by Brandes and Cornelsen [13]. Also, a new generation of Web search engines is being designed, where the data collected from the World Wide Web are clustered into meaningful semantic categories. The user interacts with the Web data by means of a graphical user interface that displays these categories organized as a tree structure or organized as a graph (see, e.g., [48, 49, 59, 120]).

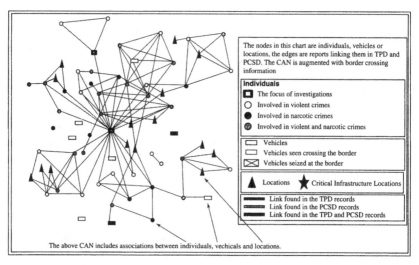

The nodes in this chart are individuals, vehicles or locations, the edges are reports linking them in TPD and PCSD. The CAN is augmented with border crossing information

Individuals
▣ The focus of investigations
○ Involved in violent crimes
● Involved in narcotic crimes
◉ Involved in violent and narcotic crimes

▭ Vehicles
▭ Vehicles seen crossing the border
⊠ Vehicles seized at the border

▲ Locations   ★ Critical Infrastructure Locations

▬▬ Link found in the TPD records
▬▬ Link found in the PCSD records
▬▬ Link found in the TPD and PCSD records

The above CAN includes associations between individuals, vechicals and locations.

Figure 3.1. A criminal activity network. (Courtesy of H. Chen, H. Atabakhsh, S. Kaza, B. Marshall, J. Xu, G. A. Wang, T. Petersen, and C. Violette.)

Computational Biology. Biologists often model the data collected from biochemical reactions of organisms by means of graphs. As a consequence, there are many types of complex biological networks, including neural networks, genome networks, protein networks, and networks of metabolic reactions. Automatic visualization of such graphs helps in understanding the complex relations among the chemical components in these networks and in extracting various information from the represented data. A limited list of publications in the vast literature of visualization systems and graph drawing techniques designed for the computational biology field includes [1, 11, 14, 15, 22, 33, 34, 52, 70, 89, 104, 110]. For example, Figure 3.2 depicts a $2\frac{1}{2}$-dimensional visualization of a metabolic network computed by the system WilmaScope [53, 110].

The remainder of this chapter is organized as follows. In Section 3.2 some of the most used graph drawing techniques are briefly illustrated. Examples of visualization systems for Web searching and for Internet computing are the subject of Section 3.3. Conclusions and a list of online graph visualization resources are given in Section 3.4.

3.2 GRAPH DRAWING TECHNIQUES

In this section we overview three of the most popular graph drawing approaches, namely, the *force-directed*, the *hierarchical*, and the *topology-shape-metrics* approach. Since an exhaustive analysis of the many algorithms that have been published in the literature based on these three approaches exceeds the scope of this chapter, we restrict our attention to the basic principles that are behind them. For more details, the interested reader is referred to [41, 75, 78, 95, 107]. As shown

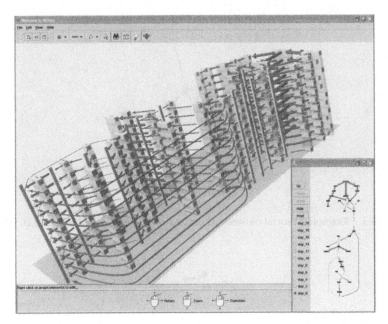

Figure 3.2. A $2\frac{1}{2}$D picture that shows experimental time-series data in the context of a metabolic network; the diagram has been computed by the system WilmaScope. (Courtesy of T. Dwyer, H. Rolletschek, and F. Schreiber.)

in the next section, algorithms that follow these approaches are often used in the drawing engine of systems that mine data by means of graph visualization.

3.2.1 Force-Directed Methods

Force-directed methods are quite popular for network analysis since they are easy to implement and often produce aesthetically pleasing results. For example, Brandes et al. [17] use a variant of force-directed methods to visually convey centrality in policy networks. Figure 3.3 describes informal communication among organizations involved in drug policy making; the dark vertices represent organizations that have a repressive attitude toward drug users, the light vertices represent organizations with a supportive attitude.

As another example, Batagelj and Mrvar [2] use different drawing methods, including force-directed ones, within Pajek, a system for visualizing and analyzing large networks. Figure 3.4 is a force-directed drawing computed by Pajek.

A milestone of the force-directed graph drawing literature is the work by Eades [54]. Indeed, most force-directed algorithms described in the literature adopt the physical model by Eades and propose variants of the forces definition (see, e.g., [31, 62, 63, 77, 108]). An experimental comparison of graph drawing algorithms based on force-directed methods is described in the work by Brandenburg et al. [12].

Eades' idea is to model a graph G as a mechanical system where the vertices of G are steel rings and the edges of G are springs that connect pairs of rings. To

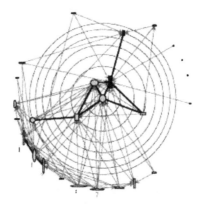

Figure 3.3.   Example of social network. (Courtesy of U. Brandes, P. Kenis, and D. Wagner.)

Figure 3.4.   Portion of a Geomlib network computed by the system Pajek. (Courtesy of V. Batajeli and A. Mrvar.)

compute a drawing of G, the rings are placed in some initial layout and let go until the spring forces move the system to a local minimum of energy. Clearly, different definitions of the repulsive and attractive forces that characterize the physical model of G give rise to different drawings of G. For example, Eades' implementation of the physical model does not reflect Hooke's law for the forces but obeys a logarithmic function for attractive forces; also, while repulsive forces act between every pair of vertices, attractive forces are calculated only between pairs of adjacent vertices. More precisely, in the model of Eades [54] the repulsive force between any pair (u, v) of vertices is

$$f_{\text{rep}}(u, v) = \frac{k \cdot \mathbf{uv}}{d^2(u, v)}$$

where $d(u, v)$ is the Euclidean distance between the points representing u and v, $\mathbf{uv}$ is the unit length vector oriented from the point representing u to the point representing v, and k is a repulsion constant. The attractive force of two adjacent vertices u and v is

$$f_{\text{attr}}(u, v) = \chi \, \log \frac{d(u, v) \, \mathbf{vu}}{l}$$

where l is the zero energy length of the spring and χ is the stiffness constant of the spring. To compute a drawing, one can start with an initial random placement of the vertices and proceed iteratively to find an equilibrium configuration. For example, a heuristic that converges to an equilibrium configuration consists of computing at each iteration and for each vertex v the force $F(v)$, that is, the sum of the repulsive and attractive forces acting on v; v is then moved in the direction of $F(v)$ by a small amount proportional to the magnitude of $F(v)$.

3.2.2 Hierarchical Methods

Hierarchies arise in a variety of applications, for example, Project Evaluation Review Technique (PERT) diagrams in project planning, class hierarchies in software engineering, and is-a relationships in knowledge representation systems. It is customary to represent hierarchical graphs so that all the edges flow in the same direction, for example, from top to bottom or from left to right. Algorithms that compute drawings of hierarchies generally follow a methodology first introduced by Sugiyama et al. [109] that accepts, as input, a directed graph G without any particular restriction (G can be planar or not, acyclic or cyclic) and that produces as output a layered drawing of G, that is, a representation where the vertices are on horizontal layers and the edges are monotone polygonal chains connecting their end vertices.

Figure 3.5 shows the heritage of the Unix operating system and is computed by the system Graphviz [58], developed at the AT&T Labs. The methodology of Sugiyama, Tagawa, and Toda [109] is also used by the system Ptolomaeus [42] to draw maps of Web navigations; see [109] for example, Figure 3.6.

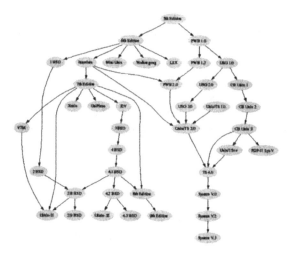

Figure 3.5. Hierarchical drawing showing the heritage of the Unix operating system, computed by Graphviz. The figure uses data from a hand-drawn graph by Ivan Darwin and Geoff Collyer from the mid-1980s. (Courtesy of J. Ellson, E. R. Gansner, E. Koutsofios, S. C. North, and G. Woodhull.)

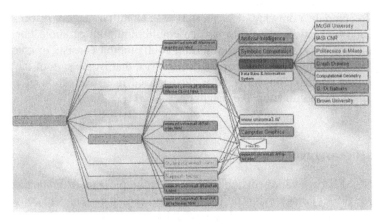

Figure 3.6. Web map computed by the system Ptolomaeus. (Courtesy of G. Di Battista, R. Lillo, and F. Vernacotola.)

The methodology of Sugiyama, Tagawa, and Toda [109] consists of the following steps:

Cycles Removal. This step removes existing cycles in the input graph by reversing the direction of some edges. Minimizing the number of reversed edges is an NP-hard problem, and different heuristics have been designed to solve this problem by reversing a small number of edges. A trivial heuristic is based on performing a Depth-First-Search (DFS) of the graph and reversing all back edges; however, in the worst case this may reverse as many as $m - n - 1$ edges, where m is the number of edges and n is the number of vertices of G. A better performing heuristic was proposed by Eades et al. [55] who present a greedy strategy that guarantees the reversal of at most $m/2 - n/6$ edges. An exact solution based on branch-and-cut is also described by Jünger et al. [76].

Layer Assignment. Each vertex is assigned to a horizontal line (layer) such that no vertices in the same layer are adjacent and if (u, v) is a directed edge of G, then the layer of u is above the layer of v in the drawing. Several objective functions can be taken into account in the layer assignment step. For example, minimizing the number of layers can be achieved by a simple linear-time longest-path algorithm: All sources are assigned to the lowest level L_0; each remaining vertex v is assigned to level L_p where p is the length of the longest path from any source to v. If one wants to minimize both the number of layers and the number of vertices on each layer, however, the problem becomes NP-hard. The polynomial time algorithm by Coffman and Graham [28] receives as input a reduced digraph G and a positive integer W; it returns a layer assignment where each layer consists of at most W vertices and the number h of layers is $h \le (2/W)h_{min}$ where h_{min} is the minimum possible number of layers. Gansner et al. [64] present a polynomial time algorithm, based on a relaxed integer linar programming (ILP) formulation, to compute a layer assignment that minimizes the total number of crossings between layers and edges (thus minimizing the sum of the so-called *edge spans*). The layer assignment step is concluded by inserting dummy vertices along edges whose end vertices are not on consecutive layers. The result is a proper k-level graph, that is, a graph where an edge connects only vertices on consecutive layers.

Crossing Reduction. This step aims at reducing the crossings among edges by permuting the order of the vertices on each layer. The number of edge crossings in a layered drawing does not depend on the precise position of the vertices but only on the ordering of the vertices within each layer. Thus the problem of reducing edge crossings is the combinatorial one of choosing an appropriate vertex ordering for each layer, not the geometric one of choosing an x coordinate for each vertex. This problem has been shown to be NP-hard by Garey and Johnson [65]; Eades and Wormald [56] showed that it remains NP-hard even if there are only two layers and the permutation of the vertices on one of them is fixed. A rich body of literature presenting heuristics and fixed parameter tractable approaches for the crossing minimization problem have been published. The reader interested in entry points to this literature can, for example, look at [74, 90, 91, 105, 106].

Horizontal Coordinate Assignment. Bends along the edges occur at the dummy vertices introduced in the layer assignment step. The horizontal coordinate assignment reduces the number of bends by readjusting the position of vertices on each layer without perturbing the order established in the crossing reduction step. The problem can be expressed as one of minimizing the total amount by which the edges deviate from being a straight line [41, 75, 78, 107]. This leads to the optimization

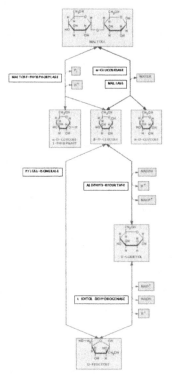

Figure 3.7. Layout of a reaction network computed by the system BioPath. (Courtesy of F. J. Brandenburg, M. Forster, A. Pick, M. Raitner, and F. Schreiber.)

of a quadratic objective function, and then the problem can be solved optimally only for small instances. In practice, a variant of the problem is often considered that attempts at drawing the edges as close to vertical lines as possible (see, e.g., Gansner et al. [64], Buchheim et al. [21], and Brandes and Köpf [18]).

We conclude this section with two additional examples of layered drawings computed by systems that adopt the above-described methodology. The drawing in Figure 3.7 shows a reaction network, containing structural formulas for substances. The drawing is computed by the layout algorithm of the system BioPath [11, 103]. The drawing in Figure 3.8 represents an example of a flow chart and is computed by the system Govisual [68].

3.2.3 Topology-Shape-Metrics Approach

The *topology-shape-metrics* approach is one of the most effective techniques for computing highly readable drawings of graphs. It represents graphs in the so-called *orthogonal drawing* standard, that is, each vertex of the graph is drawn as a point or a box and each edge is drawn as a polygonal chain of horizontal and vertical segments. Orthogonal drawings are widely used in many application fields, like very

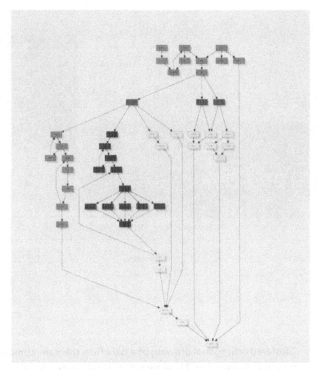

Figure 3.8. Layout of a flowchart computed by the system Govisual. (Courtesy of C. Gutwenger, M. Jünger, K. Klein, J. Kupke, S. Leipert, and P. Mutzel.)

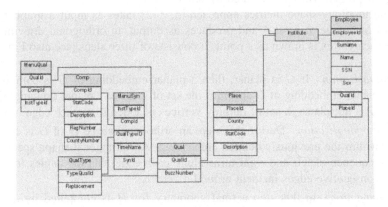

Figure 3.9. Orthogonal layout of the relational schema of a database, computed by the system DBDraw.

large scale integration (VLSI), software engineering, information systems, internet, and Web computing. For example, Figure 3.9 shows an orthogonal layout of the relational schema of a database, computed by the system DBDraw [36, 38]; Figure 3.10 depicts a clustered data flow diagram computed by the system Govisual [68]. Further

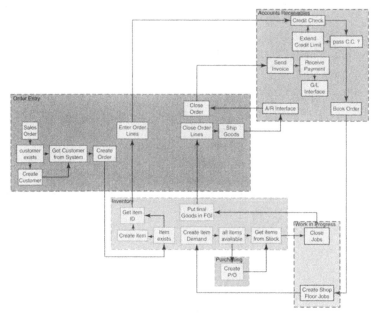

Figure 3.10. Clustered orthogonal drawing of a data flow diagram, computed by the system Govisual. (Courtesy of C. Gutwenger, M. Jünger, K. Klein, J. Kupke, S. Leipert, and P. Mutzel.)

examples of orthogonal drawings computed using variants of the topology-shape-metrics approach will be illustrated in Section 3.3.

The topology-shape-metrics approach [5, 112] takes as input a graph G with vertices of degree at most 4 and produces as output an orthogonal drawing of G where each vertex is drawn as a point. It consists of three steps (see also Fig. 3.11):

Planarization   If G is planar, then a planar embedding of G is computed. A planar embedding of G specifies the set of faces of a planar drawing of G. If G is not planar, a set of dummy vertices is added to replace edge crossings.

Orthogonalization   During this step, an orthogonal shape H of G is computed within the previously defined embedding. An orthogonal shape specifies the sequence of left and right turns along each edge, and the angles formed by consecutive edges incident around a vertex.

Compaction   In this step a final geometry for H is computed by assigning coordinates and dimensions to vertices and edge bends.

As described above, the topology-shape-metrics approach deals with the topology, shape, and geometry of the drawing separately; in each step of the approach, one or more optimization goals are considered, which are related to well-known drawing aesthetic criteria [41]. Namely, during planarization, since each dummy vertex represents an edge crossing, the goal is the minimization of the number of inserted dummy vertices. During orthogonalization the objective is typically to

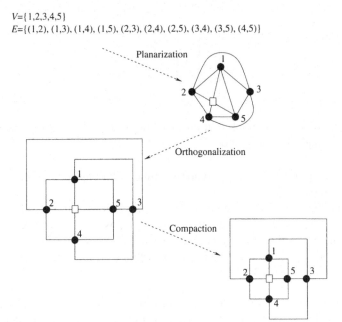

$V=\{1,2,3,4,5\}$
$E=\{(1,2),(1,3),(1,4),(1,5),(2,3),(2,4),(2,5),(3,4),(3,5),(4,5)\}$

Figure 3.11.   Illustration of the topology-shape-metrics approach.

determine a shape with the minimum number of edge bends. Finally, in the compaction step, the goal can be either the minimization of the area of the drawing or the minimization of the total edge length.

The distinct phases of the topology-shape-metrics approach have been extensively studied in the literature. If G is planar, which can be tested in linear time [73, 81], a planar embedding of G is also computable in linear time [26, 87]. If G is not planar, the minimum number of edge crossings required by the drawing may be $\Omega(n^4)$, although in practice this number is much smaller. Since, minimizing the number of crossings is an NP-hard problem [65], several heuristic planarization techniques have been proposed in the literature; the reader is referred to the paper by Liebers [82] for an annotated bibliography.

A popular and very elegant algorithm for constructing an orthogonal shape of an embedded planar graph with vertices having at most four incident edges was presented by Tamassia [112]. This algorithm uses a flow network approach to compute an orthogonal shape that has the minimum number of edge bends, while preserving the given embedding. Garg and Tamassia [66] proved that the problem of computing an orthogonal shape with the minimum number of bends in a variable embedding setting is NP-hard. Polynomial time solutions for specific classes of graphs are studied by Di Battista et al. [43]; exponential algorithms have been proposed by Bertolazzi et al. [7] and by Mutzel and Weiskircher [94].

Patrignani [99] showed that the problem of compacting an orthogonal shape minimizing the area or the total edge length of the drawing is NP-hard, while

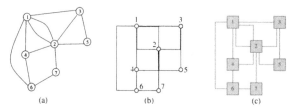

Figure 3.12. (a) A planar graph G. (b) A Kandinsky orthogonal representation of G. (c) A Kandinsky orthogonal drawing of G.

Bridgeman et al. [20] showed polynomial time solutions for particular classes of orthogonal shapes [20].

Several standards have been proposed in the literature to extend the topology-shape-metrics approach so as to work with graphs of any vertex degree. Here we recall the *Kandisky* drawing convention, originally described by Fößmeier and Kaufmann [60]. A Kandisky drawing is an orthogonal drawing with the following properties (see also Fig. 3.12):

- Segments representing edges cannot cross, with the exception that two segments that are incident of the same vertex may overlap (the angle between such segments has $0°$).
- All the polygons representing the faces have area strictly greater than 0.
- Vertices are drawn as boxes with equal size and overlapping segments are drawn as very near segments.

The Kandisky model has been further extended by Di Battista et al. [37] to deal with drawings in which the size (width and height) of each single vertex is assigned by the user. Having the possibility of customizing the size of each vertex is important in many application contexts; for example, this makes it possible to place a textual or graphical label in a vertex without intersecting any other element of the drawing (see also [8, 9]). Algorithms for nonplanar orthogonal drawings in the Kandisky model have been investigated in [27, 61].

3.3 EXAMPLES OF VISUALIZATION SYSTEMS

In this section we give examples of recent systems that use graph drawing techniques to discover properties, analyze phenomena, and extract knowledge in the context of Web searching and Internet computing. The graph drawing engines of these systems are based on variants of the visualization techniques described in the previous section.

3.3.1 Web Searching

The output of a classical Web search engine consists of an ordered list of links uniform resource locators (URLs) that are selected and ranked according to the user's

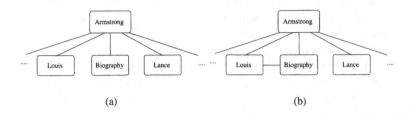

Figure 3.13. (a) Portion of a tree of categories for the query "Armstrong." (b)–(c) The same tree, plus edges that highlight cluster relationships.

query, the documents content, and (in some cases, like Google) the popularity of the links in the World Wide Web. The returned list can, however, consist of several hundreds of URLs and users may omit to check some URLs, that can be relevant for them just because these links do not appear in the first positions of the list. This problem is even more evident when the user's query presents polisemy, that is, words with different meanings. For example, suppose that the user submits the query "Jaguar." Is she interested in the "car" or in the "animal"?

A *Web meta-search clustering engine* is a system conceived to support the user in retrieving data on the Web by overcoming some of the limitations of traditional search engines. A Web meta-search clustering engine provides a visual interface to the user who submits a query; it forwards the query to (one or more) traditional search engines, and returns a set of clusters, also called *categories*, which are typically organized in a hierarchy. Each category contains URLs of documents that are semantically related with each other and is labeled with a string that describes its content. As a consequence, the user of a meta-search clustering engine has a global view of the different semantic areas involved by her query and can more easily retrieve the Web data relative to those topics in which she is interested. Clearly, the graphical user interface plays a fundamental role for Web meta-search clustering engines. Indeed, an effective representation of the categories and of their semantic relationships is essential for efficiently retrieving the wanted information.

Most Web meta-search clustering engines (see, e.g., Vivísimo[1], iBoogie[2], SnakeT[3]) have a graphical user interface (GUI) in which the clusters hierarchy is displayed as a tree. However, this type of representation may not be fully satisfactory for a complex analysis of the returned Web data. Suppose, for example, that the user's query is "Armstrong" and that the clusters hierarchy returned by a Web meta-search clustering engine is the tree depicted in Figure 3.13(a). Is the category "Biography" related to "Louis" or to "Lance" or to both (or to neither but to the astronaut Neil Armstrong)? If instead of a tree the systems returned a graph as the one in Figure 3.13(b), the user would be facilitated in deciding whether the category "Biography" is of her interest.

[1] http://vivisimo.com/.
[2] http://www.iboogie.com/.
[3] http://snaket.di.unipi.it/.

WhatsOnWeb[4] [48, 49] is a meta-search clustering engine that makes it possible to retrieve data on the Web by using drawings of graphs. The nodes represent clusters of semantically coherent URLs and the edges describe relationships between pairs of clusters. The graphical environment of WhatsOnWeb consists of two frames [see, e.g., Fig. 3.14(a)]. In the left frame the clusters hierarchy is represented as a classical directories tree. In the right frame, the user interacts with the drawing of a clustered graph. The drawing is computed with an engineered version of the topology-shape-metrics approach.

The user can expand/contract clusters in the graph and the drawing changes accordingly. Each cluster is drawn as a box having the minimum size required to host just its label (if the cluster is contracted) or a drawing of its subclusters (if the cluster is expanded). Figure 3.14(a) shows a snapshot of the interface, where the results for the query "Armstrong" are presented; in the figure, the category "Louis Armstrong" has been expanded by the user. During the browsing, the system automatically keeps consistent the map and the tree, that is, if a category is expanded in the map, it also appears expanded in the tree, and vice versa. Also, to preserve the user mental map, WhatsOnWeb preserves the orthogonal shape of the drawing during any expansion or contraction operation. For example, Figure 3.14(b) shows the map obtained by expanding the categories "Jazz," "School," and "Louis Armstrong Stamp" in the map of Figure 3.14(a).

Besides these browsing functionalities, WhatsOnWeb is equipped with several facilities that support the user's analysis of the Web data. For example, different colors are used to convey the "importance" that the system gives to every category. Red categories are the most important for the system, while yellow categories are the least important. The user can force the system to automatically prune categories on the basis of their importance and on the basis of their connectivity level with the rest of the graph. Furthermore, the user can explore the information associated with the edges of the map. Namely, if the mouse is positioned on an edge (u, v), a tool-tip is displayed that shows both the weight and the label of the relationship between u and v. Clicking on the edge, a two-layered drawing is shown to the user that displays the URLs in u, the URLs in v, and the relationships between these two sets of URLs. This feature makes it possible to isolate all those documents that give rise to a specific relationship and can be also used as a tool to evaluate and tune the mechanism adopted by the system for creating cluster relationships. For example, Figure 3.15(a) shows the information (weight and label strings) related to the edge between "Other Topics" and "University." Figure 3.15(b) depicts the two-layered drawing that shows the relationships between the URLs of "Louis Armstrong Stamp" and those of "Jazz."

3.3.2   Internet Analysis

At a high level of abstraction, the Internet can be seen as a network of so-called *autonomous systems*. An autonomous system (AS in the following) is a group of

---

[4]http://whatsonweb.diei.unipg.it/.

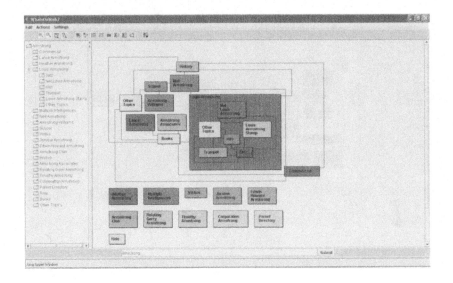

(a)

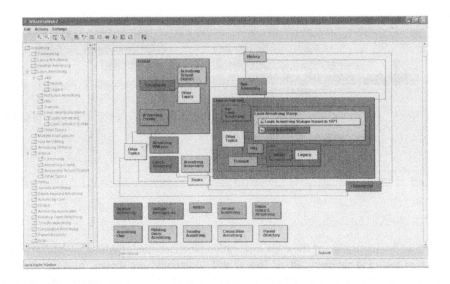

(b)

Figure 3.14. Snapshots of the user interface of WhatsOnWeb: (a) A map for the query "Armstrong"; in the map the user performed the expansion of the category "Louis Armstrong." (b) A successive map obtained by expanding the categories "Jazz," "School," and "Louis Armstrong Stamp"; this last category contains two URLs, represented by using their title.

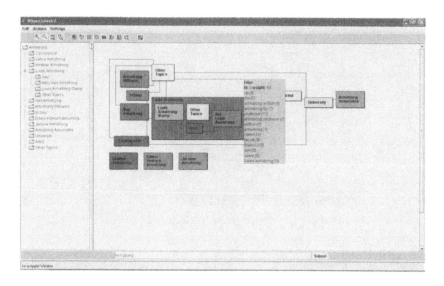

(a)

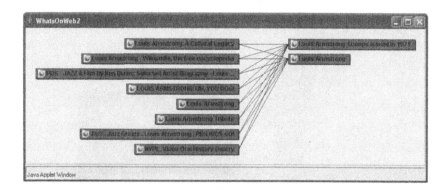

(b)

Figure 3.15.   Illustration of the edge exploration functionalities of WhatsOnWeb: (a) Information of the edge between "Other Topics" and "University" are shown. (b) A two-layered drawing that shows the relationships between the URLs of "Louis Armstrong Stamp" and those of "Jazz." The drawing is computed with a hierarchical method.

subnetworks under the same administrative authority, and it is identified by a unique integer number. In this sense, an AS can be seen as a portion of the Internet, and the Internet can be seen as the totality of the ASs. To maintain the reachability of any portion of the Internet, each AS exchanges routing information with a subset of other ASs, mainly selected on the basis of economic and social considerations. To exchange information, the ASs adopt a routing protocol called *BGP* (border gateway protocol). This protocol is based on a distributed architecture where *border routers* that belong to distinct ASs exchange information about the routes they know. Two border routers that directly communicate are said to perform a *peering session*, and the ASs they belong to are said to be *adjacent*. The *ASs graph* is the graph having a vertex for each AS and one edge between each pair of adjacent ASs. The ASs graph consists of more than 10,000 vertices, and then it is not reasonable to visualize it completely on a computer screen.

Internet service providers are often interested in visualizing and analyzing the structure of the ASs graph and the related connection policies to extract valuable information on the position of their partners and competitors, capture recurrent patterns in the Internet traffic, and detect routing instabilities. Several tools have been designed for this purpose (see, e.g., [51] for pointers). We describe here the system Hermes[5] [23] that allows users to incrementally explore the Internet topology by means of automatically computed maps. The graph drawing engine of Hermes has the following characteristics.

- Its basic drawing convention is the Kandinsky model for orthogonal drawings. However, since the handled graphs often have many vertices of degree one connected with the same vertex, the Kandinsky model is enriched with new features for effectively representing such vertices.
- It is equipped with two different graph drawing algorithms. In fact, at each exploration step of the user, the current map is enriched with new vertices and hence it has to be redrawn. Depending on the situation, the system (or the user) might want to use a static or a dynamic algorithm. Both the static and the dynamic algorithms are mainly based on variations of the topology-shape-metrics approach. The dynamic algorithm is faster than the static one, and it is designed to preserve the user's mental map during the exploration; however, it can lead, after a certain number of exploration steps, to drawings that are much less readable than those constructed with a static algorithm.

To determine the topology of the ASs graph, Hermes collects the data taken from several public routing databases, including ANS, APNIC, ARIN, BELL, CABLE&WIRELESS, CANET, MCI, RADB, RIPE, VERIO, and the routing BGP data provided by the route views project of Oregon University [88]. All data, which originally have heterogenous formats and often contain inconsistency, are filtered and represented by Hermes using the RPSL language.

The graphical user interface of Hermes offers several exploration facilities. The user can search for a specific AS and can start the exploration of the Internet from that AS. At each successive step, the user can display information about the routing policies of the ASs contained in the current map, or she can expand the map by exploring one of these ASs. For example, Figure 3.16 shows a snapshot of the system where the AS10474 (NETACTIVE, Tiscali South Africa) is searched and

[5]http://www.dia.uniroma3.it/~hermes.

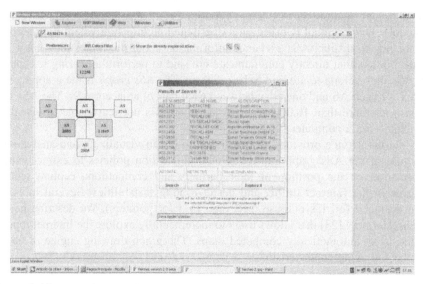

Figure 3.16. Map showing the ASs adjacent to AS10474 (NETACTIVE, Tiscali South Africa).

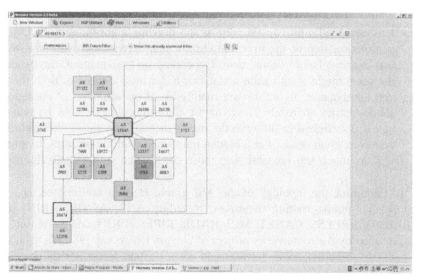

Figure 3.17. New map obtained from the previous map by exploring AS11845.

selected by the user for exploration; a first map that consists of the ASs adjacent to
AS10474 is then automatically computed and displayed by the system. Figure 3.17
shows how the map is expanded when the user decides to explore the AS11845.
Figure 3.18 depicts a more complex map obtained by performing three additional
exploration steps.

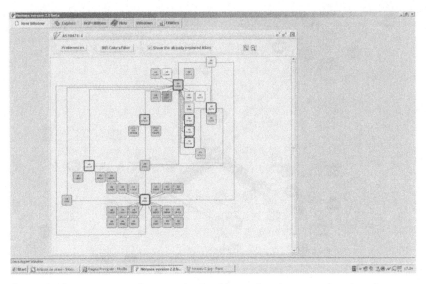

Figure 3.18. More complex map obtained by performing several exploration steps.

Although Hermes is an interesting and useful tool to analyze the topology of the Internet at the autonomous system level, it does not provide full support to understand Internet routing and its changes over time. The system BGPlay [29, 44] uses specifically tailored techniques and algorithms to display the state of routing at specific points in time and to animate its changes. The system obtains routing data from well-known routing archives of routing information, which are constantly kept up-to-date.

In some details, a route in the BGP protocol is a (directed) path in the Internet that can be used to reach a specific set of Internet protocal (IP) addresses, representing a set of subnetworks. A route is completely described by its destination IP address, its cost, and by the sequence of ASs that it traverses; this sequence is called an *AS path*. A *routing graph* is a network in which vertices are ASs and each edge defines a pair of ASs that appear consecutively in at least one of the AS paths. BGPlay is able to visualize the evolution of the AS paths used to reach a target AS from a subset of other ASs in the routing graph. To make the visualization and the animation effective, BGPlay uses a constrained force-directed algorithm, where the geometric distance between every AS and the target AS in the drawing reflects the number of (AS-) hops separating them. Figure 3.19 shows a drawing of the AS paths used to reach AS137 at a specific time.

3.4 CONCLUSIONS

This chapter summarized basic graph drawing techniques and software systems that can be used to process and analyze relational data by visually exploring the

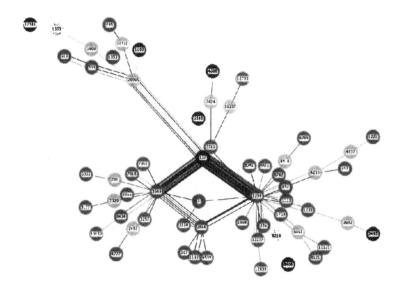

Figure 3.19. Portion of the routing graph computed by BGPlay. (Courtesy of L. Colitti, G. Di Battista, F. Mariani, M. Patrignani, and M. Pizzonia.)

related networks. We presented examples of visualization tools for data mining applications in the fields of Internet computing, social sciences, software engineering, information systems, homeland security, Web searching, and computational biology. We also described the main theoretic principles behind some of the most popular graph drawing approaches. We end the chapter with a list of online graph drawing resources.

- The graph drawing e-print archive at the University of Köln: http://gdea.informatik.uni-koeln.de/
- The Web page maintained by Tamassia: http://www.cs.brown.edu/people/rt/gd.html
- The Web page maintained by Brandes: http://graphdrawing.org/
- The Wiki for Open Problems page maintained by Marcus Raitner: http://problems.graphdrawing.org/index.php/
- The section of the "Geometry in Action" pages devoted to graph drawing and maintained by David Eppstein: http://www.ics.uci.edu/ eppstein/gina/gdraw.html
- The information visualization material in the Web pages of Hanrahan: http://www-graphics.stanford.edu/courses/cs448b-04-winter/
- The information visualization material in the Web pages of Munzner: http://www.cs.ubc.ca/ tmm/courses/infovis/

- The data mining and visualization page of Vladimir Batagelj:
  http://vlado.fmf.uni-lj.si/vlado/vladodam.htm
- The Atlas of Cyberspace at:
  http://www.cybergeography.org/atlas/atlas.html
- The graph drawing page maintained by Stephen Koubourov:
  http://www.cs.arizona.edu/people/kobourov/gdiv.html
- The social networks and visualization page maintained by Jonathon N. Cummings:
  http://www.netvis.org/resources.php

3.4.1 Acknowledgment

Work partially supported by the MIUR Project ALGO-NEXT: Algorithms for the Next Generation Internet and Web: Methodologies, Design and Applications.

REFERENCES

1. N. Amenta and J. Klingner. Case study: Visualizing sets of evolutionary trees. In *IEEE Information Visualization*, pp. 71–74, IEEE Comp. Society, New York, 2002.
2. V. Batagelj and A. Mrvar. Pajek—analysis and visualization of large networks. In M. Jünger and P. Mutzel, eds. *Graph Drawing Software*, pp. 77–103. Springer, 2003.
3. C. Batini, L. Furlani, and E. Nardelli. What is a good diagram? A pragmatic approach. In *Proceedings of 4th International Conference on the Entity-Relationship Approach*, pp. 312–319, IEEE Comp. Society, New York, 1985.
4. C. Batini, E. Nardelli, M. Talamo, and R. Tamassia. GINCOD: A graphical tool for conceptual design of data base applications. In A. Albano, V. De Antonellis, and A. Di Leva, eds. *Computer Aided Data Base Design*, pp. 33–51. North-Holland, New York, 1985.
5. C. Batini, E. Nardelli, and R. Tamassia. A layout algorithm for data flow diagrams. *IEEE Transactions on Software Engineering*, SE-12(4):538–546, 1986.
6. C. Batini, M. Talamo, and R. Tamassia. Computet aided layout of entity-relationship diagrams. *Journal of Systems and Software*, (4):163–173, 1984.
7. P. Bertolazzi, G. Di Battista, and W. Didimo. Computing orthogonal drawings with the minimum numbr of bends. *IEEE Transactions on Computers*, 49(8), 2000.
8. C. Binucci, W. Didimo, G. Liotta, and M. Nonato. Computing orthogonal drawings of graphs with vertex and edge labels. *Computational Geometry: Theory and Applications.* Vol 32(2), pp. 71–114, 2005.
9. C. Binucci, W. Didimo, G. Liotta, and M. Nonato. Computing labeled orthogonal drawings. In *10th International Symposium on Graph Drawing, GD 2002*, Vol. 2528 of *Lecture Notes in Computer Science*, pp. 66–73, Springer, Berlin, 2002.
10. F. J. Brandenburg, ed. *Graph Drawing (Proc. GD'95)*, Vol. 1027 of *Lecture Notes in Computer Science*. Springer, Berlin, 1995.
11. F. J. Brandenburg, M. Forster, A. Pick, M. Raitner, and F. Schreiber. Bio-Path—exploration and visualization of biochemical pathways. In M. Jünger and P. Mutzel, eds. *Graph Drawing Software*, pp. 215–235. Springer, Berlin, 2003.

12. F. J. Brandenburg, M. Himsolt, and C. Rohrer. An experimental comparison of force-directed and randomized graph drawing algorithms. In F. J. Brandenburg, ed. *Graph Drawing (Proc. GD '95)*, Vol. 1027 of *Lecture Notes in Computer Science*, pp. 76–87. Springer, Berlin, 1996.

13. U. Brandes and S. Cornelsen. Visual ranking of link structures. *Journal of Graph Algorithms and Applications*, 7(2):40–50, 2003.

14. U. Brandes, T. Dwyer, and F. Schreiber. Visual triangulation of network-based phylogenetic trees. In Proceedings of 6th Joint Eurographics—IEEE TCVG Symposium on Visualization (VisSym '04), pp. 75–83, IEEE Comp. Society, New York, 2004.

15. U. Brandes, T. Dwyer, and F. Schreiber. Visual understanding of metabolic pathways across organisms using layout in two and a half dimensions. *Journal of Integrative Bioinformatics*, 2:996–1003, 2004.

16. U. Brandes and T. Erlebach, eds. *Network Analysis: Methodological Foundations*, Vol. 3418 of *Lecture Notes in Computer Science*. Springer, Berlin, 2005.

17. U. Brandes, P. Kenis, and D. Wagner. Communicating centrality in policy network drawings. *IEEE Transactions on Visualization and Computer Graphics*, 9(2):241–253, 2003.

18. U. Brandes and B. Köpf. Fast and simple horizontal coordinate assignment. In S. Leipert M. Jünger, and P. Mutzel, eds. *9th International Symposium on Graph Drawing, GD 2001*, Vol. 2265 of *Lecture Notes in Computer Science*, pp. 31–44, Springer, Berlin, 2002.

19. U. Brandes and D. Wagner. Visone—analysis and visualization of social networks. In M. Jünger and P. Mutzel, eds. *Graph Drawing Software*, pp. 321–340. Springer, Berlin, 2003.

20. S. S. Bridgeman, G. Di Battista, W. Didimo, G. Liotta, R. Tamassia, and L. Vismara. Turn-regularity and optimal area drawings of orthogonal representations. *Computational Geometry: Theory and Applications*, 16(1):53–93, 2000.

21. C. Buchheim, M. Jünger, and S. Leipert. A fast layout algorithm for *k*-level graphs. In Joe Marks, ed. *8th International Symposium on Graph Drawing, GD 2000*, Vol. 1984 of *Lecture Notes in Computer Science*, pp. 229–240, Springer, Berlin, 2000.

22. Y. Byun, E. Jeong, and K. Han. A partitioned approach to protein interaction mapping. In *10th International Symposium on Graph Drawing, GD 2002*, Vol. 2528 of *Lecture Notes in Computer Science*, pp. 370–371, Springer, Berlin, 2002.

23. A. Carmignani, G. Di Battista, F. Matera, W. Didimo, and M. Pizzonia. Visualization of the high level structure of the internet with HERMES. *Journal of Graph Algorithms and Applications*, 6(3):281–311, 2002.

24. H. Chen, H. Atabakhsh, S. Kaza, B. Marshall, J. Xu, G. A. Wang, T. Petersen, and C. Violette. Bordersafe: Cross-jurisdictional information sharing, analysis, and visualization. In DG.O2005: Proceedings of the 2005 National Conference on Digital Government Research, pp. 241–242. Digital Government Research Center, Atlanta, Georgia, USA, May 15–18, 2005.

25. H. Chen, D. Zeng, H. Atabakhsh, W. Wyzga, and J. Schroeder. COPLINK: Managing law enforcement data and knoweldge. *Communication of the ACM*, 46(1):28–34, 2003.

26. N. Chiba, T. Nishizeki, S. Abe, and T. Ozawa. A linear algorithm for embedding planar graphs using PQ-trees. *Journal of Computer and System Sciences*, 30(1):54–76, 1985.

27. M. Chimani, G. Klau, and R. Weiskircher. Non-planar orthogonal drawings with fixed topology. In *Proceedings of SOFSEM'05: Theory and Practice of Computer Science*, Vol. 3381 of *Lecture Notes in Computer Science*, pp. 96–105, Springer, Berlin, 2005.

28. E. G. Coffman and R. L. Graham. Optimal scheduling for two processor systems. *Acta Informatica*, 1:200–213, 1972.

29. L. Colitti, G. Di Battista, F. Mariani, M. Patrignani, and M. Pizzonia. Visualizing Internet domain routing with BGPlay. *Journal of Graph Algorithms and Applications*, Vol. 9(1), pp. 117–148, 2005.

30. I. F. Cruz and P. Eades, eds. Special issue on graph visualization. *Journal of Visual Languages and Computing*, 6(3), 1995.

31. R. Davidson and D. Harel. Drawing graphs nicely using simulated annealing. *ACM Transactions on Graphics*, 15(4):301–331, 1996.

32. H. de Fraysseix and J. Kratochvil, eds. Graph drawing and representations: Special issue on selected papers from the 1999 Symposium on Graph Drawing. *Journal of Graph Algorithms and Applications*, 6(1), 2002.

33. E. Demir, O. Babur, U. Dogrusoz, A. Gursoy, A. Ayaz, G. Gulesir, G. Nisanci, and R. Cetin-Atalay. An ontology for collaborative construction and analysis of cellular pathways. *Bioinformatics*, 20(3):349–356, 2004.

34. E. Demir, O. Babur, U. Dogrusoz, A. Gursoy, G. Nisanci, R. Cetin-Atalay, and M. Ozturk. PATIKA: An integrated visual environment for collaborative construction and analysis of cellular pathways. *Bioinformatics*, 18(7):996–1003, 2002.

35. G. Di Battista, ed. *Graph Drawing (Proc. GD'97)*, Vol. 1353 of *Lecture Notes in Computer Science*. Springer, Berlin, 1997.

36. G. Di Battista, W. Didimo, M. Patrignani, and M. Pizzonia. Drawing database schemas. *Software—Practice and Experience*, 32:1065–1098, 2002.

37. G. Di Battista, W. Didimo, M. Patrignani, and M. Pizzonia. Orthogonal and quasi-upward drawings with vertices of prescribed sizes. In Jan Kratochvil, ed. *Graph Drawing (Proc. GD '99)*, Vol. 1731 of *Lecture Notes in Computer Science*, pp. 297–310. Springer, Berlin, 1999.

38. G. Di Battista, W. Didimo, M. Patrignani, and M. Pizzonia. DBDraw—automatic layout of relational database schemas. In M. Jünger and P. Mutzel, eds. *Graph Drawing Software*, pp. 237–256. Springer, Berlin, 2003.

39. G. Di Battista, P. Eades, H. de Fraysseix, and P. Rosenstiehl, eds. *Graph Drawing (Proc. GD'93)*, Paris, France, September 26–29, 1993.

40. G. Di Battista, P. Eades, R. Tamassia, and I. G. Tollis. Algorithms for drawing graphs: An annotated bibliography. *Computational Geometry: Theory and Applications*, 4(5):235–282, 1994.

41. G. Di Battista, P. Eades, R. Tamassia, and I. G. Tollis. *Graph Drawing*. Prentice Hall, Upper Saddle River, NJ, 1999.

42. G. Di Battista, R. Lillo, and F. Vernacotola. Ptolomaeus: The web cartographer. In *6th International Symposium on Graph Drawing, GD 1998*, Vol. 1547 of *Lecture Notes in Computer Science*, pp. 444–445, Springer, Berlin, 1998.

43. G. Di Battista, G. Liotta, and F. Vargiu. Spirality and optimal orthogonal drawings. *SIAM Journal on Computing*, 27(6):1764–1811, 1998.

44. G. Di Battista, F. Mariani, M. Patrignani, and M. Pizzonia. BGPlay: A system for visualizing the interdomain routing evolution. In *11th International Symposium on Graph Drawing, GD 2003*, Vol. 2912 of *Lecture Notes in Computer Science*, pp. 295–306, Springer, Berlin, 2003.

45. G. Di Battista and R. Tamassia, eds. Special issue on graph drawing. *Algorithmica*, 16(1), 1996.

46. G. Di Battista and R. Tamassia, eds. Special issue on geometric representations of graphs. *Computational Geometry: Theory and Applications*, 9(1–2), 1998.

47. G. Di Battista, ed. New trends in graph drawing: Special issue on selected papers from the 1997 Symposium on Graph Drawing. *Journal of Graph Algorithms and Applications*, 3(4), 1999.
48. E. Di Giacomo, W. Didimo, L. Grilli, and G. Liotta. Using graph drawing to search the web. In *13th International Symposium on Graph Drawing, GD 2005, Lecture Notes in Computer Science*, Vol 3843, pp. 480–491, Springer, Berlin.
49. E. Di Giacomo, W. Didimo, L. Grilli, and G. Liotta. A topology-driven approach to the design of web meta-search clustering engines. In *Proceedings of SOFSEM'05: Theory and Practice of Computer Science*, Vol. 3381 of *Lecture Notes in Computer Science*, pp. 106–116, Springer, Berlin, 2005.
50. J. Diaz, J. Petit, and M. Serna. A survey of graph layout problems. *ACM Computing Surveys*, 34(3):313–356, 2002.
51. M. Dodge and R. Kitchin. *Atlas of Cyberspace*. Addison Wesley, Reading, MA, 2001.
52. A. W. M. Dress and D. H. Huson. Constructing splits graphs. *IEEE Transactions on Computational Biology and Bioinformatics*, 1(3):109–115, 2004.
53. T. Dwyer, H. Rolletschek, and F. Schreiber. Representing experimental biological data in metabolic networks. In *2nd Asia-Pacific Bioinformatics Conference (APBC'04)*, Vol. 29 of *CRPIT*, pp. 13–20, ACS, Sydney, 2004.
54. P. Eades. A heuristic for graph drawing. *Congr. Numer.*, 42:149–160, 1984.
55. P. Eades, X. Lin, and W. F. Smyth. A fast and effective heuristic for the feedback arc set problem. *Information Processing Letters*, 47:319–323, 1993.
56. P. Eades and N. C. Wormald. Edge crossings in drawings of bipartite graphs. *Algorithmica*, 11(4):379–403, 1994.
57. M. Eiglsperger, C. Gutwenger, M. Kaufmann, J. Kupke, M. Jünger, S. Leipert, K. Klein, P. Mutzel, and M. Siebenhaller. Automatic layout of UML class diagrams in orthogonal style. *Information Visualization*, 3(3):189–208, 2004.
58. J. Ellson, E. R. Gansner, E. Koutsofios, G. Woodhull, and S. C. North. Graphviz and Dynagraph—static and dynamic graph drawing tools. In M. Jünger and P. Mutzel, eds. *Graph Drawing Software*, pp. 127–148. Springer, Berlin, 2003.
59. P. Ferragina and A. Gullí. A personalized search engine based on web-snippet hierarchical clustering. In 14th International Conference on World Wide Web, pp. 801–8106, Chiba, Japan, May 10–14, 2005.
60. U. Fößmeier and M. Kaufmann. Drawing high degree graphs with low bend numbers. In F. J. Brandenburg, ed. *3rd International Symposium on Graph Drawing (GD '95)*, Vol. 1027 of *Lecture Notes in Computer Science*, pp. 254–266, Springer, Berlin, 1996.
61. U. Fößmeier and M. Kaufmann. Algorithms and area bounds for non-planar orthogonal drawings. In G. Di Battista, ed. *5th International Symposium on Graph Drawing (GD 1997)*, Vol. 1353 of *Lecture Notes in Computer Science*, pp. 134–145, Springer, Berlin, 1997.
62. A. Frick, A. Ludwig, and H. Mehldau. A fast adaptive layout algorithm for undirected graphs. In R. Tamassia and I. G. Tollis, eds. *Graph Drawing (Proc. GD '94)*, Vol. 894 of *Lecture Notes in Computer Science*, pp. 388–403. Springer, Berlin, 1995.
63. T. Fruchterman and E. Reingold. Graph drawing by force-directed placement. *Software—Practice and Experience*, 21(11):1129–1164, 1991.
64. E. R. Gansner, E. Koutsofios, S. C. North, and K. P. Vo. A technique for drawing directed graphs. *IEEE Transactions on Software Engineering*, SE-19(3):214–230, 1993.
65. M. R. Garey and D. S. Johnson. Crossing number is NP-complete. *SIAM Journal of Algebraic Discrete Methods*, 4(3):312–316, 1983.

66. A. Garg and R. Tamassia. On the computational complexity of upward and rectilinear planarity testing. *SIAM Journal on Computing*, 31(2):601–625, 2001.

67. M. T. Goodrich and S. G. Kobourov, eds. *Graph Drawing (Proc. GD 2002)*, Vol. 2528 of *Lecture Notes in Computer Science*. Springer, Berlin, 2002.

68. C. Gutwenger, M. Jünger, K. Klein, J. Kupke, S. Leipert, and P. Mutzel. GoVisual—a diagramming software for UML class diagrams. In M. Jünger and P. Mutzel, eds. *Graph Drawing Software*, pp. 257–278. Springer, Berlin, 2003.

69. C. Gutwenger, J. Kupke, K. Klein, and S. Leipert. GoVisual for CASE tools Borland Together Control Center and Gentleware Poseidon—system demonstration. In *11th International Symposium on Graph Drawing, GD 2003*, Vol. 2912 of *Lecture Notes in Computer Science*, pp. 123–128, Springer, Berlin, 2003.

70. K. Han, B-H. Ju, and J. H. Park. InterViewer: Dynamic visualization of protein-protein interaction. In *10th International Symposium on Graph Drawing, GD 2002*, Vol. 2528 of *Lecture Notes in Computer Science*, pp. 364–365, Springer, Berlin, 2002.

71. X. He, ed. Special issue on selected papers from the Tenth International Symposium on Graph Drawing, GD 2002. *Journal of Graph Algorithms and Applications*, 8(2), 2004.

72. I. Herman, G. Melancon, and M. S. Marshall. Graph visualization and navigation in information visualizaion: A survey. *IEEE Transactions on Visualization and Computer Graphics*, 6(1):24–43, 2000.

73. J. Hopcroft and R. E. Tarjan. Efficient planarity testing. *Journal of the ACM*, 21(4): 549–568, 1974.

74. M. Jünger and P. Mutzel. 2-layer straightline crossing minimization: Performance of exact and heuristic algorithms. *Journal of Graph Algorithms and Applications*, 1(1):1–25, 1997.

75. M. Jünger and P. Mutzel, editors. *Graph Drawing Software*. Springer, Berlin, 2003.

76. M. Jünger, G. Reinelt, and S. Thienel. Practical problem solving with cutting plane algorithms in combinatorial optimization. In P. Seymour W. Cook, L. Lovász, ed. *DIMACS Series in Discrete Mathematics and Theoretical Computer Science*, pp. 111–152, Piscataway, New Jersey, 1995.

77. T. Kamada and S. Kawai. An algorithm for drawing general undirected graphs. *Information Processing Letters*, 31(1):7–15, 1989.

78. M. Kaufmann and D. Wagner, eds. *Drawing Graphs*, Vol. 2025 of *Lecture Notes in Computer Science*. Springer, Berlin, 2001.

79. M. Kaufmann, ed. Special issue on selected papers from the 2000 Symposium on Graph Drawing. *Journal of Graph Algorithms and Applications*, 6(3), 2002.

80. J. Kratochvil, ed. *Graph Drawing (Proc. GD'99)*, Vol. 1731 of *Lecture Notes in Computer Science*. Springer, Berlin, 1999.

81. A. Lempel, S. Even, and I. Cederbaum. An algorithm for planarity testing of graphs. In *Theory of Graphs: Internat. Symposium*, pp. 215–232. Gordon and Breach, 1967.

82. A. Liebers. Planarizing graphs—a survey and annotated bibliography. *Journal of Graph Algorithms and Applications*, 5(1):1–74, 2001.

83. G. Liotta, ed. *Graph Drawing (Proc. GD 2003)*, Vol. 2912 of *Lecture Notes in Computer Science*. Springer, Berlin, 2003.

84. G. Liotta and R. Tamassia. Drawings of graphs. In J. L. Gross and J. Yellen, eds. *Handbook of Graph Theory*, pp. 1015–1045. CRC Press, Boca Raton, FL, 2004.

85. G. Liotta and S. H. Whitesides, eds. Special issue on selected papers from the 1998 Symposium on Graph Drawing. *Journal of Graph Algorithms and Applications*, 4(3), 2000.

86. J. Marks, ed. *Graph Drawing (Proc. GD 2000)*, Vol. 1984 of *Lecture Notes in Computer Science*. Springer, Berlin, 2000.

87. K. Mehlhorn and P. Mutzel. On the embedding phase of the Hopcroft and Tarjan planarity testing algorithm. *Algorithmica*, 16:233–242, 1996.

88. D. Meyer. University of oregon route views project. On line: http://www. antc.uoregon.edu/route-views.

89. T. Munzner, F. Guimbretiere, S. Tasiran, L. Zhang, and Y. Zhou. TreeJuxtaposer: Scalable tree comparison using focus+context with guaranteed visibility. *ACM Transactions on Graphics*, 22(3):453–462, 2003.

90. P. Mutzel. An alternative method to crossing minimization on hierarchical graphs. *SIAM Journal of Optimization*, 11(4):1065–1080, 2001.

91. P. Mutzel. Optimization in leveled graphs. In P. M. Pardalos and C. A. Floudas, eds. *Encyclopedia of Optimization*, pp. 189–196. Berlin, Kluwer Academic, 2001.

92. P. Mutzel, M. Jünger, and S. Leipert, eds. *Graph Drawing (Proc. GD 2001)*, Vol. 2265 of *Lecture Notes in Computer Science*. Springer, Berlin, 2001.

93. P. Mutzel and M. Jünger, eds. Advances in graph drawing. Special issue on selected papers from the Ninth International Symposium on Graph Drawing, GD 2001. *Journal of Graph Algorithms and Applications*, 7(4), 2003.

94. P. Mutzel and R. Weiskircher. Bend minimization in orthogonal drawings using integer programming. In Proceedings of COCOON 2002, pp. 484–493, Singapore, 15–17 August, 2002.

95. T. Nishizeki and Md. Saidur Rahman. *Planar Graph Drawing*. Word Scientific, Singapore, 2004.

96. S. North, ed. *Graph Drawing (Proc. GD'96)*, Vol. 1190 of *Lecture Notes in Computer Science*. Springer, Berlin, 1996.

97. J. Pach, ed. *Graph Drawing (Proc. GD 2004)*, Vol. 3383 of *Lecture Notes in Computer Science*. Springer, Berlin, 2004.

98. J. Pach, ed. *Towards a Theory of Geometric Graphs*. American Mathematical Society, Providence, 2004.

99. M. Patrignani. On the complexity of orthogonal compaction. *Computational Geometry: Theory and Applications*, 19(1):47–67, 2001.

100. H. C. Purchase, D. Carrington, and J-A. Allder. Graph layout aesthetics in UML diagrams: User preferences. *Journal of Graph Algorithms and Applications*, 6(3):131–147, 2002.

101. H. C. Purchase, D. Carrington, and J-A. Allder. Evaluating graph drawing aesthetics: Defining and exploring a new empirical research area. In J. Di Marco, ed. *Computer Graphics and Multimedia: Applications, Problems and Solutions*, pp. 145–178. Hershey, PA, USA, Idea Group Publishing, 2004.

102. G. Sander. Graph layout for applications in compiler construction. *Theoretical Computer Science*, 217:175–214, 1999.

103. F. Schreiber. High quality visualization of biochemical pathways in BioPath. *In Cilico Biology*, 2(2):59–73, 2002.

104. F. Schreiber. Visual comparison of metabolic pathways. *Journal of Visual Languages and Computing*, 14(4):327–340, 2003.

105. M. Suderman and S. H. Whitesides. Experiments with the fixed-parameter approach for two-layer planarization. *Journal of Graph Algorithms and Applications*, Vol. 9(1). pp. 149–163, 2005.

106. M. Suderman and S. H. Whitesides. Experiments with the fixed-parameter approach for two-layer planarization. In G. Liotta, ed. *11th International Symposium on Graph*

Drawing, Vol. 2912 of *Lecture Notes in Computer Science*, pp. 345–356, Springer, Berlin, 2003.

107. K. Sugiyama. *Graph Drawing and Applications for Software and Knowledge Engineers*. World Scientific, Singapore, 2002.

108. K. Sugiyama and K. Misue. Graph drawing by the magnetic spring model. *Journal of Visual Languages and Computing*, 6(3):217–231, 1995.

109. K. Sugiyama, S. Tagawa, and M. Toda. Methods for visual understanding of hierarchical systems. *IEEE Transactions on Systems, Man, and Cybernetics*, SMC-11(2):109–125, 1981.

110. F. Schreiber, T. Dwyer, and H. Rolletschek. Representing experimental biological data in metabolic networks. In Second Asia-Pacific Bioinformatics Conference (APBC 2004), pp. 13–20, Vol. 29 of CRPIT, ACS, Sydney, 2004.

111. R. Tamassia. New layout techniques for entity-relationship diagrams. In Proceedings of 4th International Conference on Entity-Relationship Approach, pp. 304–311, IEEE Comp. Society, New York, 1985.

112. R. Tamassia. On embedding a graph in the grid with the minimum number of bends. *SIAM Journal on Computing*, 16(3):421–444, 1987.

113. R. Tamassia. Advances in the theory and practice of graph drawing. *Theoretical Computer Science*, 17:235–254, 1999.

114. R. Tamassia, G. Di Battista, and C. Batini. Automatic graph drawing and readability of diagrams. *IEEE Transactions on Systems, Man, and Cybernetics*, SMC-18(1):61–79, 1988.

115. R. Tamassia and G. Liotta. Graph drawing. In J. E. Goodman and J. O'Rourke, eds. *Handbook of Discrete and Computational Geometry, 2nd ed*, pp. 1015–1045. New York, Chapman & Hall/CRC, 2004.

116. R. Tamassia and I. G. Tollis, eds. *Graph Drawing (Proc. GD'94)*, Vol. 894 of *Lecture Notes in Computer Science*. Springer, Berlin, 1994.

117. L. Vismara, G. Di Battista, A. Garg, G. Liotta, R. Tamassia, and F. Vargiu. Experimental studies on graph drawing algorithms. *Software—Practice and Experience*, 30(11):1235–1284, 2000.

118. S. H. Whitesides, ed. *Graph Drawing (Proc. GD'98)*, Vol. 1547 of *Lecture Notes in Computer Science*. Springer, Berlin, 1998.

119. J. Xu and H. Chen. Criminal network analysis and visualization. *Communication of the ACM*, 48(6):101–107, 2005.

120. O. Zamir and O. Etzioni. Grouper: A dynamic clustering interface to web search results. *Computer Networks*, 31(11–16):1361–1374, 1999.

GRAPH PATTERNS AND THE R-MAT GENERATOR

DEEPAYAN CHAKRABARTI[1] AND CHRISTOS FALOUTSOS[2]

[1]*Yahoo! Research, Sunnyvale, California*
[2]*School of Computer Science, Carnegie Mellon University,
Pittsburgh, Pennsylvania*

4.1 INTRODUCTION

"How can we quickly generate a synthetic yet realistic graph? How can we spot fake graphs and outliers?"

Graphs, networks, and their surprising regularities/laws have been attracting significant interest recently. The World Wide Web, the Internet topology, and peer-to-peer networks follow surprising power laws [12, 22, 34], exhibit strange "bow-tie" or "jellyfish" structures [22, 63], while still having a small diameter [6]. Finding patterns, laws, and regularities in large real networks has numerous applications from criminology and law enforcement [28] to analyzing virus propagation patterns [53] and understanding networks of regulatory genes and interacting proteins [12], and so on.

How could we visualize the information contained in such large graphs efficiently? Which patterns should we be looking for? How can we *generate* "realistic" graphs that match these patterns? These are the questions that our NetMine system focuses on, and they are important for many applications:

- *Detection of Abnormal Subgraphs/Edges/Nodes:* Abnormalities should deviate from the "normal" patterns, so understanding the patterns of naturally occurring graphs is a prerequisite for detection of such outliers.

Mining Graph Data, Edited by Diane J. Cook and Lawrence B. Holder

- *Simulation Studies:* Algorithms meant for large real-world graphs can be tested on synthetic graphs that "look like" the original graphs. This is particularly useful if collecting the real data is hard or costly.
- *Realism of Samples:* Most graph algorithms are superlinear on the node count and thus prohibitive for large graphs. We might want to build a small sample graph that is "similar" to a given large graph. In other words, this smaller graph needs to match the "patterns" of the large graph to be realistic.
- *Extrapolation of Data:* Given a real, evolving graph, we expect it to have $x\%$ more nodes next year; how will it then look like, assuming that our "laws" are still obeyed? For example, to test the next-generation Internet protocol, we would like to simulate it on a graph that is "similar" to what the Internet will look like a few years into the future.
- *Graph Compression:* Graph patterns represent regularities in the data. Such regularities can be used to better compress the data.

Our NetMine system combines the computation of patterns and the generation of realistic graphs into a single package. The primary contributions of NetMine are:

1. *Scalable Pattern Computation:* Given a large graph, NetMine runs through a list of patterns commonly found in real-world graphs. These include the in-degree and out-degree distributions, singular value distributions, "hop-plots," distributions of "singular vector value," and "stress," all of which will be discussed later in this chapter. The pattern searchers are implemented scalably, enabling NetMine to quickly handle graphs with hundreds of thousands of nodes.
2. *The New "Min-cut Plot":* In addition, NetMine also generates the recently proposed "min-cut plots," an interesting pattern to check for while analyzing a graph.
3. *Interpretation of Plots:* We show how to interpret these plots and their implications regarding the datasets.
4. *The R-MAT Graph Generator:* Finally, we describe a recent, promising graph generator called R-MAT, which can be used to create synthetic yet realistic graphs, that is, graphs that match almost all of the patterns mentioned in scalable pattern computation above. NetMine also includes algorithms to automatically estimate the parameters of R-MAT given the graph whose patterns we wish to match.

For the pattern searches, the only desiderata are that they should provide some new information about the graph and that we should be able to compute them efficiently. A good graph generator, on the other hand, should satisfy many of the following properties, if not all:

(P1) *Realism*: It should only generate graphs that obey all (or at least several) of the above "laws," and it would match the properties of real graphs (degree exponents, diameters etc., which we shall discuss later) with the appropriate values of its parameters.

(P2) *Procedural Generation*: Instead of creating graphs to specifically match some patterns, the generator should offer some *process* of graph genera-tion, which automatically leads to the said patterns. This is necessary to gain insight into the process of graph generation in the real world: if a process cannot not generate synthetic graphs with the required patterns, it is probably not the underlying process (or at least the sole process) for graph creation in the real world.

(P3) *Parsimony*: It should have only a few parameters.

(P4) *Fast Parameter Fitting*: Given any real-world graph, the model parameters should be easily tunable by some published algorithms to generate graphs similar to the input graph.

(P5) *Generation Speed*: It should generate the graphs quickly, ideally, linearly on the number of nodes and edges.

(P6) *Extensibility*: The same method should be able to generate directed, undi-rected, and bipartite graphs, both weighted or] unweighted.

We will show how R-MAT matches all of these, including most of the patterns required for (P1).

The rest of this chapter is organized as follows: Section 4.2 surveys the existing graph laws and generators. Section 4.3 presents our proposed methods and algo-rithms for mining large graphs. Section 4.4 demonstrates the usefulness of NetMine using experiments on several real-world datasets. We conclude in Section 4.5. The Appendix provides the details of a proof.

4.2 BACKGROUND AND RELATED WORK

Let us recall the terminology first. An *unlabeled graph* $g = (\mathcal{V}, \mathcal{E})$ is a set $\mathcal{V}$ of N nodes and a set $\mathcal{E}$ of E edges between them. The edges may be undirected (like the network of Internet routers and their physical links) or directed (like the network of who-trusts-whom in the epinions.com database [57]). A special case is that of *bipartite graphs* with two sets of nodes, $\mathcal{V}_1$ (say, "actors") and $\mathcal{V}_2$ (say, "movies"), with edges between them (which actor played in which movie). Unless otherwise specified, all edges are unweighted: In a *weighted* graph, each edge has a (positive) weight. For example, the Internet router graph has physical links between routers as the edges of the graph, and the bandwidth of each link in megabits per second (Mbps) is its weight.

The twin problems of finding graph patterns and building graph generators have attracted a lot of recent interest, and a large body of work has been done on both, not only by computer scientists but also by physicists, mathematicians, sociologists, and others. However, there is little interaction among these fields, with the result that they often use different terminology and do not benefit from each other's advances.

In this section, we attempt to give a brief overview of the main ideas, with a focus on combining sources from all the different fields, to gain a coherent picture of the current state-of-the-art. The interested reader is also referred to some excellent and entertaining books on the topic [12, 64].

4.2.1   Graph Patterns and "Laws"

While there are many differences between graphs, some patterns show up regularly. Work has focused on finding several such patterns, which *together* characterize naturally occurring graphs. The main ones appear to be:

- Power laws
- Small diameters
- Community effects

Power Laws. While the Gaussian distribution is common in nature, there are many cases where the probability of events far to the right of the mean is significantly higher than in Gaussians. In the Internet, for example, most routers have a very low degree (perhaps "home" routers), while a few routers have extremely high degree (perhaps the "core" routers of the Internet backbone) [34]. Power law distributions attempt to model this.

Definition 4.1 (Power Law)   Two variables x and y are related by a power law when their scatter plot is linear on a log–log scale:

$$y(x) = cx^{-\gamma} \tag{4.1}$$

where c and γ are positive constants. The constant γ is often called the power law exponent.

Definition 4.2 (Degree Distribution)   The degree distribution of an undirected graph is a plot of the count c_k of nodes with degree k, versus the degree k, typically on a log–log scale. Occasionally, the fraction $\frac{c_k}{N}$ is used instead of c_k; however, this merely translates the log–log plot downwards. For directed graphs, out-degree and in-degree distributions are defined similarly.

Definition 4.3 (Scree Plot)   This is a plot of the eigenvalues (or singular values) of the adjacency matrix of the graph, versus their rank, using a log–log scale.

Both degree distributions and scree plots of many real-world graphs have been found to obey power laws. Examples include the Internet Autonomous System (AS) and router graphs [34, 36], the World Wide Web [10, 22, 40, 41], citation graphs [56], online social networks [25], and many others. Power laws also show up in the distribution of "bipartite cores" ($\approx$ communities) and the distribution of PageRank values [21, 52]. Indeed, power laws appear to be a defining characteristic of almost all large real-world graphs.

SIGNIFICANCE OF POWER LAWS. The significance of power law distributions lies in the fact that they are *heavy-tailed*, meaning that they decay more slowly than exponential or Gaussian distributions. Thus, a power law degree distribution would be much more likely to have nodes with a very high degree (much larger than the mean) than the other two distributions.

DEVIATIONS FROM POWER LAWS. Pennock et al. [54] and others have observed deviations from a pure power law distribution in several datasets. Two of the more common deviations are exponential cutoffs and lognormals. The exponential cutoff models distributions look like power laws over the lower range of values on the x axis but decay exponentially quickly for higher values; examples include the network of airports [9]. Lognormal distributions look like truncated parabolas on log–log scales and model situations where the plot "dips" downwards in the lower range of values on the x axis; examples include degree distributions of subsets of the World Wide Web (WWW), and many others [14, 54].

Small Diameters. Several definitions of the term "graph diameter" exist: the definition we use is called the "effective diameter" or "eccentricity" and is closely related to the "hop-plot" of a graph; both of these are defined below. The advantages are twofold: (a) the hop-plot and effective diameter can be computed in linear time and space using a randomized algorithm [50], and (b) this particular definition is robust in the presence of outliers.

Definition 4.4 (Hop-plot) Starting from a node u in the graph, we find the number of nodes $N_h(u)$ in a neighborhood of h hops. We repeat this starting from each node in the graph and sum the results to find the total neighborhood size N_h for h hops [$N_h = \sum_u N_h(u)$]. The hop-plot is just the plot of N_h versus h.

Definition 4.5 (Effective Diameter) This is the minimum number of hops in which some fraction (say, 90%) of all connected pairs of nodes can reach each other [63].

SIGNIFICANCE OF GRAPH DIAMETER. The diameters of many real-world graphs are very small compared to the graph size [6]: only around 4 for the Internet AS-level graph, 12 for the Internet router-level graph, 16 for the WWW, and the famous "six degrees of separation" in the social network. Any realistic graph generator needs to match this criterion.

Community Effects. Informally, a community is a set of nodes where each node is "closer" to the other nodes within the community than to nodes outside it. This effect has been found (or is believed to exist) in many real-world graphs, especially social networks [47, 60].

Community effects have typically been studied in two contexts: (a) local one-hop neighborhoods, as characterized by the *clustering coefficient* and (b) node groups with possibly longer paths between members, such as *graph partitions* and *bipartite cores*. All of these are discussed below.

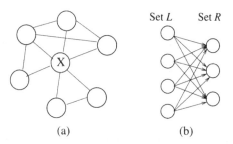

Figure 4.1. Indicators of community structure: (a) Node X has 6 neighbors (clustering coefficient). These neighbors could have been connected by $\binom{6}{2} = 15$ edges, but only 5 such edges exist. So, the local clustering coefficient of node X is $5/15 = 1/3$. (b) A 4×3 bipartite core, with each node in set L being connected to each node in set R.

Definition 4.6 (Clustering Coefficient)   For a node v with edges (u, v) and (v, w), the clustering coefficient of v measures the probability of existence of the third edge (u, w) [Fig. 4.1(a)]. The clustering coefficient of the entire graph is found by averaging over all nodes in the graph[1].

Definition 4.7 (Graph Partitioning)   Graph partitioning techniques typically break the graph into two disjoint partitions (or communities) while optimizing some measure; these two communities may then be repartitioned separately.

The popular METIS software tries to find the best separator, minimizing the number of edges cut in order to form two disconnected components of relatively similar sizes [37]. Many other measures and techniques exist [7, 20]. The recent cross-associations method [26, 27] finds such partitions in a completely *parameter-free* manner, using the Minimum Description Length (MDL) principle to choose the number of partitions as well as their memberships.

Definition 4.8 (Bipartite Core)   A bipartite core in a graph consists of two (not necessarily disjoint) sets of nodes L and R such that every node in L links to every node in R; links from R to L are optional [Fig. 4.1(b)].

SIGNIFICANCE OF GRAPH COMMUNITIES. Most real-world graphs exhibit strong community effects. Moody [47] found groupings based on race and age in a network of friendships in one American school, Schwartz and Wood [60] group people with shared interests from email logs; Borgs et al. [19] find communities from "cross posts" on Usenet; and Flake et al. [35] discover communities of Web pages in the WWW. This is also reflected in the clustering coefficients of real-world graphs: They are almost always much larger than in random graphs of the same size [65].

---

[1]We note here that there is at least one other definition based on counting triangles in the graph; both definitions make sense, but we use this one.

Other Patterns. Many other graph patterns have also been studied in the literature. We mention two of these that we will use later: (a) edge betweenness or stress, (b) "singular vector value" versus rank plots, and (c) resilience under attack.

Definition 4.9 (Edge Betweenness or Stress)   Consider all shortest paths between all pairs of nodes in a graph. The edge betweenness or stress of an edge is the number of these shortest paths that the edge belongs to and is thus a measure of the "load" on that edge.

Definition 4.10 (Stress Plot)   This is a plot of the number of edges s_k with stress k, versus k.

Definition 4.11 (Singular Vector Value versus Rank Plots)   The singular vector value of a node is the absolute value of the corresponding component of the first singular vector of the graph. It can be considered to be a measure of the "importance" of the node, and as we will see later, is closely related to the widely used concept of "Bonacich centrality" in social network analysis [18].

Definition 4.12 (Resilience)   A resilient graph is one whose diameter does not increase when its nodes or edges are removed according to some "attack" process [5, 50]. Most real-world graphs are very resilient against random failures but susceptible to targeted attacks (such as removal of nodes of the highest degree) [62].

As against these previous patterns, which appear in many real-world graphs, several patterns *specific* to the Internet and the WWW have been discovered. Broder et al. [22] show that the WWW has a "bow-tie" structure, with four roughly equal parts. There is a central strongly connected component (SCC), a set of pages (IN) pointing into this center, a set of pages (OUT) that the pages in the center point to, and a set of TENDRILS hang off the IN and OUT sections. Dill et al. [29] extend this by finding self-similarity in this structure. For the Internet topology, Tauro et al. [63] observe a "jellyfish" structure, with nodes organized in a set of concentric circles around a small core.

4.2.2 Graph Generators

Graph generators allow us to create synthetic graphs, which can then be used for, say, simulation studies. However, to be realistic, the generated graph must match all (or at least several) of the patterns mentioned above. By telling us which processes can (or cannot) lead to the development of these patterns, graph generators can provide insight into the creation of real-world graphs.

Graph models and generators can be broadly classified into four categories:

1. *Random Graph Models:* The graphs are generated by a random process. The basic random graph model has attracted a lot of research interest due to its phase transition properties.

2. *Preferential Attachment Models:* In these models, the "rich" get "richer" as the network grows, leading to power law effects. Some of today's most popular models belong to this class.

3. *Optimization-Based Models:* Here, power laws are shown to evolve when risks are minimized using limited resources. Together with the preferential attachment models, they try to provide mechanisms that automatically lead to power laws.

4. *Geographical Models:* These models consider the effects of geography (i.e., the *positions* of the nodes) on the growth and topology of the network. This is especially important for modeling router or power grid networks, which involve laying wires between points on the globe.

We will briefly touch upon some of these generators below. Tables 4.1 and 4.2 provide a taxonomy of generators.

Random Graph Models. In the earliest random graph model [32], we start with N nodes, and for every pair of nodes, an edge is added between them with probability p. This simple model leads to a surprising list of properties, including phase transitions in the size of the largest component and diameter logarithmic in graph size. Its ease of analysis has proven to be very useful in the early development of the field. However, its degree distribution is Poisson, whereas most real-world graphs seem to exhibit power law distributions. Also, the graphs lack community effects: The clustering coefficient is usually far smaller than that in comparable real-world graphs.

The basic Erdös–Rényi model has been extended in several ways, typically to match the power law degree distribution pattern [2, 49, 51]. These methods retain the simplicity and ease of analysis of the original model, while removing one of its weaknesses: the unrealistic degree distribution. However, these models do not describe any *process* by which power laws may arise automatically, which makes them less useful in understanding the internal processes behind graph formation in the real world (property P2). Also, most models make no attempt to match any other patterns (property P1), and further work is needed to incorporate community effects into the model.

Preferential Attachment Models. First developed in the mid-1950s [61], the idea of preferential attachment models has been rediscovered recently due to their ability to generate skewed distributions such as power laws [10, 55]. Informally, they use the concept of the "rich getting richer" over time: New nodes join the graph each time step and preferentially connect to existing nodes with high degree. This basic idea has been very influential and has formed the basis of a large body of further work [3, 4, 6, 11, 15, 17, 23, 52, 54, 67].

The preferential attachment models have several interesting properties:

- *Power Law Degree Distributions:* These models lead to power laws as a *by-product* of the graph generation method and not as a specific designed-in feature.

- *Low Diameter:* The generated graphs have $O(\log N)$ diameter. Thus, the increase in diameter is slower than the growth in graph size.
- *Resilience:* The generated graphs are resilient against random node/edge removals but quickly become disconnected when nodes are removed in descending order of degree [5, 50]. This matches the behavior of the Internet.
- *A Procedural Method:* Perhaps most importantly, these models describe a *process* that can lead to realistic graphs and power laws (matching property P2). The two main ideas are those of (a) growth and (b) preferential attachment; variants to the basic model add other ideas such as node "fitness." The ability of these models to match many real world graphs implies that graphs in the real world might indeed have been generated by similar processes.

Preferential attachment models are probably the most popular models currently, due to their ability to match power law degree distributions by such a simple set of steps. However, they typically do not exhibit community effects, and, apart from the paper of Pennock et al. [54], little effort has gone into finding reasons for deviations from power laws in real-world graphs (property P1).

One set of related models has shown promise recently: These are the "edge-copying" models [40, 41], where a node (such as a website) acquires edges by *copying links* from other nodes (websites). This is similar to preferential attachment because pages with high degree will be linked to by many other pages and so have a greater chance of getting copied. However, such graphs can be expected to have a large number of bipartite cores (which leads to the community effect). This makes the edge-copying technique a promising research direction.

Geographical Models. Several models introduce the constraints of geography into network formation. For example, it is easier (cheaper) to link two routers that are physically close to each other; most of our social contacts are people we meet often, and who consequently probably live close to us (say, in the same town or city), and so on. In this area, there are two important models: the *small-world* model, and the *Waxman* model.

The small-world model [65] starts with a regular lattice and adds/rewires some edges randomly. The original lattice represents ties between close friends, while the random edges link "acquaintances" and serve to connect different social circles. Thus, the resulting graph has low diameter but a high clustering coefficient—two patterns common to real-world graphs. However, the degree distribution is not a power law, and the basic model needs extensions to match this (property P1).

The Waxman model [66] probabilistically links two nodes in the graph, based on their geographical distance (in fact, the probability decreases exponentially with distance). The model is simple yet attractive and has been incorporated into the popular BRITE [45] generator used for graph generation in the networking community. However, it does not yield a power law degree distribution, and further work is needed to analyze the other graph patterns for this generator (property P1).

TABLE 4.1 Taxonomy of Graph Generators[a]

| Generator | Graph Type | | | | | | Degree Distributions | | | |
| | Undir. | Dir. | Bip. | Self-Loops | Mult. Edges | Geog. Info | Power Law | | | Exponential |
							Plain	Exp. Cutoff	Devia-tion	
Erdös–Rényi [32]	✓									✓
PLRG [2], PLOD [51]	✓			✓	✓		✓			
Exponential cutoff [49]	✓			✓	✓		✓	✓		
BA [10]	✓						✓ (γ = 3)			
AB [4]	✓			✓	✓		✓			
Edge copying [40, 41]		✓		✓	✓		✓			✓
GLP [23]	✓			✓	✓		✓			
Accelerated growth [11]	✓						Power-law mixture of γ = 2 and γ = 3			
Fitness model [15]	✓						✓ (modified)			

Aiello et al. [3]
Pandurangan et al. [52]
Inet [67]
Pennock et al. [54]
Small world [65]
Waxman [66]
BRITE [45]
Yook et al. [68]
Fabrikant et al. [33]
R-MAT [25] (DGX)

[a]This table shows the graph types and degree distributions that different graph generators can create. The graph type can be undirected, directed, bipartite, allowing self-loops or multigraph (multiple edges possible between nodes). The degree distributions can be power law (with possible exponential cutoffs, or other deviations such as lognormal/DGX) or exponential decay. Empty cells indicate that the corresponding property does not occur in the corresponding model.

TABLE 4.2 Taxonomy of Graph Generators (Continued.)[a]

Generator	Diameter or Avg Path Length	Community Bip. Core vs. Size	Community $C(k)$ vs. k	Clustering Coefficient	Remarks
Erdős–Rényi [32]	$O(\log N)$		Indep.	Low, CC $\propto N^{-1}$	
PLRG [2], PLOD [51]	$O(\log N)$	Indep.		CC $\to 0$ for large N	
Exponential cutoff [49]	$O(\log N)$			CC $\to 0$ for large N	
BA [10]	$O(\log N)$ or $O\left(\frac{\log N}{\log \log N}\right)$			CC $\propto N^{-0.75}$	
AB [4]					
Edge copying [40, 41]		Power law			
GLP [23]				Higher than AB, BA, PLRG	Internet only
Accelerated growth [11]				Non monotonic with N	
Fitness model [15]					

Generator	Diameter	Clustering coefficient CC	Notes
Aiello et al. [3]			
Pandurangan et al. [52]			
Inet [67]			Specific to the AS graph
Pennock et al. [54] Small world [65]	$O(N)$ for small N, $O(\ln N)$ for large N, depends on p	$CC(p) \propto (1 - p)^3$, Indep of N	N = num nodes p = rewiring prob
Waxman [66] BRITE [45]	Low (like in BA)	Like in BA	BA + Waxman with additions
Yook et al. [68] Fabrikant et al. [33]			Tree, density 1
R-MAT [25]	Low (empirically)		

[a]The comparisons are made for graph diameter, existence of community structure [number of bipartite cores versus core size, or clustering coefficient CC(k) of all nodes with degree k vs. k], and clustering coefficient. N is the number of nodes in the graph. The empty cells represent information unknown to the authors and require further research.

Optimization-Based Models. Optimization-based models provide another process leading to power laws. Carlson and Doyle [24, 30] propose in their *highly optimized tolerance (HOT)* model that power laws may arise in systems due to *trade-offs* between yield (or profit), resources (to prevent a risk from causing damage), and tolerance to risks. Apart from power laws, HOT also matches the resilience pattern of many real-world graphs (see Definition 4.12): it is robust against "designed-for" uncertainties but very sensitive to design flaws and unanticipated perturbations.

Several variations of the basic model have also been proposed: COLD [48] truncates the power law tails, while the *heuristically optimized trade-offs* model [33] needs only locally optimal decisions instead of global optimality.

Optimization-based models provide a very intuitive process that leads to both power laws and resilience (matching property P2). However, further research needs to be conducted on other patterns for such graphs (property P1). One step in this direction is the work of Berger et al. [13], who generalize the heuristically optimized trade-offs model and show that it is equivalent to a form of preferential attachment; thus, competition between opposing forces can give rise to preferential attachment, and we already know that preferential attachment can, in turn, lead to power laws and exponential cutoffs.

Summary of Graph Generators. Thus, we see that these are four general categories of graph generators, along with several "cross-category" generators too. Most of the generators do well with properties P3–P6 but need further research to determine their realism (property P1, that is, which patterns they match, and which they do not). Next, we will discuss our R-MAT graph generator, which attempts to match all of these properties, including the degree distribution (with both power laws and deviations), community structure, singular vector value versus rank patterns, and so on.

4.2.3   Other Related Concepts

There has been a lot of recent interest [31, 44] in the field of relational learning: learning structure from datasets that are structured as graphs. The primary difference between this exciting body of research and ours is that relational learning focuses on finding structure at the local level (e.g., if a movie is related to another movie that was nominated for/won an Academy Award, then it is likely to be nominated itself), whereas we are attempting to find patterns at the global level (e.g., power laws in degree distributions). Also, relational learning typically involves labeled data (e.g., the nodes in a graph could have different labels) and structure/predictions are made on those labels; our work attempts to find and model patterns that are solely properties of the graph topology.

Other topics of interest involving graphs include graph partitioning [59], frequent subgraph discovery (see [42, 43] and Chapter 7), finding cycles in graphs [8], navigation in graphs [39], and many others. All of these address interesting problems, and we are investigating their use in our work.

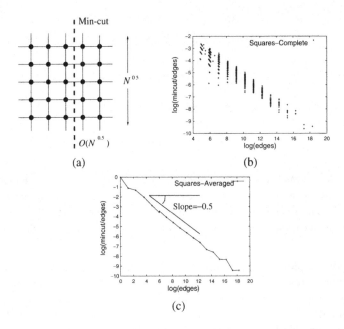

Figure 4.2. Example of min-cut plot: Plot (a) shows a portion of a regular 400 × 400 2D grid and a possible min-cut. Plot (b) shows the full 2D min-cut plot, and plot (c) shows the averaged plot. If the number of nodes is N, the length of each side is $\sqrt{N}$. Then the size of the min-cut is $O(\sqrt{N})$, which leads to a slope of -0.5, which is exactly what we observe.

4.3 NetMine AND R-MAT

Given a large graph, NetMine computes several patterns that often show up in real-world graphs. These include in- and out-degree distributions, singular value distributions, hop-plots, and distributions of singular vector value and stress. In the previous section, we defined all of these and referenced research that allows fast computation of these patterns. In this section, we will discuss two recent developments: the min-cut plots and the R-MAT graph generator.

4.3.1 Min-cut plots

A min-cut of an unlabeled graph $g = (\mathcal{V}, \mathcal{E})$ is a partition of the set of vertices $\mathcal{V}$ into two sets $\mathcal{V}_1$ and $\mathcal{V} - \mathcal{V}_1$ such that both partitions are of approximately the same size, and the number of edges crossing partition boundaries is minimized. The number of such crossing edges is called the min-cut size. Min-cut sizes of various classes of graphs have been studied extensively, and are known to have important effects on other properties of the graphs [58]. For example, Figure 4.2(a) shows a regular two-dimensional (2D) grid graph and one possible min-cut of the graph. We see that if the number of nodes is N, then the size of the min-cut (in this case) is $O(\sqrt{N})$.

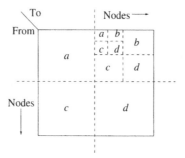

Figure 4.3. The R-MAT model: The adjacency matrix is broken into four equal-sized partitions, and one of those four is chosen according to a (possibly nonuniform) probability distribution. This partition is then split recursively till we reach a single cell, where an edge is placed. Multiple such edge placements are used to generate the full synthetic graph.

The min-cut plot is built as follows: Given a graph, its min-cut is found, and the set of edges crossing partition boundaries is deleted. This divides the graph into two disjoint graphs; the min-cut algorithm is then applied recursively to each of these subgraphs. This continues till the size of the graph reaches a small value (set to 20 in our case). Each application of the min-cut algorithm becomes a point in the min-cut plot. The graphs are drawn on a log–log scale. The x axis is the number of edges in a given graph. The y axis is the fraction of that graph's edges that were included in the edge-cut for that graph's separator.

Figure 4.2(b) shows the min-cut plot for the 2D grid graph. In plot (c), the value on the y axis is averaged over all points having the same x coordinate. The min-cut size is $O(\sqrt{N})$, so this plot should have a slope of -0.5, which is exactly what we observe.

4.3.2 The R-MAT Graph Generator

We have seen that most of the current graph generators focus on only one graph pattern—typically the degree distribution—and give low importance to all the others. There is also the question of how to fit model parameters to match a given graph. What we would like is a trade-off between parsimony (property P3), realism (property P1), and efficiency (properties P4 and P5). In this section, we present the R-MAT generator, which attempts to address all of these concerns.

Description and Properties. R-MAT is based on the well-known "80–20 rule" in one dimension (80% of the data falls on 20% of the data range), which is known to result in self-similarity and power laws; R-MAT extends this rule to a two-dimensional graph adjacency matrix. The R-MAT generator creates directed graphs with 2^n nodes and E edges, where both values are provided by the user. We start with an empty adjacency matrix and divide it into four equal-sized partitions. One of the four partitions is chosen with probabilities a, b, c, d, respectively ($a + b + c + d = 1$), as in Figure 4.3 The chosen partition is again subdivided into four smaller partitions, and the procedure is repeated until we reach a simple cell (= 1×1 partition). The nodes (i.e., row and column) corresponding to this cell are linked by an edge in the graph. This process is repeated E times to generate the full graph. There is a subtle point here: We may have *duplicate* edges (i.e., edges that fall into the same cell in the adjacency matrix), but we only keep one of them when generating an unweighted graph. To smooth out fluctuations in the degree distributions, some noise is added to the (a, b, c, d) values at each stage of the recursion, followed by renormalization (so that $a + b + c + d = 1$). Typically, $a \geq b$, $a \geq c$, $a \geq d$.

PARSIMONY. The algorithm needs only three parameters: the partition probabilities a, b, and c; $d = 1 - a - b - c$. Thus, the models is parsimonious.

DEGREE DISTRIBUTION. The following theorem gives the expected degree distribution of an R-MAT generated graph.

THEOREM 4.1 (Count vs. Degree)   For a pure R-MAT generated graph (i.e., without any smoothing factors), the expected number of nodes c_k with out-degree k is given by

$$c_k = \binom{E}{k} \sum_{i=0}^{n} \binom{n}{i} \left[p^{n-i}(1-p)^i \right]^k \left[1 - p^{n-i}(1-p)^i \right]^{E-k} \qquad (4.2)$$

where 2^n is the number of nodes in the R-MAT graph (typically $n = \lceil \log_2 N \rceil$ and $p = a + b$.

Proof. Theorem 4.1 is repeated and proven in the Appendix.

This is well modeled by a *discrete lognormal* [14], which looks like a truncated parabola on the log–log scale. By setting the parameters properly, this can successfully match both power law and "unimodal" distributions [54].

COMMUNITIES. Intuitively, R-MAT is generating "communities" in the graph:

- The partitions a and d represent separate groups of nodes that correspond to communities (say, Linux and Windows users).
- The partitions b and c are the *cross-links* between these two groups; edges there would denote friends with separate preferences.
- The recursive nature of the partitions means that we automatically get sub-communities within existing communities (say, RedHat and Mandrake enthusiasts within the Linux group).

DIAMETER, SINGULAR VALUES, AND OTHER PROPERTIES. We show experimentally that graphs generated by R-MAT have small diameters and match several other criteria as well.

EXTENSIONS TO UNDIRECTED, BIPARTITE, AND WEIGHTED GRAPHS. The basic model generates directed graphs; all the other types of graphs can be easily generated by minor modifications of the model. For undirected graphs, a directed graph is generated and then made symmetric. For bipartite graphs, the same approach is used; the only difference is that the adjacency matrix is now rectangular instead of square. For weighted graphs, the number of *duplicate* edges in each cell of the adjacency matrix is taken to be the weight of that edge. More details may be found in [25].

PARAMETER FITTING ALGORITHM. We are given some input graph and need to fit the R-MAT model parameters so that the generated graph matches the input graph in terms of graph patterns. Using Theorem 4.1 we can fit the in-degree and out-degree distributions; this gives us two equations (specifically, we get the values of $p = a + b$ and $q = a + c$). We need one more equation to fit the three model parameters.

We tried several experiments where we fit the *scree plot* (see Definition 4.3). However, we obtained comparable (and much faster) results by conjecturing that the $a : b$ and $a : c$ ratios are approximately $75 : 25$ (as seen in many real-world scenarios) and using these to fit the parameters. Hence, this is our current parameter-fitting method for R-MAT.

4.4 EXPERIMENTS

The questions we wish to answer are:

(Q1) How do the min-cut plots look for real-world graphs, and does R-MAT match them?

(Q2) How well can R-MAT match real-world graphs? How does it compare against other graph generators?

The datasets we use for our experiments are:

- *Epinions:* A directed graph of who-trusts-whom from epinions.com [57]: $N = 75,879$; $E = 508,960$.

- *Epinions-U:* An undirected version of the Epinions graph: $N = 75,879$; $E = 811,602$.

- *Clickstream:* A bipartite graph of Internet users' browsing behavior [46]. An edge (u, p) denotes that user u accessed page p. It has $23,396$ users, $199,308$ pages, and $952,580$ edges.

- *Lucent* is an undirected graph of network routers, obtained from www.isi.edu/scan/mercator/maps.html. $N = 112,969$; $E = 181,639$.

- *Router* is a larger graph (the SCAN + Lucent map) from the same uniform resource locator (URL), which subsumes the *Lucent* graph. $N = 284,805$; $E = 898,492$.

- *Google* is a graph of webpage connectivity from the Google [1] programming contest. $N = 916,428$; $E = 5,105,039$.

4.4.1 (Q1) Min-cut Plots

We plotted min-cut sizes for a variety of graphs. For each graph listed we used the Metis graph partitioning library [38] to generate a separator, as described by Blandford et al. [16].

Figure 4.4 shows min-cut sizes of some real-world graphs. For random graphs, we expect about half the edges to be included in the cut. Hence, the min-cut plot of a random graph would be a straight horizontal line with a y coordinate of about $\log(0.5) = -1$. A very separable graph (e.g., a line graph) might have only one edge in the cut; such a graph with N edges would have a y coordinate of $\log(1/N) = -\log(N)$, and its min-cut plot would thus be on the line $y = -x$. As we can see from Figure 4.4, the plots for real-world graphs do not match either of these situations, meaning that real-world graphs are quite far from either random graphs or simple line graphs.

Observation 4.1 (Noise)   We see that real-world graphs seem to have a lot of "noise" in their min-cut plots, as shown by the first row of Figure 4.4

Observation 4.2 ("Lip")   The ratio of min-cut size to number of edges decreases with increasing edges, except for graphs with large number of edges, where we observe a "lip" in the min-cut plot.

The min-cut plot contains important information about the graph [58]. Hence, any synthetically generated graph meant to simulate a real-world graph should match the min-cut plot of the real-world graph. In Figure 4.5, we compare the min-cut plots for the Epinions graph with a graph generated R-MAT. As can be seen, the basic shape of the plot is the same in both cases, though the R-MAT plot appears to be shifted slightly from the original.

Observation 4.2   The graph generated by R-MAT appears to match the basic shape of the min-cut plot.

4.4.2 (Q2) Comparing R-MAT to Real-World Graphs

As described in Section 4.3, R-MAT is applicable to directed, undirected, as well as bipartite graphs. We demonstrate results on one example graph of each type: the Epinions directed graph, the Epinions-U undirected graph, and the Clickstream bipartite graph. For each dataset, we fit the R-MAT parameters, generate a synthetic graph with these parameters, and compare it to the true graph. We also compare R-MAT with three existing generators chosen for their popularity or recency:

- The *AB* model [4]: This is a preferential attachment model with additional process to rewire edges and add links between existing nodes (instead of only adding links to new nodes).

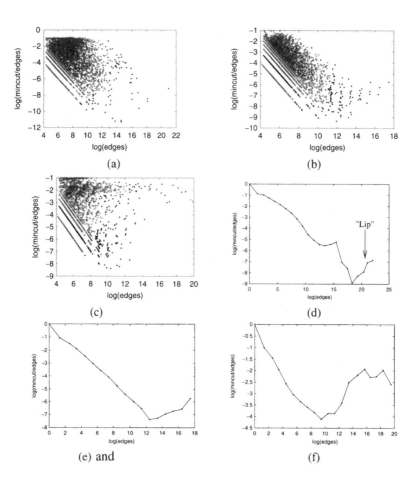

Figure 4.4. Min-cut plots for real-world graphs: Min-cut plots are shown for several real-world datasets. We plot the ratio of min-cut size to edges versus number of edges on a log–log scale. The first row shows the actual plots; in the second row, the cutsize-to-edge ratio is averaged over all points with the same number of edges. (a) Google min-cut plot, (b) Lucent min-cut plot, (c) Clickstream min-cut plot, (d) Google averaged, (e) Lucent averaged, and (f) Clickstream averaged.

- The *Generalized Linear Preference (GLP)* model [23]: This modifies the original preferential attachment equation with an extra parameter and is highly regarded in the networking community.
- The *PG* model [54]: This model has a parameter to traverse the continuum from pure preferential attachment to pure random attachment.

There are two important points to note:

- All of the above models are used to generate undirected graphs, and thus, we can compare them to R-MAT only on Epinions-U.

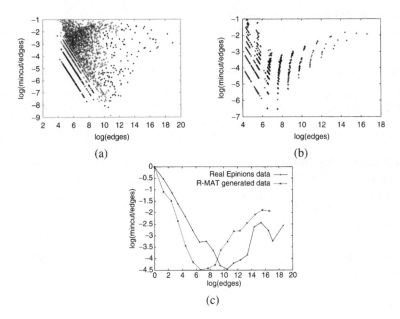

Figure 4.5. Min-cut plots for RMAT: We compare min-cut plots for the Epinions dataset and a dataset generated by RMAT, using properly chosen parameters. We see from plot (c) that the shapes of the min-cut plots are similar. (a) Epinions min-cuts, (b) RMAT min-cuts, and (c) averaged min-cuts.

- We were unaware of any method to fit the parameters of these models, so we fit them using a brute-force method. We use AB+, PG+, and GLP+ for the original algorithms augmented by our parameter fitting.

The graph patterns we look at are:

1. Both in-degree and out-degree distributions (Definition 4.2).
2. Hop-plot and effective diameter (Definitions 4.4 and 4.5).
3. Singular value vs. rank plot (also known as the scree plot; see Definition 4.3).
4. Singular vector value versus rank plots (Definition 4.11).
5. Stress distribution (Definition 4.10).

We have already compared R-MAT min-cut plots to those in the real graph, earlier in this section.

R-MAT on Directed Graphs. Figure 4.6 shows results on the Epinions directed graph. The R-MAT fit is very good; the other models considered are not applicable. The corresponding R-MAT parameters are shown in Table 4.3

TABLE 4.3   R-MAT Parameters for the Datasets

| | | | | R-MAT Parameters | | | |
Dataset	Graph Type	Dimensions	Edges	a	b	c	d
Epinions	Directed	$75,879 \times 75,879$	508,960	0.56	0.19	0.18	0.07
Clickstream	Bipartite	$23,396 \times 199,308$	952,580	0.50	0.15	0.19	0.16
Epinions-U	Undirected	$75,879 \times 75,879$	811,602	0.55	0.18	0.18	0.09

R-MAT on Bipartite Graphs. Figure 4.7 shows results on the Clickstream bipartite graph. As before, the R-MAT fit is very good. In particular, note that the in-degree distribution is a power law while the out-degree distribution deviates significantly from a power law; R-MAT matches *both* of these very well. Again, the other models are not applicable.

R-MAT on Undirected Graphs. Figure 4.8 shows the comparison plots on the Epinions-U undirected graph. R-MAT gives the closest fits. Also, note that all the y scales are logarithmic, so small differences in the plots actually represent significant deviations.

4.5   CONCLUSIONS

We described our NetMine toolkit for mining and visualizing the information in real-world graphs. The emphasis is on scalability, so that our algorithms can handle arbitrarily large graphs. When applied on real graphs, NetMine can check for many of the patterns that frequently occur in real-world graphs, and fit the R-MAT model parameters so that "similar" graphs can be generated. Deviations from the usual patterns could signify abnormalities or outliers in the data, and the R-MAT model with the fitted parameters can be used for, say, extrapolation of the data and "what-if" scenarios.

Specifically, the major contributions of this work are:

- **Min-cut Plots:** They show the relative size of the minimum cut in a graph partition. For regular 2D and 3D grid-style networks (like Delaunay triangulations for finite element analysis), these plots have a slope that depends on the intrinsic dimensionality of the grid. However, for real graphs, these plots show significantly more "noise" as well as a "lip". Their slope offers insight into their intrinsic structure.

- **R-MAT**: This simple, parsimonious graph model is able to match almost all the patterns of real-world graphs and is more widely applicable and accurate than other existing graph generators. We also showed how the parameters of R-MAT can be efficiently set to mimic any given graph; this allows the user to quickly generate similar graphs that are synthetic yet *realistic*. We were pleasantly surprised when R-MAT also matched the min-cut plots of real-world graphs.

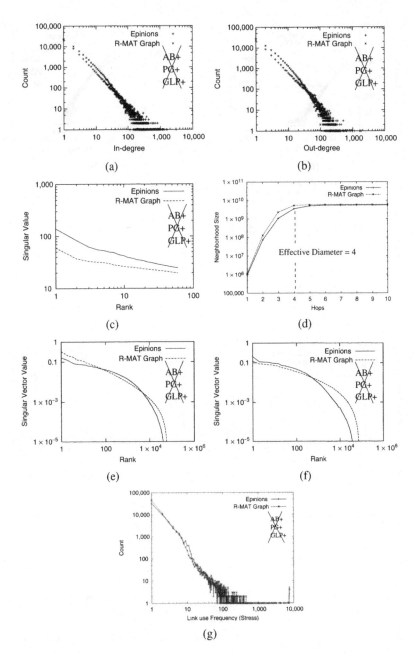

Figure 4.6. Results on the Epinions directed graph: The AB+, PG+ and GLP+ methods do not apply. The crosses and dashed lines represent the R-MAT-generated graphs, while the pluses and strong lines represent the real graph. (a) In-degree, (b) out-degree, (c) scree plot, (d) hop-plot, (e) first left singular vector, (f) first right singular vector, and (g) stress.

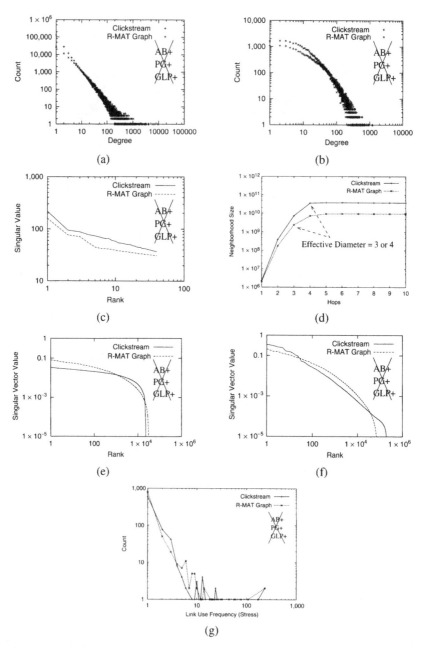

Figure 4.7. Results on the Clickstream bipartite graph: The AB+, PG+, and GLP+ methods do not apply. The crosses and dashed lines represent the R-MAT-generated graphs, while the pluses and strong lines represent the real graph. (a) In-degree, (b) out-degree, (c) scee plot, (d) hop-plot, (e) first left singular vector, (f) first right singular plot, and (g) stress.

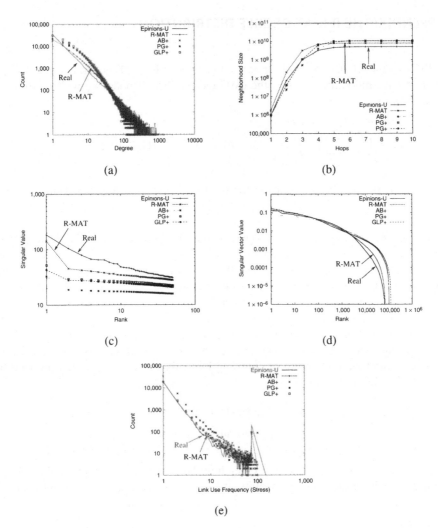

Figure 4.8. Results on the Epinions-U undirected graph: We show (a) degree, (b) hop-plot, (c) singular value, (d) singular vector value, and (e) stress distributions for the Epinions-U dataset. R-MAT gives the best matches to the Epinions-U graph, among all the generators. In fact, for the stress distribution, the R-MAT and Epinions-U plots are almost indistinguishable.

Moreover, we propose a list of natural tests that hold for a variety of real graphs: matching the power law/DGX distribution for the in- and out-degree, the hop-plot and the diameter of the graph, the singular value distribution, the values of the first singular vector ("Google-score"), and the "stress" distribution over the edges of the graph. All of these are packaged into the NetMine package, which provides a complete toolkit for analyzing and visualizing large graphs in a scalable manner, as shown by our experiments.

APPENDIX: THE R-MAT DEGREE DISTRIBUTION

THEOREM 4.A1 (Count vs. Degree)   For a pure R-MAT generated graph (i.e., without any smoothing factors), the expected number of nodes c_k with out-degree k is given by

$$c_k = \binom{E}{k} \sum_{i=0}^{n} \binom{n}{i} \left[p^{n-i}(1-p)^i \right]^k \left[1 - p^{n-i}(1-p)^i \right]^{E-k} \qquad (4.A.1)$$

where 2^n is the number of nodes in the R-MAT graph (typically $n = \lceil \log_2 N \rceil$ and $p = a + b$.

Proof. In the following discussion, we neglect the elimination of duplicate edges. This is a reasonable assumption: In most of our experiments, we found that the number of duplicate edges is far less than the total number of edges. Each edge that is "dropped" on to the adjacency matrix takes a specific path: At each stage of the recursion, the edge chooses either Up (corresponding to partitions a or b) or Down (corresponding to partitions c or d). There are n such stages, where 2^n is the number of nodes. Row X can be reached only if the edge follows a *unique* ordered sequence of Up and Down choices. Since the probability of choosing Up is $p = a + b$ and the probability of choosing Down is $1 - p = c + d$, the probability of the edge falling to row X is

$$P(X) = p^{\text{num(Up)}}(1 - p)^{\text{num(Down)}}$$
$$= p^{\text{num(Up)}}(1 - p)^{n-\text{num(Up)}} \qquad (4.A.2)$$

This equation means that any other row that requires the same number of Up choices will have the same probability as row X. The number of such rows can be easily seen to be $\binom{n}{\text{num(Up)}}$. Thus, we can think of different *classes* of rows:

Class	Probability of getting an edge	Num(rows)
0	p^n	$\binom{n}{0}$
1	$p^{n-1}(1-p)^1$	$\binom{n}{1}$
$\vdots$	$\vdots$	$\vdots$
$i - 1$	$p^{n-i}(1-p)^i$	$\binom{n}{i}$
$\vdots$	$\vdots$	$\vdots$
n	$(1-p)^n$	$\binom{n}{n}$

Now, we can calculate the count-degree plot. Let

$$NR_k = \text{number of rows with outdegree } k$$

$$= NR_{0,k} + NR_{1,k} + \cdots + NR_{n,k} \qquad (4.A.3)$$

where $NR_{i,k}$ = number of rows *of class i* with out-degree k. Thus, the expected number of rows with out-degree k is

$$E[NR_k] = \sum_{i=0}^{n} E[NR_{i,k}] \qquad (4.A.4)$$

Now, for a row in class i, each of the E edges can either drop into it [with probability $p^{n-i}(1-p)^i$] or not [with probability $1 - p^{n-i}(1-p)^i$]. Thus, the number of edges falling into this row is a binomially distributed random variable: $\text{Bin}[E, p^{n-i}(1-p)^i]$. Thus, the probability that it has exactly k edges is given by

$$P_{i,k} = \binom{E}{k} \left[p^{n-i}(1-p)^i\right]^k \left[1 - p^{n-i}(1-p)^i\right]^{E-k}$$

Thus, the expected number of such rows from class i is

$$E[NR_{i,k}] = \text{number of rows in class } i \times P_{i,k} = \binom{n}{i} P_{i,k}$$

$$= \binom{n}{i}\binom{E}{k} \left[p^{n-i}(1-p)^i\right]^k \left[1 - p^{n-i}(1-p)^i\right]^{E-k}$$

Using this in Equation (4.A.3) gives us

$$E[NR_k] = \binom{E}{k} \sum_{i=0}^{n} \left\{ \binom{n}{i} \left[p^{n-i}(1-p)^i\right]^k \left[1 - p^{n-i}(1-p)^i\right]^{E-k} \right\} \qquad (4.A.5)$$

Equation (4.A.5) gives us the count of nodes with out-degree k; thus we can plot the count-vs.-outdegree plot using this equation. *Q.E.D.*

Acknowledgment

This work is partially supported by the National Science Foundation under Grants No. CCR-0208853, ANI-0326472, IIS-0083148, IIS-0113089, IIS-0209107, IIS-0205224, INT-0318547, SENSOR-0329549, EF-0331657, IIS-0326322, and CNS-0433540 and by the Pennsylvania Infrastructure Technology Alliance, a partnership of Carnegie Mellon, Lehigh University, and the Commonwealth of Pennsylvania's Department of Community and Economic Development (DCED). Additional funding was provided by donations from Intel and by a gift from Northrop-Grumman Corporation. Any opinions, findings, and conclusions or recommendations expressed in this material are those of the author(s) and do not necessarily reflect the views of the National Science Foundation or other funding parties.

REFERENCES

1. Google programming contest (2002). http://www.google.com/programming-contest/.
2. W. Aiello, F. Chung, and L. Lu. A random graph model for massive graphs. In *ACM Symposium on Theory of Computing*, pp. 171–180, ACM Press, NY, NY, 2000.
3. W. Aiello, F. Chung, and L. Lu. Random evolution in massive graphs. In *IEEE Symposium on Foundations of Computer Science*, IEEE Computer Society Press, Los Alamitos, CA, 2001.
4. R. Albert and A.-L. Barabási. Topology of complex networks: Local events and universality. *Physical Review Letters*, 85(24), 2000.
5. R. Albert, H. Jeong, and A.-L. Barabási. Error and attack tolerance of complex networks. *Nature*, 406:378–381, 2000.
6. R. Albert and A.-L. Barabási. Statistical mechanics of complex networks. *Reviews of Modern Physics*, 74(1):47–97, 2002.
7. N. Alon. Spectral techniques in graph algorithms. In C. L. Lucchesi and A. V. Moura, eds. *Lecture Notes in Computer Science 1380*, pp. 206–215. Springer, Berlin, 1998.
8. N. Alon, R. Yuster, and U. Zwick. Finding and counting given length cycles. *Algorithmica*, 17(3):209–223, 1997.
9. L. A. N. Amaral, A. Scala, M. Barthélémy, and H. E. Stanley. Classes of behavior of small-world networks. In *Proceedings of National Academy of Science, U.S.A.*, 97, 2000.
10. A.-L. Barabási and R. Albert. Emergence of scaling in random networks. *Science*, 286:509–512, 1999.
11. A. L. Barabási, H. Jeong, Z. Néda, E. Ravasz, A. Schubert, and T. Vicsek. Evolution of the social network of scientific collaborations. *Physica A*, 311:590–614, 2002.
12. A.-L. Barabási. *Linked: The New Science of Networks*. Perseus Publishing, May, New York, NY, 2002.
13. N. Berger, C. Borgs, J. Chayes, R. M. D'Souza, and R. D. Kleinberg. Degree distribution of competition-induced preferential attachment graphs. *Combinatorics, Probability and Computing*, 14, pp. 697–721, 2005.
14. Z. Bi, C. Faloutsos, and F. Korn. The DGX distribution for mining massive, skewed data. In *Knowledge Discovery and Datamining*, pp. 17–26, ACM Press, New York, NY, 2001.
15. G. Bianconi and A.-L. Barabási. Competition and multiscaling in evolving networks. *Europhysics Letters*, 54:436–442, 2001.
16. D. Blandford, G. E. Blelloch, and I. Kash. Compact representations of separable graphs. In *ACM-SIAM Symposium on Discrete Algorithms*, SIAM, Philadelphia, PA, 2003.
17. B. Bollobás, C. Borgs, J. Chayes, and O. Riordan. Directed scale-free graphs. In *SIAM Symposium on Discrete Algorithms*, pp. 132–139, 2003.
18. P. Bonacich. Power and centrality: A family of measures. *American Journal of Sociology*, 92:1170–1182, 1987.
19. C. Borgs, J. Chayes, M. Mahdian, and A. Saberi. Exploring the community structure of newsgroups (Extended Abstract). In *Knowledge Discovery and Datamining*, ACM Press, New York, NY, 2004.
20. U. Brandes, M. Gaertler, and D. Wagner. Experiments on graph clustering algorithms. In European Symposium on Algorithms, 568–579, Springer Verlag, Berlin, Germany, 2003.
21. S. Brin and L. Page. The anatomy of a large-scale hypertextual Web search engine. *Computer Networks and ISDN Systems*, 30(1–7):107–117, 1998.

22. A. Broder, R. Kumar, F. Maghoul, P. Raghavan, S. Rajagopalan, R. Stata, A. Tomkins, and J. Wiener. Graph structure in the web: Experiments and models. In *World Wide Web Conference*, 2000.

23. T. Bu and D. Towsley. On distinguishing between Internet power law topology generators. In *INFOCOM*, pp. 638–647, IEEE Computer Society Press, Los Alamitos, CA, 2002.

24. J. M. Carlson and J. Doyle. Highly optimized tolerance: A mechanism for power laws in designed systems. *Physics Review E*, 60(2):1412–1427, 1999.

25. D. Chakrabarti, Y. Zhan, and C. Faloutsos. R-MAT: A recursive model for graph mining. In *SIAM Data Mining*, 2004.

26. D. Chakrabarti, S. Papadimitriou, D. Modha, and C. Faloutsos. Fully automatic Cross-associations. In *Knowledge Discovery and Datamining*, ACM Press, NY, NY, 2004.

27. D. Chakrabarti. AutoPart: Parameter-free graph partitioning and outlier detection. In *Principles & Practice of Knowledge Discovery in Databases*, Springer, Berlin, Germany, 112–124, 2004.

28. H. Chen, J. Schroeder, R. Hauck, L. Ridgeway, H. Atabaksh, H. Gupta, C. Boarman, K. Rasmussen, and A. Clements. COPLINK Connect: Information and knowledge management for law enforcement. *CACM*, 46(1):28–34, 2003.

29. S. Dill, R. Kumar, K. S. McCurley, S. Rajagopalan, D. Sivakumar, and A. Tomkins. Self-similarity in the Web. In *Very Large Data Bases*, Morgan Kaufmann, San Francisco, CA, 2001.

30. J. Doyle and J. M. Carlson. Power laws, highly optimized tolerance, and generalized source coding. *Physical Review Letters*, 84(24):5656–5659, June 2000.

31. S. Džeroski and N. Lavrač, eds. *Relational Data Mining*. Springer, Berlin, Germany, 2001.

32. P. Erdös and A. Rényi. On the evolution of random graphs. *Publication of the Mathematical Institute of the Hungarian Acadamy of Science*, 5:17–61, 1960.

33. A. Fabrikant, E. Koutsoupias, and C. H. Papadimitriou. Heuristically optimized trade-offs: A new paradigm for power laws in the Internet (extended abstract), International Colloquium on Automata, Languages, and Programming, 110–122, Springer Verlag, Berlin, 2002.

34. M. Faloutsos, P. Faloutsos, and C. Faloutsos. On power-law relationships of the Internet topology. In *Conf. of the ACM Special Interest Group on Data Communications*, pp. 251–262, ACM Press, NY, NY, 1999.

35. G. W. Flake, S. Lawrence, and C. Lee Giles. Efficient identification of Web communities. In *Knowledge Discovery & Datamining*, ACM Press, NY, NY, 2000.

36. R. Govindan and H. Tangmunarunkit. Heuristics for Internet map discovery. In *IEEE INFOCOM 2000*, pp 1371–1380, Tel Aviv, Israel, March 2000.

37. G. Karypis and V. Kumar. Multilevel algorithms for multi-constraint graph partitioning. Technical Report 98-019, University of Minnesota, Twin Cities, 1998.

38. G. Karypis and V. Kumar. A fast and high quality multilevel scheme for partitioning irregular graphs. SIAM Journal on Scientific Computing, 20(1):359–392, 1999.

39. J. Kleinberg. The small-world phenomenon: An algorithmic perspective. Technical Report 99-1776, Cornell Computer Science Department, Cornell University, Ithaca, NY, 1999.

40. J. Kleinberg, S. R. Kumar, P. Raghavan, S. Rajagopalan, and A. Tomkins. The web as a graph: Measurements, models and methods. In Proceedings of the International Conference on Combinatorics and Computing, Tokyo, Japan, pp. 1–17, 1999.

41. S. R. Kumar, P. Raghavan, S. Rajagopalan, and A. Tomkins. Extracting large-scale knowledge bases from the web. In *Very Large Data Bases*, 639–650, Edinburgh, Scotland, 1999.

42. M. Kuramochi and G. Karypis. Frequent subgraph discovery. In *IEEE International Conference on Data Mining*, Nov 29–Dec 2, pp. 313–320, 2001.

43. M. Kuramochi and G. Karypis. Discovering frequent geometric subgraphs. In *IEEE International Conference on Data Mining*, Taebashi City, Japan; Dec 9–12, 2002.

44. A. McGovern and D. Jensen. Identifying predictive structures in relational data using multiple instance learning. In *International Conference on Machine Learning*, Aug 21–24, Washington, DC, 2003.

45. A. Medina, I. Matta, and J. Byers. On the origin of power laws in Internet topologies. In *Conf. of the ACM Special Interest Group on Data Communications*, Aug 28–Sep 1, Stockholm, Sweden, pp. 18–34, 2000.

46. A. L. Montgomery and C. Faloutsos. Identifying Web browsing trends and patterns. *IEEE Computer*, 34(7):94–95, 2001.

47. J. Moody. Race, school integration, and friendship segregation in America. *American Journal of Sociology*, 107(3):679–716, 2001.

48. M. E. J. Newman, M. Girvan, and J. D. Farmer. Optimal design, robustness and risk aversion. *Physical Review Letters*, 89(2), 028301 1–4, 2002.

49. M. E. J. Newman, S. H. Strogatz, and D. J. Watts. Random graphs with arbitrary degree distributions and their applications. *Physical Review E*, 64, 026118 1–17, 2001.

50. C. R. Palmer, P. B. Gibbons, and C. Faloutsos. ANF: A fast and scalable tool for data mining in massive graphs. In *Knowledge Discovery & Datamining*, Edmonton, AB, Canada, July 23–26, 2002.

51. C. R. Palmer and J. Gregory Steffan. Generating network topologies that obey power laws. In *GLOBECOM*, November, San Francisco, CA, 2000.

52. G. Pandurangan, P. Raghavan, and E. Upfal. Using PageRank to characterize Web structure. In International Computing and Combinatorics Conference, 330–339, Aug 15–17, Singapore, 2002.

53. R. Pastor-Satorras and A. Vespignani. Epidemic spreading in scale-free networks. *Physical Review Letters*, 86(14):3200–3203, 2001.

54. D. M. Pennock, G. W. Flake, S. Lawrence, E. J. Glover, and C. Lee Giles. Winners don't take all: Characterizing the competition for links on the Web. *Proceedings of the National Academy of Sciences*, 99(8):5207–5211, 2002.

55. D. S. De S. Price. A general theory of bibliometric and other cumulative advantage processes. *Journal of American Society of Information Science*, 27:292–306, 1976.

56. S. Redner. How popular is your paper? An empirical study of the citation distribution. *European Physical Journal B*, 4:131–134, 1998.

57. M. Richardson and P. Domingos. Mining knowledge-sharing sites for viral marketing. In *Knowledge Discovery & Datamining*, pp. 61–70, Edmonton, Canada, 2002.

58. A. L. Rosenberg and L. S. Heath. *Graph Separators, with Applications*. Kluwer Academic/Plenum, London, 2001.

59. K. Schloegel, G. Karypis, and V. Kumar. Graph partitioning for high performance scientific simulations. In *CRPC Parallel Computing Handbook*, Morgan Kaufmann, San Francisco, CA, 2000.

60. M. F. Schwartz and D. C. M. Wood. Discovering shared interests using graph analysis. *Communications of the ACM*, 40(3):78–89, 1993.

61. H. Simon. On a class of skew distribution functions. *Biometrika*, 42:425–440, 1955.

62. H. Tangmunarunkit, R. Govindan, S. Jamin, S. Shenker, and W. Willinger. Network topologies, power laws, and hierarchy. Technical Report 01-746, University of Southern California, Los Angeles, 2001.

63. S. L. Tauro, C. Palmer, G. Siganos, and M. Faloutsos. A simple conceptual model for the Internet topology. In Global Internet, San Antonio, Texas, 2001.

64. D. J. Watts. *Six Degrees: The Science of a Connected Age.* W. W. Norton, New York, NY, 2003.

65. D. J. Watts and S. H. Strogatz. Collective dynamics of "small-world" networks. *Nature*, 393:440–442, 1998.

66. B. M. Waxman. Routing of multipoint connections. *IEEE Journal on Selected Areas in Communications*, 6(9):1617–1622, December 1988.

67. J. Winick and S. Jamin. Inet-3.0: Internet Topology Generator. Technical Report CSE-TR-456-02, University of Michigan, Ann Arbor, MI, 2002.

68. S.-H. Yook, H. Jeong, and A.-L. Barabási. Modeling the Internet's large-scale topology. *Proceedings of the National Academy of Sciences*, 99(21):13382–13386, 2002.

Part II

MINING TECHNIQUES

5

DISCOVERY OF FREQUENT SUBSTRUCTURES

XIFENG YAN AND JIAWEI HAN

University of Illinois at Urbana-Champaign, Urbana-Champaign, Illinois

5.1 INTRODUCTION

Graphs become increasingly important in modeling complicated structures, such as circuits, images, chemical compounds, protein structures, biological networks, the Web, work flows, and XML documents. We have witnessed many graph-related algorithms developed in chemical informatics [3, 4, 26, 27], computer vision [29], video indexing [25], and text retrieval [19].

Among various kinds of graph patterns, frequent substructures are very basic ones that can be discovered in a set of graphs. They are useful at characterizing graph sets, discriminating different groups of graphs, classifying and clustering graphs, and building graph indices. Borgelt and Berthold [6] illustrated the discovery of active chemical structures in a dataset that screened for the human immunodeficiency virus (HIV) by contrasting the support of frequent graphs between different classes. Deshpande et al. [8] used frequent structures as features to classify chemical compounds. Huan et al. [13] successfully applied the frequent graph mining technique to study protein structural families. Frequent graph patterns were also used as indexing features by Yan et al. [33] to perform fast graph search. Their method outperforms the traditional path-based indexing approach significantly. Koyuturk et al. [16] proposed a method to detect frequent subgraphs in biological networks.

Mining Graph Data, Edited by Diane J. Cook and Lawrence B. Holder
Copyright © 2007 John Wiley & Sons, Inc.

For example, they observed considerably large frequent subpathways in metabolic networks.

The discovery of frequent substructures usually consists of two steps. In the first step, it generates frequent substructure candidates while the frequency of each candidate is checked in the second step. Most studies of frequent substructure discovery focus on the first step since the second step involves subgraph isomorphism that is NP-complete.

The initial frequent substructure mining algorithm, called AGM, was proposed by Inokuchi et al. (see [15] and Chapter 9), which shares similar characteristics with the Apriori-based itemset mining [1]. The Apriori property is also used by other frequent substructure discovery algorithms such as FSG (see [17] and Chapter 6) and the path-join algorithm [28]. All of them require a join operation to merge two (or more) frequent substructures into one larger substructure candidate. They distinguish themselves by using different building blocks: vertices, edges, and edge-disjoint paths. In the context of frequent substructure mining, Apriori-based algorithms have two kinds of considerable overheads: (1) joining two size-k frequent graphs (or other structures like paths in [28]) to generate size-$(k + 1)$ graph candidates, and (2) checking the frequency of these candidates separately. These overheads constitute the performance bottleneck of Apriori-based algorithms.

To avoid the overheads incurred in Apriori-based algorithms, non-Apriori-based algorithms such as gSpan [30], MoFa [6], FFSM [14], SPIN [24], and Gaston [20] have been developed recently. These algorithms are inspired by PrefixSpan [23], TreeMinerV (see [35] and Chapter 15), and FREQT [2] at mining sequences and trees, respectively. All of these algorithms adopt the pattern growth methodology [10], which intends to extend patterns from a single pattern directly.

The Apriori-based approach has to use the breadth-first search (BFS) strategy because of its level-wise candidate generation. To determine whether a size-$(k + 1)$ graph is frequent, it has to check all of its corresponding size-k subgraphs to obtain an upper bound of its frequency. Thus, before mining any size-$(k + 1)$ subgraph, the Apriori-based approach usually has to complete the mining of size-k subgraphs. Therefore, BFS is necessary in the Apriori-like approach. In contrast, the pattern growth approach is more flexible on the search method. Both breadth-first search and depth-first search (DFS) can work.

In this chapter, we will examine Apriori-based and pattern growth algorithms to give an overview of frequent substructure discovery algorithms. The variants of frequent substructures and their applications will be presented afterwards.

5.2 PRELIMINARY CONCEPTS

We denote the vertex set of a graph g by $V(g)$ and the edge set by $E(g)$. A label function, L, maps a vertex or an edge to a label. A graph g is a subgraph of another graph g' if there exists a subgraph isomorphism from g to g'.

Definition 5.1 (Frequent Graph)   Given a labeled graph dataset, $D = \{G_1, G_2, \ldots, G_n\}$, support($g$) [or frequency($g$)] is the percentage (or number) of graphs

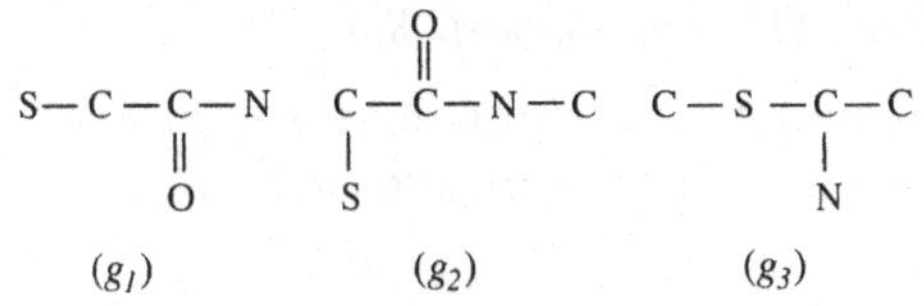

$$
\begin{array}{ccc}
\underset{\underset{\displaystyle O}{\overset{\displaystyle \|}{}}}{S-\underset{\underset{O}{\overset{\|}{}}}{C}-C-N} &
\begin{array}{c} O \\ \| \\ C-C-N-C \\ | \\ S \end{array} &
C-S-\underset{\underset{N}{|}}{C}-C
\end{array}
$$

$$(g_1) \qquad\qquad (g_2) \qquad\qquad (g_3)$$

Figure 5.1. Sample graph dataset.

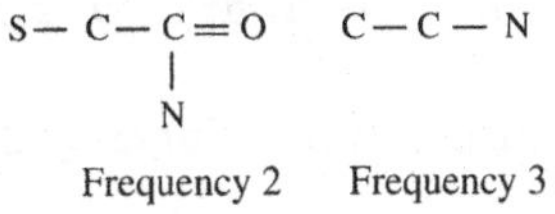

$$
\begin{array}{cc}
S-C-\underset{\underset{N}{|}}{C}=O & \qquad C-C-N
\end{array}
$$

Frequency 2   Frequency 3

Figure 5.2. Frequent graphs.

in D where g is a subgraph. A graph is frequent if its support is no less than a minimum support threshold, min_support.

EXAMPLE 5.1   Figure 5.1 shows a sample chemical structure dataset. Figure 5.2 depicts two frequent subgraphs in this dataset if the minimum support is set at 66.6%.

5.3 APRIORI-BASED APPROACH

The initial frequent substructure mining algorithm proposed by Inokuchi et al. [15] shares similar characteristics with Apriori-based frequent itemset mining algorithms developed by Agrawal and Srikant [1]. The frequent graphs having larger sizes are searched in a bottom-up manner by generating candidates having an extra vertex, edge, or path.

The general framework of Apriori-based methods is outlined in Algorithm 5.1. S_k is the frequent substructure set of size k. We will clarify the definition of graph size when the concrete mining algorithms are presented. Algorithm 5.1 adopts a level-wise mining methodology. At each iteration, the size of newly discovered frequent substructures is increased by one. These new substructures are first generated by joining two similar but slightly different frequent subgraphs that are discovered in the last call of Algorithm 5.1. This candidate generation procedure is shown on line 4. The newly formed graphs are then checked for their frequency. The frequent ones are detected and used to generate larger candidates in the next round.

The main design complexity of Apriori-based substructure mining algorithms is the candidate generation step. The candidate generation for frequent itemset mining is straightforward. For example, suppose we have two frequent itemsets of size 3: (abc) and (bcd), the candidate frequent itemset of size 4 generated from these two is $(abcd)$. That is, after we find two itemsets (abc) and (bcd) are frequent, we should check the frequency of $(abcd)$. However, the candidate generation problem

Algorithm 5.1   Apriori(D, *min_support*, S_k)

Input: A graph dataset D and *min_support*.
Output: A frequent substructure set S_k.

```
1: S_{k+1} ← ∅;
2: for each frequent g_i ∈ S_k do
3:    for each frequent g_j ∈ S_k do
4:       for each size (k+1) graph g formed by the merge of
         g_i and g_j do
5:          if g is frequent in D and g ∉ S_{k+1} then
6:             insert g to S_{k+1};
7: if S_{k+1} ≠ ∅ then
8:    call Apriori(D, min_support, S_{k+1});
9: return;
```

in frequent substructure mining becomes much harder than the case in frequent itemset mining since there are a lot of ways to join two substructures.

Researchers have proposed various kinds of candidate generation strategies. AGM [15] proposed a vertex-based candidate generation method that increases the substructure size by one vertex at each iteration of Algorithm 5.1. Two size-k frequent graphs are joined only when the two graphs have the same size-$(k-1)$ subgraph. Here the size of a graph means the number of vertices in a graph. The newly formed candidate includes the common size-$(k-1)$ subgraph and the additional two vertices from the two size-k patterns. Since it is undetermined whether there is an edge connecting the additional two vertices, we actually can form two candidates. Figure 5.3 depicts the two substructures joined by two chains.

FSG proposed by Kuramochi and Karypis [17] adopts an edge-based method that increases the substructure size by one edge in each call of Algorithm 5.1. In FSG, two size-k patterns are merged if and only if they share the same subgraph that has $k-1$ edges, which is called the *core*. Here the size of a graph means the number of edges in a graph. The newly formed candidate includes the core and the additional two edges from the size-k patterns. Figure 5.4 shows potential candidates formed by two structure patterns. Each candidate has one more edge than these two patterns. This example illustrates the complexity of joining two structures to form a large pattern candidate. In [17], Kuramochi and Karypis illustrate two other candidate generation cases.

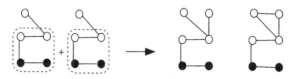

Figure 5.3.   AGM [15].

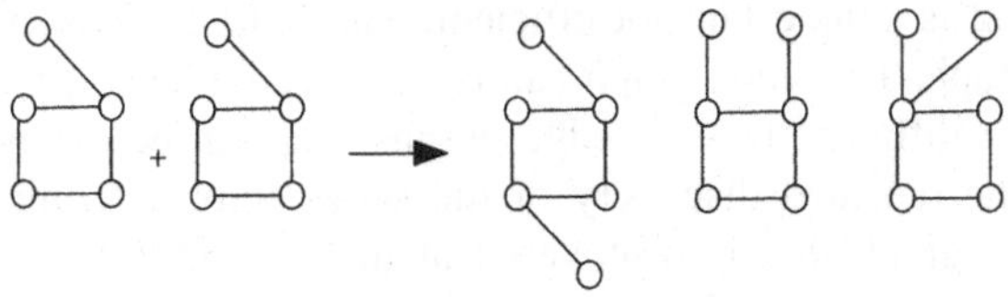

Figure 5.4. FSG [17].

Other Apriori-based methods such as the disjoint-path method proposed by Vanetik [28] use more complicated candidate generation procedures. For example, in [28], graphs are classified by the number of disjoint paths they have. A substructure pattern with $k+1$ disjoint paths is generated by joining substructures with k disjoint paths.

Apriori-based algorithms have considerable overheads at joining two size-k frequent substructures to generate size-$(k+1)$ graph candidates. To avoid such overheads, non-Apriori-based algorithms have been developed recently, most of which adopt the pattern growth methodology [10], which intends to extend patterns from a single pattern directly. In the next section, we are going to introduce the concept of a pattern growth approach.

5.4 PATTERN GROWTH APPROACH

A graph g can be extended by adding a new edge e. The newly formed graph is denoted by $g \diamond_x e$. Edge e may or may not introduce a new vertex to g. If e introduces a new vertex, we denote the new graph by $g \diamond_{xf} e$, otherwise, $g \diamond_{xb} e$, where f or b indicates that the extension is in a *forward* or *backward* direction.

Algorithm 5.2 (PatternGrowth) illustrates a general framework of pattern growth-based frequent substructure mining algorithm. For each discovered graph g, it performs extensions recursively until all the frequent graphs with g embedded are discovered. The recursion stops once no frequent graph can be generated any more.

Algorithm **5.2**   PatternGrowth(g, D, *min_support*, S)

```
Input: A frequent graph g, a graph dataset D, and
       min_support.
Output: A frequent substructure set S.
1: if g ∈ S then return;
2: else insert g to S;
3: scan D once, find all the edges e such that g can be
   extended to g ◊ₓ e ;
4: for each frequent g ◊ₓ e do
5:    Call PatternGrowth(g ◊ₓ e, D, min_support, S);
6: return;
```

Algorithm 5.2 is simple but not efficient. The bottleneck is at the inefficiency of extending a graph. The same graph can be discovered many times. For example, there may exist n different $(n-1)$-edge graphs that can be extended to the same n-edge graph. The repeated discovery of the same graph is computationally inefficient. We call a graph that is discovered at the second time a *duplicate graph*. Although line 1 of Algorithm 5.2 gets rid of duplicate graphs, the generation and detection of duplicate graphs may cause additional workloads. To reduce the generation of duplicate graphs, each frequent graph should be extended as conservatively as possible. This principle leads to the design of several new algorithms such as gSpan [30], MoFa [6], FFSM [14], SPIN [24], and Gaston [20]. We take gSpan as an example to illustrate the concept of a pattern-growth approach.

gSpan is designed to reduce the generation of duplicate graphs; it need not search previous discovered frequent graphs for duplicate detection; and it does not extend any duplicate graph but still guarantees the discovery of the complete set of frequent graphs.

Depth-first search is adopted by gSpan to traverse graphs. Initially, a starting vertex is randomly chosen, and the vertices in a graph are marked so that one can tell which vertices are visited. The visited vertex set is expanded repeatedly until a full DFS tree is built. One graph may have various DFS trees depending on how the depth-first search is performed, that is, the vertex visiting order. The darkened edges in Figures 5.5(b)–5.5(d) show three DFS trees for the same graph of Figure 5.5(a) (the vertex labels are x, y, and z; the edge labels are a and b; the alphabetic order is taken as the default order in the labels). When building a DFS tree, the visiting sequence of vertices forms a linear order. We use subscripts to record this order, where $i < j$ means v_i is visited before v_j. T is named a *DFS subscripting* of G.

Given a DFS tree T, we call the starting vertex in T, v_0, the *root*, and the last visited vertex, v_n, the *rightmost vertex*. The straight path from v_0 to v_n is called the *rightmost path*. In Figures 5.5(b)–5.5(d), three different subscriptings are generated based on the corresponding DFS trees. The rightmost path is (v_0, v_1, v_3) in Figures 5.5(b) and 5.5(c), and (v_0, v_1, v_2, v_3) in Figure 5.5(d).

PatternGrowth extends a frequent graph in every possible position, which may generate a large number of duplicate graphs. In gSpan, we introduce a more sophisticated extension method. The new method restricts the extension as follows: Given a graph G and a DFS tree T in G, a new edge e can be added between the rightmost vertex and other vertices on the rightmost path (*backward extension*); or it can introduce a new vertex and connect to vertices on the rightmost path (*forward*

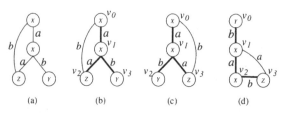

(a)   (b)   (c)   (d)

Figure 5.5.   DFS subscripting.

extension). Since both kinds of extensions take place on the rightmost path, we call them *rightmost extension*, denoted by $G \diamond_r e$ (for brevity, T is omitted here).

EXAMPLE 5.2 If we want to extend the graph in Figure 5.5(b), the backward extension candidates can be (v_3, v_0). The forward extension candidates can be edges extending from v_3, v_1, or v_0 with a new vertex introduced.

    Figures 5.6(b)–5.6(g) show all the potential rightmost extensions of Figure 5.6(a) (the darkened vertices represent the rightmost path). Among them, Figures 5.6(b)–5.6(d) grow from the rightmost vertex while Figures 5.6(e)–5.6(g) grow from other vertices on the rightmost path. Figures 5.6(b.0)–5.6(b.4) are children of Figure 5.6(b), and Figures 5.6(f.0)–5.6(f.3) are children of Figure 5.6(f). In summary, backward extension only takes place on the rightmost vertex while forward extension introduces a new edge from vertices on the rightmost path. This restricted extension is similar to TreeMinerV's equivalence class extension [35] and FREQT's rightmost expansion [2] in frequent tree discovery.

    Since many DFS trees/subscriptings may exist for the same graph, we choose one of them as the *base subscripting* and only conduct rightmost extensions on that DFS tree/subscripting. Otherwise, rightmost extensions cannot reduce the generation of duplicate graphs because we have to extend the same graph for every DFS subscripting.

    We transform each subscripted graph to an edge sequence, called DFS code, so that we can build an order among these sequences. The goal is to select the subscripting that generates the minimum sequence as its base subscripting. There are two kinds of orders in this transformation process: (1) edge order, which maps edges in a subscripted graph into a sequence; and (2) sequence order, which builds an order among edge sequences, that is, graphs. In [31] Yan and Han illustrate how to build an order among all of DFS codes, either from the same graph or from different graphs. We are not going to elaborate the ordering technique. In summary, one graph has at least one DFS code. Some graphs may have multiple DFS codes, among which one code is minimum.

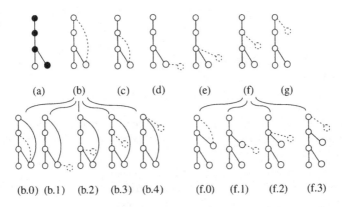

Figure 5.6. Rightmost extension.

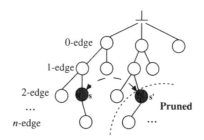

Figure 5.7. Lexicographic search tree.

Figure 5.7 shows how to arrange all DFS codes in a search tree through right-most extensions. The root is an empty code. Each node is a DFS code encoding a graph. Each edge represents a rightmost extension from a $(k-1)$-length DFS code to a k-length DFS code. The tree itself is ordered: Left siblings are smaller than right siblings in the sense of DFS lexicographic order. Since any graph has at least one DFS code, the search tree can enumerate all possible subgraphs in a graph dataset. However, one graph may have several DFS codes, minimum and nonminimum. The search of nonminimum DFS codes does not produce useful results. *Is it necessary to perform rightmost extensions on nonminimum DFS codes?* The answer is *"no."* If codes s and s' in Figure 5.7 encode the same graph and s' is not a minimum code, the search space under s' can be safely pruned [31].

The details of gSpan are depicted in Algorithm 5.3. gSpan is called recursively to extend a graph pattern until the support of the newly formed graph is lower than *min_support* or its code is not minimum any more. The difference between gSpan and PatternGrowth is at the rightmost extension and extension termination of nonminimum DFS codes (Algorithm 5.3 lines 1 and 2). We replace the existence condition in Algorithm 5.2 lines 1 and 2 with the inequation $s \neq$ dfs(s). Actually,

Algorithm **5.3**   gSpan(s, D, *min_support*, S)

Input: A DFS code s, a graph dataset D, and *min_support*.
Output: A frequent substructure set S.

1: **if** $s \neq$ dfs(s), **then**
2:      **return**;
3: insert s into S;
4: set C to $\emptyset$;
5: scan D once, find all the edges e such that s can be
   rightmost extended to $s \diamond_r e$; insert $s \diamond_r e$ into C and
   count its frequency;
6: sort C in DFS lexicographic order;
7: **for each** frequent $s \diamond_r e$ in C **do**
8:    Call gSpan($s \diamond_r e$, D, *min_support*, S);
9: **return**;

$s \neq \text{dfs}(s)$ is more efficient to calculate. Line 5 requires exhaustive enumeration of s in D in order to count the frequency of all the possible rightmost extensions of s.

Algorithm 5.3 implements a depth-first search version of gSpan. Actually, breadth-first search works too: for each newly discovered frequent subgraph in line 8, instead of directly calling gSpan, we insert it into a global FIFO queue Q, which records all subgraphs that have not been extended. Then, we gSpan each subgraph in Q one by one. We implemented a breadth-first search version of gSpan. Its performance is very close to that of the depth-first search one though the latter usually consumes less memory.

5.5 VARIANT SUBSTRUCTURE PATTERNS

5.5.1 Closed Frequent Substructure

According to the Apriori property, all the subgraphs of a frequent substructure must be frequent. A large graph pattern may generate an exponential number of frequent subgraphs. For example, among 423 confirmed active chemical compounds in an acquired immunodeficiency syndrome (AIDS) antiviral screen dataset, there are nearly 1,000,000 frequent graph patterns whose support is at least 5%. This renders the further analysis on frequent graphs nearly impossible.

The same issue also exists in frequent itemset mining and sequence mining. So far, two solutions have been proposed. One is closed frequent pattern mining [7, 21, 22, 32, 36]. The other is maximal frequent pattern mining [5, 7, 9, 18]. A frequent pattern is *closed* if and only if there does not exist a superpattern that has the same support. A frequent pattern is *maximal* if and only if it does not have a frequent superpattern.

EXAMPLE 5.3 The two graphs in Figure 5.2 are closed frequent graphs while only the first graph is a maximal frequent graph. The second graph is not maximal because it has a frequent supergraph.

Since the maximal pattern set is a subset of the closed pattern set, usually it is more compact than the closed pattern set. However, it cannot reconstruct the whole set of frequent patterns while the closed frequent pattern set can. The pruning techniques developed in closed frequent pattern mining can also be applied to maximal frequent pattern mining. For the AIDS antiviral dataset mentioned above, among the one million frequent graphs, only about 2000 are closed frequent graphs. If further analysis, such as classification or clustering, is performed on closed frequent graphs instead of frequent graphs, it will achieve similar accuracy with less redundancy and higher efficiency.

5.5.2 Approximate Substructure

An alternative way to reduce the number of patterns is to mine approximate frequent substructures that allow minor structural variations. With this technique, one can represent several slightly different frequent substructures using one approximate substructure.

Holder et al. (see [11] and Chapter 7) adopt the principle of minimum description length (MDL) in their substructure discovery system, called "SUBDUE," for mining approximate frequent substructures. It looks for a substructure pattern that can best compress a graph set based on the MDL principle. SUBDUE adopts a constrained beam search method. It grows a single vertex incrementally by expanding a node in it. At each expansion it searches for the best total description length: the description length of a pattern and the description length of the graph set with all the instances of the pattern condensed into single nodes. SUBDUE performs approximate matching to allow slight variations of substructures, thus supporting the discovery of approximate substructures.

5.5.3   Contrast Substructure

Between two predefined sets of graphs, contrast patterns are substructures that are frequent in one set but infrequent in the other. The search of contrast patterns requires two parameters: the *minimum support* of a substructure in the positive set and the *maximum support* in the negative set. Borgelt and Berthold discussed the mining of contrast substructures using their MoFa algorithm [6]. The mining is carried using a pattern growth approach. The pruning is done on the search of substructures in the positive set while the maximum support in the negative set is used to filter out unqualified substructures.

5.5.4   Coherent Substructure

A frequent substructure G is a coherent subgraph if the mutual information between G and each of its subgraphs is above some threshold. The number of coherent substructures is significantly smaller than that of frequent substructures. Thus, mining coherent substructures can efficiently prune redundant patterns—the patterns that are similar to each other and have the similar supports. Furthermore, as demonstrated by Huan et al. [13] in mining spatial motifs from protein structure graphs, the discovered coherent substructures are usually statistically significant. Their experimental study shows that coherent substructure mining selects a small subset of features that have high distinguishing power between protein classes.

5.5.5   Discriminative Frequent Substructure

Given a graph query, it is desirable to retrieve graphs quickly from a large graph database via graph-based indices. Yan et al. [33] developed a frequent substructure-based graph index method, called gIndex, that is significantly different from the existing path-based methods. Frequent substructures are ideal candidates since they explore the shared structures in the data and are relatively stable to database updates. To reduce the index size, that is, the number of frequent substructures that are used in the indices, a new concept called *discriminative frequent substructure* is introduced in [33]. A frequent substructure is *discriminative* if its support cannot be approximated well by its subgraphs that are being indexed. For the AIDS antiviral dataset we tested, the index built on discriminative frequent substructures is 10 times smaller but achieves similar performance in comparison with the index built on frequent substructures directly.

5.5.6 Dense Substructure

There exists a specific kind of graph structure, called *relational graph*, where each node label is used only once per graph. Relational graphs are widely used in modeling and analyzing massive networks, for example, biological networks, social networks, transportation networks, and the World Wide Web. In biological networks, nodes represent objects like genes, proteins, and enzymes, whereas edges encode the relationships, such as control, reaction, and correlation between these objects. In social networks, each node represents a unique entity, and an edge describes a kind of relationship between entities. One particular interesting patterns are *frequent highly connected* or *dense* subgraphs in large relational graphs. In social networks, this kind of pattern can help identify groups where people are strongly associated. In computational biology, a highly connected subgraph could represent a set of genes within the same functional module, that is, a set of genes participating in the same biological pathways. We define the connectivity by the minimum cut size for a given graph. In [34], Yan et al. proposed two algorithms, CloseCut and Splat, to discover exact dense frequent substructures in a set of relational graphs. Hu et al. [12] developed an algorithm called CoDense to find dense subgraphs across multiple biological networks.

5.6 EXPERIMENTS AND PERFORMANCE STUDY

In this section, we compare the performance between Apriori-based and pattern growth based methods. The real dataset used in the experiments is an AIDS antiviral screen chemical compound dataset[1]. The synthetic data generator is provided by Kuramochi and Karypis [17].

The AIDS antiviral screen compound dataset from Developmental Therapeutics Program in NCI/NIH is available publicly. We select the most up-to-date release, March 2002 Release. The dataset contains 43,905 chemical compounds. The screening tests categorized the compounds into three classes: CA (confirmed active), CM (confirmed moderately active), and CI (confirmed inactive). Among these 43,905 compounds, 423 of them belong to CA, 1083 are of CM, and the remainings are in class CI.

All the experiments are done on a 1.7-GHz Intel Pentium-4 PC with 1 GB main memory, running RedHat 7.3. In each experiment, we show the performance of an Apriori-based approach, a revised version of FSG, and a pattern growth based

---

[1] http://dtp.nci.nih.gov/docs/aids/aids_data.html.

approach, gSpan. We also depict the performance of a closed frequent graph mining algorithm, CloseGraph.

We run these three algorithms to find frequent structures in class CA and CM compounds. All the hydrogens in these compounds are removed. The most popular atoms in these two datasets are C, O, N, and S. There are 21 kinds of atoms in class CA compounds, whereas 25 kinds can be found in class CM. Three kinds of bonds are popular in these compounds: single bond, double bond, and aromatic bond. On average, each compound in class CA has 40 vertices and 42 edges. The maximum one has 188 vertices and 196 edges. Each compound in class CM has 32 vertices and 34 edges on average. The maximum one has 221 vertices and 234 edges.

Figure 5.8(a) shows the runtime with *min_support* varying from 5 to 10%. Figure 5.8(b) shows the memory consumption of these three algorithms. CloseGraph and gSpan consume less main memory than FSG. The reduction is between 1 and 2 orders of magnitude. The number of frequent graphs and frequent closed graphs is shown in Figure 5.8(c). CloseGraph generates fewer patterns than gSpan and FSG. The ratio between the number of frequent graphs and closed ones is close to 100 : 1. It demonstrates that closed graph mining can deliver compact mining results.

Figure 5.9 shows the largest graph patterns we discovered in three different minimum support thresholds: 20% in Figure 5.9(a) (14 edges), 10% in Figure 5.9(b) (21 edges), and 5% in Figure 5.9(c) (42 edges). The second structure in Figure 5.9

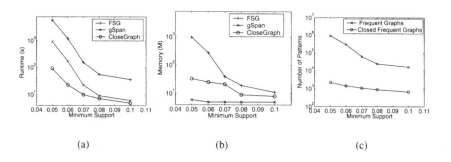

(a)   (b)   (c)

Figure 5.8. Mining patterns in class CA compounds: (a) runtime, (b) memory, and (c) number of patterns.

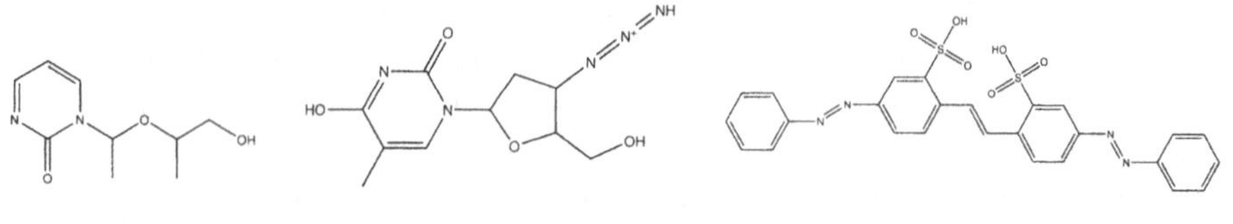

(a)   (b)   (c)

Figure 5.9. Discovered patterns in class CA: (a) 20%, (b) 10%, and (c) 5%.

is a compound of class azido pyrimidines, a known inhibitor of HIV-1, as reported by Borgelt and Berthold [6].

Next, we conduct experiments on class CM compounds. Figure 5.10 shows the performance and the number of discovered patterns. The ratio between frequent graphs and closed ones is around 10 : 1. The largest pattern with 5% support has 23 edges. That means the compounds in class CA share a small group of larger chemical fragments whereas the compounds in class CM are very diverse. This explains why CloseGraph does not achieve a similar speedup in this dataset.

We then test these algorithms on a series of synthetic graph datasets. Table 5.1 shows the major parameters and their meanings, as described in [17]. The synthetic data generator works as follows: First, it generates a set of L potential frequent graphs as seeds. They have I edges on average. Then, it randomly picks several seeds and merges them (overlaps these seeds as much as possible) to construct a new graph in the dataset. A user can set parameters to decide the number of graphs (D) wanted and their average size (T). For example, we may have the following setting in a synthetic dataset: It has 10,000 graphs, each graph has 20 edges on average, each potential frequent graph has 10 edges on average, and there are in total 200 potential frequent graphs and 40 available labels. We denote this dataset by $D10kN40I10T20L200$.

Figure 5.11 shows the runtime and the mining result for dataset $D10kN40I12T20$. The synthetic graphs have 27 edges and 16 vertices on average. Compared with the real dataset, CloseGraph has a similar performance gain in this

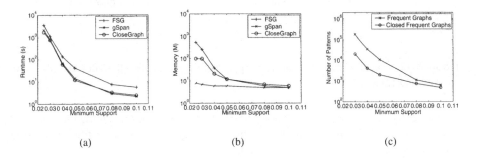

(a)   (b)   (c)

Figure 5.10. Mining patterns in class CM compounds: (a) performance, (b) memory, and (c) number of patterns.

TABLE 5.1   Parameters of Synthetic Graph Generator

Notation	Description
D	Total number of graphs in a dataset
N	Number of possible labels
T	Average size of graphs in terms of edges
I	Average size of potentially frequent graphs
L	Number of potentially frequent graphs

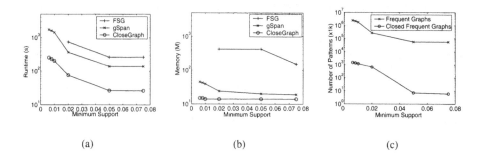

Figure 5.11. Varying support for D10kN40I12T20L200: (a) performance, (b) memory, and (c) number of patterns.

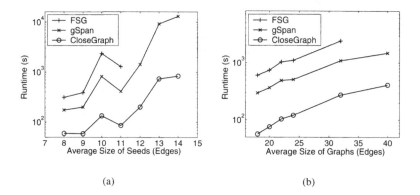

Figure 5.12. Performance vs. varying parameters: (a) D10kN40I?T20L200: 1% and (b) D10kN40I12T?L200: 2%.

synthetic dataset. The revised version of FSG now works well when the *min_support* is above 1%. At 1% *min_support*, we aborted computation because FSG exhausted the main memory, which is around 1 GB.

We then test the performance by changing some major parameters in the synthetic data. Two parameters in Table 5.1 are selected: One is the average size of potentially frequent graphs (seeds), and the other is the average size of graphs in the dataset. For each experiment, only one parameter varies. Figure 5.12 shows the experimental results, where FSG was aborted when $I \geq 12$ or $T \geq 40$ because it used up the 1-GB main memory.

5.7 CONCLUSIONS

Discovery of frequent substructures is one of the fundamental tasks in structured data mining since the discovered patterns can be used for characterizing structure datasets, classifying and clustering complex structures, building graph indices, and performing similarity search in large graph databases.

In this chapter, we overview the methodologies for efficient mining of frequent substructures and present a pattern growth approach for frequent substructure mining. Our experimental study on two typical such methods, gSpan and CloseGraph, demonstrates the efficiency and scalability of the pattern growth approach. Moreover, we discussed methods for discovery of variant substructure patterns, including closed frequent substructure, approximate substructure, contrast substructure, coherent substructure, discriminative frequent substructure, and dense substructure.

There are still many interesting issues to be further studied in graph and complex structured data mining, including mining of frequent subgraphs in a large single graph (such as Web or large social network), further development of graph classification, clustering and multidimensional analysis methods based on the mined patterns, and mining of the interactions among multiple social and biological networks.

REFERENCES

1. R. Agrawal and R. Srikant. Fast algorithms for mining association rules. In Proceedings of 1994 International Conference Very Large Data Bases (VLDB'94), pp. 487–499, Santiago, Chile, Sept. 1994.

2. T. Asai, K. Abe, S. Kawasoe, H. Arimura, H. Satamoto, and S. Arikawa. Efficient substructure discovery from large semi-structured data. In Proceedings of 2002 SIAM International Conference Data Mining (SDM'02), Arlington, VA, April 2002.

3. R. Attias and J. E. Dubois. Substructure systems: Concepts and classifications. *Journal of Chemical Information and Computer Sciences*, 30:2–7, 1990.

4. D. M. Bayada, R. W. Simpson, and A. P. Johnson. An algorithm for the multiple common subgraph problem. *Journal of Chemical Information and Computer Sciences*, 32:680–685, 1992.

5. R. J. Bayardo. Efficiently mining long patterns from databases. In Proceedings of 1998 ACM-SIGMOD International Conference on Management of Data (SIGMOD'98), pp. 85–93, Seattle, WA, June 1998.

6. C. Borgelt and M. R. Berthold. Mining molecular fragments: Finding relevant substructures of molecules. In Proceedings of 2002 International Conference on Data Mining (ICDM'02), pp. 211–218, Maebashi, Japan, Dec. 2002.

7. D. Burdick, M. Calimlim, and J. Gehrke. MAFIA: A maximal frequent itemset algorithm for transactional databases. In Proceedings of 2001 International Conference on Data Engineering (ICDE'01), pp. 443–452, Heidelberg, Germany, April 2001.

8. M. Deshpande, M. Kuramochi, and G. Karypis. Automated approaches for classifying structures. In Proceedings of 2002 Workshop on Data Mining in Bioinformatics (BIOKDD'02), Edmonton, Canada, July 2002.

9. K. Gouda and M. J. Zaki. Efficiently mining maximal frequent itemsets. In Proceedings of 2001 International Conference on Data Mining (ICDM'01), pp. 163–170, San Jose, CA, Nov. 2001.

10. J. Han, J. Pei, and Y. Yin. Mining frequent patterns without candidate generation. In Proceedings of 2000 ACM-SIGMOD International Conference on Management of Data (SIGMOD'00), pp. 1–12, Dallas, TX, May 2000.

11. L. B. Holder, D. J. Cook, and S. Djoko. Substructure discovery in the subdue system. In Proceedings of AAAI'94 Workshop Knowledge Discovery in Databases (KDD'94), pp. 169–180, Seattle, WA, July 1994.

12. H. Hu, X. Yan, H. Yu, J. Han, and X. J. Zhou. Mining coherent dense subgraphs across massive biological networks for functional discovery. In Proceedings of 2005 International Conference on Intelligent Systems for Molecular Biology (ISMB'05), pp. 213–221, Ann Arbor, MI, June 2005.

13. J. Huan, W. Wang, D. Bandyopadhyay, J. Snoeyink, J. Prins, and A. Tropsha. Mining spatial motifs from protein structure graphs. In Proceedings of 8th International Conference on Research in Computational Molecular Biology (RECOMB), pp. 308–315, San Diego, CA, March 2004.

14. J. Huan, W. Wang, and J. Prins. Efficient mining of frequent subgraph in the presence of isomorphism. In Proceedings of 2003 International Conference on Data Mining (ICDM'03), pp. 549–552, Melbourne, FL, Nov. 2003.

15. A. Inokuchi, T. Washio, and H. Motoda. An apriori-based algorithm for mining frequent substructures from graph data. In Proceedings of 2000 European Symposium Principle of Data Mining and Knowledge Discovery (PKDD'00), pp. 13–23, Lyon, France, Sept. 2000.

16. M. Koyuturk, A. Grama, and W. Szpankowski. An efficient algorithm for detecting frequent subgraphs in biological networks. *Bioinformatics*, 20:I200–I207, 2004.

17. M. Kuramochi and G. Karypis. Frequent subgraph discovery. In Proceedings of 2001 International Conference on Data Mining (ICDM'01), pp. 313–320, San Jose, CA, Nov. 2001.

18. D. Lin and Z. Kedem. Pincer-search: A new algorithm for discovering the maximum frequent set. In Proceedings of 1998 International Conference on Extending Database Technology (EDBT'98), pp. 105–119, Valencia, Spain, Mar. 1998.

19. B. Liu, G. Cong, L. Yi, and K. Wang. Discovering frequent substructures from hierarchical semi-structured data. In Proceedings of 2002 SIAM International Conference on Data Mining (SDM'02), Arlington, VA, April 2002.

20. S. Nijssen and J. Kok. A quickstart in frequent structure mining can make a difference. In Proceedings of 2004 ACM SIGKDD International Conference on Knowledge Discovery in Databases (KDD'04), pp. 647–652, Seattle, WA, Aug. 2004.

21. N. Pasquier, Y. Bastide, R. Taouil, and L. Lakhal. Discovering frequent closed itemsets for association rules. In Proceedings of 7th International Conference on Database Theory (ICDT'99), pp. 398–416, Jerusalem, Israel, Jan. 1999.

22. J. Pei, J. Han, and R. Mao. CLOSET: An efficient algorithm for mining frequent closed itemsets. In Proceedings of 2000 ACM-SIGMOD International Workshop Data Mining and Knowledge Discovery (DMKD'00), pp. 11–20, Dallas, TX, May 2000.

23. J. Pei, J. Han, B. Mortazavi-Asl, H. Pinto, Q. Chen, U. Dayal, and M.-C. Hsu. PrefixSpan: Mining sequential patterns efficiently by prefix-projected pattern growth. In Proceedings of 2001 International Conference on Data Engineering (ICDE'01), pp. 215–224, Heidelberg, Germany, April 2001.

24. J. Prins, J. Yang, J. Huan, and W. Wang. Spin: Mining maximal frequent subgraphs from graph databases. In Proceedings of 2004 ACM SIGKDD International Conference on Knowledge Discovery in Databases (KDD'04), pp. 581–586, Seattle, WA, Aug. 2004.

25. K. Shearer, H. Bunke, and S. Venkatesh. Video indexing and similarity retrieval by largest common subgraph detection using decision trees. *Pattern Recognition*, 34:1075–1091, 2001.

26. S. Su, D. J. Cook, and L. B. Holder. Knowledge discovery in molecular biology: Identifying structural regularities in proteins. *Intelligent Data Analysis*, 3:413–436, 1999.

27. Y. Takahashi, Y. Satoh, and S. Sasaki. Recognition of largest common fragment among a variety of chemical structures. *Analytical Sciences*, 3:23–28, 1987.

28. N. Vanetik, E. Gudes, and S. E. Shimony. Computing frequent graph patterns from semistructured data. In Proceedings of 2002 International Conference on Data Mining (ICDM'02), pp. 458–465, Maebashi, Japan, Dec. 2002.

29. E. K. Wong. Model matching in robot vision by subgraph isomorphism. *Pattern Recognition*, 25:287–304, 1992.

30. X. Yan and J. Han. gSpan: Graph-based substructure pattern mining. In Proceedings of 2002 International Conference on Data Mining (ICDM'02), pp. 721–724, Maebashi, Japan, Dec. 2002.

31. X. Yan and J. Han. CloseGraph: Mining closed frequent graph patterns. In Proceedings of 2003 ACM SIGKDD International Conference on Knowledge Discovery and Data Mining (KDD'03), pp. 286–295, Washington, D.C., Aug. 2003.

32. X. Yan, J. Han, and R. Afshar. CloSpan: Mining closed sequential patterns in large datasets. In Proceedings of 2003 SIAM International Conference on Data Mining (SDM'03), pp. 166–177, San Fransisco, CA, May 2003.

33. X. Yan, P. S. Yu, and J. Han. Graph indexing: A frequent structure-based approach. In Proceedings of 2004 ACM-SIGMOD International Conference on Management of Data (SIGMOD'04), pp. 335–346, Paris, France, June 2004.

34. X. Yan, X. J. Zhou, and J. Han. Mining closed relational graphs with connectivity constraints. In Proceedings of 2005 ACM SIGKDD International Conference on Knowledge Discovery in Databases (KDD'05), pp. 324–333, Chicago, IL, Aug. 2005.

35. M. J. Zaki. Efficiently mining frequent trees in a forest. In Proceedings of 2002 ACM SIGKDD International Conference on Knowledge Discovery in Databases (KDD'02), pp. 71–80, Edmonton, Canada, July 2002.

36. M. J. Zaki and C. J. Hsiao. CHARM: An efficient algorithm for closed itemset mining. In Proceedings of 2002 SIAM International Conference on Data Mining (SDM'02), pp. 457–473, Arlington, VA, April 2002.

6

FINDING TOPOLOGICAL FREQUENT PATTERNS FROM GRAPH DATASETS

MICHIHIRO KURAMOCHI AND GEORGE KARYPIS

Department of Computer Science & Engineering
University of Minnesota

6.1 INTRODUCTION

Efficient algorithms for finding frequent patterns—both sequential and nonsequential—in very large datasets have been one of the key success stories of data mining research [1, 2, 26, 54, 60, 69]. Nevertheless, as data mining techniques have been increasingly applied to nontraditional domains, there is a need to develop efficient and general-purpose frequent pattern discovery algorithms that are capable of capturing the spatial, topological, geometric, and/or relational nature of the datasets that characterize these domains. In many application domains, there exist datasets that possess inherently structural or relational characteristics, which are suitable for graph-based representations, and can greatly benefit from graph-based data mining algorithms [e.g., network topology, very large scale integration (VLSI) circuit design, protein–protein interactions, biological pathways, Web graph, etc.].

The power of graphs to model complex datasets has been recognized by various researchers [4, 9, 13, 18, 24, 31, 38, 46, 55, 62, 66] as it allows us to represent arbitrary relations among entities and solve problems that we could not previously solve. One way of formulating the frequent pattern discovery problem for graph datasets is that of discovering subgraphs occurring frequently in the given input graph dataset. For instance, consider the problem of mining chemical compounds to find recurrent substructures. We can achieve that using a graph-based pattern discovery algorithm by creating a graph for each one of the compounds whose vertices correspond to different atoms, and whose edges correspond to bonds between them. We can assign to each vertex a label corresponding to the atom involved (and potentially its charge) and assign to each edge a label corresponding to the type of the bond [and potentially information about their relative three-dimensional (3D) orientation]. Once these graphs have been created, recurrent substructures across different compounds become frequently occurring subgraphs. In fact, within the context of chemical compound classification, such techniques have been used to mine chemical compounds and identify the substructures that best discriminate between the different classes [8, 14, 40, 61], and were shown to produce superior classifiers than more traditional methods [27].

In this chapter we focus on the problem of finding frequently occurring topological subgraphs. We explore various ways of formally defining the problems that are suited for different types of graphs and applications and have varying complexity–completeness trade-offs. Our exposition of the rather large problem formulation space is followed by a description of a number of illustrative algorithms developed by our group for finding such subgraphs and a brief survey of the related research in this field.

6.2 BACKGROUND DEFINITIONS AND NOTATION

A graph $G = (V, E)$ is made of two sets, the set of vertices V and the set of edges E. Each edge itself is a pair of vertices, and throughout this chapter we assume that the graph is undirected, that is, each edge is an unordered pair of vertices. Furthermore, we will assume that the graph is *labeled*. That is, each vertex and edge has a label associated with it that is drawn from a predefined set of vertex labels (L_V) and edge labels (L_E). Each vertex (or edge) of the graph is not required to have a unique label, and the same label can be assigned to many vertices (or edges) in the same graph.

Given a graph $G = (V, E)$, a graph $G_s = (V_s, E_s)$ will be a *subgraph* of G if and only if $V_s \subseteq V$ and $E_s \subseteq E$, and it will be an *induced subgraph* of G if $V_s \subseteq V$ and E_s contains all the edges of E that connect vertices in V_s. A graph is *connected* if there is a path between every pair of vertices in the graph. Two graphs $G_1 = (V_1, E_1)$ and $G_2 = (V_2, E_2)$ are *isomorphic* if they are topologically identical to each other, that is, there is a mapping from V_1 to V_2 such that each edge in E_1 is mapped to a single edge in E_2 and vice versa. In the case of labeled graphs, this mapping must also preserve the labels on the vertices and edges. An *automorphism* is an isomorphism mapping where $G_1 = G_2$. Given two graphs $G_1 = (V_1, E_1)$ and $G_2 = (V_2, E_2)$, the problem of *subgraph isomorphism* is to find an isomorphism between G_2 and a subgraph of G_1, that is, to determine whether or not G_2 is included in G_1. The *canonical label* of a graph $G = (V, E)$, cl(G), is defined to be a unique *code* (i.e., a sequence of bits, a string, or a sequence of numbers) that is invariant on the ordering of the vertices and edges in the graph [20]. As a result, two graphs will have the same canonical label if they are isomorphic. Examples of

TABLE 6.1 Notation Used Throughout Chapter

Notation	Description
k subgraph	A connected subgraph with k edges (also written as a size-k subgraph)
G^k, H^k	(Sub)graphs of size k
$E(G)$	Edges of a (sub)graph G
$V(G)$	Vertices of a (sub)graph G
$cl(G)$	A canonical label of a graph G
a, b, c, e, f	edges
u, v	Vertices
$d(v)$	Degree of a vertex v
$l(v)$	The label of a vertex v
$l(e)$	The label of an edge e
$H = G - e$	H is a graph obtained by the deletion of edge $e \in E(G)$
$\mathcal{D}$	A dataset of graph transactions
$\{\mathcal{D}_1, \mathcal{D}_2, \ldots, \mathcal{D}_N\}$	Disjoint N partitions of $\mathcal{D}$ (for i and j, $i \neq j$, $\mathcal{D}_i \cap \mathcal{D}_j = \emptyset$ and $\bigcup_i \mathcal{D}_i = \mathcal{D}$)
T	A graph transaction
$\mathcal{G}$	A single input graph
C	A candidate subgraph
C^k	A set of candidates with k edges
$\mathcal{C}$	A set of all candidates
F	A frequent subgraph
$\mathcal{F}^k$	A set of frequent k subgraphs
$\mathcal{F}$	A set of all frequent subgraphs
k^*	The size of the largest frequent subgraph in $\mathcal{D}$
L_E	A set of all edge labels in $\mathcal{D}$
L_V	A set of all vertex labels in $\mathcal{D}$

different canonical label codes and details on how they are computed are presented in Section 6.2.1. Both canonical labeling and determining graph isomorphism are not known to be either in P or NP-complete [20].

The size of a graph $G = (V, E)$ is the number of edges ($|E|$). Given a size-k connected graph $G = (V, E)$, by *adding an edge* we will refer to the operation in which an edge $e = (u, v)$ is added to the graph so that the resulting size-$(k + 1)$ graph remains connected. Similarly, by *deleting an edge* we refer to the operation in which $e = (u, v)$ such that $e \in E$ is deleted from the graph and the resulting size-$(k - 1)$ graph remains connected. Note that depending on the particular choice of e, the deletion of the edge may result in deleting at most one of its incident vertices if that vertex has only e as its incident edge. Finally, the notation that we will be using throughout the chapter is shown in Table 6.1.

6.2.1 Canonical Labeling

One of the key operations required by any frequent subgraph discovery algorithm is a mechanism by which to check whether two subgraphs are identical or not.

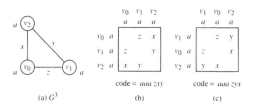

Figure 6.1. Simple examples of codes and canonical adjacency matrices.

One way of performing this check is to perform a graph isomorphism operation. However, in cases in which many such checks are required among the same set of subgraphs, a better way of performing this task is to assign to each graph a unique *code* (i.e., a sequence of bits, a string, or a sequence of numbers) that is invariant on the ordering of the vertices and edges in the graph. Such a code is referred to as the *canonical label* of a graph $G = (V, E)$ [20, 56], and we will denote it by cl(G). By using canonical labels, we can check whether or not two graphs are identical by checking to see whether they have identical canonical labels. Moreover, by comparing the canonical labels we can obtain a complete ordering of a set of graphs in a unique and deterministic way, regardless of the original vertex and edge ordering.

A simple way of defining the canonical label of an unlabeled graph is to use the string obtained by concatenating the upper triangular elements[1] of the graph's adjacency matrix when this matrix has been symmetrically permuted such that this string becomes the lexicographically largest (or smallest) among the strings obtained from all such permutations. This is illustrated in Figure 6.1 that shows a graph G^3 and the permutation of its adjacency matrix[2] that leads to its canonical label *aaazyx*. In this code, *aaa* was obtained by concatenating the vertex labels in the order that they appear in the adjacency matrix and *zyx* was obtained by concatenating the columns of the upper triangular portion of the matrix. Note that any other permutation of G^3's adjacency matrix will lead to a code that is lexicographically smaller (or equal) to *aaazyx*. If a graph has $|V|$ vertices, the complexity of determining its canonical label using this scheme is in $O(|V|!)$, making it impractical even for moderate size graphs. Note that the problem of determining the canonical label of a graph is equivalent to determining isomorphism between graphs, because if two graphs are isomorphic with each other, their canonical labels must be identical. Both canonical labeling and determining graph isomorphism are not known to be either in P or NP-complete [20].

In practice, the complexity of finding a canonical labeling of a graph can be reduced by using various heuristics to narrow down the search space or by using alternate canonical label definitions that take advantage of special properties that may exist in a particular set of graphs [20, 48, 49]. As part of our earlier research

---

[1]For a directed graph, the string obtained by using all the elements in the adjacency matrix can be used, instead of the elements in the upper triangle.

[2]The symbol v_i in the figure is a vertex ID, not a vertex label, and blank elements in the adjacency matrix means there is no edge between the corresponding pair of vertices.

we have developed such a canonical labeling algorithm that fully makes use of edge and vertex labels for fast processing and various vertex invariants to reduce the complexity of determining the canonical label of a graph [41, 42]. Our algorithm can compute the canonical label of graphs containing up to 50 vertices extremely fast and will be the algorithm used to compute the canonical labels of the different subgraphs in this chapter.

6.2.2 Maximum Independent Set

Some of our frequent subgraph discovery algorithms that will be discussed later in this chapter focus on finding subgraphs whose embeddings are edge-disjoint. A critical step in obtaining this set of edge-disjoint embeddings for a particular subgraph is to find the maximum independent set of its overlap graph. Given a graph $G = (V, E)$, a subset of vertices $I \subset V$ is called *independent* if no two vertices in I are connected by an edge in E. An independent set I is called a *maximal independent set* if for every vertex v in I there is an edge in E that connects v to a vertex in $V \setminus I$. A maximal independent set I is called the *maximum independent set* (MIS) if I contains as many vertices of V as possible.

The problem of finding the MIS of a graph was among the first problems proved to be NP-complete [21] and remains so even for bounded degree graphs. Moreover, it has been shown that the size of MIS cannot be approximated even within a factor of $n^{1-o(1)}$ in polynomial time [19]. However, the importance of the problem and its applicability to a wide range of domains has attracted a considerable amount of research. This research has been focused on developing both faster exact algorithms as well as approximate algorithms. The fastest exact algorithm to date is the algorithm by Robson [57] that solves the MIS problem in time $O(1.211^n)$, making it possible to solve in reasonable amount of time problem instances containing up to around 100 vertices. In this thesis, we used a fast implementation of the exact *maximum clique* (MC) problem solver *wclique* [51] instead of those fast exact MIS algorithms. Because the MIS problem on a graph G is equivalent to the MC problem on G's complement graph $\overline{G}$, we can use *wclique* as a fast exact MIS (EMIS) algorithm. Heuristic algorithms focus on finding maximal independent sets whose size is bounded in terms of the size of the optimal solution, and a number of such methods have been developed [6, 25, 29, 39].

One of the most widely used heuristics is the *greedy algorithm* (GMIS), which selects a vertex of the minimum degree, deletes that vertex and all of its neighbors from the graph, and repeats this process until the graph becomes empty. A recent detailed analysis of the GMIS algorithm has shown that it produces reasonably good approximations of the MIS for bounded and low-degree graphs [25]. In particular, for a graph G with a maximum degree Δ and an average degree $\overline{d}$, the size $|I|$ of the MIS satisfies the following:

$$|I| \leq \min\left(\frac{\Delta + 2}{3}|\text{GMIS}(G)|, \frac{\overline{d} + 2}{2}|\text{GMIS}(G)|\right) \qquad (6.1)$$

where $|\text{GMIS}(G)|$ is the size of the approximate MIS found by the GMIS algorithm.

6.3 FREQUENT PATTERN DISCOVERY FROM GRAPH
DATASETS—PROBLEM DEFINITIONS

There are three key axes regarding the problem definitions, and depending on the combinations of the axes, different problems may exist. Those axes are the type of graphs that these algorithms operate on, the type of subgraphs that we would like to find, and whether or not we want to find the complete set of patterns or just a subset of them.

There are two distinct forms of the input to a frequent subgraph discovery algorithm, which are referred to as the *graph-transaction setting* and the *single-graph setting*. In the graph-transaction setting, the input to a pattern mining algorithm is a set of relatively small graphs (called transactions), whereas in the single-graph setting the input data is a single large graph. The difference affects the way the frequency of the various patterns is determined. For the graph-transaction setting, the frequency of a pattern is determined by the number of graph transactions that the pattern occurs in, irrespective of how many times a pattern occurs in a particular transaction, whereas in the single-graph setting the frequency of a pattern is based on the number of its occurrences (i.e., embeddings) in the single graph. Due to the inherent differences of the characteristics of the underlying dataset and the problem formulation, algorithms developed for the graph-transaction setting cannot be used to solve the single-graph setting, whereas the latter algorithms can be easily adapted to solve the former problem.

There are a number of different ways of defining the types of subgraphs that we would like to discover. These definitions themselves are primarily divided along three dimensions. The first has to do with the topology of the subgraph itself and deals with patterns of specified topology. Examples include arbitrary subgraphs, induced subgraphs, trees, and paths. The second dimension has to do with their relation within the context of the frequent pattern lattice and contain subgraphs that are frequent, maximal, or closed. Finally, the third dimension has to do with whether or not there is a relation between the various embeddings of the discovered patterns and contain subgraphs whose embeddings can have arbitrary overlaps, partial overlaps, and being vertex- and/or edge-disjoint.

Frequent pattern discovery algorithms can be classified into two categories depending on the their completeness. Complete algorithms find all frequent patterns that satisfy a given specification (typically the minimum support threshold). On the other hand, heuristic algorithms return only a subset of all frequent patterns. The reason we are interested in heuristic algorithms is because depending on the problem formulation, complete algorithms may simply become unfeasible. For those problems, heuristic algorithms can be useful practical solutions.

6.3.1 Several Possible Problem Definitions

Complete and Exact Discovery from Graph Transactions. The first problem formulation is for finding frequently occurring connected subgraphs in a set of graphs, which is defined as follows:

Definition 6.1 (Complete Subgraph Discovery from a Set of Graph Transactions)   Given a set of graphs $\mathcal{D}$ each of which is an undirected labeled graph, and a parameter σ such that $0 < \sigma \leq 1$, find all connected undirected graphs that are subgraphs in at least $\sigma |\mathcal{D}|$ of the input graphs.

We will refer to each of the graphs in $\mathcal{D}$ as a *graph-transaction* or simply *transaction* when the context is clear, to $\mathcal{D}$ as the *graph-transaction dataset*, and to σ as the *support* threshold.

There are two key aspects in the above problem statement. First, we are only interested in subgraphs that are connected. This is motivated by the fact that the resulting frequent subgraphs will be encapsulating relations (or edges) between some of the entities (or vertices) of various objects. Within this context, connectivity is a natural property of frequent patterns. An additional benefit of this restriction is that it reduces the complexity of the problem, as we do not need to consider disconnected combinations of frequent connected subgraphs. Second, we allow the graphs to be labeled, and input graph transactions and discovered frequent patterns can contain multiple vertices and edges carrying the same label. This greatly increases our modeling ability, as it allow us to find patterns involving multiple occurrences of the same entities and relations, but at the same time makes the problem of finding such frequently occurring subgraphs nontrivial. This is because in such cases, any frequent subgraph discovery algorithm needs to correctly identify how a particular subgraph maps to the vertices and edges of each graph transaction, that can only be done by solving many instances of the subgraph isomorphism problem, which has been shown to be NP-complete [21].

Three Exact Discovery Problems from a Single Graph.   A fundamental issue that needs to be considered by any frequent subgraph discovery problem formulation similar to the single-graph setting is the counting method of the occurrence frequency. In general, there are two possible methods of frequency counting. According to the first method, two embeddings of a subgraph are considered different, as long as they differ by at least one edge (i.e., nonidentical). As a result, arbitrary overlaps of embeddings of the same subgraph are allowed. On the other hand, by the second method, two embeddings are considered different, only if they do not share edges (i.e., they are edge-disjoint). These two methods are illustrated in Figure 6.2. In this example, there are three possible embeddings of the subgraph shown in Figure 6.2(a) in the input graph of Figure 6.2(b). Two of these embeddings [Figs. 6.2(c) and 6.2(e)] do not share any edges, whereas the third embedding

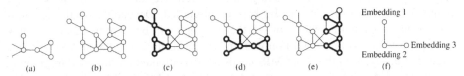

Figure 6.2. Overlapped embeddings: (a) subgraph, (b) input graph, (c) embedding 1, (d) embedding 2, (e) embedding 3, and (f) overlaps.

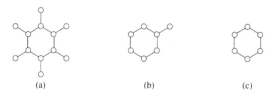

(a)	(b)	(c)

Figure 6.3. Patterns with the nonmonotonic frequency: (a) size-12 graph $\mathcal{D}$, (b) size-7 graph G^7, and (c) size-6 subgraph G^6.

[Fig. 6.2(d)] shares edges with the other two. Thus, if we allow overlaps, the frequency of the subgraph is (c), and if we do not it is (b).

These two ways of counting the frequency of a subgraph lead to problems with dramatically different characteristics. If we allow arbitrary overlaps between nonidentical embeddings, then the resulting frequency is not any longer downward closed (i.e., the frequency of a subgraph does not monotonically decrease as a function of its length). This is illustrated in Figure 6.3. Both G^7 and G^6 are subgraphs of $\mathcal{D}$. Although the smaller subgraph G^6 has only one nonidentical embedding, the larger G^7 has six nonidentical embeddings. On the other hand, if we determine the frequency of each subgraph by counting the maximum number of its edge-disjoint embeddings, then the resulting frequency is downward closed [63].

Being able to take advantage of a frequency counting method that is downward closed is essential for the computational tractability of most frequent pattern discovery algorithms. For this reason, our problem formulation uses edge-disjoint embeddings. Given this, one way of formulating the frequent subgraph discovery problem for the single-graph setting is defined as follows [63]:

Definition 6.2 (Complete Discovery from a Single Graph)   Given an input graph $\mathcal{D}$ that is undirected and labeled, and a parameter f, find all connected undirected labeled subgraphs that have at least f edge-disjoint embeddings in $\mathcal{D}$.

Unfortunately quite often this problem can be intractable. By this definition, to determine if a subgraph is frequent or not, we need to find whether the overlap graph of its nonidentical embeddings contain an independent set whose size is at least f. When a subgraph is relatively frequent compared to the frequency threshold f, by using approximate MIS algorithms we can quickly tell that such a subgraph is actually frequent. However, in the cases in which the approximate MIS algorithm does not find a sufficiently large independent set, the exact MIS needs to be computed before a pattern will be kept or discarded. Depending on the resulting size of the maximum independent set, the subgraph will be identified as frequent or infrequent. Also, if we need not only to find frequent subgraphs but also to find their exact frequency, then the exact MIS needs to be computed on the overlap graph of every pattern. In both cases, because solving the exact MIS problem is NP-complete (see Section 6.2.2), the above definition of the frequent subgraph discovery problem cannot be tractable, even for a relatively simple input graph.

To make the problem more practical, we propose two alternative formulations that can find frequent subgraphs without solving the exact MIS problem.

Definition 6.3 (Subset Discovery from a Single Graph)   Given an input graph $\mathcal{D}$ that is undirected and labeled, and a parameter f, find as many connected undirected labeled subgraphs as possible that have at least f edge-disjoint embeddings in $\mathcal{D}$.

Definition 6.4 (Superset Discovery from a Single Graph)   Given an input graph $\mathcal{D}$ that is undirected and labeled, and a parameter f, find all connected undirected labeled subgraphs such that an upper bound on the number of its edge-disjoint embeddings is above the threshold f.

Essentially the solutions for these two problems become a subset and a superset of the solution for Definition 6.2, respectively. The first formulation, Definition 6.3, which asks for a subset of the solution of Definition 6.2, requires that the embeddings of each subgraph form an overlap graph that has an *approximate MIS* whose size is greater than or equal to f. The second formulation, Definition 6.4, which asks for a superset of the solution of Definition 6.2, requires that an upper bound on the size of the exact MIS of this overlap graph is greater than or equal to f. Note that as discussed in Section 6.2.2, such upper bounds can be easily obtained for both the GMIS algorithm as well as for other approximate algorithms.

Heuristic Discovery from a Single Graph. Although existing algorithms [8, 30, 33, 35–37, 41, 44, 64] for the graph-transaction problem formulation have improved the performance and scalability over recent years, the fact still remains that those complete algorithms based on exact graph matching can handle only a limited class of problems in practice. For example, in any of those studies, the average size of input graph transactions tested never exceeds 1000 edges and the average vertex degree is almost always around two. In other words, input graph transactions and frequent subgraphs from them are sparse and quite often they are even trees. While it is still useful, for example, to find common substructures from a set of chemical compounds as that type of datasets is repeatedly tested in those studies, it is also clear that algorithms with such scalability limitations cannot process larger or denser datasets.

Nevertheless, there are many graph or relational datasets that are either large or dense, or both, and existing complete and exact algorithms are useless. In fact, even SUBDUE [31] cannot handle dense graphs at all. For example, given a random graph of 1000 vertices and 49,852 edges (the average vertex degree 100) with 10 distinct vertex labels and no edge labels, the only patterns SUBDUE could find were single vertices, after spending 755 ss. At this moment, regardless of being complete or heuristic, there is no generic algorithm that can find frequently occurring patterns from a massive graph dataset.

Another serious issue with those complete frequent subgraph algorithms is their use of exact matching. Those algorithms can find a large number of frequent subgraphs efficiently. Yet, all the embeddings of a frequent subgraph must be exactly identical. Even though there are two frequent subgraphs that are similar to each other, they are treated as two distinct ones. This makes those algorithms difficult to use. For example, if there is noise in an input dataset, or if a frequent pattern has minor variation in topology, according to those exact-matching-based algorithms, each of the variations is treated as a distinct pattern and therefore its frequency decreases and eventually we may miss the pattern. As another example, consider finding topological structures of amino acids in protein sequences. We can model a protein sequence as a graph where a vertex corresponds to an amino acid and an edge shows the proximity between the two amino acids as the endpoints of the edge. It is common that for a well-known frequently occurring substructure, some amino acids are missing or superfluous depending on a protein sequence. Exact-matching-based algorithms have difficulty to find frequent substructures under this type of situations. Users of a pattern finding algorithm are typically interested in a relatively small number of nontrivial frequent patterns that represent or character-ize input datasets. If two subgraphs look alike, they should be treated as the same pattern.

To overcome the shortcomings of the existing frequent subgraph discovery algorithms, our proposed research will focus on the following two problems.

Definition 6.5 (Heuristic and Exact Subgraph Discovery from a Single Graph) Given an input undirected graph D with vertex and edge labels, and a parameter f as the minimum frequency threshold, find frequent subgraphs that sat-isfy the threshold so that the discovery process maximizes a predefined objective function $f_{obj}(D, \mathcal{F}, t)$ where t is the runtime of the discovery process.

The objective function is the measure of the quality and the efficiency of a discovery algorithm. Needless to say it is the ultimate goal of our research for this problem to find as many frequent subgraphs as possible. Because of the NP-hardness of approximating the subgraph isomorphism problem, we cannot expect any sort of guarantee on the coverage of the solution (i.e., how much fraction of truly frequent subgraphs can be discovered). We will evaluate our methods by comparing them against the existing algorithms such as SUBDUE [31] or the complete algorithms such as SEuS [22] and SiGraM [43] in the quality of the output (i.e., the coverage) and the efficiency in runtime and memory. In the proposed research, in addition to the design of discovery algorithms for this problem definition, we will also investigate various objective function definitions including the minimum description length (MDL) principle.

The goal of the second problem is not only to discover frequent subgraphs approximately but also to generalize the frequent subgraphs embedded in the given graph.

6.3.2 Representative Algorithms for Different Problem Formulations

In the following section, we will briefly describe three algorithms (FSG, SiGraM, and Grew), which are designed for the different problem formulations explained above. FSG is an algorithm for finding all frequent connected subgraphs from a set of labeled graphs, which is, in other words, a complete algorithm in the graph-transaction setting. FSG works in the level-by-level fashion starting from smaller patterns toward larger ones. It traverses the search space of frequent subgraphs in the breadth-first manner, by generating candidates with joining and counting the frequencies of the generated candidate patterns. SiGraM, which is a pair of a horizontal (hSiGraM) and a vertical (vSiGraM) algorithms, is designed for finding all frequent connected subgraphs from single-label sparse graphs. hSiGraM can be considered an extension of FSG to the single-graph setting, while vSiGraM is a novel depth-first algorithm using efficient pruning strategies and heuristics. As an example of an efficient incomplete algorithm, we will present Grew in Section 6.6.

6.4 FSG FOR THE GRAPH-TRANSACTION SETTING

In developing our frequent subgraph discovery algorithm, we decided to follow the level-by-level structure of the Apriori [2] algorithm used for finding frequent item-sets. The motivation behind this choice is the fact that the level-by-level structure of Apriori requires the smallest number of subgraph isomorphism computations during frequency counting, as it allows it to take full advantage of the downward closed property of the minimum support constraint and achieves the highest amount of prun-ing when compared with the most recently developed depth-first-based approaches such as dEclat [69], Tree Projection [1], and FP-growth [26]. In fact, despite the extra overhead due to candidate generation incurred by the level-by-level approach, recent studies have shown that because of its effective pruning, it achieves compara-ble performance with that achieved by the various depth-first-based approaches, as long as the dataset is not dense or the support value is not extremely small [23, 28]. Table 6.2 illustrates the performance of FSG when it is applied to the PTC dataset (see Section 6.8 for the dataset description). FSG is able to operate efficiently up to 0.75% of the minimum support threshold, which corresponds to the absolute

TABLE 6.2   FSG Results for the PTC Dataset[a]

Support[b]	t (s)[c]	$|\mathcal{F}|$[d]	k^{*e}	$|\mathcal{F}|/t$[f]
5.0	0.2	310	9	1,550
2.0	3.9	7,811	20	2,003
1.0	26	49,475	24	1,374
0.75	333	404,028	36	1,213

[a]Measured on a dual Intel Xeon 3.2-GHz machine with 2 Gbytes main memory, running the Linux operating system.
[b]The minimum support threshold (%).
[c]The runtime in seconds.
[d]The total number of frequent subgraphs discovered.
[e]The size of the largest frequent subgraphs discovered.
[f]The number of frequent subgraphs found per second.

frequency of 3. As shown in this table, the number of frequent subgraphs discovered in a unit time decreases moderately as the minimum support is lowered, which shows that the scalability of FSG is well maintained even for very low support threshold.

FSG starts by enumerating all frequent single- and double-edge subgraphs. Then, it enters its main computational phase, which consists of a main iteration loop. During each iteration, FSG first generates all candidate subgraphs whose size is greater than the previous frequent ones by one edge and then counts the frequency for each of these candidates and prunes subgraphs that do no satisfy the support constraint. FSG stops when no frequent subgraphs are generated for a particular iteration. Details on how FSG generates the candidates subgraphs and on how it computes their frequency are provided in Sections 6.4.1 and 6.4.2, respectively.

To ensure that the various graph-related operations are performed efficiently, FSG stores the various input graphs and the various candidate and frequent subgraphs that it generates using an adjacency list representation.

6.4.1 Candidate Generation

FSG generates candidate subgraphs of size $k + 1$ by joining two frequent size-k subgraphs. In order for two such frequent size-k subgraphs to be eligible for joining, they must contain the same size-$(k - 1)$ connected subgraph. The simplest way to generate the complete set of candidate subgraphs is to join all pairs of size-k frequent subgraphs that have a common size-$(k - 1)$ subgraph. Unfortunately, the problem with this approach is that a particular size-k subgraph can have up to k different size-$(k - 1)$ subgraphs. As a result, if we consider all such possible subgraphs and perform the resulting join operations, we will end up generating the same candidate pattern multiple times, and generating a large number of candidate patterns that are not downward closed. The net effect of this is that the resulting algorithm spends a significant amount of time identifying unique candidates and eliminating non-downward-closed candidates (both of which are nontrivial operations as they require to determine the canonical label of the generated subgraphs). Note that candidate generation approaches in the context of frequent itemsets (e.g., Apriori [2]) do not suffer from this problem because they use a consistent way to order the items within an itemset (e.g., lexicographically). Using this ordering, they only join two size-k itemsets if they have the same $(k - 1)$-prefix. For example, a particular itemset $\{A, B, C, D\}$ will only be generated once (by joining $\{A, B, C\}$ and $\{A, B, D\}$), and if that itemset is not downward closed, it will never be generated if only its $\{A, B, C\}$ and $\{B, C, D\}$ subsets were frequent.

Fortunately, the situation for subgraph candidate generation is not as severe as the above discussion seems to indicate, and FSG addresses both of these problems by only joining two frequent subgraphs if and only if they share a certain, properly selected, size-$(k - 1)$ subgraph. Specifically, for each frequent size-k subgraph F_i, let $\mathcal{P}(F_i) = \{H_{i,1}, H_{i,2}\}$ be the two size-$(k - 1)$-connected subgraphs of

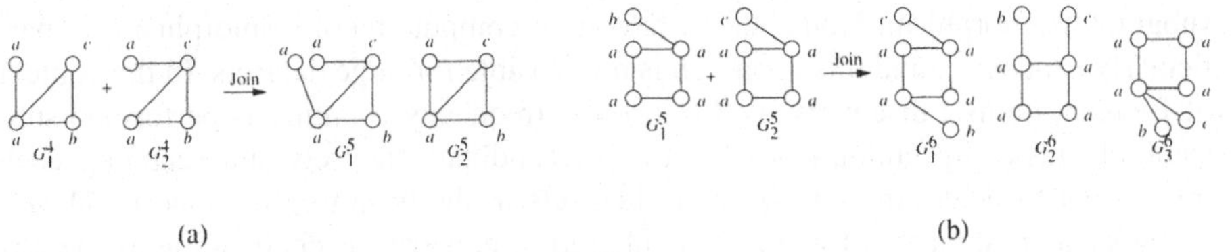

(a)   (b)

Figure 6.4. Two cases of joining: (a) by vertex labeling and (b) by multiple automorphisms of a single core

F_i such that $H_{i,1}$ has the smallest canonical label and $H_{i,2}$ has the second small-est canonical label among the various connected size-$(k-1)$ subgraphs of F_i. We will refer to these subgraphs as the *primary subgraphs* of F_i. Note that if every size-$(k-1)$ subgraph of F_i is isomorphic to each other, $H_{i,1} = H_{i,2}$ and $|\mathcal{P}(F_i)| = 1$. FSG will only join two frequent subgraphs F_i and F_j, if and only if $\mathcal{P}(F_i) \cap \mathcal{P}(F_j) \neq \emptyset$, and the join operation will be done with respect to the common size-$(k-1)$ subgraph(s). The completeness of this candidate generation approach is shown in [44]. This candidate generation approach dramatically reduces the number of redundant and non-downward-closed patterns that are generated and leads to sig-nificant performance improvements over the naive approach (originally implemented in [41]).

The actual join operation of two frequent size-k subgraphs F_i and F_j that have a common primary subgraph H is performed by generating a candidate size-$(k+1)$ subgraph that contains H plus the two edges that were deleted from F_i and F_j to obtain H. However, unlike the joining of itemsets in which two frequent size-k itemsets lead to a unique size-$(k+1)$ itemset, the joining of two size-k subgraphs may produce multiple distinct size-$(k+1)$ candidates. This happens for the follow-ing two reasons. First, the difference between the common primary subgraph and the two frequent subgraphs can be a vertex that has the same label. In this case, the joining of such size-k subgraphs will generate two distinct subgraphs of size $k+1$. This is illustrated in Figure 6.4(a) where the pair of graphs G_a^4 and G_b^4 generates two different candidates G_a^5 and G_b^5. Second, the primary subgraph itself may have multiple automorphisms and each of them can lead to a different size-$(k+1)$ can-didate. In the worst case, when the primary subgraph is an unlabeled clique, the number of automorphisms is $k!$. An example for this case is shown in Figure 6.4(b) in which the primary subgraph—a square of four vertices labeled with a—has four automorphisms resulting in three different candidates of size 6. Finally, in addition to joining two different subgraphs, FSG also needs to perform self-join in order to correctly generate a size-$(k+1)$ candidate subgraph whose all size-k-connected subgraphs are isomorphic to each other.

6.4.2 Frequency Counting

The simplest way to determine the frequency of each candidate subgraph is to scan each one of the dataset transactions and determine if it is contained or not using subgraph isomorphism. Nonetheless, having to compute these isomorphisms is particularly expensive, and this approach is not feasible for large datasets. In the context of frequent itemset discovery by Apriori, the frequency counting is performed substantially faster by building a hash tree of candidate itemsets and scanning each transaction to determine which of the itemsets in the hash tree it supports. Developing such an algorithm for frequent subgraphs, however, is challenging as there is no natural way to build the hash tree for graphs.

For this reason, FSG instead uses transaction identifier (TID) lists, proposed by [17, 59, 68]. In this approach for each frequent subgraph FSG keeps a list of transaction identifiers that support it. Now when FSG needs to compute the frequency of G^{k+1}, it first computes the intersection of the TID lists of its frequent k subgraphs. If the size of the intersection is below the support, G^{k+1} is pruned, otherwise FSG computes the frequency of G^{k+1} using subgraph isomorphism by limiting the search only to the set of transactions in the intersection of the TID lists. The advantages of this approach are twofold. First, in the cases where the intersection of the TID lists is below the minimum support level, FSG is able to prune the candidate subgraph without performing any subgraph isomorphism computations. Second, when the intersection set is sufficiently large, FSG only needs to compute subgraph isomorphisms for those graphs that can potentially contain the candidate subgraph and not for all the graph transactions.

Reducing Memory Requirements of TID Lists. The computational advantages of TID lists come at the expense of higher memory requirements for maintaining them. To address this limitation we implemented a database-partitioning-based scheme that was motivated by a similar scheme developed for mining frequent itemsets [58]. In this approach, the database is partitioned into N disjoint parts $\mathcal{D} = \{\mathcal{D}_1, \mathcal{D}_2, \ldots, \mathcal{D}_N\}$. Each of these subdatabases $\mathcal{D}_i$ is mined to find a set of frequent subgraphs $\mathcal{F}_i$, called *local frequent subgraphs*. The union of the local frequent subgraphs $\bar{\mathcal{C}} = \bigcup_i \mathcal{F}_i$, called *global candidates*, is determined and their frequency in the entire database is computed by reading each graph transaction and finding the set of subgraphs that it supports. The subset of $\bar{\mathcal{C}}$ that satisfies the minimum support constraint is output as the final set of frequent patterns $\mathcal{F}$. Since the memory required for storing the TID lists depends on the size of the database, their overall memory requirements can be reduced by partitioning the database in a sufficiently large number of partitions.

One of the problems with a naive implementation of the above algorithm is that it can dramatically increase the number of subgraph isomorphism operations that are required to determine the frequency of the global candidate set. To address this problem, FSG incorporates three techniques: (i) a priori pruning the number of candidate subgraphs that need to be considered; (ii) using bitmaps to limit the frequency counting of a particular candidate subgraph to only those partitions that this frequency has not already being determined locally; and (iii) taking advantage of the lattice structure of $\bar{\mathcal{C}}$ to check each graph transaction only against the subgraphs that are descendants of patterns that are already being supported by that transaction.

The a priori pruning of the candidate subgraphs is achieved as follows. For each partition $\mathcal{D}_i$, FSG finds the set of local frequent subgraphs and the set of local negative border subgraphs[3] and stores them into a file S_i along with their associated frequencies. Then, it organizes the union of the local frequent and local negative border subgraphs across the various partitions into a lattice structure (called a *pattern lattice*), by incrementally incorporating the information from each file S_i. Then, for each node v of the pattern lattice it computes an upper bound $f^*(v)$ of its occurrence frequency by adding the corresponding upper bounds for each one of the N partitions, $f^*(v) = f_1^*(v) + \cdots + f_P^*(v)$. For each partition $\mathcal{D}_i$, $f_i^*(v)$ is determined using the following equation:

$$f_i^*(v) = \begin{cases} f_i(v) & \text{if } v \in S_i \\ \min_u \left(f_i^*(u) \right) & \text{otherwise} \end{cases}$$

where $f_i(v)$ is the actual frequency of the pattern corresponding to node v in $\mathcal{D}_i$, and u is a connected subgraph of v that is smaller from it by one edge (i.e., it is its parent in the lattice). Note that the various $f_i^*(v)$ values can be computed in a bottom-up fashion by a single scan of S_i, and used directly to update the overall $f^*(v)$ values. Now, given this set of frequency upper bounds, FSG proceeds to prune the nodes of the pattern lattice that are either infrequent or fail the downward closure property.

6.5 SiGRaM FOR THE SINGLE-GRAPH SETTING

We developed two algorithms, called HSiGRaM and vSiGRaM, which find all frequent subgraphs according to Definitions 6.2–6.4 described in Section 6.3.1.[4] In both algorithms, the frequent patterns are conceptually organized in a form of a lattice that is referred to as the *lattice of frequent subgraphs*. The kth level of this lattice contains all frequent subgraphs with k edges (i.e., size-k subgraphs), and a node at level k representing a subgraph G^k is connected to at most k nodes at level $k - 1$, each corresponding to a distinct (i.e., nonisomorphic) connected size-$(k - 1)$ subgraph of G^k. The goal of both HSiGRaM and vSiGRaM is to identify the various nodes of this lattice and the frequency of the associated subgraphs.

The difference between the two algorithms is the method they use to discover (i.e., generate) the nodes of the lattice. HSiGRaM follows a horizontal approach and discovers the nodes in a breadth-first fashion, whereas vSiGRaM follows a vertical approach and discovers the nodes in a depth-first fashion. Both horizontal and vertical approaches have been previously used to find frequent subgraphs in the graph-transaction setting [8, 37, 44, 64] and have their origins on algorithms developed for finding frequent itemsets and sequences [2, 3, 26, 70].

---

[3]A local negative border subgraph is the one generated as a local candidate subgraph but does not satisfy the minimum threshold for the partition.
[4]SiGRaM stands for SINGLE-GRAPH MINER.

A detailed description of HSIGRAM and VSIGRAM is provided in the rest of this section.

6.5.1 Horizontal Algorithm: HSIGRAM

The general structure of HSIGRAM is shown in Algorithm 6.1 (the notation used in the pseudocode is shown in Table 6.1). HSIGRAM takes as input the graph G, the minimum frequency threshold f, and the parameter *Problem_Type* that specifies the particular problem definition (as discussed in Section 6.3.1). It starts by enumerating all frequent single- and double-edge subgraphs in G and then enters its main computational loop (lines 7–13). During each iteration, HSIGRAM first generates all candidate subgraphs of size $k + 1$ by joining pairs of size-k frequent subgraphs (line 8) and then computes their frequency (HSIGRAM-COUNT in line 11). The candidate subgraphs whose frequency is lower than the minimum threshold f are discarded, and the remaining are kept for the next level of the algorithm. The computation terminates when no frequent subgraphs are generated during a particular iteration.

The two key components of the HSIGRAM algorithm that significantly affect its overall computational complexity are the methods used to perform candidate generation and to compute the frequency of the candidate subgraphs. The candidate generation is basically identical to the method used for FSG, which was described in Section 6.4.1 in detail. In the rest of this section we provide additional details on how the frequency counting is performed and describe various optimizations that are designed to reduce their runtime.

Algorithm **6.1** HSIGRAM(G,***Problem_Type***,f)

```
 1: ▷ f is the minimum frequency threshold.
 2: ▷ Problem_Type is either subset, complete, or
    superset.
 3: F ← ∅
 4: F¹ ← all frequent size-1 subgraphs in G
 5: F² ← all frequent size-2 subgraphs in G
 6: k ← 2
 7: while Fᵏ ≠ ∅ do
 8:     Cᵏ⁺¹ ← HSIGRAM-GEN(Fᵏ⁻¹, Fᵏ, f)
 9:     Fᵏ⁺¹ ← ∅
10:     for each candidate C in Cᵏ⁺¹ do
11:         C.freq ← HSIGRAM-COUNT(C, G, Problem_Type)
12:         if C.freq ≥ f then
13:             add C to Fᵏ⁺¹
14:         end if
15:     end for
16:     F ← F ∪ Fᵏ⁺¹
17:     k ← k + 1
18: end while
19: return F
```

Frequency Counting. HSiGraM-Count in Algorithm 6.2 computes the frequency of a candidate subgraph C by first identifying all of its embeddings, constructing the overlap graph of these embeddings, and then, based on the *Problem_Type* parameter, finding an approximate or exact MIS of this overlap graph. The outline of this process is shown in Algorithms 6.2 and 6.3. In the rest of this section we first describe how the various embeddings are identified. Then, we explain the method used to efficiently compute the desired maximal independent sets.

Embedding Identification. To identify all the embeddings of a candidate C, HSiGraM-Embed shown in Algorithm 6.3 needs to solve the subgraph isomorphism problem. Performing the subgraph isomorphism for every candidate from scratch may be expensive, especially when an input graph is large. HSiGraM-Embed reduces this computational requirement by using *anchor edges*. An anchor edge is a partial embedding of a candidate C and works as a constraint of the subgraph isomorphism problem, which narrows down the search space only around the anchor edge.

More specifically, HSiGraM-Embed creates and uses anchor edges as follows. First, the list of anchor edges are created right after frequency counting for size-$(k - 1)$ frequent subgraphs, by converting the list of its nonidentical embeddings. These edges will be used later for counting a candidate of size k. Let F_i denote a frequent subgraph of size $k - 1$ and suppose F_i has N nonidentical embeddings in total. After the frequency counting, F_i has a list of all its embeddings $\mathcal{M}(F_i) = \{m_1, \ldots, m_N\}$. An anchor edge e of an embedding m_i of F is an edge in $E(\mathcal{G})$ that

Algorithm **6.2** HSiGraM-Count($C^{k+1}, \mathcal{G}, Problem\_Type$)

```
 1: (M(C^{k+1}), A(C^{k+1})) ← HSiGraM-Embed(C, G)
 2: G ← build an overlap graph from M(C^{k+1}, G)
 3: {G_1, G_2, ..., G_m} ← decompose G
 4: f_MIS ← 0
 5: for each G_i in {G_1, G_2, ..., G_m} do
 6:     if G_i is easy to handle then
 7:         f_MIS ← f_MIS + |EMIS(G_i)|
 8:     else if Problem_Type = approximate then
 9:         f_MIS ← f_MIS + |GMIS(G_i)|
10:     else if Problem_Type = exact then
11:         f_MIS ← f_MIS + |EMIS(G_i)|
12:     else if Problem_Type = upper bound then
13:         f_MIS ← f_MIS + |GMIS(G_i)|min((Δ + 2)/3, (d̄ + 2)/2)
14:     end if
15: end for
16: ▷ S(C^{k+1}) is a set of all connected size-k subgraphs
       in C^{k+1}
17: f_p ← the lowest frequency among S(C^{k+1})
18: return min(f_MIS, f_p)
```

Algorithm **6.3** HSIGRAM-EMBED$(C, \mathcal{G})$

```
 1: ▷ A: a set of all anchor edges of C
 2: A ← intersection of anchor edges across S(C)
 3: ▷ collect all unique embeddings of C into M
 4: M ← ∅
 5: for each anchor edge e in A do
 6:     Mₑ ← all embeddings of C that includes the edge e
 7:     M ← M ∪ Mₑ
 8: end for
 9: ▷ collect all unique anchor edges of C into A
10: A ← ∅
11: for each embedding m in M do
12:     e ← choose one edge from m arbitrarily
13:     add e to A
14: end for
15: return  (M, A)
```

is also a part of m_i. For every m_i, HSIGRAM-EMBED arbitrarily chooses an edge and adds it to $\mathcal{A}(F_i)$ (line 11 in Algorithm 6.3). Because of overlapped embeddings, some embeddings may lead to the same anchor edge.

Now, in the next iteration, suppose a k-candidate C contains a frequent $(k-1)$-subgraph F_i. Because there are k edges in $E(C)$, C may have up to k distinct such frequent subgraphs of size $k-1$, and each F_i holds the anchor edge list. Before starting the frequency counting of C, first HSIGRAM-EMBED selects one of F_i whose frequency is the lowest among $\{F_i\}$. For each $e_n \in \mathcal{A}(F_i)$, HSIGRAM-EMBED checks if there is an edge $e_m \in \mathcal{A}(F_j)$ for all $j \neq i$ such that the shortest path length between e_n and e_m, denoted by d, is within the diameter of C, denoted by dia(C). If there is such an edge e_m from every $\mathcal{A}(F_j)$ for $j \neq i$, e_n may be a part of an embedding of C, because if C is a frequent subgraph of size k, there must be a set of frequent subgraphs of size $k-1$ inside the same embedding of C. To compute the exact path length between edges e_n and e_m in $\mathcal{G}_i$ requires all pairs shortest paths, which may be computationally expensive when $|E(\mathcal{G}_i)|$ is large. HSIGRAM-EMBED bounds this length d by the difference between two lengths, $|d_n - d_m|$, where d_n and d_m are the shortest path lengths from an arbitrarily chosen vertex $v \in V(\mathcal{G}_i)$ to e_n and e_m, respectively. If e_n and e_m are in the same embedding of C_i, always $d \leq$ dia(C) holds and $d_n \leq d_m + d$. Thus, if $|d_n - d_m| \leq$ dia(C) is true, then e_n and e_m may belong to the same embedding of C, otherwise e_n and e_m cannot be in the same embedding (see Fig. 6.5). If e_n cannot find such e_m from every $\mathcal{A}(F_j)$ for $j \neq i$, e_m is removed from $\mathcal{A}(F_i)$ (line 2). Because the subgraph isomorphism will be performed for each e_n, this pruning procedure can effectively reduce the runtime.

Finally, after removing unnecessary anchor edges, for each of the remaining anchor edges, all the subgraph isomorphisms of C are repeatedly identified and the set of embeddings $\mathcal{M}$ is built (line 6).

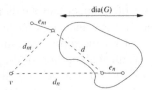

Figure 6.5. Distance estimation between two edges.

Computing the Frequency. The frequency of each subgraph C^{k+1} is computed by the HSiGraM-COUNT function shown in Algorithm 6.2. In particular, HSiGraM-COUNT computes two different frequencies. The first, denoted by f_{MIS}, is computed based on the size of the MIS of the overlap graph created from the embeddings of C^{k+1}. The second, denoted by f_p, is the least frequency of all the connected size-k subgraphs of C^{k+1} (line 15), which represents an upper bound on C^{k+1}'s frequency derived entirely from the lattice of frequent subgraphs. In the case in which f_{MIS} is computed using Definition 6.4 in Section 6.3.1, the frequency bound provided by f_p may actually be tighter, and thus may lead to more effective pruning. For this reason, the overall frequency of C^{k+1} is obtained by taking the minimum of f_{MIS} and f_p.

The frequency f_{MIS} is computed as follows (lines 2–13). Given a pattern and all of its nonidentical embeddings, HSiGraM-COUNT generates its overlap graph G. Then, HSiGraM-COUNT decomposes G into its connected components $G_1, G_2, \ldots, G_m$ ($m \geq 1$). Next, for each connected component G_i, it checks the maximum degree of its vertices, and, if it is less that or equal to 2 (a cycle or a path), it computes its maximum independent set directly by the EMIS algorithm because it is trivial to compute the exact MIS for this class of graphs (line 7). If the maximum degree is greater than 2, HSiGraM-COUNT uses either the result of the GMIS algorithm (line 9), the result of the EMIS algorithm (line 11), or the upper bound on the size of the exact MIS [Eq. (6.1)]. The summation of those MIS sizes for the components is the final value of f_{MIS}. Note that the decomposition of the overlap graph into its connected components allow us to take advantage of the properties of the special graphs and also obtain tighter bounds for each component as the maximum degree for some of them will be lower than the maximum degree of the entire overlap graph.

In addition, every edge is marked if it is included in any embedding of a frequent subgraph. Unmarked edges are removed before proceeding to the next iteration.

6.5.2 Vertical Algorithm: vSiGraM

The most computationally expensive step in the HSiGraM algorithm is frequency counting as it needs to repeatedly perform subgraph isomorphism computations. The overall time can be greatly reduced if instead of storing only the anchor edges we store the complete set of embeddings across successive levels of the algorithm. Because of HSiGraM's level-by-level structure, these complete embeddings need to be stored for the entire set of frequent and candidate patterns of each successive pair of levels. This substantially increases the memory requirements of this approach, making it impractical for the most interesting of datasets. On the other hand, within the context of a vertical algorithm, storing the complete set of embeddings is feasible since we need to do that only for the subgraphs along the path from the current node to the root. Thus, a vertical algorithm has potentially a computational advantage over a horizontal algorithm, which motivated the development of vSiGraM.

However, before developing efficient algorithms that generate the lattice of frequent subgraphs in a depth-first fashion, two critical steps need to be addressed. The first step is the method that is used to ensure that the same node of the lattice and the depth-first subtree rooted at that node should not be discovered and explored multiple times. This is important because each node at level k will be connected to up to k different nodes at level $(k-1)$. As a result, if there are no mechanisms by which to prevent the repeated generation of the same node, a depth-first algorithm will end up performing redundant computations (i.e., generating the same nodes multiple times), adversely impacting the overall performance of the algorithm. vSiGraM eliminates these redundant computations by assigning each node at level k (corresponding to a subgraph F^k) to a unique parent node at level $k-1$ (corresponding to a subgraph F^{k-1}), such that only F^{k-1} is allowed to create F^k. The subgraph F^{k-1} is called *the generating parent* of F^k. Details on how this is achieved is provided in Section 6.5.2.

The second step is the method that is used to create successor nodes in the course of the traversal. In the case of hSiGraM, this corresponds to the candidate generation phase and is performed by joining the frequent subgraphs of the previous level. However, since the lattice is explored in a depth-first fashion, such a joining-based approach will not work, as the algorithm may not have yet discovered the required frequent subgraphs. To address this problem, vSiGraM creates the successor nodes (i.e., extended subgraphs) by analyzing all the embeddings of the current subgraph F^k, and identifying the distinct one-edge extensions to these embeddings that are sufficiently frequent. The frequent extensions for which F^k is the generating parent are then used as the successor nodes during the depth-first traversal.

The general structure of vSiGraM is shown in Algorithm 6.4. vSiGraM starts by determining all frequent size-1 patterns and then uses each one of them as the starting point of a recursive depth-first extension (vSiGraM-Extend function). vSiGraM-Extend takes as input a size-k frequent subgraph F^k and all of its embeddings $\mathcal{M}(F^k)$ in $\mathcal{G}$ and proceeds as follows. For each size-k embedding $m \in \mathcal{M}(F^k)$, it identifies and stores every possible size-$(k+1)$ subgraph in $\mathcal{G}$ that contains m. From this set of subgraphs, it extracts all size-$(k+1)$ subgraphs that are not isomorphic to each other and stores them in C^{k+1}. Then, vSiGraM-Extend eliminates from C^{k+1} all the subgraphs that do not have F^k as their generating parent (lines 5 and 6) or are infrequent (lines 7 and 8). The subgraphs remaining in C^{k+1} are the frequent subgraphs of size $(k+1)$ obtained by a one-edge-extension of F^k and are used as input for the next recursive call. The recursion terminates when $C^{k+1} = 0$, and the depth-first search backtracks.

Algorithm **6.4** vSiGraM

vSiGraM ($\mathcal{G}$, *Problem_Type*, f)

```
1: F ← ∅
2: F¹ ← all frequent size-1 subgraphs in G
3: for each F¹ in F¹ do
4:     M(F¹) ← all embeddings of F¹
5: end for
6: for each F¹ in F¹ do
7:     F ← F ∪ vSiGraM-Extend(F¹, G, Problem_Type, f)
8: end for
9: return F
```

vSiGraM-Extend (F^k, $\mathcal{G}$, *Problem_Type*, f)

```
1:  F ← ∅
2:  for each embedding m in M(Fᵏ) do
3:      Cᵏ⁺¹ ← Cᵏ⁺¹ ∪ {all (k + 1)-subgraphs of G containing m}
4:  end for
5:  for each Cᵏ⁺¹ in Cᵏ⁺¹ do
6:      if Fᵏ is not the generating parent of Cᵏ⁺¹ then
7:          continue
8:      end if
9:      compute Cᵏ⁺¹.freq from M(Cᵏ⁺¹)
10:     if Cᵏ⁺¹.freq < f then
11:         continue
12:     end if
13:     add Cᵏ⁺¹ to F
14: end for
15: return F
```

In the rest of this section we provide additional details on how the various operations are performed and describe various optimizations that are designed to reduce vSiGraM's runtime.

Generating Parent Identification. The scheme that vSiGraM uses to determine the generating parent of a particular subgraph is as follows. Suppose a size-$(k + 1)$ frequent subgraph F^{k+1} is just created by extension from a size-k frequent subgraph F^k. By the canonical labeling, the order of edges and vertices in F^{k+1} is uniquely determined. vSiGraM removes the last edge that does not disconnect F^{k+1} and obtains another size-k subgraph F.

If F is isomorphic to F^k then F^k becomes the generating parent of F^{k+1}, and vSiGraM keeps the further exploration from F^{k+1}. Similar types of approaches have been used earlier in the context of vertical algorithms for the graph-transaction setting [63, 64]. All of these share the same idea, which avoids redundant frequent pattern generation and traverses the lattice of patterns as if it were a tree.

Efficient Subgraph Extension. Starting from a frequent size-k subgraph, vSI-GRAM obtains the extended subgraphs of size $k + 1$ by adding an additional edge (while preserving connectivity) to all of its possible embeddings. Specifically, for each embedding m of a frequent k-subgraph F, vSIGRAM enumerates all the edges that can be added to m to form a size-$(k + 1)$ extended subgraph. Each of those edges is represented by a tuple of five elements $s = (x, y, u, v, e)$, called a *stem*, where x and y are the vertex IDs of the edge in $\mathcal{G}$, u and v, $u < v$, are the corresponding vertex IDs in F, and e is the label of the edge. For u and v, if there is no corresponding vertex in F, -1 is used to show that it is outside the subgraph F.

However, because of the automorphism of the subgraph F, we cannot use this stem representation directly. For a particular embedding m of a frequent subgraph F in $\mathcal{G}$, there may be more than one vertex mapping of the subgraph onto the embedding. If we simply used a pair of vertex IDs of the subgraph to represent a stem, depending on the mapping, the same edge addition might be considered a different stem, which would result in the wrong frequency of the subgraph. To avoid this problem, every time a stem is generated, its representation is normalized as follows: vSIGRAM enumerates all possible automorphisms of F, denoted by $\{\phi_i\}$. By an appropriate ϕ_i, we obtain the canonical vertex ID for every vertex $v \in V(F)$. The *canonical ID* of a vertex v, denoted by cvid(v), is defined as

$$\text{cvid}(v) = \min_i \phi_i(v)$$

The automorphism with the least subscript that gives the canonical ID for v is called the *canonical automorphism*, denoted by ϕ_v^*.

$$\phi_v^* = \arg \min_{\phi_i} \phi_i(v) \qquad i < j \text{ if } \phi_i(v) = \phi_j(v)$$

For example, given the size-6 graph G shown in Figure 6.6(a), cvid(v_3) = v_1 and $\phi_{v_3}^* = \phi_2$. Figure 6.6(b) shows cvid and ϕ^* for every vertex in G. Note that although $\phi_3(v_3)$ is also v_1, because ϕ_2 has the smaller subscript, 2, $\phi_{v_3}^*$ is ϕ_2.

Now for each stem $s = (x, y, u, v, e)$, $\phi^*(u, v) = (u', v')$ are defined as follows.

$$u' \equiv \text{cvid}(u) \qquad v' \equiv \phi_u^*(v) \qquad \text{if cvid}(u) \leq \text{cvid}(v)$$

$$u' \equiv \phi_v^*(u) \qquad v' \equiv \text{cvid}(v) \qquad \text{otherwise}$$

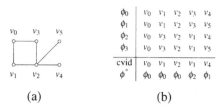

(a)   (b)

Figure 6.6. (a) Size-6 graph G and (b) canonical vertex IDs and canonical automorphism.

Then, stem s is rewritten as (x, y, u', v', e), which is an automorphism-invariant representation of s and is used by vSiGraM to properly determine the frequency of size-$(k + 1)$ extended subgraphs.

Frequency Counting. In the vertical algorithm, when a size-$(k + 1)$ extension is processed, there is only one size-k frequent subgraph visible, the generating parent. vSiGraM's frequency counting is similar to hSiGraM-Count, except for the computation of f_p (see line 15 in Algorithm 6.2). hSiGraM enforces the downward closure property on the frequency of a size-$(k + 1)$ candidate, by using the least frequency of all size-k subgraphs of the candidate. vSiGraM cannot take the same step because vSiGraM does not hold all size-k frequent subgraphs at the time a size-$(k + 1)$ extended subgraph is created. Instead vSiGraM simply uses the frequency of the size-k generating parent from which the current size-$(k + 1)$ extension is obtained. As a result, vSiGraM's pruning is looser than that of hSiGraM.

6.5.3 Interplay of the Two Search Paradigms and Three Problem Formulations

In this section, we show and compare the performance of the two algorithms with various parameters and real datasets. All experiments were done on dual AMD Athlon MP 1800+ (1.53 GHz) machines with 2 GB main memory, running the Linux operating system. All the runtimes reported are in seconds.

Intractability of Solving the Exact MIS. Table 6.3 shows the results obtained by the hSiGraM and vSiGraM algorithms for the VLSI dataset. For all the frequency thresholds, both hSiGraM and vSiGraM could not finish the computation because solving the exact MIS became intractable. On the other hand, both the subset and superset discovery based on the approximate and upper bound MIS can be solved efficiently.

Superset Discovery and Pruning. In the superset discovery problem (see Definition 6.4), the key is how effectively we can reduce the number of potential frequent patterns because some of the potential patterns may not be actually frequent enough by definition. As shown in Table 6.4, often hSiGraM achieves superior performance over vSiGraM because of its tighter pruning strategies based on the downward closure property. In other words, for certain datasets, when the superset formulation is used, vSiGraM ends up generating significantly more patterns than those generated by hSiGraM. For example, in the case of the DTP dataset and $f = 20$, vSiGraM generates almost 16 times more patterns than hSiGraM. In such cases, the amount of time required by vSiGraM is substantially greater than that required by hSiGraM (32.4 times greater in the DTP example). The reason for that is the fact that because of its depth-first nature, vSiGraM cannot take advantage of the frequent subgraph lattice to get a tight upper bound on the frequency of a subgraph based on the frequency of all of its subgraphs, and it bases its upper bound only on the frequency of the generating parent. On the other hand, because of its level-by-level

TABLE 6.3   Results for the VLSI Dataset[a]

| | Runtime t (s) | | | | Number of Patterns $|\mathcal{F}|$ | | | | Largest Size k^* | | | |
| | Subset | | Superset | | Subset | | Superset | | Subset | | Superset | |
f	H	V	H	V	H	V	H	V	H	V	H	V
300	0	0	1	1	10	10	19	22	1	1	5	5
250	1	0	1	1	17	17	25	33	2	2	5	5
200	11	3	37	8	137	137	347	415	5	5	5	5
150	13	4	46	9	156	156	437	503	5	5	5	5
100	42	7	54	10	379	379	519	609	5	5	5	5
75	49	8	56	10	409	409	571	679	5	5	5	5
50	236	15	282	17	683	683	946	1051	5	5	5	5
25	428	18	469	20	1452	1452	1907	2131	5	5	5	5

[a] f: the minimum frequency threshold. $|\mathcal{F}|$: the total number of frequent subgraphs discovered. k^*: the size of the largest frequent subgraphs discovered. H and V: HSIGRAM and vSIGRAM, respectively.
Subset (using the approximate MIS), Complete (using the exact MIS), and Superset (using the upper bound MIS) correspond to problem definitions 6.3 (subset discovery) 6.2 (complete discovery), and 6.4 (superset discovery) described in Section 6.3.1, respectively.

TABLE 6.4   Superset Discovery Results[a]

| | | Runtime t (s) | | No. of Patterns $|\mathcal{F}|$ | | k^* | |
Dataset	f	H	V	H	V	H	V
Citation	100	0.1	0.0	7	11	2	5
	50	0.6	—	113	—	7	—
	20	139	—	12,203	—	16	—
Contact Map	500	3	—	106	—	7	—
	400	10	—	246	—	8	—
	300	183	—	2,358	—	10	—
DTP	1,000	78	26	48	80	10	11
	500	96	30	153	226	12	13
	200	115	38	641	916	15	15
	100	169	64	2,484	3,788	16	18
	50	247	103	8,295	13,622	18	21
	20	616	19,998	52,180	824,702	20	81
	10	2,018	—	232,810	—	21	—

[a] —: the computation was aborted because of the too long runtime or memory exhaustion. f: the minimum frequency threshold. t: the runtime in seconds. $|\mathcal{F}|$: the total number of potentially frequent patterns found. H and V: HSIGRAM and vSIGRAM, respectively. k^*: the size of the largest potentially frequent patterns found.

nature, HSIGRAM can use the information from all its subpatterns and obtains better upper bounds (see discussion in Section 6.5.1).

Subset and Complete Discovery.   In the subset and complete discovery problems (see Definitions 6.3 and 6.2 in Section 6.3.1), as shown in Table 6.5, generally vSI-GRAM outperforms HSIGRAM with respect to runtime. In fact, as the value of the frequency threshold decreases, vSIGRAM is up to five times faster than HSIGRAM. This is true across all datasets for the subset and complete problem formulations,

TABLE 6.5 Subset and Complete Discovery Results[a]

Dataset	f	Runtime t (s)				Number of Patterns $|\mathcal{F}|$				Largest Size k^*			
		Subset		Complete		Subset		Complete		Subset		Complete	
		H	V	H	V	H	V	H	V	H	V	H	V
Aviation	2,000	308	130	306	130	833	833	833	833	8	8	8	8
	1,750	779	342	787	342	2,249	2,249	2,249	2,249	9	9	9	9
	1,500	1,603	743	1,674	745	5,207	5,207	5,207	5,207	10	10	10	10
	1,250	2,726	1,461	2,720	1,496	11,087	11,087	11,087	11,087	12	12	12	12
	1,000	5,256	3,667	5,158	3,683	30,331	30,331	30,331	30,331	13	13	13	13
Citation	100	0.1	0.0	0.1	0.0	6	6	6	6	1	1	1	1
	50	0.1	0.1	0.1	0.1	39	39	39	39	2	2	2	2
	20	0.6	0.3	0.9	0.5	266	266	266	266	3	3	3	3
	10	4.0	1.5	4.2	1.9	986	986	988	988	5	5	5	5
Contact map	500	1	1	1	1	62	62	62	62	2	2	2	2
	400	3	2	3	2	100	100	100	100	2	2	2	2
	300	10	3	10	3	186	186	186	186	2	2	2	2
	200	44	9	45	9	505	505	505	505	3	3	3	3
	100	362	63	356	71	3,183	3,183	3,186	3,186	5	5	5	5
	50	3,505	607	3,532	632	29,237	2,9237	29,298	29,298	6	6	6	6
Credit	500	0	0	0	0	24	24	24	24	3	3	3	3
	200	10	4	10	4	1,325	1,325	1,325	1,325	7	7	7	7
	100	49	20	45	21	11,696	11,696	11,696	11,696	9	9	9	9
	50	169	78	172	80	73,992	73,992	73,992	73,992	11	11	11	11
	20	2,019	461	1,855	468	613,884	613,884	613,884	613,884	13	13	13	13
DTP	1,000	70	19	71	20	31	31	31	31	6	6	6	6
	500	92	20	86	21	109	109	109	109	7	7	7	7
	200	101	23	100	24	414	414	415	415	9	9	9	9
	100	113	27	114	27	1,244	1,244	1,244	1,244	12	12	12	12
	50	145	34	134	35	4,028	4,028	4,028	4,028	14	14	14	14
	20	243	86	249	83	21,477	21,477	21,478	21,478	16	16	16	16
	10	813	311	882	294	112,535	112,535	112,539	112,539	21	21	21	21

[a] —: the computation was aborted because of the too long runtime or memory exhaustion. f: the minimum frequency threshold. t: the runtime in seconds. $|\mathcal{F}|$: the total number of potentially frequent patterns found. k^*: the size of the largest potentially frequent patterns found. Subset (using the approximate MIS) and Complete (using the exact MIS) correspond to problem definitions 6.3 (subset discovery) and 6.2 (complete discovery) described in Section 6.3.1, respectively. H and V: hSiGraM and vSiGraM, respectively.

and for those datasets for which the superset formulation leads to the same number of frequent patterns for both algorithms. As discussed in Section 6.5.2, the reason for that performance advantage is the fact that by keeping track the embeddings of the frequent subgraphs along the depth-first path, vSiGraM spends significantly less time in subgraph isomorphism-related computations than hSiGraM does.

6.6 Grew—SCALABLE FREQUENT SUBGRAPH DISCOVERY ALGORITHM

Grew is a heuristic algorithm designed to operate on a large graph and to find patterns corresponding to connected subgraphs that have a large number of vertex-disjoint embeddings. Specifically, the patterns that Grew finds satisfy the following two properties:

Property 6.1   The number of vertex-disjoint embeddings of each pattern is guaranteed to be at least as high as the user-supplied minimum frequency threshold.

Property 6.2   If a vertex contributes to the support of multiple patterns $\{G_1, G_2, \ldots, G_k\}$ of increasing size, then G_i is a subgraph of G_{i+1} for $i = 1, \ldots, k - 1$.

The first property ensures that the patterns discovered by GREW will be frequent. However, GREW is not guaranteed to find all the vertex-disjoint embeddings of each pattern that it reports, nor is it guaranteed to find all the patterns that have a sufficiently large number of vertex-disjoint embeddings. As a result, GREW will tend to undercount the frequency of the patterns that it discovers and will miss some patterns. Moreover, the second property imposes some additional constraints on the types of patterns that it can discover, as it does not allow each vertex to contribute to the support of patterns that do not have a subgraph/supergraph relationship. As a result of these properties, the number of patterns that GREW discovers is significantly smaller to those discovered by complete algorithms such as SiGRaM [45].

GREW discovers frequent subgraphs in an iterative fashion. During each iteration, GREW identifies vertex-disjoint embeddings of subgraphs that were determined to be frequent in previous iterations and merges certain subgraphs that are connected to each other via one or multiple edges. This iterative frequent subgraph merging process continues until there are no such candidate subgraphs whose combination will lead to a larger frequent subgraph. Note that unlike existing subgraph growing methods used by complete algorithms [36, 44, 45, 64], which increase the size of each successive subgraph by one edge or vertex at a time, GREW, in each successive iteration, can potentially double the size of the subgraphs that it identifies.

The key feature that contributes to GREW's efficiency is that it maintains the location of the embeddings of the previously identified frequent subgraphs by rewriting the input graph. As a result of this graph rewriting, the vertices involved in each particular embedding are *collapsed* together to form a new vertex (referred to as *multivertex*), whose label uniquely identifies the particular frequent subgraph that is supported by them. Within each multivertex, the edges that are not part of the frequent subgraph are added as loop edges. To ensure that the rewritten graph contains all the information present in the original graph, these newly created loop edges, as well as the edges of the original graph that are incident to a multivertex, are augmented to contain information about (i) the label of the incident vertices and (ii) their actual endpoint vertices within each multivertex (with respect to the original graph). Using the above representation, GREW identifies the sets of embedding pairs to be merged by simply finding the frequent edges that have the same augmented edge label. In addition, GREW obtains the next level rewritten graph by simply contracting together the vertices that are incident to the selected edges.

6.6.1 Graph Representation

GREW represents the original input graph $\mathcal{G}$ as well as the graphs obtained after each successive rewriting operation in a unified fashion. This representation, referred to as the *augmented graph*, is designed to contain all necessary information by which we can recover the original input graph $\mathcal{G}$ from any intermediate graph obtained after a sequence of rewriting operations.

Each vertex v and edge e of the augmented graph $\hat{G}$ has a label associated with it, which is denoted by $\hat{l}(v)$ and $\hat{l}(e)$, respectively. In addition, unlike the original graph that is simple, the augmented graph can contain loops. Furthermore, there can be multiple loop edges associated with each vertex, and there can be multiple edges connecting the same pair of vertices. However, whenever there exist such multiple loops or edges, the augmented label of each individual edge will be different from the rest.

The label of each vertex v in the augmented graph depends on whether or not it corresponds to a single vertex or a multivertex obtained after collapsing together a set of vertices that are used by an embedding of a particular subgraph. In the former case the label of the vertex is identical to its label in the original graph, whereas in the latter case its label is determined by the canonical labeling (Section 6.2.1) of its corresponding subgraph. This canonical-labeling-based approach ensures that the multivertices representing the embeddings of the same subgraphs will be uniquely assigned the same label.

To properly represent edges that are connected to multivertices, the augmented graph representation assigns a label to each edge that is a tuple of five elements. For an edge $e = uv$ in an augmented graph $\hat{G}$, this tuple is denoted by $(\hat{l}(u), \hat{l}(v), l(e), e.\text{epid}(u), e.\text{epid}(v))$, where $\hat{l}(u)$ and $\hat{l}(v)$ are the labels of the vertices u and v in $\hat{G}$, $l(e)$ is the original label of the edge e, and $e.\text{epid}(u)$ and $e.\text{epid}(v)$ are two numbers, referred to as *endpoint identifiers*, that uniquely identify the specific pair of $\mathcal{G}$'s vertices within the subgraphs encapsulated by u and v to which e is incident. The endpoint identifiers are determined by first ordering the original vertices in u and v according to their respective canonical labeling and then using their rank as the endpoint identifier. If an endpoint is not a multivertex, but just a plain vertex, the endpoint identifier is always set to zero.

Since the endpoint identifiers are derived from the canonical labels of u and v, it is easy to see that this approach will correctly assign the same five elements to all the edges that have the same original label and connect the same pair of subgraphs at exactly the same vertices. However, to ensure that topologically equivalent edges can be quickly identified by comparing their tuple representation, the order of the tuple's elements must be determined in a consistent fashion. For this reason, given an edge $e = uv$, the precise tuple representation is defined as follows:

$$(\hat{l}(u), \hat{l}(v), l(e), e.\text{epid}(u), e.\text{epid}(v)) \quad \text{if } \hat{l}(u) < \hat{l}(v), \quad \text{or}$$

$$\text{if } \hat{l}(u) = \hat{l}(v) \quad \text{and} \quad e.\text{epid}(u) \le \text{epid}(v)$$

or

$(\hat{l}(v), \hat{l}(u), l(e), e.\,\text{epid}(v), e.\,\text{epid}(u))$ if $\hat{l}(u) > \hat{l}(v)$,   or

$$\text{if } \hat{l}(u) = \hat{l}(v) \quad \text{and} \quad e.\,\text{epid}(u) > \text{epid}(v)$$

This consistent tuple representation ensures that all the edges that share the same label in the augmented graph correspond to identical subgraphs in the original graph.

Note that loops and multiple edges can also be represented by these five-element tuples, and the augmented graph representation treats them like ordinary edges.

6.6.2 Grew-se —Single-Edge Collapsing

The simplest version of GREW, which is referred to as GREW-SE, operates on the augmented graph and repeatedly identifies frequently occurring edges and contracts them in a heuristic fashion.

The overall structure of GREW-SE is shown in Algorithm 6.5. It takes as input the original graph G and the minimum frequency threshold f, and on completion, it returns the set of frequent subgraphs $\mathcal{F}$ that it identified. During each iteration (loop starting at line 5), it scans the current augmented graph $\hat{G}$ and determines the set of edge types $\mathcal{E}$ that occur at least f times in $\hat{G}$. This is achieved by comparing the labels of the various edges in $\hat{G}$ and determining those edge types that occur at least f times.

From the discussion in Section 6.6.1, we know that each of these edge types represents identical subgraphs, and as a result each edge type in $\mathcal{E}$ can lead to a frequent subgraph. However, because some vertices can be incident to multiple embeddings of the same (or different) frequent edge types, the frequencies obtained at this step represent upper bounds, and the actual number of the vertex-disjoint embeddings can be smaller. For this reason, GREW-SE further analyzes the embeddings of each edge type to select a maximal set of embeddings that do not share any vertices with each other or with embeddings selected previously for other edge types. This step (loop starting at line 8) is achieved by constructing the overlap graph G_o for the set of embeddings of each edge type e and using a greedy maximal independent set algorithm [25] to quickly identify a large number of vertex-disjoint embeddings. If the size of this maximal set is greater than the minimum frequency threshold, this edge type survives the current iteration and the embeddings in the independent set are marked. Otherwise the edge type is discarded as it does not lead to a frequent subgraph in the current iteration. After processing all the edge types, the contraction operations are performed, graph $\hat{G}$ is updated, and the next iteration begins.

To illustrate some of the steps performed by GREW-SE, let us take the simple example shown in Figure 6.7 in which the original graph [Fig. 6.7(a)] contains two squares connected to each other by an edge. Assume all the edges have the same label at the beginning, while there are two distinct vertex labels (the white and the slightly shaded ones). Edge contraction process proceeds as illustrated in

Algorithm **6.5** GREW-SE($\mathcal{G}, f$)

```
 1: ▷ G is the input graph.
 2: ▷ f is the minimum frequency threshold.
 3: F ← ∅
 4: Ĝ ← augmented graph representation of G
 5: while true do
 6:     E ← all edge-types in Ĝ that occur at least f times
 7:     order E in decreasing frequency
 8:     for each edge-type e in E do
 9:         Gₒ ← overlap graph of e
10:         ▷ each vertex in Gₒ corresponds to an embedding
                of e in Ĝ
11:         M_MIS ← obtain MIS for Gₒ
12:         e.f ← |M_MIS|
13:         if e.f ≥ f then
14:             F ← F∪{e}
15:             for each embedding m in M_MIS do
16:                 mark m
17:             end for
18:         end if
19:     end for
20:     if no marked edge in Ĝ then
21:         break
22:     end if
23:     update Ĝ by rewriting all of its marked edges
24: end while
25: return  F
```

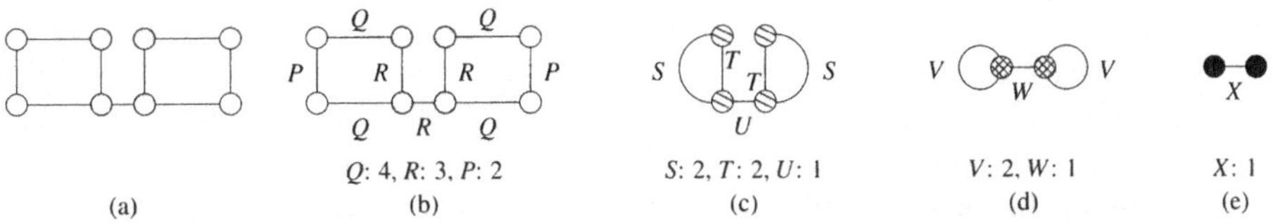

Figure 6.7. Sequence of edge collapsing operations performed by GREW-SE.

Figures 6.7(b)–(e), assuming the edge types are selected in decreasing order of the raw frequency (each edge type is represented by a capital letter and the raw frequency of each edge type is also shown in the figure). Every time an edge is contracted, a new multivertex is created whose label identifies a subgraph that the multivertex represents (shown by the difference in shading and filling pattern of vertices). Note that at the end of this sequence of edge collapsing, the two squares originally existed in Figure 6.7(a) are represented by two black vertices in Figure 6.7(e).

6.6.3 GREW-ME—Multiedge Collapsing

As discussed in Section 6.6.1, a result of successive graph rewriting operations is the creation of multiple loops and multiple edges in $\hat{G}$. In many cases, there may be the same set of multiple edges connecting similar pairs of vertices in $\hat{G}$, all of which can be collapsed together to form a larger frequent subgraph. GREW-SE can potentially identify such a subgraph by collapsing a sequence of single edges (which after the first iteration, each successive iteration will involve loop edges). However, this will require multiple iterations, and, owing to the heuristic nature of the overall algorithm, it may fail to *orchestrate* the proper sequence of steps.

To address this problem, we developed the GREW-ME algorithm that in addition to collapsing vertices connected via a single edge, it also analyzes the sets of multiple edges connecting pairs of vertices to identify any frequent subsets of edges. This is achieved by using a traditional frequent closed itemset mining algorithm (e.g., [52, 53, 71]) as follows. For each pair of vertices that are connected via multiple edges (or a single vertex with multiple loops), GREW-ME creates a list that contains the multiple edge types that are involved and treats each list as a transaction whose items correspond to the multiple edges. Then, by running a closed frequent itemset mining algorithm, GREW-ME finds all the frequent sets of edges whose raw frequency is above the minimum threshold. Each of these multiple sets of edges is treated as a different edge type, and GREW-ME proceeds in a fashion identical to GREW-SE.

6.6.4 Mining a Large Graph Using GREW—Two Randomized Schemes

A key limitation of GREW is that it tends to find a very small number of frequent subgraphs compared to complete algorithms [45, 47]. This is primarily due to the fact that its graph rewriting-based approach substantially constraints the sets of frequent subgraphs that are allowed to have overlapping embeddings (Property 6.2 in Section 6.6), and secondary due to the fact that it underestimates the frequency of each subgraph and consequently it may miss subgraphs that are actually frequent.

However, because of its low computational requirements, GREW can be used as a building block to develop *meta-strategies*, which can be used to effectively mine very large graphs in a reasonable amount of time. The key to such schemes is the observation that the particular set of frequent subgraphs that GREW discovers is determined by the order in which the different edge types are listed up, and by the maximal independent set algorithm. The order of the edge-types is important because during each iteration, a vertex of $\hat{G}$ can only participate in a single contraction operation. As a result, edge types examined earlier have a greater chance in having a larger number of vertex-disjoint embeddings and thus resulting in a frequent subgraph. At the same time, a particular set of maximal independent embeddings can directly or indirectly impact the discovered subgraphs. The direct impact occurs when the size of the selected maximal independent set happens to be smaller than the minimum frequency threshold (when the maximum size independent set is above the threshold), in which case the subgraph will not be identified as frequent. The indirect impact occurs when the chosen independent set negatively affects the number of vertex-disjoint embeddings of some of the edge types that are examined afterward.

GREW performs each of these steps by employing relatively simple heuristics that are designed to maximize the frequency of the discovered patterns. However, each of these two steps can be performed using different heuristics that can guide/bias GREW to find/prefer one set of subgraphs over the others. This ability makes it possible to develop approaches that invoke GREW multiple times, each time biasing it toward discovering a different set of frequent subgraphs, and return the union of the frequent subgraphs that were found across the multiple runs.

We developed two different schemes for biasing each successive run. Both of them rely on the observations made in Section 6.6.4, which identified GREW's key operations that affect the set of subgraphs that it identifies. The first scheme, instead of considering the different edge types in decreasing order of their raw frequency, it considers them in a random order, which is different for each successive run. Since each run of GREW will tend to favor different sets of edge types, the frequent subgraphs identified in successive runs will be somewhat different. We will refer to this as the *simple randomization scheme*. The second scheme uses the same random traversal strategy as before but also tries to bias the patterns to different edges from those used earlier. This is done by maintaining statistics as to which and how many times each edge was used to support a frequent subgraph identified by earlier iterations, and biases the algorithm that selects the maximal independent set of embeddings so that it will prefer edges that have not been used (or used infrequently) in previously discovered subgraphs. Thus, this scheme tries to learn from previous invocations, and for this reason it will be referred to as the *adaptive randomization scheme*.

Table 6.6 shows the performance and characteristics of the subgraphs discovered by multiple runs (ranging from 1 to 10) of GREW-SE and GREW-ME for the cases in which these multiple runs were performed using the simple and the adaptive randomization schemes.

From these results we can see that the two randomization schemes are quite effective in allowing GREW to find a larger number of frequent subgraphs. As the number of runs increases, both the number of frequent patterns and the size of the largest frequent patterns increase monotonically. As expected, there is a certain degree of overlap between the patterns found by different runs, and for this reason the distinct number of subgraphs does not increase linearly. In addition, the set of vertices and edges of the original graph that are covered by the discovered frequent subgraphs also increases with the number of runs. For all the datasets, after invoking GREW-SE and GREW-ME 10 times, the resulting set of frequent subgraphs *cover* more than 50% of the vertices and/or edges of the input graph. This suggests that GREW is able to find a diverse set of frequent subgraphs that captures a good fraction of the input graph.

Comparing the relative performance of GREW-SE and GREW-ME within the context of this randomization framework, we can see that GREW-ME tends to find a

TABLE 6.6   Simple and Adaptive Randomization[a]

Dataset	f	Method	Runs	Simple Randomization					Adaptive Randomization				
				t (s)	$|\mathcal{F}|$	k^*	R_V	R_E	t (s)	$|\mathcal{F}|$	k^*	R_V	R_E
Aviation	100	GREW-SE	1	23	245	14	54	67	18	245	14	54	67
			2	46	386	14	44	59	35	386	14	44	59
			5	261	795	17	28	47	166	793	19	30	48
			10	423	1426	22	20	39	260	1476	19	19	39
		GREW-ME	1	49	233	14	55	68	37	233	14	55	68
			2	193	363	22	44	59	144	363	22	44	59
			5	345	754	22	28	47	252	754	22	23	47
			10	615	1422	22	19	40	634	1434	40	18	39
Citation	10	GREW-SE	1	39	659	5	54	94	41	659	5	54	94
			2	82	881	5	45	92	86	881	5	44	92
			5	231	1340	7	36	88	224	1365	5	35	87
			10	461	1912	7	31	82	453	1940	8	30	80
		GREW-ME	1	188	658	5	56	94	189	658	5	56	94
			2	367	940	6	46	92	358	936	7	46	92
			5	899	1527	7	37	88	916	1519	9	36	87
			10	1843	2311	8	32	82	2683	2319	9	30	80
VLSI	10	GREW-SE	1	12	394	20	6	73	12	389	21	6	73
			2	25	712	20	2	61	24	674	34	1	58
			5	74	1372	20	1	42	62	1452	34	0	34
			10	146	2335	21	0	27	140	2416	34	0	16
		GREW-ME	1	37	509	20	10	73	37	509	18	10	74
			2	83	959	22	3	58	74	933	21	3	57
			5	235	2049	30	1	38	202	1925	22	0	31
			10	440	3547	30	0	25	362	3403	27	0	14
Web	10	GREW-SE	1	298	2716	9	74	86	199	2716	9	74	86
			2	393	3268	9	69	82	395	3273	13	67	80
			5	992	4095	15	62	74	994	4155	13	58	70
			10	1970	4871	15	56	67	1974	4881	13	51	61
		GREW-ME	1	805	2719	14	74	86	550	2719	14	74	86
			2	1084	3249	14	69	82	978	3257	14	67	80
			5	2578	4138	16	62	74	2464	4158	14	58	70
			10	5074	4945	16	57	67	5175	4979	15	51	61

[a] f: the minimum frequency threshold. Runs: the number of randomized runs. t: the runtime in seconds. $|\mathcal{F}|$: the number of frequent patterns discovered. k^*: the size of the largest frequent patterns found. R_V, R_E: the fraction of vertices and edges (%), respectively, in the input graph that are not covered by any of the frequent patterns discovered.

larger number of distinct frequent subgraphs whose maximum size is larger than the subgraphs discovered by GREW-SE. This indicates that GREW-ME's ability to identify multiple edges and collapse vertices that are connected by them becomes more effective in increasing the diversity of the discovered patterns in the context of this randomization strategy.

Finally, comparing the relative performance of the two randomization schemes, we can see that, as expected, the adaptive randomization scheme improves the pattern discovery process as it finds a larger number of distinct patterns than the simple randomization. However, the overall size of the patterns identified by both schemes remains the same, as they both cover similar fractions of the vertices and/or edges of the input graph.

6.7 RELATED RESEARCH

6.7.1 Related Work on Algorithms for the Graph-Transaction Setting

In recent years, a number of efficient and scalable algorithms have been developed to find patterns in the graph-transaction setting [8, 32, 33, 44, 64, 65]. These algorithms are complete in the sense that they are guaranteed to discover all frequent subgraphs and were shown to scale to very large graph datasets. However, developing algorithms that are capable of finding patterns in the single-graph setting has received much less attention, despite the fact that this problem setting is more generic and applicable to a wider range of datasets and application domains than the other. Moreover, existing algorithms that are guaranteed to find all frequent patterns [22, 63] or algorithms that are heuristic, such as GBI [67] and SUBDUE [31], which tend to miss a large number of frequent patterns, are computationally expensive and do not scale to large datasets.

Different algorithms have been developed capable of finding all frequently occurring subgraphs in a database of graphs with reasonable computational efficiency. Some of these are the AGM algorithm developed by Inokuchi et al. [35, 37], the FSG algorithm developed by Kuramochi and Karypis [41, 42], the chemical substructure discovery algorithm developed by Borgelt and Berthold [8], the gSpan algorithm developed by Yan and Han [64], FFSM by Huan et al. [33], and GASTON by Nijssen and Kok [50].

The AGM algorithm was initially developed to find frequently induced subgraphs [35]. After Kuramochi and Karypis proposed FSG [41], Inokuchi et al. [37] extended AGM to find arbitrary frequent subgraphs and discovered the frequent subgraphs using a breadth-first approach and grew the frequent subgraphs one vertex at a time. To distinguish a subgraph from another, it uses a canonical labeling scheme based on the adjacency matrix representation.

The FSG algorithm initially presented in [41], with subsequent improvements presented in [42], is the first work on finding frequent connected subgraphs from a set of labeled graph transactions. Many algorithms proposed afterward are following the problem formulation first presented in [41]. FSG uses a breadth-first approach to discover the lattice of frequent subgraphs. The size of these subgraphs is grown by adding one edge at a time, and the frequent pattern lattice is used to prune non-downward-closed candidate subgraphs. FSG employs a number of techniques to achieve high computational performance including efficient canonical labeling, efficient candidate subgraph generation algorithms, and various optimizations during frequency counting.

The chemical substructure mining algorithm developed by Borgelt and Berthold [8] finds frequent substructures (connected subgraphs) using a depth-first approach similar to that used by dEclat [69]. To reduce the number of subgraph isomorphism operations, it keeps the embeddings of previously discovered subgraphs and tries to extend the embeddings by one edge. However, despite these optimizations, the reported speed of the algorithm is slower than that achieved by FSG and gSpan. This is mainly due to the fact that their candidate subgraph generation scheme does not ensure that the same subgraph is generated only once, and the algorithm generates and counts the frequency of the same subgraph multiple times.

The gSpan algorithm finds the frequently occurring subgraphs also following a depth-first approach. However, unlike the previous algorithm, every time a candidate subgraph is generated, it is checked against previously discovered subgraphs to determine whether or not it has already been explored. To ensure that these subgraph comparisons are done efficiently, they use a canonical labeling scheme based on depth-first traversals. According to the reported performance in [42, 64], gSpan and FSG are comparable on real datasets corresponding to chemical compounds (the PTE chemical dataset[5]), whereas gSpan performs better than FSG on synthetic datasets. FFSM incorporates the join operation that is commonly used in horizontal frequent mining algorithms in the depth-first search scheme for efficient execution. GASTON focuses on enumerating simple frequent structures first, such as paths and trees, and then moves to identifying general graphs. Yan and Han [65] and Huan et al. [34] proposed closed and maximal frequent subgraph mining algorithms, respectively.

6.7.2 Related Work on Algorithms for the Single-Graph Setting

In the rest of this section, we will describe related research for the single-graph setting as it is directly related to the topic of this chapter.

The most well-known algorithm for finding recurring subgraphs in a single large graph is the SUBDUE system, originally developed in 1994 and improved over the years [10–12, 31]. SUBDUE is an approximate algorithm and finds patterns that can compress the original input graph by substituting those patterns with a single vertex. In evaluating the extent to which a particular pattern can compress the original graph, it uses the minimum description length (MDL) principle and employs a heuristic beam search to narrow the search space. These approximations improve its computational efficiency but at the same time it prevents it from finding subgraphs that are indeed frequent. GBI [67] is another greedy heuristics-based algorithm similar to SUBDUE.

Ghazizadeh and Chawathe [22] developed an algorithm called SEuS that uses a data structure called *summary* to construct a lossy compressed representation of the input graph. This summary is obtained by collapsing together all the vertices of the input graph that have the same label and is used to quickly prune infrequent candidates. As the authors indicate, this summary data–structure is useful only when the input graph contains a relatively small number of frequent subgraphs with high frequency and is not effective if there are a large number of frequent subgraphs with low frequency.

Vanetik et al. [63] presented an algorithm for finding all frequently occurring subgraphs from a single labeled undirected graph using the maximum number of edge-disjoint embeddings of a graph as a measure of its frequency. Each subgraph is represented by its minimum number of edge-disjoint paths (*path number*) and

[5]ftp://ftp.comlab.ox.ac.uk/pub/Packages/ILP/Datasets/carcinogenesis/progol/carcinogenesis.tar.Z.

uses a level-by-level approach to grow the patterns based on their path number. Their emphasis is on efficient candidate generation and no special attention is paid for frequency counting.

Motoda et al. developed an algorithm called GBI [66] that is similar to SUB-DUE and later proposed the improved version called B-GBI [47] adopting the beam search. B-GBI is the closest algorithm to our study, in the sense that both perform the same basic operation to identify frequent patterns based on edge contraction. However, while B-GBI focuses on one edge type at a time when collapsing the embeddings of the edge type in a greedy manner, GREW identifies and contracts more than one edge type concurrently using a greedy MIS algorithm. Because B-GBI works on a single edge type at at time, it uses the beam search to compensate the greedy nature of the algorithm. On the other hand, we adopted the randomized process to increase the diversity of frequent patterns to be found, on top of the concurrent edge collapsing scheme of GREW. Furthermore, our algorithm employs various heuristics such as multiedge collapsing and the adaptive and sequential schemes to ensure the coverage of the input graph by the frequent patterns. Unfortunately, because the current implementation of B-GBI is mainly designed for a set of graphs, not for a single large graph, it cannot be directly compared with GREW.

6.8 CONCLUSIONS

In this chapter, we described different problem definitions and formulations for finding frequent topological patterns from graph datasets. We also showed three representative approaches for both the graph-transaction setting and the single-graph setting, illustrating their performance and scalability using experimental results with real datasets from various application domains.

In recent years, frequent subgraph mining has been attracting strong attention and various algorithms have been proposed mainly for the graph-transaction setting, and the efficiency and scalability have been improved significantly (e.g., [33, 36, 44, 50, 64]). For a certain type of graph datasets such as chemical graphs, those algorithms have shown excellent performance. Furthermore, frequent subgraph methodology has been shown to be promising when the frequent patterns discovered by those methods are used as features in the chemical compound classification or virtual screening tasks. For example, in [15, 16], frequent subgraphs are used as features for chemical compound classification, and the resulting classification accuracy is equivalent or even superior to the accuracy based on fingerprints. We should note that those fingerprints can be considered a collection of domain expertise, whereas frequent subgraphs do not require much domain-specific knowledge. However, it is yet unclear whether those algorithms can operate efficiently on other types of graph datasets as they do on chemical graphs. Chemical graphs are *easy* to handle in the sense that vertex degree is bound to a small constant (normally at most 4), that there are relatively diverse vertex and edge labels, and that the size of each graph is reasonably small (typically around 20–25 edges). It is possible that those algorithms have overfit for chemical graph datasets.

The issue of overfitting is also related to the lack of significant applications. Other than chemical compound classification or virtual screening, there have not been important real applications that require frequent subgraph mining. This is partly because those existing frequent subgraph mining methods may not be able to scale efficiently on different kinds of graph datasets (e.g., larger or denser graphs with less variety of edge/vertex labels). Without needs for other practical applications, it may be inevitable to concentrate on the only killer application domain.

One possible direction for future research is to focus on a certain type of subgraph and develop complete algorithms that can find all frequent subgraphs of that kind, although it may be difficult to rationalize focusing on a particular class of subgraphs from an application user standpoint. Another approach will be to find reasonably large sets of frequent patterns heuristically so that one can handle more difficult graphs (i.e., larger, denser, or less labeled). In this direction, it will be crucial to define theoretically well the characteristics of the frequent subgraphs to be discovered.

Datasets

We briefly describe the source and characteristics of the datasets used in the experiments shown in this chapter. The PTC dataset[6] used with FSG contains 417 chemical compounds. From the description of chemical compounds in the dataset, we created a transaction for a compound, a vertex for an atom, and an edge for a bond. Each vertex has a label assigned for its atom type and each edge has an edge label assigned for its bond type. There are 27 atom types and 4 bond types. On average the size of the graph transactions in the dataset is 15 edges.

The basic characteristics of the datasets used with HSIGRAM and vSIGRAM and with GREW are shown in Tables 6.7 and 6.8, respectively. Note that we created different graph datasets from the same source for SIGRAM and for GREW because SIGRAM can handle only sparse graph datasets.

The *Aviation* and *Credit* datasets are obtained from the website of SUBDUE.[7] The Aviation dataset is originally from the Aviation Safety Reporting System Database and the Credit dataset is from the UCI machine learning repository [7]. The directed edges in the original graph data were converted into undirected ones. For the experiments using HSIGRAM and vSIGRAM, we removed undirected edges to show "near_to" relation between two vertices because those edges form cliques that make this graph difficult to mine, while for GREW the graph is used as is.

The *Citation* dataset was created from the citation graph used in KDD Cup 2003.[8] For HSIGRAM and vSIGRAM, each vertex in this graph corresponds to a document and each edge corresponds to a citation. Because our algorithms are for undirected graphs, the direction of these citations was ignored. Since the original dataset does not have any meaningful label for vertices, we generated vertex labels as follows. We first used a clustering algorithm to form clusters of the document

---

[6]http://www.predictive-toxicology.org/ptc/.
[7]http://cygnus.uta.edu/subdue/databases/index.html.
[8]KDD Cup 2003, http://www.cs.cornell.edu/projects/kddcup/datasets.html.

TABLE 6.7   Datasets Used with HSiGRAM and vSiGRAM[a]

Dataset	$|V(G)|$	$|E(G)|$	$|L_V|$	$|L_E|$	C
Aviation	101,185	196,964	6,173	51	2,703
Credit	14,700	28,000	59	20	700
Citation	29,014	42,064	50	12	16,999
Contact map	33,443	224,488	21	2	170
DTP	41,190	86,140	58	3	2,319
VLSI	12,752	23,084	23	1	2,633

[a]$|V(G)|$: the total number of vertices. $|E(G)|$: the total number of edges. $|L_V|$: the total number of distinct vertex labels. $|L_E|$: the total number of distinct edges labels. C: the total number of connected components.

TABLE 6.8   Datasets Used with GREW[a]

Dataset	$|V(G)|$	$|E(G)|$	$|L_V|$	$|L_E|$	C
Aviation	101,185	133,113	6,173	52	1,049
Citation	29,014	294,171	742	1	3,018
VLSI	29,347	81,353	11	1	1
Web	255,798	317,247	3,438	1	25,685

[a]$|V(G)|$: the total number of vertices. $|E(G)|$: the total number of edges. $|L_V|$: the total number of distinct vertex labels. $|L_E|$: the total number of distinct edges labels. C: the total number of connected components.

abstracts into 50 thematically coherent topics and then assigned the cluster ID as the label to the corresponding vertices. For the edges, we used as labels the difference in the publication year of the two papers. For example, if two papers were published in 1997 and 2002, an edge is created between those two document vertices with the label "5." Finally, because some of the vertices in the resulting graph had a very high degree (i.e., authorities and hubs), we kept only the vertices whose degree was less or equal to 15. On the other hand, for GREW, we assigned only vertices labels obtained from the subdomains of the first author's email address, since the original graph has no meaningful labels on either the vertices or the edges. Self-citations based on this vertex assignment were removed.

The *Contact Map* dataset is made of 170 proteins from the Protein Data Bank [5] with pairwise sequence identity lower than 25%. The vertices in these graphs correspond to the different amino acids, and the edges connect two amino acids if they are either at consecutive sequence positions or they are in contact in their 3D structure. Amino acids are considered to be in contact if the distance between their C_α atoms is less than 8 Å. Furthermore, while creating the graphs we only considered nonlocal contacts that are defined as the contacts between amino acids whose sequence separation is at least six amino acids.

The *DTP* dataset is a collection of 2319 chemical compounds randomly selected from the dataset of 223,644 chemical compounds provided by the Developmental

Therapeutics Program (DTP) at National Cancer Institute.[9] Note that each chemical compound forms a connected component and there are 2319 such components in this dataset. Each vertex corresponds to an atom and its label represents the atom type. An edge is formed between two vertices if the corresponding two atoms are connected by a bond. The type of a bond is used as an edge label, and there are three distinct edge labels.

The *VLSI* dataset was obtained from the International Symposium on Physical Design '98 (ISPD98) benchmark suite[10] and corresponds to the netlist of a real circuit. For HSiGraM and vSiGraM, the netlist was converted into a graph by first removing any nets that are longer than four and then using a star-based approach to replace each net (i.e., hyperedge) by a set of edges. We also limited the size of the largest discovered pattern to five edges. This is because for the values of the frequency threshold used in our experiments, the only frequent patterns that contained more than five edges were paths, and because of the highly connected nature of the underlying graph, there were a very large number of such paths, making it hard to find these longer path patterns in a reasonable amount of time. For Grew, the netlist was converted into a graph by using a star-based approach to replace each net (i.e., hyperedge) by a set of edges.

The *Web* dataset was obtained from the 2002 Google Programming Contest.[11] The original dataset contains various Web pages and links from various "edu" domains. We converted the dataset into an undirected graph in which each vertex corresponds to a Web page and an edge to a hyperlink between Web pages. In creating this graph, we kept only the links between "edu" domains that connected sites from different subdomains. Every edge has an identical label (i.e., unlabeled), whereas each vertex was assigned a label corresponding to the subdomain of the Web server.

REFERENCES

1. R. C. Agarwal, C. C. Aggarwal, and V. V. V. Prasad. A tree projection algorithm for generation of frequent item sets. *Journal of Parallel and Distributed Computing*, 61(3):350–371, 2001.
2. R. Agrawal and R. Srikant. Fast algorithms for mining association rules. In J. B. Bocca, M. Jarke, and C. Zaniolo, eds. *Proceedings of the 20th International Conference on Very Large Data Bases (VLDB)*, pp. 487–499. Morgan Kaufmann, September 1994.
3. R. Agrawal and R. Srikant. Mining sequential patterns. In P. S. Yu and A. L. P. Chen, eds. *Proceedings of the 11th International Conference on Data Engineering (ICDE)*, pp. 3–14. IEEE Press, 1995.
4. Y. Amit and A. Kong. Graphical templates for model registration. *IEEE Transactions on Pattern Analysis and Machine Intelligence*, 18(3):225–236, 1996.

[9]DTP 2D and 3D Structural Information. http://dtp.nci.nih.gov/docs/3d_database/structural_information/structural_data.html.
[10]http://vlsicad.cs.ucla.edu/~cheese/ispd98.html.
[11]http://www.google.com/programming-contest/.

5. H. M. Berman, J. Westbrook, Z. Feng, G. Gilliland, T. N. Bhat, H. Weissig, I. N. Shindyalov, and P. E. Bourne. The protein data bank. *Nucleic Acids Research*, 28:235–242, 2000.

6. P. Berman and T. Fujito. On the approximation properties of independent set problem in degree 3 graphs. In Proceedings of Workshop on Algorithms and Data Structures, pp. 449–460, 1995.

7. C. L. Blake and C. J. Merz. UCI repository of machine learning databases. 1998.

8. C. Borgelt and M. R. Berthold. Mining molecular fragments: Finding relevant substructures of molecules. In Proceedings of 2002 IEEE International Conference on Data Mining (ICDM), pp. 51–58, 2002.

9. C.-W. K. Chen and D. Y. Y. Yun. Unifying graph-matching problem with a practical solution. In Proceedings of International Conference on Systems, Signals, Control, Computers, September 1998.

10. D. J. Cook and L. B. Holder. Substructure discovery using minimum description length and background knowledge. *Journal of Artificial Intelligence Research*, 1:231–255, 1994.

11. D. J. Cook and L. B. Holder. Graph-based data mining. *IEEE Intelligent Systems*, 15(2):32–41, 2000.

12. D. J. Cook, L. B. Holder, and S. Djoko. Knowledge discovery from structural data. *Journal of Intelligent Information Systems*, 5(3):229–245, 1995.

13. L. Dehaspe, H. Toivonen, and R. D. King. Finding frequent substructures in chemical compounds. In R. Agrawal, P. Stolorz, and G. Piatetsky-Shapiro, eds, *Proceedings of the 4th ACM SIGKDD International Conference on Knowledge Discovery and Data Mining (KDD-98)*, pp. 30–36. AAAI Press, 1998.

14. M. Deshpande, M. Kuramochi, and G. Karypis. Automated approaches for classifying structures. In Proceedings of the 2nd Workshop on Data Mining in Bioinformatics (BIOKDD '02), 2002.

15. M. Deshpande, M. Kuramochi, and G. Karypis. Frequent sub-structure based approaches for classifying chemical compounds. In Proceedings of 2003 IEEE International Conference on Data Mining (ICDM), pp. 35–42, 2003.

16. M. Deshpande, M. Kuramochi, N. Wale, and G. Karypis. Frequent sub-structure-based approaches for classifying chemical compounds. *IEEE Transactions on Knowledge and Data Engineering (TKDE)*, Special Issue on Mining Biological Data, 2005.

17. B. Dunkel and N. Soparkar. Data organization and access for efficient data mining. In Proceedings of the 15th IEEE International Conference on Data Engineering, pp. 522–529, March 1999.

18. D. Dupplaw and P. H. Lewis. Content-based image retrieval with scale-spaced object trees. In M. M. Yeung, B.-L. Yeo, and C. A. Bouman, eds. Proceedings of SPIE: Storage and Retrieval for Media Databases, Vol. 3972, pp. 253–261, 2000.

19. U. Feige, S. Goldwasser, L. Lovasz, S. Safra, and M. Szegedy. Approximating clique is almost NP-complete. In Proceedings of the 32nd IEEE Symposium on Foundations of Computer Science (FOCS), pp. 2–12, 1991.

20. S. Fortin. The graph isomorphism problem. Technical Report TR96-20, Department of Computing Science, University of Alberta, Edmonton, Canada, 1996.

21. M. R. Garey and D. S. Johnson. *Computers and Intractability: A Guide to the Theory of NP-Completeness*. W. H. Freeman, New York, 1979.

22. S. Ghazizadeh and S. Chawathe. SEuS: Structure extraction using summaries. In Proceedings of the 5th International Conference on Discovery Science, pp. 71–85, 2002.

23. B. Goethals. *Efficient Frequent Pattern Mining*. Ph.D. thesis, University of Limburg, Diepenbeek, Belgium, December 2002.

24. J. Gonzalez, L. B. Holder, and D. J. Cook. Application of graph-based concept learning to the predictive toxicology domain. In Proceedings of the Predictive Toxicology Challenge Workshop, 2001.

25. M. M. Halldórsson and J. Radhakrishnan. Greed is good: Approximating independent sets in sparse and bounded-degree graphs. *Algorithmica*, 18(1):145–163, 1997.

26. J. Han, J. Pei, and Y. Yin. Mining frequent patterns without candidate generation. In Proceedings of ACM SIGMOD International Conference on Management of Data, pp. 1–12, Dallas, TX, May 2000.

27. C. Hansch, P. P. Maolney, T. Fujita, and R. M. Muir. Correlation of biological activity of phenoxyacetic acids with hammett substituent constants and partition coefficients. *Nature*, 194:178–180, 1962.

28. J. Hipp, U. Güntzer, and G. Nakhaeizadeh. Algorithms for association rule mining––a general survey and comparison. *SIGKDD Explorations*, 2(1):58–64, July 2000.

29. D. S. Hochbaum. Efficient bounds for the stable set, vertex cover, and set packing problems. *Discrete Applied Mathematics*, 6:243–254, 1983.

30. H. Hofer, C. Borgelt, and M. R. Berthold. Large scale mining of molecular fragments with wildcards. In *Proceedings of the 5th International Symposium on Intelligent Data Analysis (IDA 2003)*, Vol. 2810 of *Lecture Notes in Computer Science*, pp. 376–385, 2003.

31. L. B. Holder, D. J. Cook, and S. Djoko. Substructure discovery in the SUBDUE system. In Proceedings of the AAAI Workshop on Knowledge Discovery in Databases, pp. 169–180, 1994.

32. M. Hong, H. Zhou, W. Wang, and B. Shi. An efficient algorithm of frequent connected subgraph extraction. In *Proceedings of the 7th Pacific-Asia Conference on Knowledge Discovery and Data Mining (PAKDD-03)*, Vol. 2637 of *Lecture Notes in Computer Science*, pp. 40–51. Springer, 2003.

33. J. Huan, W. Wang, and J. Prins. Efficient mining of frequent subgraph in the presence of isomophism. In Proceedings of 2003 IEEE International Conference on Data Mining (ICDM'03), pp. 549–552, 2003.

34. J. Huan, W. Wang, J. Prins, and J. Yang. Spin: Mining maximal frequent subgraphs from graph databases. In Proceedings of the 10th ACM SIGKDD International Conference on Knowledge Discovery and Data Mining (KDD-2004), pp. 581–586, 2004.

35. A. Inokuchi, T. Washio, and H. Motoda. An apriori-based algorithm for mining frequent substructures from graph data. In Proceedings of the 4th European Conference on Principles and Practice of Knowledge Discovery in Databases (PKDD'00), pp. 13–23, Lyon, France, September, 2000.

36. A. Inokuchi, T. Washio, and H. Motoda. Complete mining of frequent patterns from graphs: Mining graph data. *Machine Learning*, 50(3):321–354, March 2003.

37. A. Inokuchi, T. Washio, K. Nishimura, and H. Motoda. A fast algorithm for mining frequent connected subgraphs. Technical Report RT0448, IBM Research, Tokyo Research Laboratory, 2002.

38. H. Kälviäinen and E. Oja. Comparisons of attributed graph matching algorithms for computer vision. In Proceedings of STEP-90, Finnish Artificial Intelligence Symposium, pp. 354–368, Oulu, Finland, June, 1990.

39. S. Khanna, R. Motwani, M. Sudan, and U. V. Vazirani. On syntactic versus computational views of approximability. In Proceedings of IEEE Symposium on Foundations of Computer Science, pp. 819–830, 1994.

40. R. D. King, S. H. Muggleton, A. Srinivasan, and M. J. E. Sternberg. Structure-activity relationships derived by machine learning: The use of atoms and their bond connectivities

to predict mutagenicity by inductive logic programming. In Proceedings of the National Academy of Sciences, Vol. 93, pp. 438–442, 1996.

41. M. Kuramochi and G. Karypis. Frequent subgraph discovery. In Proceedings of 2001 IEEE International Conference on Data Mining (ICDM), pp. 313–320, November 2001.

42. M. Kuramochi and G. Karypis. An efficient algorithm for discovering frequent subgraphs. Technical Report 02-026, University of Minnesota, Department of Computer Science, 2002.

43. M. Kuramochi and G. Karypis. Finding frequent patterns in a large sparse graph. Technical Report 03-039, University of Minnesota, Department of Computer Science, 2003.

44. M. Kuramochi and G. Karypis. An efficient algorithm for discovering frequent subgraphs. *IEEE Transactions on Knowledge and Data Engineering*, 16(9):1038–1051, 2004.

45. M. Kuramochi and G. Karypis. Finding frequent patterns in a large sparse graph. In Proceedings of the 2004 SIAM International Conference on Data Mining (SDM04), 2004.

46. T. K. Leung, M. C. Burl, and P. Perona. Finding faces in cluttered scenes using random labeled graph matching. In Proceedings of the 5th IEEE International Conference on Computer Vision, June 1995.

47. T. Matsuda, H. Motoda, T. Yoshida, and T. Washio. Mining patterns from structured data by beam-wise graph-based induction. In *Proceedings of the 5th International Conference on Discovery ScienceDiscoveery (DS 2002)*, Vol. 2534 of *Lecture Notes in Computer Science*, pp. 422–429. Springer, 2002.

48. B. D. McKay. Nauty users guide. http://cs.anu.edu.au/~bdm/nauty/.

49. B. D. McKay. Practical graph isomorphism. *Congressus Numerantium*, 30:45–87, 1981.

50. S. Nijssen and J. N. Kok. A quickstart in frequent structure mining can make a difference. In Proceedings of the 10th ACM SIGKDD International Conference on Knowledge Discovery and Data Mining (KDD-2004), pp. 647–652, 2004.

51. P. R. J. Östergård. A fast algorithm for the maximum clique problem. *Discrete Applied Mathematics*, 120:195–205, 2002.

52. N. Pasquier, Y. Bastide, R. Taouil, and L. Lakhal. Discovering frequent closed itemsets for association rules. In Proceedings of International Conference on Database Theory (ICDT), pp. 398–416, 1999.

53. J. Pei, J. Han, and R. Mao. CLOSET: An efficient algorithm for mining frequent closed itemsets. In ACM SIGMOD Workshop on Research Issues in Data Mining and Knowledge Discovery, pp. 21–30, 2000.

54. J. Pei, J. Han, B. Mortazavi-Asl, H. Pinto, Q. Chen, U. Dayal, and M.-C. Hsu. PrefixSpan: Mining sequential patterns efficiently by prefix-projected pattern growth. In Proceedings of 2001 International Conference on Data Engineering (ICDE'01), pp. 215–226, 2001.

55. E. G. M. Petrakis and C. Faloutsos. Similarity searching in medical image databases. *Knowledge and Data Engineering*, 9(3):435–447, 1997.

56. R. C. Read and D. G. Corneil. The graph isomorph disease. *Journal of Graph Theory*, 1:339–363, 1977.

57. J. M. Robson. Algorithms for maximum independent sets. *Journal of Algorithms*, 7:425–440, 1986.

58. A. Savasere, E. Omiecinski, and S. B. Navathe. An efficient algorithm for mining association rules in large databases. In Proceedings of the 21st International Conference on Very Large Data Bases (VLDB), pp. 432–444, 1995.

59. P. Shenoy, J. R. Haritsa, S. Sundarshan, G. Bhalotia, M. Bawa, and D. Shah. Turbo-charging vertical mining of large databases. In Proceedings of ACM SIGMOD International Conference on Management of Data, pp. 22–33, May 2000.

60. R. Srikant and R. Agrawal. Mining sequential patterns: Generalizations and performance improvements. In Proceedings of the 5th International Conference on Extending Database Technology (EDBT), Vol. 1057, pp. 3–17, 1996.

61. A. Srinivasan and R. D. King. Feature construction with inductive logic programming: A study of quantitative predictions of biological activity aided by structural attributes. *Data Mining and Knowledge Discovery*, 3(1):37–57, 1999.

62. A. Srinivasan, R. D. King, S. H. Muggleton, and M. Sternberg. The predictive toxicology evaluation challenge. In Proceedings of the 15th International Joint Conference on Artificial Intelligence (IJCAI), pp. 1–6. Morgan-Kaufmann, 1997.

63. N. Vanetik, E. Gudes, and S. E. Shimony. Computing frequent graph patterns from semistructured data. In Proceedings of 2002 IEEE International Conference on Data Mining (ICDM), pp. 458–465, 2002.

64. X. Yan and J. Han. gSpan: Graph-based substructure pattern mining. In Proceedings of 2002 IEEE International Conference on Data Mining (ICDM), pp. 721–724, 2002.

65. X. Yan and J. Han. CloseGraph: Mining closed frequent graph patterns. In Proceedings of the 9th ACM SIGKDD International Conference on Knowledge Discovery and Data Mining (KDD-2003), pp. 286–295, 2003.

66. K. Yoshida and H. Motoda. CLIP: Concept learning from inference patterns. *Artificial Intelligence*, 75(1):63–92, 1995.

67. K. Yoshida, H. Motoda, and N. Indurkhya. Graph-based induction as a unified learning framework. *Journal of Applied Intelligence*, 4:297–328, 1994.

68. M. J. Zaki. Scalable algorithms for association mining. *IEEE Transactions on Knowledge and Data Engineering*, 12(2):372–390, 2000.

69. M. J. Zaki and K. Gouda. Fast vertical mining using diffsets. Technical Report 01-1, Department of Computer Science, Rensselaer Polytechnic Institute, 2001.

70. M. J. Zaki and K. Gouda. Fast vertical mining using diffsets. In Proceedings of the 9th ACM SIGKDD International Conference on Knowledge Discovery and Data Mining (KDD-2003), 2003.

71. M. J. Zaki and C.-J. Hsiao. CHARM: An efficient algorithm for closed association rule mining. Technical Report 99-10, Department of Computer Science, Rensselaer Polytechnic Institute, October 1999.

7

UNSUPERVISED AND SUPERVISED PATTERN LEARNING IN GRAPH DATA

DIANE J. COOK, LAWRENCE B. HOLDER, AND NIKHIL KETKAR

*School of Electrical Engineering & Computer Science Washington State University Pullman, Washington**

7.1 INTRODUCTION

The success of machine learning and data mining for business and scientific purposes has fueled the expansion of its scope to new representations and techniques. Much collected data is structural in nature, containing entities as well as relationships between these entities. Compelling data in bioinformatics [32], network intrusion detection [15], Web analysis [2, 8], and social network analysis [7, 27] has become available that requires effective handling of structural data. The ability to learn concepts from relational data has also become a crucial challenge in many security-related domains. For example, the U.S. House and Senate Intelligence Committees' report on their inquiry into the activities of the intelligence community before and after the September 11, 2001, terrorist attacks revealed the necessity for "connecting the dots" [28], that is, focusing on the relationships between entities in the data rather than merely on an entity's attributes.

In this chapter we describe an approach to mining concepts from graph-based data. We provide an algorithmic overview of the approach implemented in the Subdue system and describe its uses for unsupervised discovery and supervised

*Part of this work was done while the authors were at the University of Texas at Arlington, Arlington, Texas

Mining Graph Data, Edited by Diane J. Cook and Lawrence B. Holder
Copyright © 2007 John Wiley & Sons, Inc.

concept learning. This book details several different approaches for performing these same tasks. To assist in understanding the relationships between these approaches, we also perform experimental comparisons between Subdue and frequent subgraph miners and inductive logic programming (ILP) learning methods. We observe that each type of approach offers unique advantages for certain classes of applications and all face challenges that will drive future research.

7.2 MINING GRAPH DATA USING SUBDUE

The Subdue graph-based relational learning system[1] [3, 4] encompasses several approaches to graph-based learning, including discovery, clustering, and supervised learning, which will be described in this section. Subdue uses a labeled graph $G = (V, E, \mu, \nu)$ as both input and output, where $V = \{v_1, v_2, \ldots, v_n\}$ is a set of vertices, $E = \{(v_i, v_j)|v_i, v_j \in V\}$ is a set of edges, and μ and ν represent node and edge labeling functions, as described in Chapter 2 by Bunke and Neuhaus. The graph G can contain directed edges, undirected edges, self-edges [i.e., $(v_i, v_i) \in E$], and multiedges (i.e., more than one edge between vertices v_i and v_j). The input graph need not be connected, but the learned patterns must be connected subgraphs (called substructures) of the input graph. The input to Subdue can consist of one large graph or several individual graph transactions, and in the case of supervised learning, the individual graphs are classified as positive or negative examples.

7.2.1 Substructure Discovery

As an unsupervised discovery algorithm, Subdue searches for a substructure, or subgraph of the input graph, that best compresses the input graph. Subdue uses a variant of beam search for its main search algorithm, as summarized in Figure 7.1. A substructure in Subdue consists of a subgraph definition and all its instances throughout the graph. The initial state of the search is the set of substructures consisting of all uniquely labeled vertices. The only operator of the search is the ExtendSubstructure operator. As its name suggests, it extends a substructure in all possible ways by a single edge and a vertex, or by only a single edge if both vertices are already in the subgraph.

The search progresses by applying the ExtendSubstructure operator to each substructure in the current state. The resulting state, however, does not contain all the substructures generated by the ExtendSubstructure operator. The substructures are kept on a queue and are ordered based on their compression (or sometimes referred to as value) as calculated using the minimum description length (MDL) principle described below. Only the top *beam* substructures remain on the queue for expansion during the next pass through the main discovery loop.

The search terminates upon reaching a limit on the number of substructures extended (this defaults to half the number of edges in the graph) or upon exhaustion of the search space. Once the search terminates and Subdue returns the list of

---

[1]Subdue source code, sample datasets, and publications are available at http://ailab.uta.edu/subdue.

```
SUBDUE(Graph,BeamWidth,MaxBest,MaxSubSize, Limit)
    ParentList = Null;
    ChildList = Null;
    BestList = Null;
    ProcessedSubs = 0;
    Create a substructure from each unique vertex label
        and its single-vertex instances;
    Insert the resulting substructures in ParentList;
    while ProcessedSubs ≤ Limit
        and ParentList not empty
    do
        while ParentList is not empty
        do
            Parent = RemoveHead(ParentList);
            Extend each instance of Parent in all possible ways;
            Group the extended instances into Child substructures;
            for each Child
            do
                if SizeOf(Child) less than MaxSubSize
                then
                    Evaluate Child;
                    Insert Child in ChildList in order by value;
                    if BeamWidth < Length(ChildList)
                    then
                        Destroy substructure at end of ChildList;
            Increment ProcessedSubs;
            Insert Parent in BestList in order by value;
            if MaxBest < Length(BestList)
            then
                Destroy substructure at end of BestList;
        Switch ParentList and ChildList;
    return BestList;
```

Figure 7.1.   Subdue's discovery algorithm.

best substructures, the graph can be compressed using the best substructure. The compression procedure replaces all instances of the substructure in the input graph by single vertices, which represent the substructure definition. Incoming and outgoing edges to and from the replaced instances will point to or originate from the new vertex that represents the instance. The Subdue algorithm can be invoked again on this compressed graph. This procedure can be repeated multiple times and is referred to as an iteration.

Subdue's search is guided by the MDL [26] principle given in Eq. (7.1), where $DL(S)$ is the description length of the substructure being evaluated, $DL(G|S)$ is the description length of the graph as compressed by the substructure, and $DL(G)$ is the description length of the original graph. The best substructure is the one that minimizes this compression ratio:

$$\text{Compression} = \frac{DL(S) + DL(G|S)}{DL(G)} \tag{7.1}$$

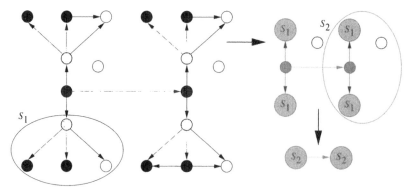

Figure 7.2. Example of Subdue's substructure discovery capability.

As an example, Figure 7.2 shows the four instances that Subdue discovers of a pattern S_1 in the example input graph and the resulting compressed graph, as well as the pattern S_2 found in this new graph and the resulting compressed graph. To allow slight variations between instances of a discovered pattern (as is the case in Fig. 7.2), Subdue applies an error-tolerant graph match between the substructure definition and potential instances (see Chapter 2 by Bunke and Neuhaus for more details on this topic). The discovery algorithm can be biased by incorporating prior knowledge in the form of predefined substructures or preference weights on desirable (undesirable) edge and vertex labels. As shown by Rajappa, Subdue's run time is polynomial in the size of the input graph [25]. Substructure discovery using Subdue has yielded expert-evaluated significant results in domains including predictive toxicology, network intrusion detection, earthquake analysis, Web structure mining, and protein data analysis [16, 23, 30].

7.2.2 Graph-Based Clustering

Given the ability to find a prevalent subgraph pattern in a larger graph and then compress the graph with this pattern, iterating over this process until the graph can no longer be compressed will produce a hierarchical, conceptual clustering of the input data. On the ith iteration, the best subgraph S_i is used to compress the input graph, introducing new vertices labeled S_i in the graph input to the next iteration. Therefore, any subsequently discovered subgraph S_j can be defined in terms of one or more S_i, where $i < j$. The result is a lattice, where each cluster can be defined in terms of more than one parent subgraph. For example, Figure 7.3 shows such a clustering extracted from the graph representation of a DNA (deoxyribonucleic acid) molecule (visualized in Fig. 7.4) [10]. Note that in this cluster hierarchy, for example, subgraph S_4 is composed of an S_1 and an S_2 together with a CH_2.

7.2.3 Supervised Learning

Extending a graph-based discovery approach to perform supervised learning intro-duces the need to handle negative examples (focusing on the two-class scenario). In

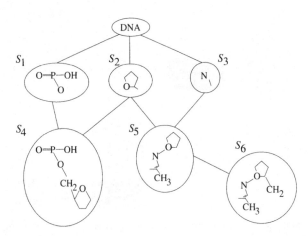

Figure 7.3. Subdue's hierarchical cluster generated from DNA data.

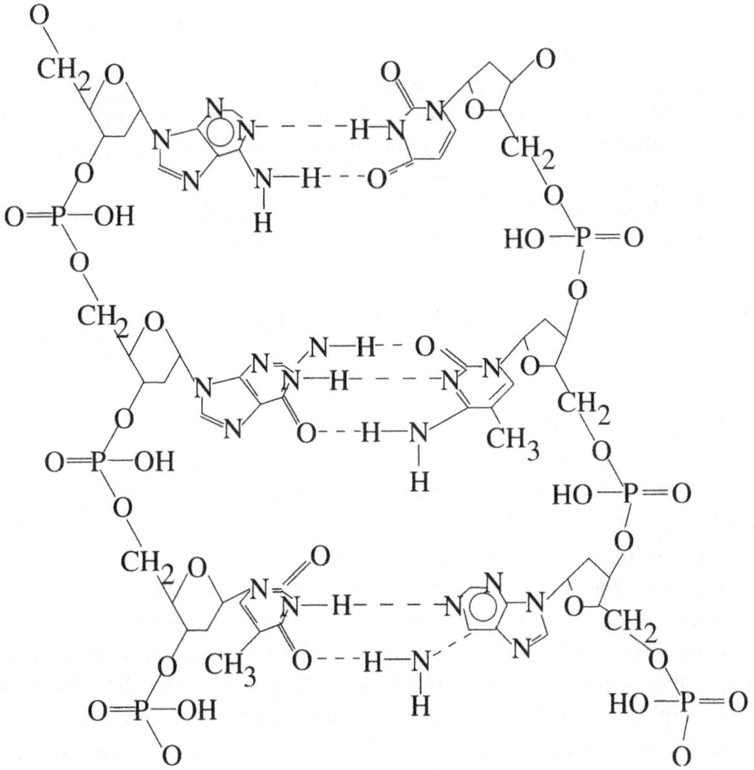

Figure 7.4. Visualization of a portion of DNA.

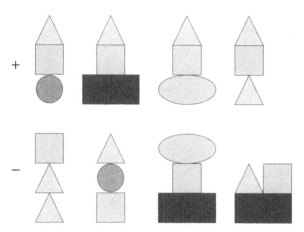

Figure 7.5. Visualization of graph-based data with four positive and four negative examples.

the case of a graph the negative information can come in two forms. First, the data may be in the form of numerous small graphs, or graph transactions, each labeled either positive or negative. Second, data may be composed of two large graphs: one positive and one negative.

The first scenario is closest to the standard supervised learning problem in that we have a set of clearly defined examples. Figure 7.5 depicts a simple set of positive (G^+) and negative (G^-) examples. One approach to supervised learning is to find a subgraph that appears in many positive graphs but in few negative graphs. This amounts to replacing the compression-based measure with an error-based measure. For example, we would find a subgraph S that minimizes

$$\frac{|\{g \in G^+ | S \not\subseteq g\}| + |g \in G^- | S \subseteq g\}|}{|G^+| + |G^-|} = \frac{FN + FP}{P + N} \tag{7.2}$$

where $S \subseteq g$ means S is isomorphic to a subgraph of g (although we do not actually perform a subgraph isomorphism test during learning). The first term of the numerator is the number of false negatives, and the second term is the number of false positives.

This approach will lead the search toward a small subgraph that discriminates well, for example, the subgraph on the left in Figure 7.6. However, such a subgraph does not necessarily compress well, nor represent a characteristic description of the target concept. We can bias the search toward a more characteristic description by using the compression-based measure to look for a subgraph that compresses the positive examples, but not the negative examples. If $DL(G)$ represents the description length (in bits) of the graph G, and $DL(G|S)$ represents the description length of graph G compressed by subgraph S, then we can look for an S that minimizes $DL(G^+|S) + DL(S) + DL(G^-) - DL(G^-|S)$, where the last two terms represent the portion of the negative graph incorrectly compressed by the subgraph. This

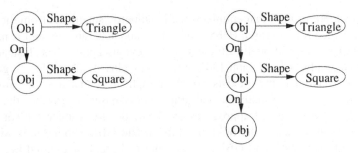

Figure 7.6. Two possible graph concepts learned from example data.

approach will lead the search toward a larger subgraph that characterizes the positive examples, but not the negative examples, for example, the subgraph on the right in Figure 7.6.

Finally, this process can be iterated in a set-covering approach to learn a disjunctive hypothesis. Using the error measure, any positive example containing the learned subgraph would be removed from subsequent iterations. Using the compression-based measure, instances of the learned subgraph in both the positive and negative examples (even multiple instances per example) are compressed to a single vertex.

7.3 COMPARISON TO OTHER GRAPH-BASED MINING ALGORITHMS

Because graph-based data mining has demonstrated success for a variety of tasks in structural domains (see Chapters 14–17 of this book for examples), a number of varied techniques and methodologies have arisen for mining interesting subgraph patterns from graph datasets. These include mathematical graph theory-based approaches like FSG ([14] and Chapter 6 of this book), gSpan ([33] and Chapter 5 of this book), greedy search-based approaches like Subdue or GBI [17], ILP approaches such as used by WARMR [6], and kernel function-based approaches ([11] and Chapter 11 of this book). In this chapter we contrast the methodologies employed by Subdue with frequent substructure and ILP approaches and attempt to highlight differences in discoveries that can be expected from the alternative techniques.

7.4 COMPARISON TO FREQUENT SUBSTRUCTURE MINING APPROACHES

Mathematical graph theory-based approaches mine a complete set of subgraphs mainly using a support or frequency measure. The initial work in this area was the AGM [9] system, which uses the Apriori levelwise approach. FSG takes a similar approach and further optimizes the algorithm for improved running times. gFSG

[12] is a variant of FSG that enumerates all geometric subgraphs from the database. gSpan uses depth-first search (DFS) codes for canonical labeling and is much more memory and computationally efficient than previous approaches. Instead of mining all subgraphs, CloseGraph [34] only mines closed subgraphs. A graph G is closed in a dataset if there exists no supergraph of G that has the same support as G. In comparison to mathematical graph theory-based approaches that are complete, greedy search-based approaches use heuristics to evaluate the solution. The two pioneering works in the field are Subdue and GBI. Subdue uses MDL-based compression heuristics, and GBI uses an empirical graph size-based heuristic. The empirical graph size definition depends on the size of the extracted patterns and the size of the compressed graph.

Methodologies that focus on complete, frequent subgraph discovery, such as FSG and gSpan, are guaranteed to find all subgraphs that satisfy the user-specified constraints. Although completeness is a fundamental and desirable property, a side effect of the complete approaches is that these systems typically generate a large number of substructures, which by themselves provide relatively less insight about the domain. As a result, interesting substructures have to be identified from the large set of substructures either by domain experts or by other automated methods to achieve insights into this domain. In contrast, Subdue typically produces a smaller number of substructures that best compress the graph dataset and that can provide important insights about the domain.

A thorough comparison of Subdue with these approaches is difficult because most frequent subgraph discovery approaches are designed to look for patterns that are frequent across a disjoint set of graph transactions, rather than to find patterns that are frequent or interesting within a single graph (see [13, 31] for some representation-limited exceptions). Another key distinction of Subdue from many other graph-based methods is that Subdue can accommodate a free-form graph representation, where others often have a specific representation tailored to transactional data, which is a restricted form of relational data. Table 7.1 lists for Subdue, FSG, and gSpan, whether these systems can process directed graphs, multigraphs, graphs with self-edges, as well as restrictions on labels or graph size. These restrictions are based on the original algorithm and could be superseded by modifications.

TABLE 7.1   Acceptable Input Graphs

	Types of Graphs				
	Directed	Labels	Multigraph	Self-edges	Number of Vertices/ Edges
Subdue	√	Alphanumeric	√	√	Unlimited
FSG	√	Alphanumeric	×	×	Unlimited
gSpan	×	Numeric	×	×	Max 254/254

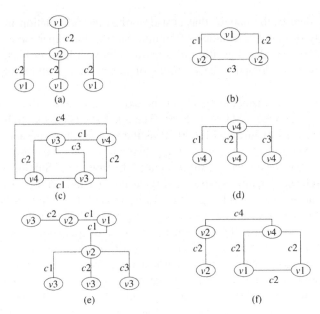

Figure 7.7. Embedded substructures.

7.4.1 Experiments Using Synthetic Datasets

We demonstrate the advantage of Subdue's compression-based methodology by performing an experimental comparison of Subdue with the graph-based data mining systems FSG and gSpan on a number of artificial datasets. We generate graph datasets with 500, 1000, 1500 and, 2000 transactions by embedding one of the 6 substructures shown in Figure 7.7. Each of the generated 24 datasets (6 different embedded substructures, each with 500, 1000, 1500, and 2000 transactions) have the following properties:

1. Of the transactions 60% have the embedded substructure: The rest of the transactions are generated randomly.
2. For all the transactions that contain the embedded substructure, 60% of the transaction is the embedded substructure, that is, coverage of the embedded substructure is 60%.

Since each of the 24 datasets have both properties listed above, it is clear that the embedded substructure is the most interesting substructure in each of the datasets. Using these datasets, we now compare the performance of Subdue, FSG, and gSpan. For each of the 24 datasets, Subdue was run with the default parameters and FSG and gSpan were run at a 10% support level. We want to determine if Subdue can find the best substructure despite its greedy search approach. We also want to compare quality of the reported substructure based on an interestingness factor, for which we use compression. This is motivated by work in information theory that uses this measure to identify the pattern that best describes the information in the data. We use a slightly different calculation of compression in the experiments than the one employed by the Subdue algorithm in order to remove a level of bias in the results. Lastly, we want to compare the runtimes of the three algorithms on a variety of datasets.

Table 7.2 summarizes the results of the experiments and Table 7.3 summarizes the runtimes of Subdue, FSG, and gSpan. The results indicate that Subdue discovers the embedded substructure and reports it as the best substructure about 80% of the time. Both FSG and gSpan generate approximately 200,000 substructures among which there exists the embedded substructure. The runtime of Subdue is intermediate between FSG and gSpan. Subdue clearly discovers and reports fewer but more interesting substructures. Although it could be argued that setting a higher support value for FSG and gSpan can lead to fewer generated substructures, it should be noted that this can cause FSG and gSpan to miss the interesting pattern. We observed that the increase in runtime for Subdue is nonlinear when we increase the size of the dataset. Increase for FSG and gSpan was observed to be linear (largely because of various optimizations). The primary reason for this behavior is the less efficient implementation of graph isomorphism in Subdue than in FSG and gSpan. A more efficient approach for graph isomorphism, with the use of canonical labeling, needs to be developed for Subdue.

7.4.2 Experiments Using Real Datasets

In addition, we performed an experimental comparison of Subdue with gSpan and FSG on the chemical toxicity and the chemical compounds datasets that are provided with gSpan. The chemical toxicity dataset has a total of 344 transactions. There are

TABLE 7.2   Results (Average) on Six Artificial Datasets

Number of Transactions	Cases Subdue Reported Embedded Substructure as Best (%)	Number of Substructures Generated by FSG/gSpan
500	66	233495
1000	83	224174
1500	83	217477
2000	78	217479
Average	79	223156

TABLE 7.3   Runtimes (Seconds) on Six Artificial Datasets

Number of Transactions	FSG	gSpan	Subdue
500	734	61	51
1000	815	107	169
1500	882	182	139
2000	1112	249	696
Average	885	150	328

TABLE 7.4   Results from the Chemical Toxicity Dataset

Compression from Subdue's best substructure	16%
Best compression from any FSG/gSpan substructure	8%
Number of substructures reported by FSG/gSpan	844
Runtime Subdue (s)	115
Runtime FSG (s)	8
Runtime gSpan (s)	7

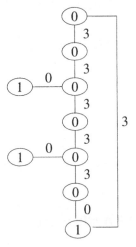

Figure 7.8.   Best compressing substructure discovered by Subdue on the chemical toxicity dataset.

66 different edge and vertex labels in the dataset. FSG and gSpan results were recorded based on a 5% support threshold. From our experiments we determined the support of any best compressing substructure should be greater than 10% for the dataset of size greater than 50. Therefore, we selected 5% support. If support is set to a lesser value, large numbers of random and insignificant patterns are generated.

Table 7.4 summarizes the results of the experiment, and Figure 7.8 shows the best compressing substructure discovered by Subdue. Because gSpan supports only numeric labels, atom and bond labels have been replaced by unique identifying numbers in this database. As the results indicate, Subdue discovered frequent patterns missed by FSG and gSpan that resulted in greater compression. However, the runtime of Subdue is much larger than that of FSG and gSpan.

The chemical compounds dataset has a total of 422 transactions. There are 21 different edge and vertex labels in the dataset. The results for FSG and gSpan were recorded based on a 10% support threshold. We selected 10% support because if support is set to a lesser value, large numbers of random and insignificant patterns are generated. Table 7.5 summarizes the results of the experiment, and Figure 7.9 shows the best compressing substructure discovered by Subdue. Again, Subdue found better-compressing substructures but required a greater runtime.

TABLE 7.5   Results from the Chemical Compounds Dataset

Compression from Subdue's best substructure	19%
Best compression from any FSG/gSpan substructure	7%
Number of substructures reported by FSG/gSpan	15966
Runtime Subdue (s)	142
Runtime FSG (s)	21
Runtime gSpan (s)	4

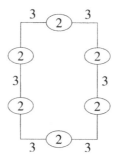

Figure 7.9.   Best compressing substructure discovered by Subdue on the chemical compound dataset.

7.5   COMPARISON TO ILP APPROACHES

Logic-based mining, popularly known as inductive logic programming (ILP) [18], is characterized by the use of logic for the representation of structural data. ILP systems represent examples, background knowledge, hypotheses, and target concepts in Horn clause logic. The core of ILP is the use of logic for representation and the search for syntactically legal hypotheses constructed from predicates provided by the background knowledge. ILP systems such as FOIL [24], CProgol [20], Golem [22], and WARMR [6] have been extensively applied to supervised learning and to a certain extent to unsupervised learning.

In contrast, graph-based approaches are characterized by representation of structural data in the form of graphs. Graph-based approaches represent examples, background knowledge, hypotheses, and target concepts as graphs. Here we perform a qualitative comparison of Subdue and logic-based mining approaches.

By performing a comparison of the graph-based and logic-based approaches, we intended to analyze the ability of the approaches to efficiently discover complex structural concepts and to effectively utilize background knowledge. To do so, we must establish some notions on the complexity of a structural concept and identify the types of background knowledge generally available for a mining task.

The complexity of a multirelational, or structural, concept is a direct consequence of the number of relations in the concept. For example, learning the concept of arene (a chemical compound with a six-member ring as in benzene), which comprises learning six relations, involves the exploration of a larger portion of the hypothesis space than learning the concept of hydroxyl (oxygen connected to hydrogen as in methanol), which comprises learning one relation. The concept of arene is thus more complex than that of hydroxyl.

Although the number of relations in the concept is a key factor in the complexity of the concept, there are also other factors such as the number of relations in the examples from which the concept is to be learned. For example, learning the concept of hydroxyl from a set of phenols (hydroxy group attached to an arene) involves the exploration of a larger hypothesis space than learning the same hydroxyl concept from a set of alcohols (hydroxy group attached to an alkyl). The concept of a hydroxyl group is thus more complex to learn from phenols than it is from a set of alcohols. We identify this complexity as *structural complexity.*

To learn a particular concept, it is essential that the representation used by a data mining system be able to express that particular concept. For a representation to express a particular concept, it is beneficial to have both the syntax that expresses the concept and the semantics that associates meaning to the syntax. A relational concept can be said to have a *greater complexity* than some other relational concept if it requires a more expressive representation. To learn numerical ranges, for example, it is essential to have the syntax and the semantics to represent notions such as "less than," "greater than," and "equal to." We identify this complexity as *semantic complexity.*

A relational learner can be provided background knowledge that condenses the hypothesis space. For example, if the concept to be learned is "compounds with three arene rings" and the concept of an arene ring is provided as a part of the background knowledge, then the arene rings in examples could be condensed to a single entity. This would cause a massive reduction in the hypothesis space required to be explored to learn the concept, and the mining algorithm would perform more efficiently than without the background knowledge. We identify such background knowledge as background knowledge intended to condense the hypothesis space.

A mining algorithm can also be provided background knowledge that augments the hypothesis space. For example, consider that the algorithm is provided with background knowledge that allows it to learn concepts such as "less than," "greater than," and "equal to." In this case, the algorithm would explore a hypothesis space larger than what it would explore without the background knowledge. Thus, introducing background knowledge has augmented the hypothesis space and has facilitated the learning of concepts that would not be learned without the background knowledge. We identify such background knowledge as background knowledge intended to augment the hypothesis space.

Using these notions, we now identify the factors on the basis of which the graph-based and logic-based approaches can be compared. They are:

1. Ability to learn structurally large relational concepts
2. Ability to learn semantically complex relational concepts or the ability to effectively use background knowledge that augments the hypothesis space to learn semantically complex relational concepts

3. Ability to effectively use background knowledge that condenses the hypothesis space

7.5.1 CProgol

The representative for ILP systems that we will use for our experiments is CProgol [19], which is characterized by the use of mode-directed inverse entailment and a hybrid search mechanism. Inverse entailment is a procedure that generates a single, most specific clause that, together with the background knowledge, entails the observed data. The inverse entailment in CProgol is mode-directed, that is, uses mode declarations. A mode declaration is a constraint that imposes restrictions on the predicates and their arguments appearing in a hypotheses clause. These constraints specify which predicates can occur in the head and the body of hypotheses. They also specify which arguments can be input variables, output variables, or constants, as well as the number of alternative solutions for instantiating the predicate.

In CProgol, user-defined mode declarations aid the generation of the most specific clause. CProgol first computes the most specific clause that covers the seed example and belongs to the hypothesis language. The most specific clause can be used to bound the search from below. The search is now bounded between the empty clause and the most specific clause. The search proceeds within the bounded-subsumption lattice in a general-to-specific manner, bounded from below with respect to the most specific clause. The search strategy is A* guided by a weighted compression and accuracy measure. The A* search returns a clause that covers the most positive examples and maximally compresses the data. Any arbitrary Prolog program can serve as background knowledge for CProgol.

7.5.2 Experiments Using Structurally Large Concepts from the Mutagenesis Dataset

The Mutagenesis dataset [29] has been collected to identify mutagenic activity in a compound based on its molecular structure and is considered to be a benchmark dataset for multirelational data mining. The Mutagenesis dataset consists of the molecular structure of 230 compounds, of which 138 are labeled as mutagenic and 92 as nonmutagenic. The mutagenicity of the compounds has been determined by the Ames test. The task is to distinguish mutagenic compounds from nonmutagenic ones based on their molecular structure. The Mutagenesis dataset basically consists of atoms, bonds, atom types, bond types, and partial charges on atoms. The dataset also consists of the hydrophobicity of the compound ($\log P$), the energy level of the compound's lowest unoccupied molecular orbital (LUMO), a Boolean attribute identifying compounds with 3 or more benzyl rings (I1), and a Boolean attribute identifying compounds that are acenthryles (Ia). Ia, I1, $\log P$, and LUMO are relevant properties in determining mutagenicity. When run on the entire Mutagenesis dataset, as described above, using 10-fold cross validation, Subdue achieved 63% accuracy with an average learning time per fold of 94 min, while CProgol achieved 60% accuracy with an average learning time per fold of 24 min. The difference in accuracy was not statistically significant.

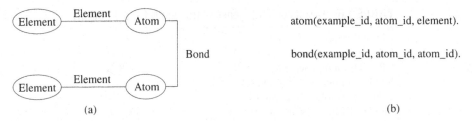

Figure 7.10. Representation for structurally large concepts: (a) graph representation and (b) logic representation.

To compare the performance of the approaches while learning structurally large concepts, we ran Subdue and CProgol on a variant of the Mutagenesis dataset. Since we intended to compare the ability of the approaches to learn large structural concepts, both the mining algorithms were provided only with the basic information of the atoms, the elements, and the bonds without any other information or background knowledge. This is shown in Figure 7.10. The algorithms are not provided with any additional information or any form of background knowledge because we intended to compare the ability to learn large structural concepts. The introduction of any additional information or background knowledge would prevent this from happening. If systems were provided with the partial charge on the atoms and background knowledge to learn ranges, the systems would learn ranges on partial charges that would contribute to the accuracy. This would make it difficult to analyze how the approaches compared while learning structurally large concepts. Hence the partial charge information and the background knowledge to learn ranges was not given to either system.

The atom type and bond type information was also not provided to either system. The reasoning behind doing so is that we view the atom-type and bond-type information as a propositional representation of relational data. Such information allows the relational learners to learn propositional representations of relational concepts rather than the true relational concept. Consider, for example, the rule found by CProgol on the Mutagenesis dataset [29], atom(A,B,c,195,C). This rule denotes that compounds with a carbon atom of type 195 are mutagenic. The atom type 195 occurs in the third adjacent pair of 3 fused 6-member rings. Therefore all compounds with 3 fused 6-member rings are labeled active. Thus, a rule involving 15 relations (3 fused 6-member rings) has been discovered by learning a single relation. Discovering such a rule has allowed CProgol to learn a propositional representation of a relational concept rather than the true relational concept. Providing atom-type and bond-type information would allow both systems to learn propositional representations of structurally large relational concepts rather than the true relational concepts. We do not consider the learning of such concepts equivalent to the learning of structurally large relational concepts, and therefore do not provide either system with the atom-type and bond-type information.

The results of the experiment are shown in Table 7.6. For the training set, the accuracy for one run on the entire dataset and the learning time are shown. For the

TABLE 7.6   Results on Mutagenesis Structurally Large Concepts

CProgol training set accuracy	60.00%
Subdue training set accuracy	86.00%
CProgol training set runtime (s)	2010
Subdue training set runtime (s)	1876
CProgol 10-fold CV accuracy	61.74%
Subdue 10-fold CV accuracy	81.58%
CProgol 10-fold CV runtime (average, s)	1940
Subdue 10-fold CV runtime (average, s)	2100
CProgol—Subdue, Δerror $\pm\ \delta$	20.84% $\pm$ 12.78%
CProgol—Subdue, confidence	99.94%

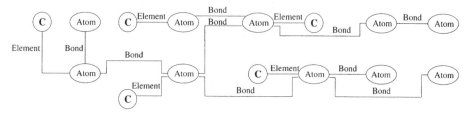

Figure 7.11.   Rule discovered by Subdue on the mutagenesis dataset while learning structurally large concepts.

10-fold cross validation (CV), average learning time over 10-fold is shown. The results show that Subdue performs significantly better than CProgol. Subdue learns 17 graphs representing 17 rules. One of the rules discovered by Subdue is shown in Figure 7.11. This rule has an accuracy of 76.72% and a coverage of 81.15%. The hypotheses learned by CProgol mostly consisted of a single atom or bond predicate. These results give a strong indication that a graph-based approach can perform better than a logic-based approach when learning structurally large concepts.

7.5.3   Experiments Using Structurally Large Concepts from Artificial Data

We performed additional experiments using artificially generated Bongard problems [1] to reinforce the insights from the experiments on the Mutagenesis dataset. Bongard problems were introduced as an artificial domain in the field of pattern recognition. A simplified form of Bongard problems has been used as an artificial domain in the field of ILP [5]. We use a similar form of Bongard problems for our artificial domain experiments. We use a Bongard problem generator to generate datasets as shown in Figure 7.12. Each dataset consists of a set of positive and negative examples. Each example consists of a number of simple geometrical objects placed inside one another. The task is to determine the particular set of objects, their shapes, and their placement, which can correctly distinguish the positive examples from the negative ones.

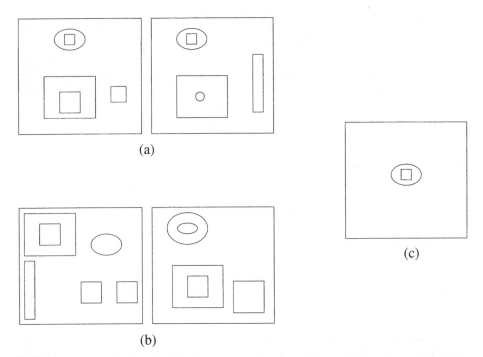

Figure 7.12. A Bongard problem: (a) positive examples; (b) negative examples; and (c) concept.

Figure 7.13 shows the representations used for Subdue and CProgol. We systematically analyzed the performance of Subdue and CProgol on artificially generated Bongard problems with increasing numbers of objects in the concept and increasing numbers of objects in the examples. In the first experiment, the number of objects in the Bongard concept was varied from 5 to 35. The number of additional objects in each example (objects that are not part of the concept) were kept constant at 5. For every concept size from 5 to 35, 10 different concepts were generated. For each of the 10 concepts a training set and a test set of 100 positive and 100 negative examples were generated. CProgol and Subdue were run on the training sets and were tested on the test sets.

Figure 7.14(a) shows the average accuracy achieved by CProgol and Subdue on 10 datasets for every concept size ranging from 5 to 35. It is observed that Subdue clearly outperforms CProgol. To further analyze the performance of the systems, we reran the same experiment, but in this case the systems were iteratively given increased resources (this was achieved by varying the "nodes" parameter in CProgol and the "limit" parameter in Subdue) so that we could determine the number of hypotheses each system explored before it learned the concept (a cutoff accuracy of 80% was decided). Figure 7.14(b) shows the number of hypotheses explored by each system to achieve an accuracy of 80% (this experiment was only performed for concept sizes varying from 5 to 18 as a significantly large

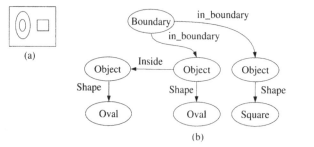

Figure 7.13. Representation for Bongard problems: (a) Bongard example; (b) graph representation; and (c) logic representation.

amount of time was required). A snapshot of the experiment for concept size 10 is shown in Figure 7.14(c). The results show that CProgol explores a larger number of hypotheses than Subdue.

An analysis of CProgol indicates that it first generates a mostspecific clause from a randomly selected example using the mode definitions. Mode definitions together with the background knowledge form a user-defined model for generation of candidate hypotheses. After generation of the mostspecific clause, CProgol performs a general-to-specific search in the bounded-subsumption lattice guided by the mode definitions. The most general hypothesis is the empty clause, and the most specific hypothesis is the clause generated in the previous step. The process of hypothesis generation is affected more by the mode definitions and the background knowledge than the examples, first because a single example is used to construct the most specific clause, and second because the mode definitions have a major effect on the process of hypothesis generation. Thus, CProgol makes more use of the mode definitions and background knowledge and less use of the examples.

This observation about CProgol can be partially generalized to other logic-based approaches such as top-down search of refinement graphs [24], inverse resolution [21], and relative least general generalization [22]. An analysis of Subdue indicates that hypotheses are generated only on the basis of the examples. The candidate hypotheses are generated by extending the subgraph by an edge and a vertex or just an edge in all possible ways as in the examples. As Subdue generates the hypotheses only on the basis of the examples, it is more example driven. This observation about Subdue can be partially generalized to other graph-based systems such as FSG, AGM, and gSpan because there is more use of the examples and less use of the model. Graph-based approaches tend to explore the hypothesis space more efficiently because they use only the examples to generate candidate hypotheses and thus can search a larger portion of the smaller hypothesis space with a given amount of resources, which is essential in learning structurally large relational concepts.

7.5.4 Experiments Using Semantically Complex Concepts from the Mutagenesis Dataset

To compare the performance of the approaches for learning semantically complex concepts, we ran Subdue and CProgol on the Mutagenesis dataset. Each system was

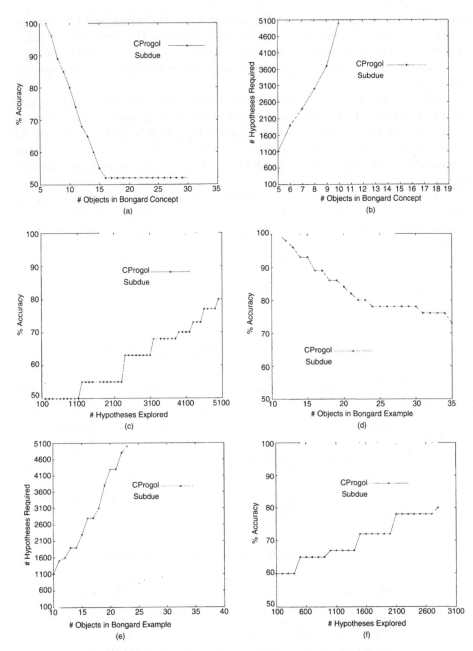

Figure 7.14.   Results for structurally large Bongard problems.

provided with background knowledge so that numerical ranges could be learned. For CProgol this was achieved by introducing Prolog-based background knowledge. For Subdue this was achieved by explicitly instantiating the background knowledge, that is, additional structure was added to the training examples. This is shown in Figure 7.15.

The results of this experiment are shown in Table 7.7. The results indicate that CProgol uses the background knowledge and shows an improved performance while Subdue has achieved a lower accuracy than what it achieved without the background knowledge. These results give a strong indication that a logic-based approach performs better than a graph-based approach when learning semantically complex concepts.

In general, graph-based mining algorithms explore only those hypotheses that are explicitly present in the examples. For hypotheses to be explicitly present in the examples, it is essential that the targeted semantically complex concepts be explicitly instantiated in the examples. The drawbacks of the data-driven approach are that explicit instantiation is cumbersome in most cases and also that explicit instantiation is not a generalized methodology to learn complex semantic concepts. For example, suppose a domain expert were to suggest that the ratio of

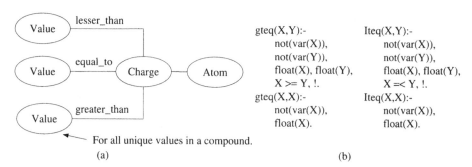

Figure 7.15. Representation for learning semantically complex concepts: (a) graph representation and (b) logic representation.

TABLE 7.7   Results on Mutagenesis Semantically Complex Concepts

CProgol training set accuracy	82.00%
Subdue training set accuracy	80.00%
CProgol training set runtime (s)	960
Subdue training set runtime (s)	848
CProgol 10-fold CV accuracy	78.91%
Subdue 10-fold CV accuracy	77.39%
CProgol 10-fold CV runtime (average, s)	810
Subdue 10-fold CV runtime (average, s)	878
CProgol—Subdue, Δerror ± δ	1.52% ± 11.54%
CProgol—Subdue, confidence	31.38%

the number of carbon atoms to the number of hydrogen atoms in a molecule has an effect on the mutagenicity. CProgol with some added background knowledge could use this information to classify the molecules. Subdue, on the other hand, would require making changes to the representation such that the pattern would be found in terms of a graph. Logic-based approaches allow the exploration of hypotheses through implicitly defined background knowledge rather than explicit instantiation in the examples. This is essential in learning semantically complex multirelational concepts.

7.6 CONCLUSIONS

The comparisons that were performed in this project highlight features of alternative approaches to graph mining and point to directions for continued research. We have observed that Subdue is preferred over FSG or gSpan when data is presented in one large graph or when a pattern is dominantly present in small or medium-size datasets. However, for large databases and those that exhibit a high degree of randomness, FSG or gSpan will likely be better choices because Subdue may not find the best pattern. Subdue can discover concepts that are less frequent but potentially of greater interest. However, Subdue needs to make use of techniques such as canonical labeling and approximate algorithms for graph isomorphism in order to scale to larger datasets.

In addition, we can conclude that graph-based mining algorithms tend to explore the concept hypothesis space more efficiently than logic-based algorithms, which is essential for mining structurally large concepts from databases. However, logic-based systems make more efficient use of background knowledge and are better at learning semantically complex concepts. These experiments point out the need for graph mining algorithms to effectively use background knowledge. Additional features such as generalizing numeric ranges from graph data will further improve these algorithms.

In addition to these comparisons, we are currently in the process of comparing Subdue to graph kernel learning algorithms. By performing these comparisons we hope to define a unifying methodology for mining graph-based data.

Acknowledgment

This work is partially supported by the National Science Foundation grants IIS-0505819 and IIS-0097517.

REFERENCES

1. M. Bongard. *Pattern Recognition*. Spartan Books, New York, New York, 1970.
2. S. Chakrabarti, M. van den Berg, and B. Dom. Focused crawling: A new approach to topic-specific web resource discovery. In Proceedings of the International World Wide Web Conference, 1999.

3. D. Cook and L. Holder. Graph-based data mining. *IEEE Intelligent Systems*, 15(2): 32–41, 2000.

4. D. J. Cook and L. B. Holder. Substructure discovery using minimum description length and background knowledge. *Journal of Artificial Intelligence Research*, 1:231–255, 1994.

5. L. de Raedt and W. V. Laer. Inductive constraint logic. In Proceedings of the Workshop on Algorithmic Learning Theory, pp. 80–94, 1995.

6. L. Dehaspe and H. Toivonen. Discovery of frequent datalog patterns. *Data Mining and Knowledge Discovery*, 3(1):7–36, 1999.

7. P. Domingos and M. Richardson. Mining the network value of customers. In Proceedings of the International Conference on Knowledge Discovery and Data Mining, p. 5766, San Francisco, California, 2001.

8. G. W. Flake, S. Lawrence, C. L. Giles, and F. Coetzee. Self-organization of the web and identification of communities. *IEEE Computer*, 35(3):66–71, 2000.

9. A. Inokuchi, T. Washio, and H. Motoda. Complete mining of frequent patterns from graphs: Mining graph data. *Machine Learning*, 50(3):321–354, 2003.

10. I. Jonyer, L. B. Holder, and D. J. Cook. Hierarchical conceptual structural clustering. *International Journal on Artificial Intelligence Tools*, 10(1–2):107–136, 2001.

11. R. I. Kondor and J. D. Lafferty. Diffusion kernels on graphs and other discrete input spaces. In Proceedings of the International Conference on Machine Learning, pp. 315–322, 2002.

12. M. Kuramochi and G. Karypis. Discovering frequent geometric subgraphs. In Proceedings of the International Conference on Data Mining, pp. 258–265, 2002.

13. M. Kuramochi and G. Karypis. Finding frequent patterns in a large sparse graph. In Proceedings of the SIAM Data Mining Conference, 2003.

14. M. Kuramochi and G. Karypis. An efficient algorithm for discovering frequent subgraphs. *IEEE Transactions on Knowledge and Data Engineering*, 16(9):1038–1051, 2004.

15. W. Lee and S. Stolfo. Data mining approaches for intrusion detection. In Proceedings of the Seventh USENIX Security Symposium, 1998.

16. N. Manocha, D. J. Cook, and L. B. Holder. Structural web search using a graph-based discovery system. *Intelligence Magazine*, 12(1), pages 20–29, 2001.

17. T. Matsuda, H. Motoda, T. Yoshida, and T. Washio. Mining patterns from structured data by beam-wise graph-based induction. In Proceedings of the Fifth International Conference on Discovery Science, pp. 422–429, 2002.

18. S. Muggleton, ed. *Inductive Logic Programming*. Academic, London, 1992.

19. S. Muggleton. Inverse entailment and Progol. *New Generation Computing*, 13:245–286, 1995.

20. S. Muggleton. Stochastic logic programs. In L. de Raedt, ed. *Advances in Inductive Logic Programming*. IOS Press, Amsterdam, 1996.

21. S. Muggleton and W. Buntine. Machine invention of first-order predicates by inverting resolution. In Proceedings of the Fifth International Conference on Machine Learning, pp. 339–352, 1988.

22. S. Muggleton and C. Feng. Efficient induction of logic programs. In Proceedings of the Workshop on Algorithmic Learning Theory, pp. 368–381, 1990.

23. C. Noble and D. Cook. Graph-based anomaly detection. In Proceedings of the International Conference on Knowledge Discovery and Data Mining, 2003.

24. J R Quinlan. Learning logical definitions from relations. *Machine Learning*, 5:239–266, 1990.

25. S. Rajappa. Data mining in nonuniform distributed databases. Master's thesis, The University of Texas, 2003.
26. J. Rissanen. *Stochastic Complexity in Statistical Inquiry.* World Scientific, Singapore, 1989.
27. M. F. Schwartz and D. C. M. Wood. Discovering shared interests using graph analysis. *Communications of the ACM,* 36(8):7889, 1993.
28. U.S. Senate and House Committees on Intelligence. Joint inquiry into intelligence community activities before and after the terrorist attacks of September 11, 2001, U.S. Government Printing Office, December 2002.
29. A. Srinivasan, S. Muggleton, M. J. E. Sternberg, and R. D. King. Theories for mutagenicity: A study in first-order and feature-based induction. *Artificial Intelligence,* 85(1–2):277–299, 1996.
30. S. Su, D. J. Cook, and L. B. Holder. Knowledge discovery in molecular biology: Identifying structural regularities in proteins. *Intelligent Data Analysis,* 3:413–436, 1999.
31. N. Vanetik, E. Gudes, and S. Shimony. Computing frequent graph patterns from semistructured data. In Proceedings of the IEEE International Conference on Data Mining (ICDM), 2002.
32. J. T. L. Wang, M. J. Zaki, H. T. T. Toivonen, and D. Shasha. *Data Mining in Bioinformatics.* Springer, New York, 2004.
33. X. Yan and J. Han. gSpan: Graph-based substructure pattern mining. In Proceedings of the International Conference on Data Mining, 2002.
34. X Yan and J Han. CloseGraph: Mining closed frequent graph patterns. In Proceedings of the Conference on Knowledge Discovery and Data Mining, pp. 286–295, 2003.

8

GRAPH GRAMMAR LEARNING

ISTVAN JONYER

Department of Computer Science, Oklahoma State University,
Stillwater, Oklahoma

8.1 INTRODUCTION

Graphs offer an effective approach to representing and mining relational and complex data sets. They have the ability to represent the most diverse and complex
relationships we may find in such databases. However, even current graph-based
data mining approaches are somewhat limited in their expressivity and ability to
generalize from the data. The majority of current approaches discover frequent
itemsets (subgraphs) from an input set of disjoint graphs (see [11, 16] as well
as Chapters 6 and 9 of this book). The result is a set of association rules linking
grounded subgraphs or subgraphs actually occurring in the input data. Variations
in the graph pattern are not allowed. One approach allows variability in the graph
structure and labels [6] by finding imperfect instances of a subgraph. The generality,
however, is lost since information stored in the imperfect instances is typically not
recovered or is impractical to examine.

Graph grammars enable the specification of elaborate graphs using simple production rules. In this work, we propose to merge the expressive power of grammars
with the representational power of graphs to provide a hierarchical decomposition
of relational databases that also allow for variations of the learned patterns. Namely,
graph grammars can provide increased generalization over extracted subgraphs

Mining Graph Data, Edited by Diane J. Cook and Lawrence B. Holder
Copyright © 2007 John Wiley & Sons, Inc.

found in the data, while representing desirable features such as recursion and disjunctive rules.

8.2 RELATED WORK

In this section we survey work related to our approach. These include graph-based discovery algorithms, grammar induction algorithms, and inductive logic programming approaches.

Many existing graph mining algorithms are largely based on the Apriori algorithm [1]. These approaches restrict themselves to *sets* of disjoint graphs called *graph transactions*, and they are not able to work with arbitrary graphs. One system by Inokuchi and colleagues [11] is called Apriori-based graph mining (AGM). Another system by Kuramochi and Karypis [16], called frequent subgraph (FSG), further develops AGM by applying an edge-growing strategy. FSG employs a number of techniques for fast graph matching, avoiding the need for graph isomorphism tests. The gSpan algorithm by Yan and Han [28] (also see Chapter 5 of this book) takes the ideas in FSG one step further by parting with the candidate generation approach of the Apriori algorithm for a faster, *generate-evaluate* approach using depth-first search. While such approaches can exploit the disconnected property of some graphs for fast frequent itemset discovery, domains involving single connected graphs are left unaddressed.

Recent work on grammar induction includes learning string grammars with a bias toward those that minimize description length [17], inferring compositional hierarchies from strings in the Sequitur system [20], and learning search control from successful parses [29]. In contrast, work on *graph grammar* induction is almost nonexistent. Computational uses of graph grammars have been reported in the annual journal *Graph Grammars and Their Application to Computer Science* [8] and more recently in the *International Conference on Graph Transactions*, which was first held in 2002. Despite the large number of articles on graph grammars, only a small number of them address the problem of inferring graph grammars from examples. An enumerative method for inferring a very limited class of context-sensitive graph grammars is due to Bartsch-Spörl [2]. Other algorithms utilize a merging technique for hyperedge replacement grammars [18] and regular tree grammars [4]. None of these are as general and comprehensive as our approach, which is based on a method for discovering interesting substructures in graphs (see [5] and Chapter 7 of this book).

Inductive logic programming (ILP) systems are important to mention as related work since they work well with complex data. They represent databases in first-order logic (FOL), not as graphs, and perform induction on the world of logic statements. Several approaches have emerged since the idea was first introduced by Plotkin [21], along with the inductive mechanism of relative least general generalization (RLGG). One prominent representative of ILP systems is First Order Inductive Logic (FOIL) [22], which learns sets of Horn clause-like first-order rules. It has been applied to learning search-control rules [29] and learning patterns in hypertext domains [25]. A different approach is realized by Progol [18], which uses inverse entailment to generate a specific bound to its general-to-specific search. Progol has been applied to chemical carcinogenicity [26] and three-dimensional protein structure [27], among others.

8.3 GRAPH GRAMMAR LEARNING

A distinct direction in graph-based knowledge discovery research is toward finding more efficient algorithms. This is driven by the high complexity associated with graph isomorphism computations. While this line of research is important, we must also pursue the expansion of the expressive power of graph-based hypotheses. We recognize that graph-based data structures exhibit characteristics that may be exploited in the effort of finding regularities in graphs, and that such regularities can be expressed using graph grammars.

8.3.1 Graph Grammars

In this research we are addressing the problem of inferring graph grammars from positive and, optionally, negative examples. In other words, we seek to design a machine learning algorithm in which the learned hypothesis is a graph grammar. Machine learning algorithms, in general, attempt to learn theories that generalize beyond the seen examples, so that new, unseen data can be accurately categorized. Translated to grammar terms, we would like to find grammars that accept more than just the training language. Therefore, we would like to learn *parser* grammars, which have the power to express more general concepts than the sum of the positive examples.

In this research we are concerned with graph grammars of the set-theoretic or expression approach [19]. In this approach a *graph* is a pair of sets $G = \langle V, E \rangle$ where V is a set of labeled *vertices* or *nodes*, and E is a set of labeled *edges*. Graph grammar *rules* are of the form $S \rightarrow P$, where S and P are graphs. Such rules can be applied in both directions, depending on whether we want to parse or generate a graph. When generating a new graph, an isomorphic copy of S is removed from the existing graph along with all its incident edges, and is replaced with a copy of P, together with edges connecting it to the graph. The new edges are given new labels to reflect their connection to the subgraph instance. When parsing, all instances of P are removed from the existing graph and each are replaced with S. Again, occurrences of S are connected to the rest of the graph according to a specific embedding function.

A special case of the set-theoretic approach is the node-label-controlled grammar, in which S consists of a single labeled node [9]. This is the type of grammar on which we are focusing. In our case, S is always a nonterminal, but P can be any graph and can contain both terminals and nonterminals. Since we are focusing on parser grammars, the embedding function is trivial: External edges that are incident on a vertex in the subgraph being replaced (P) always get reconnected to the single vertex S being inserted in its place.

This research focuses on identifying recursive and disjunctive productions. We would like to take advantage of these features both in discovery and in concept learning tasks, and on both flat domains that consist of a number of data instances and structural domains that can be represented as one big graph. As we will see, we cannot learn all possible kinds of graph grammar rules. The most important obstacle in discovering graph grammars is computational cost, which must be kept under control if our algorithm is to be practical.

8.3.2   Graph Grammar Learning

According to the intuition we gave previously, graph grammar induction (GGI) expands on the expressive power of frequent subgraphs. Therefore, a graph grammar learner algorithm can use a subgraph learning algorithm as its basis, which can be extended with additional capabilities for learning recursive rules, disjunctive rules, and logical relationships. In our research we used the Subdue graph mining algorithm and extended it as described below. We refer to our grammar learning algorithm as SubdueGL. We give the complete pseudocode for our algorithm in Figure 8.1.

```
SubdueGL ( graph G, int Beam, int Limit )
   grammar = {}
   repeat
      queue Q = { v | vertex v has a unique label in G }
      bestSub = first substructure in Q
      repeat
         newQ = {}
         for each substructure S in Q
            newSubs = ExtendSubstructure(S)
            recursiveSubs = RecursifySubstructure(S)
            newQ = newQ U newSubs U recursiveSubs
            Limit = Limit - 1
         evaluate substructures in newQ by MDL or SetCover
         Q = substructures in newQ with top Beam compression scores
         if best substructure in Q better than bestSub
         then bestSub = best substructure in Q
      until Q is empty or Limit <= 0
      grammar = grammar U bestSub
      G = G compressed by bestSub
   until bestSub cannot compress the graph G
   return grammar


ExtendSubstructure (substructure S
   newSubs = S extended by an adjacent edge in all possible ways
   // Discover variables
   varSubs = S extended by an adjacent edge in all possible ways,
                  replacing the added vertex with a non-terminal
   // Discover relationships
   for each substructure V in varSubs
      for each variable w in V
         if relationship '<=' or '=' holds between w and any other vertex in V
            add relationship to V
   return newSubs U varSubs


RecursifySubstructure (substructure S)
   recSubs = all possible chains of instances of S, linked by a
                  single edge
   return recSubs
```

Figure 8.1.   The SubdueGL algorithm.

The two learning modes supported by the algorithm are discovery mode and concept learning mode (or unsupervised learning and supervised learning, respectively). In discovery mode the algorithm attempts to find grammar rules in the input graph, which need not be a connected graph. In some domains, such as market basket analysis, a set of connected graphs may be used. In concept learning mode the input consists of two graphs. In this case the algorithm seeks to identify grammar rules in one graph, called the positive graph, such that the rules identified should not be found in the other graph, called the negative graph. Our evaluation heuristics, which are discussed next, penalize for dances of grammar rules in the negative input.

The last consideration before we give the details is the heuristic that drives the search, that is, the method of evaluation that designates a grammar rule to be better or worse than another. By default, we use the minimum description length (MDL) principle to drive the search [24]. The evaluation heuristic based on the MDL principle assumes that the best substructure is the one that minimizes the description length of the input graph when compressed by the substructure [6]. The description length of the substructure S given the input graph G is calculated as $DL(S, G) = DL(S) + DL(G|S)$, where $DL(S)$ is the description length of the subgraph, and $DL(G|S)$ is the description length of the input graph compressed by the subgraph. We are looking for a minimum set of grammar rules that compress the graph the best. The formula for computing the description length in concept learning, given positive and negative inputs, is $DL(G^+, G^-|S) = DL(G^+|S) + DL(S) + DL(G^-) - DL(G^-|S)$, where $DL(S)$ is the description length of the grammar rule, G^+ is the positive input, G^- is the negative input, $DL(G^+|S)$ is the positive input compressed by the grammar rule, $DL(G^-|S)$ is the negative input compressed by the grammar rule, and $DL(G^+, G^-|S)$ is the positive and negative inputs compressed by the grammar rule. $DL(S) + DL(G^+|S)$ expresses the number of bits needed to encode the grammar rule plus parts of the positive input that are not described by the rule. $DL(G^-) - DL(G^-|S)$ is the number of bits needed to describe the portion of the negative input that *is* described by the grammar rule.

An alternative heuristic, called set cover, drives the search toward minimal error (maximal accuracy), especially in concept learning tasks, by covering as many positive examples as possible while covering as few negative examples as possible. In this heuristic the concept value is computed as $SC(S) = E(S)/X(G)$, where $E(S)$ is the error (examples not covered) by subgraph S, and $X(G)$ is the total number of examples (both positive and negative) in graph G. Then $E(G)$ is defined $E(S) = X^+(G) - C^+(S) + C^-(S)$, where $X^+(G)$ is the total number of positive examples in G, $C^+(S)$ is the number of positive examples covered by S, and $C^-(S)$ is the number of negative examples covered by S.

Basic Rules. The most basic of grammar rules consists only of a connected, static graph. Such rules can be identified by the Subdue system. A grammar consisting only of such rules provides a hierarchical decomposition of the input graph [13]. To identify each rule, Subdue searches for interesting subgraphs using a computationally constrained beam search. It starts by identifying all one-vertex substructures in the input graph and all their occurrences. The search progresses by applying the ExtendSubgraph operator to each substructure. The resulting substructures are kept on a queue and are ordered based on their description length. The top few—4, by default—are further extended. By the repeated application of the ExtendSubgraph operator the most interesting substructure is found. The substructure identified in this way will comprise the right side of the first grammar rule.

The next step in the grammar learning process is to remove all occurrences of the previously identified substructure, and replace them with a single nonterminal vertex. The process starts over and a new grammar rule is identified. This process can be repeated until the graph is exhausted (reduced to a single nonterminal vertex), or until a user-specified limit is reached.

Figure 8.1 shows the complete SubdueGL algorithm. A description of recursive and disjunctive rules follow next.

Recursion. Before we describe the details of the algorithm for learning recursive rules, let us examine the motivation behind such rules and their possible characteristics. Recursion comes in two flavors: *structural* and *functional*. As graph representation is mainly used to take advantage of the structural relationships in data, our research focuses on structural recursion. We seek to learn graph grammars, where the left side of each production is only a single nonterminal. We do not address functional recursion in this work.

Structural recursion can also be divided into several subtypes based on the number of recursive nonterminals and their way of embedding on the right side of a production rule. We restrict our grammars to have only one recursive nonterminal on the right side of a production, although any number of nonrecursive nonterminals may appear on the right side. This restriction is due to the additional computational complexity needed to identify such grammars.

This limited type of recursion also has two variations, chained and embedded, based on the number of neighbors a recursive nonterminal has. Chained recursion occurs when a specific motif repeats in a sequence. This mostly happens with sequential data, as in a DNA (deoxyribonucleic acid) sequence, but is also possible in complex structures. Rules of the chained recursion allow only a single neighbor to the nonterminal. In graph terms this means only a single edge can connect to the nonterminal.

Embedded recursion occurs when a motif is removed from a structure and its surroundings, when moved into its place, form the same pattern as the removed one. This is perhaps most familiar to computer scientists in the parenthesis matching problem. When removing the innermost matching set of parentheses, the ones just outside of them move in to take their place. Rules of the embedded-recursion type allow two or more edges to connect to the nonterminal.

Figure 8.2 shows two examples of embedded recursion. The first one describes a graph grammar for generating DNA, and the second grammar generates matching parentheses. Figure 8.3 shows a chained recursive rule. Recognizing embedded recursion is computationally expensive, especially in graphs where the higher number of connections between objects creates a combinatorial explosion of possible

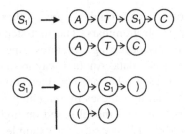

Figure 8.2. Sample embedded recursive rules.

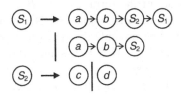

Figure 8.3. Sample chained graph grammar rules.

patterns. In this research we do not seek to find embedded recursion and focus exclusively on chained recursion.

In our specification, recursive productions are of the form $S \rightarrow PS$. The nonterminal S is on both sides of the production, and P is linked to S via a single edge. If the grammar is used for graph generation, this rule would generate an infinitely long sequence of the graph P. If the language is to be finite, a stopping alternative production is required. One such production is $S \rightarrow PS|\emptyset$, which reads "replace S with PS or nothing." For our purposes, however, we use the production $S \rightarrow PS|P$. The rule $S \rightarrow PS|\emptyset$, when used for parsing, would imply that nothing can be replaced with S, introducing an arbitrary number of S's. At the same time, it cannot parse a chain of P's of finite length as it would have no starting point, since PS does not exist in the input graph. Remember that the stopping alternative of a graph generator rule is the starting point of a parser rule.

When parsing a graph, we start from the complete graph and work toward a single nonterminal. This is done by removing subgraphs from the graph that match the right side of a production and inserting the nonterminal on the left side. In our example, replace PS with S, and finally, P with S. An example of a recursive production is shown in Figure 8.3 (S_1).

Recursive productions are created in SubdueGL by the RecursifySubstructure operator. It is applied to each substructure after the ExtendSubstructure operator. RecursifySubstructure checks each instance of the substructure to see if it is connected to any other instances by a single edge. If so, a recursive production is possible. The operator adds the connecting edge to the substructure definition and collects all possible chains of instances. If a recursive production emerges to be the best at the end of an iteration, it will compose the next grammar rule, and each recursive chain of subgraphs is replaced by a single nonterminal vertex.

Variables. Disjunctive rules—or variables—are another feature of the grammars, which we include in our graph grammars. The first step toward discovering variables is discovering commonly occurring structures, since variability in the data is defined by the surrounding static data. Another way to look at this is that structures that are very different from each other are not interesting. Most machine learning algorithms attempt to construct hypotheses by searching for regularities in the data. Accounting for slight variations in a model based on high regularity may enhance the performance of the model. ILP systems, for example, recognize this and allow the specification of variables by single-sided bounds for numeric variables. Decision tree learners take a similar approach.

Variable productions are of the form $S \rightarrow P_1|P_2|\ldots|P_n$ for discrete variables and $S \rightarrow [P_{min} \ldots P_{max}]$ for numeric variables. The nonterminal graph S can be thought of as a variable having possible discrete values $P_1, P_2, \ldots, P_n$ or numeric values in the range P_{min} to P_{max}. If S is a single vertex, and P_i are also single vertices, then S is synonymous with a regular nongraph variable whose values are the vertex labels, which can be alphanumeric values like numbers (discrete or continuous) or string descriptions. An example of a variable is shown in Figure 8.3, where S_2 has possible values c and d.

A variable S can have many values. It is possible to create a number of different substructures from a subset of these values that may evaluate better according to our evaluation measure. Many techniques exist for selecting values that are descriptive of one class and not the other. Information gain has been utilized in decision tree induction, regression, and other statistical methods are used to build statistical models, and the list goes on. Even description length has been utilized for guiding the search for hypotheses [3]. Linear and surface-based separation techniques can be useful but only in numeric domains or where categorical values can be ordered naturally. Problems can still arise if the classes cannot be separated in such a manner, as in the case of a normal distribution where the negative class consists of outliers. Neural networks and-nearest neighbor algorithms may overcome this, but human expertise is essential for good results. More sophisticated statistical methods can achieve better results, but they only work on flat, numeric domains. We show a comparison on the performance of 15 such techniques on the Wisconsin Breast Cancer domain.

SubdueGL discovers variables inside the ExtendSubstructure search operator. As mentioned before, Subdue extends each instance of a substructure in all possible ways and groups the resulting subgraphs that are alike. In addition, SubdueGL also groups subgraphs that result from the same extension (extended by the same edge from the same vertex), *regardless of what vertex they point to*. For example, let $(v1, e, v2)$ represent edge e from vertex $v1$ to vertex $v2$. Vertex $v1$ is part of the original substructure, which was extended by e and $v2$. For variable creation, instances are grouped using $(v1, e, V)$, where V is a variable (nonterminal) vertex whose values include the labels of all matching $v2$'s. The substructure so created is then evaluated and competes with others for top placement on the extension queue.

As mentioned earlier, variable V can have many values. It is possible to create other substructures from a subset of these values that may evaluate better according to a given heuristic. Even though the number of possible subsets of values is exponential, we can apply the observation that the greatest reduction in description length is provided by values that occur more frequently. Hence, we can create a ranking of values based on their frequency. This frequency is based only on occurrences in the positive input. That is, it does not take into account instances in the negative input, and does not assume anything about the global performance of the variable. This ranking, however, allows us to consider only a number of subsets of values that is linear in the number of occurrences of the values in the positive input. The substructure is evaluated by successively removing the lowest ranked variable value. The set of variable values that evaluates the best is kept.

For continuous values such a ranking may not be practical since each value may be unique. Here, we use a heuristic that assumes that numeric variables have a central tendency, and we seek to eliminate outliers. That is, we compute the mean of the variable's values and successively eliminate the value that is the farthest from the mean. We detect continuous variables by checking to see if each variable value is a number. If so, the variable is handled as continuous, otherwise as discrete. The GetBestVariableValues search operator finds the best values for a variable. When appropriate, the user may control the search process by specifying a minimum percentage of substructure instances that must contain the variable value for it to be kept as part of the variable definition.

Relationships. Relationships increase the expressive power of grammars by identifying variables in production rules that are equivalent or, in the case of numerical variables, have a *less-than or equal-to* relationship. Relationships are logical entities and are signified by special relationship edges. This is in contrast with structural components of graph grammars, which are vertices and structural edges.

Graph grammars typically do not include logical components or relationship edges specifically. At the same time we have to realize that graph grammars have not been used for machine learning purposes before or to construct models for databases represented as graphs. Relationships, however, are used widely in machine learning. ILP systems, for example, allow $\leq$ relationships between two variables and also between a variable and a static value [3]. Building on the experience of ILP systems, we allow $\leq$ and $=$ relationships. The $<$ relationship will be part of future investigation.

A precondition for the existence of relationships is variables. At least one vertex participating in a relationship has to be a variable nonterminal since relationships between nonvariables are trivial. The only relationship that can occur between a variable and a nonvariable vertex is the less-than or equal-to relationship. *Equal-to* relationships must be between two variables, as they had no value between two static vertices.

A relationship edge is identified by comparing a newly discovered variable's values in each instance of a substructure to every other one of its vertices. If the same relationship holds between the variable and another vertex in every instance of a substructure, a relationship edge is created. Figure 8.4 shows an example of a

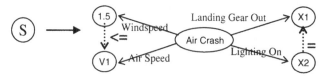

Figure 8.4. Graph grammar production with relationships.

production that contains two relationships. The relationship edges are marked with dotted arrows and are labeled ≤ and =.

Relationships only show up in concept learning mode. Although not explicitly prohibited, they do not show up in grammar rules in discovery mode because they do not help cover additional examples on their own. Relationships are logical components of hypotheses and serve only to increase their accuracy. They do add to the description length of the hypothesis, so a hypothesis without a relationship evaluates better than with one, everything else being equal. In concept learning this added cost pays off if it helps to eliminate negative examples from the coverage of the hypothesis. Therefore, relationships can only make a contribution by improving the accuracy of hypotheses in concept learning. This is done not by additional reward on the positive examples but by the reduced penalty from the negative ones.

8.3.3   Class of Grammars Learned

Earlier we discussed some aspects of the graph grammars to be learned in terms of graph transactions. That is, we know that the graph grammars are node-label controlled, where the embedding function is trivial. We have not discussed, however, the exact class of graph grammars to be learned. We have stated that the grammar is context free, but further details are in order.

According to the Chomsky hierarchy, in unrestricted grammars, both sides of production rules can be any string, including terminal and nonterminal symbols. Context-sensitive grammars are restricted such that the left side of a production must be no longer than the right side. Context-free grammars are further restricted such that there must be only one nonterminal on the left side of productions. Regular grammars introduce restrictions on the right side of productions as well. There are two types of regular grammars: right linear and left linear. If all productions of a grammar are of the form $A \rightarrow wB$ or $A \rightarrow w$, it is called a right linear regular grammar, where A and B are nonterminals and w is a string of terminals. A grammar consisting of productions of the form $A \rightarrow Bw$ or $A \rightarrow w$ is called a left-linear regular grammar.

This discussion was presented with string grammars in mind, but the concepts transfer easily to graph grammars. Examining the restrictions we are placing on the learning process, we can see that the resulting grammars fall within the class of regular grammars. In SubdueGL, which stands for Subdue Grammar Learner, recursive productions are restricted to those that include the recursive nonterminal at one end of a single edge originating from an arbitrary graph. This restriction is introduced to make the algorithm computationally tractable (and hence, practical) since identifying recursive rules is exponential in the number of edges allowed around the nonterminal. This structure restricts the production to generate only linear sequences of subgraphs, much like in string grammars. Therefore, it can be thought of as a regular graph grammar. Since we do not place any restrictions on the direction of the edge, the grammar can be both right and left linear.

The single-edge connection restriction can easily be removed to enable SubdueGL to learn context-free graph grammars. There are no fundamental limitations that prevent this. For the additional expressive power, however, we would naturally incur additional computational cost. Whether this additional cost is fundamentally prohibitive or not will have to be the subject of further investigation.

8.4   EMPIRICAL EVALUATION

This section describes the empirical evaluation of our system on real-world domains. The algorithm is also compared to existing systems both on flat and structured domains.

8.4.1   Flat Domains

Although addressing flat databases was not a primary objective of this research, a good performance on these domains demonstrates our goal of creating a general-purpose data mining tool. Even more importantly, comparing SubdueGL's performance to accepted systems on well-known domains confirms the validity of our approach.

Converting flat, or feature-vector-based datasets to graphs is straightforward. Each feature vector forms a starlike pattern, with a general central vertex (perhaps "object" or "event"), from which attributes and their values originate. Attributes correspond to edges while attribute values correspond to vertices pointed to by the attribute edge. Hence, the entire dataset is a collection of disconnected, starlike graphs.

We were particularly interested in how SubdueGL compared against other systems on flat data that were specifically designed for relational domains, just like SubdueGL. We selected FOIL [3] and Progol [18], which are prominent inductive logic learning systems. Both have had great success in a wide variety of domains. We also wanted to know how our approach measures up against an algorithm that was designed for flat domains. For this, we used C4.5 [23].

We selected the *vote, diabetes*, and *credit* domains to serve as the basis for our comparison, which are available from the UCI machine learning repository [17]. The vote domain is the Congressional Voting Records Database. It contains 16 discrete-valued attributes, having values y, n, and u (for *yes, no,* and *unknown*). The diabetes domain is the Pima Indians Diabetes Database, which contains 7 continuous-valued attributes. The credit domain is the German Credit Dataset from the Statlog Project Databases. The credit data set contains 13 discrete and 7 continuous-valued attributes.

TABLE 8.1 Comparison of FOIL, Progol, C4.5, Subdue, and SubdueGL

	Vote (%)	Diabetes (%)	Credit (%)
FOIL	93.02	70.66	68.60
Progol	94.19	63.68	63.20
C4.5	94.48	74.62	70.90
Subdue	89.07	61.71	70.50
SubdueGL	94.23	70.94	71.30

Predictive accuracies are shown in Table 8.1, which were generated using 10-fold cross validation. As is evident from the table, SubdueGL outperformed the two ILP systems and Subdue in all three domains and was the best on the credit domain. C4.5, however, did significantly better on the diabetes domain than the other three systems.

The difference in performance may lie in the way these systems handle continuous values. As the table shows, all systems did equally well on the vote domain, which has only discrete-valued attributes. Subdue is an exception, which is unable to incorporate multiple values into rules without graph grammar induction. When creating rules for continuous attributes, FOIL can only create one-sided intervals, while SubdueGL can create two-sided intervals. The success of FOIL's strategy depends on the data, since it may be able to create a perfect separation of positive and negative examples only if the values of one class is greater than those of the other for an attribute. For detecting a central tendency versus outliers, FOIL must include outliers on one side and can only make a determination as to how many negatives to cover. SubdueGL, on the other hand, is able to handle a central distribution with outliers as well as a splitting of values. Based on this reasoning, FOIL should do no better than SubdueGL, at least on domains with continuous values. This is evidenced by Table 8.1.

In the case of Progol, which is biased to learn clauses that entail a particular positive example, the bias causes Progol to make guesses and overcommit to particular numerical constants based on the evidence of a single positive example. Since numerical constants are ordered, which may be interpreted as part of an interval, the inclusion of new constant symbols, not already present in the training sample or background knowledge, is justified. At the same time this strategy is less successful on numerical constants than nonnumerical ones, which may be the reason why Progol did much worse than SubdueGL and FOIL once continuous attributes were introduced into the database.

C4.5 is a decision tree learning algorithm, which creates a hierarchical rule in the form of a tree in which items can be classified by starting at the top of the tree and answering questions posed at each of the nodes encountered. Even though C4.5 handles continuous values using one-sided intervals, like FOIL, it can employ two one-sided intervals in succeeding nodes of the decision tree to create two-sided intervals. This strategy proves to be the most successful one on the diabetes domain, which contains continuous attributes exclusively. SubdueGL outperformed C4.5 on the credit domain, which contains mixed attribute types.

8.4.2 Complex Domains

In this section we describe a series of experiments that have real-world applications and are structural, not flat, in nature. First we show experiments from biochemistry that deal with finding common patterns in myoglobin and hemoglobin primary and secondary structures. Then, we present an experiment inspired by Defense Advanced Research Project Agency (DARPA's) Evidence Extraction and Link Discovery (EELD) project.

Biochemistry. To show SubdueGL's applicability to the sequence-structured domains, we analyzed the primary and secondary structure of the proteins *myoglobin* and *hemoglobin*. Myoglobin and hemoglobin are hemeproteins whose physiological importance is principally related to their ability to bind molecular oxygen. Myoglobin is found mainly in muscle tissue where it serves as an intracellular storage site for oxygen. Its secondary structure is unusual in that it contains a very high proportion (75%) of α-helical secondary structure. Hemoglobin is found in erythrocytes where it is responsible for binding oxygen in the lung and transporting the bound oxygen throughout the body [15]. These proteins are used widely to illustrate nearly every important feature of protein structure, function, and evolution [10].

The primary structure of proteins is represented as a sequence of amino acids, which have three-letter acronyms. These compose the vertices of the input graph, which are connected by edges labeled "next." A small part of the input graph is shown in Figure 8.5.

The grammar induced from the primary sequence of myoglobin is shown in Figure 8.6, where graph vertices are only shown by their labels. The arrow ($\rightarrow$) is the production operator, while ($-$) signifies the edge labeled "next" in the graph. We show a representative number of rules while omitting others (S_7 through S_{17}). The expressive power of the grammar is apparent at the first glance. Production S

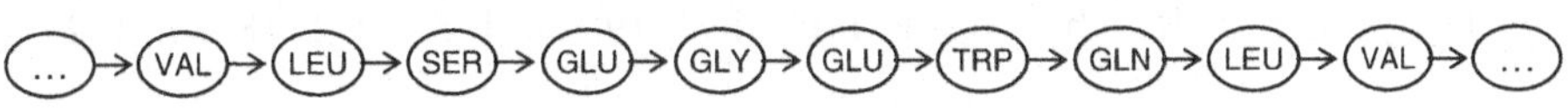

Figure 8.5.   Part of the myoglobin primary sequence.

$S \rightarrow S_2 - S_3 - S_4 - S_5 - S_6 - S_7 - S_8 - S_9 - S_{10} - S_{11} - S_{12} - S_{13} - S_{14} - S_{15} - S_{16} - S_{17} - \text{GLU} - S$
$S_2 \rightarrow \text{VAL} \mid \text{SER} \mid \text{HIS} \mid \text{LYS}$
$S_3 \rightarrow \text{LEU} \mid \text{GLY} \mid \text{HIS} \mid \text{PHE} \mid \text{PRO}$
$S_4 \rightarrow \text{GLY} \mid \text{GLN} \mid \text{ALA} \mid \text{ASP} \mid \text{THR}$
$S_5 \rightarrow \text{ALA} \mid \text{ASP} \mid \text{ARG} \mid \text{THR} \mid \text{ASN}$
$S_6 \rightarrow \text{LEU} \mid \text{LYS} \mid \text{ILE} \mid \text{PHE}$
...
$S_{20} \rightarrow S_{21} - S_{22} - S_{23} - S_{24} - S_{25} - S_{26} - S_{27} - S_{28} - S_{29} - \text{LEU}$
$S_{21} \rightarrow \text{VAL} \mid \text{GLY} \mid \text{ARG} \mid S$
$S_{22} \rightarrow \text{LEU} \mid \text{LYS} \mid \text{PRO} \mid S$
$S_{23} \rightarrow \text{LEU} \mid \text{SER} \mid \text{ASP}$
$S_{24} \rightarrow \text{LEU} \mid \text{GLU} \mid \text{ALA} \mid \text{ASP} \mid \text{ILE}$

Figure 8.6.   Partial grammar induced by SubdueGL on myoglobin primary sequence data.

h_1_15 − h_1_15 − h_1_6 − h_1_6 − h_1_19 − h_1_8 − h_1_18 − h_1_23

$$
\begin{aligned}
\mathbf{S} &\rightarrow \mathbf{S_2} - \mathbf{S_3} - \text{h\_1\_6} - \mathbf{S_4} - \text{h\_1\_19} - \text{h\_1\_8} - \text{h\_1\_18} - \mathbf{S_5} \\
\mathbf{S_2} &\rightarrow \text{h\_1\_14} \mid \text{h\_1\_15} \\
\mathbf{S_3} &\rightarrow \text{h\_1\_14} \mid \text{h\_1\_15} \\
\mathbf{S_4} &\rightarrow \text{h\_1\_6} \mid \text{h\_1\_1} \\
\mathbf{S_5} &\rightarrow \text{h\_1\_20} \mid \text{h\_1\_23}
\end{aligned}
$$

Figure 8.7. Partial grammar SubdueGL on hemoglobin secondary structure.

$$
\begin{aligned}
\mathbf{S} &\rightarrow \text{h\_1\_15} - \text{h\_1\_15} - \text{h\_1\_6} - \text{h\_1\_6} - \text{h\_1\_19} - \mathbf{S_2} - \text{h\_1\_18} - \mathbf{S_3} \\
\mathbf{S_2} &\rightarrow \text{h\_1\_9} \mid \text{h\_1\_8} \\
\mathbf{S_3} &\rightarrow \text{h\_1\_25} \mid \text{h\_1\_23}
\end{aligned}
$$

Figure 8.8. Partial grammar SubdueGL on myoglobin secondary structure.

contains 16 variables and it is recursive. Rules S_{21} and S_{22} are both variables and contain **S** as one of their values.

For the next example we use the secondary structure of hemoglobin, which is represented in graph form as a sequence of helices and sheets along the primary sequence. Each helix is a vertex connected via edges labeled "next". Each helix is encoded in the form h_t_l, where h stands for helix, t is the helix type, and l is the length of the helix. Part of the grammar identified by SubdueGL is shown in Figure 8.7. This grammar only involves helices of type 1 (right-handed α helix). This grammar can generate the most frequently occurring helix sequences that are unique to hemoglobin. It can also generate others as well, for instance, ones that also occur in myoglobin. The grammar generated for myoglobin secondary sequence is shown in Figure 8.8. We can see that the two sequences have the same length and the languages generated by them intersect in the following sequence:

h_1_15 − h_1_15 − h_1_6 − h_1_6 − h_1_19 − h_1_8 − h_1_18 − h_1_23

This is interesting since researchers have long been speculating that a common evolutionary path exists for these proteins. In February of 2000, Hou and colleagues reported on myoglobin-like proteins that are prime candidates to be common ancestors [10].

CounterTerrorism. In recent years the U.S. government has sponsored an increased number of research projects directed toward counterterrorism. Many projects center on social network analysis in which social interactions and relationships are represented and analyzed. In the following example will show one such example to highlight the ability of our system to pinpoint hard to find patterns that could not be identified previously.

Our example domain was inspired by the *Russian Contract Killing* problem in which various events are represented in a social network among which are murders. They seek to identify events leading up to a contract killing event in order to predict and prevent such events in the future. A portion of the input is depicted in Figure 8.9.

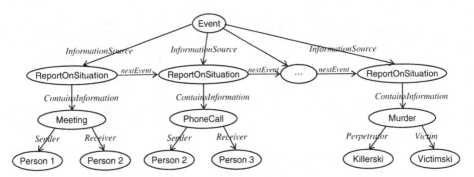

Figure 8.9. Portion of the RCK domain.

As we can see in the figure, the domain consists of a series of communication events that can be a face-to-face meeting, a phone call, or an email communication, which is followed by a murder event. The actual input contains four chains of events, two of which have three communications and two have two communications. The names of the persons involved are different in the actual input.

We applied SubdueGL to this domain. The resulting graph grammar is shown in Figure 8.10. Rule S_6 contains only one of the four subgraphs that remain in the database after the input graph is parsed by rules S through S_5. In reality, there are four of these subgraphs.

Inspecting the results, we can see that rule S describes a series of general communications. These are of different types and involve different people. Rule S_5 describes the murder event. In this example the name of the killer and victim were always the same, but if we change these, SubdueGL returns a more general rule that has variables in place of these names. Looking at rule S_6 we can see that a series of communication events (inferred from the recursive nature of rule S) are followed by a murder event (rule S_4), which is exactly the type of information that can help predict crime and, in this case, murder.

Even though the number and type of communications change, SubdueGL is able to identify a general pattern that is useful in identifying the events leading up to a murder. An algorithm that can only identify static patterns would not be able to discern any usable information from this type of data.

8.4.3 Discussion

In this section we have demonstrated the practical utility of grammar learning in graphs on biochemistry and counterterrorism domains. Our examples were specific, but other configurations of both data and the learning system are possible.

In the counterterrorism domain, for instance, we could also represent and track the money trail that is sure to trace from the contractor of the murder to the assassin. This may involve bank-to-bank transactions, Internet-assisted payments between persons, and so on. Other accessories that may be involved in the contract killing

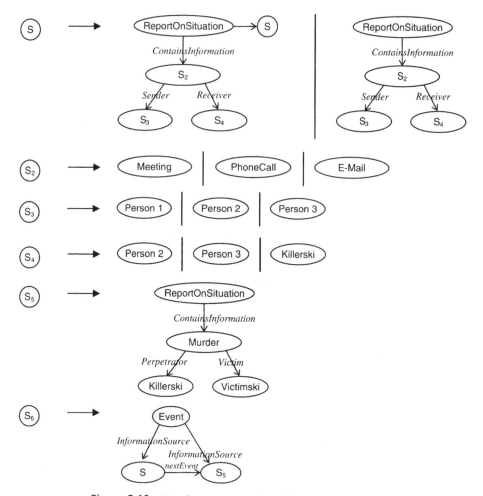

Figure 8.10. Graph grammar induced from the RCK domain.

may also be tracked, such as purchase and transport of weapons and raw materials needed to make weapons. In all cases, a series of events may be found and modeled similarly to the series of contacts shown in the example. These events may also be intermixed in occurrence and may also contain events that are irrelevant to the contract killing scenario.

Similarly, other approaches may be employed in finding useful patterns in biochemistry. For instance, experts are already aware of some patterns and the role of various amino acids, which are assigned into groups based on their role and chemical properties. It may be useful to incorporate this background information into the database to jump start or steer the discovery process. The same is true for counterterrorism domains, where certain information may be available to government agents, information that can be used as background knowledge to help the discovery algorithm.

8.5 CONCLUSION

In this work we developed a machine learning algorithm for learning graph grammars. Using the representational power of graphs and the ability of graph grammars to generalize, we developed an approach that extends the expressive power of hypotheses learned from examples. Our main goal was to achieve a better performance over algorithms that learn static patterns only, and to develop a graph-based algorithm that is competitive in performance with those of inductive logic systems. This was partially motivated by the emergence of specific problems that we determined could benefit from the increased expressive power of graph grammars. Previous works on graph-based systems lack the ability to learn structurally recursive concepts.

Our approach includes the ability to learn recursive concepts, concepts that include variable data points, and relationships between data points. We extended our approach for both discovery and concept learning modes. We found that the class of graph grammars learned by our algorithm is equivalent to the class of regular graph grammars, which is the most restricted form of grammars.

We evaluated the system on several real-world examples. We compared SubdueGL to ILP systems as well since they are known to perform well on relational domains. Hence, we performed experiments on publicly available databases using SubdueGL and two of the most prominent ILP systems: FOIL and Progol. Our approach outperformed these systems significantly about half the time and marginally in the remaining experiments.

We experimented with real-world domains as well. We found interesting results in protein sequences, which demonstrated that the additional expressive power of graph grammars have practical utility. We applied our approach to a counterterrorism domain, where the task involved recognizing a sequence of events leading up to a murder. Our system found such a pattern, which, again, verifies the practicality of the approach.

REFERENCES

1. R. Agrawal and R. Srikant. *Fast algorithms for mining association rules*. In Proceedings of the International Conference on Very Large Databases, pp. 487–499, Santiago, Chile, 1994.
2. B. Bartsch-Spörl. *Grammatical inference of graph grammars for syntactic pattern recognition. Lecture Notes in Computer Science*, 153:1–7, 1983.
3. R. M. Cameron-Jones and J. R. Quinlan. *Efficient top-down induction of logic programs. SIGART Bulletin.* 5(1):33–42, 1994.
4. R. C. Carrasco, J. Oncina, and J. Calera. *Stochastic inference of regular tree languages. Lecture Notes in Artificial Intelligence*, 1433:187–198, 1998.
5. D. J. Cook and L. B. Holder. *Graph-based data mining, IEEE Intelligent Systems*, 15:32–41, 2000.
6. D. J. Cook and L. B. Holder. *Substructure discovery using minimum description length and background knowledge. Journal of Artificial Intelligence Research*, 1: 231–255, 1994.

7. R. E. Dickerson and I. Geis. *Hemoglobin: Structure, Function, Evolution, and Pathology.* Benjamin/Cummings, Boston, 1982.

8. H. Ehrig, H.-J. Kreowski, and G. Rozenberg, eds. *Graph Grammars and Their Application to Computer Science, Lecture Notes Computer Science*, Vol. 532. Springer, Berlin, 1991.

9. J. Engelfriet and G. Rozenberg. *Graph Grammars Based on Node Rewriting: An Introduction to NLC Grammars. Lecture Notes in Computer Science*, Vol. 532, pp. 12–23. Springer, Berlin, 1991.

10. S. Hou, R. W. Larsen, D. Boudko, C. W. Riley, E. Karatan, M. Zimmer, G. W. Ordal, and M. Alam. *Myoglobin-like aerotaxis transducers in Archaea and Bacteria. Nature* 403:540–544, 2000.

11. A. Inokuchi, T. Washio, and H. Motoda. *An apriori-based algorithm for mining frequent substructures from graph data.* Proceedings of the European Conference on Principles and Practice of Knowledge Discovery in Databases, Lyon, France, 2000.

12. E. Jeltsch and H. J. Kreowski. Grammatical inference based on hyperedge replacement. *Lecture Notes in Computer Science*, 532:461–474, 1991.

13. I. Jonyer, L. B. Holder, and D. J. Cook, Graph-based hierarchical conceptual clustering, *Proceedings of the Thirteenth Annual Florida AI Research Society*, Orlando, FL, pp. 91–95, 2000.

14. E. Keogh, C. Blake, and C. J. Merz. In *UCI Repository of Machine Learning Databases*, Irvine, 1998.

15. M. W. King. *Hemoglobin and myoglobin.* The Internet. Indiana State University, Terre Haute, http://www.indstate.edu/thcme/mwking/hemoglobin-myoglobin.html.

16. M. Kuramochi and G. Karypis. *An Efficient Algorithm for Discovering Frequent Subgraphs.* Technical Report 02-026, Department of Computer Science, University of Minnesota, Twin Cities, 2002.

17. P. Langley and S. Stromsten. *Learning context-free grammars with a simplicity bias. Proceedings of the Eleventh European Conference on Machine Learning*, pp. 220–228. Springer, Barcelona, 2000.

18. S. Muggleton. *Inductive logic programming: Derivations, successes and shortcomings. SIGART Bulletin*, 5:1, 1994.

19. M. Nagl, 1987. Set theoretic approaches to graph grammars. In H. Ehrig, M. Nagl, G. Rozenberg, and A. Rozenfeld, eds., Graph grammars and their application to Computer Science, 41–54.

20. C. G. Nevill-Manning and I. H. Witten. *Identifying hierarchical structure in sequences: A linear-time algorithm. Journal of Artificial Intelligence Research*, 7:67–82, 1997.

21. G. D. Plotkin. *Automatic methods of inductive inference.* Ph.D. thesis. Edinburgh University, 1971.

22. J. R. Quinlan. *Learning logical definitions from relations. Machine Learning*, 5: 239–266, 1990.

23. J. R. Quinlan. *C4.5: Programs for Machine Learning.* Morgan Kaufmann, San Francisco, 1993.

24. J. Rissanen. *Stochastic Complexity in Statistical Inquiry.* World Scientific, New Jersey, 1989.

25. S. Slattery and M. Craven. *Combining statistical and relational methods for learning in hypertext domains.* Proceedings of the Eighth International Conference on ILP, pp. 38–52, 1998.

26. A. Srinivasan, R. D. King, S. H. Muggleton, and M. J. E. Sternberg. *Carcinogenesis predictions using ILP*. Proceedings of the Seventh International Conference on Inductive Logic Programming, Amherst, MA, pp. 273–88; 1997.

27. M. Turcotte, S. H. Muggleton, and M. J. E. Sternberg. *Application of inductive logic programming to discover rules governing the three-dimensional topology of protein structure.* Proceedings of the Eighth International Conference on Inductive Logic Programming, Madison, Wisconsin, pp. 53–64, 1998.

28. X. Yan and J. Han. *gSpan: Graph-based substructure pattern mining*. Proceedings of the International Conference on Data Mining (ICDM), Maebashi City, Japan, 2002.

29. J. M. Zelle and R. J. Mooney. *Combining FOIL and EBG to speedup logic programs*. Proceedings of the Thirteenth International Joint Conference on Artificial Intelligence, Chambery, France, pp. 1106–1111, 1993.

9

CONSTRUCTING DECISION TREE BASED ON CHUNKINGLESS GRAPH-BASED INDUCTION

KOUZOU OHARA, PHU CHIEN NGUYEN, AKIRA MOGI, HIROSHI MOTODA, AND TAKASHI WASHIO

Institute of Scientific and Industrial Research, Osaka University, Osaka, Japan

9.1 INTRODUCTION

Over the last few years there has been much research work on data mining in seeking for better performance. Better performance includes mining from structured data, which is a new challenge. Since structure is represented by proper relations and a graph can easily represent relations, knowledge discovery from graph-structured data poses a general problem for mining from structured data.

On one hand, from this background, discovering frequent patterns of graph-structured data, that is, frequent subgraph mining or simply graph mining, has attracted much research interest in recent years because of its broad application areas such as bioinformatics [2, 11, 17], cheminformatics [13, 15, 27], and the like. Apriori-based Graph Mining (AGM) [13] and a number of other methods including Apriori-based connected Graph Mining (AcGM) [12], FSG ([15] and Chapter 6 of this book), graph-based substructure pattern mining (gSpan) ([27] and Chapter 5 of this book), Fast Frequent Subgraph Mining (FFSM) [10], and the like have been developed for the purpose of enumerating all frequent subgraphs of a graph database. However, the computation time increases exponentially with input graph

Mining Graph Data, Edited by Diane J. Cook and Lawrence B. Holder
Copyright © 2007 John Wiley & Sons, Inc.

size and minimum support. This is because the kernel of frequent subgraph mining is subgraph isomorphism, which is known to be NP-complete [6].

To avoid the complex subgraph isomorphism problem, heuristic algorithms, which are not guaranteed to find the complete set of frequent subgraphs, such as SUBDUE ([5] and Chapter 7 of this book) and graph-based induction (GBI) [28] have also been proposed. They tend to find an extremely small number of patterns based on greedy search. GBI extracts typical patterns from graph-structured data by recursively chunking two adjoining nodes. Later an improved version called beam-wise graph-based induction (B-GBI) [17] adopting the beam search was proposed to increase the search space, thus extracting more discriminative patterns while keeping the computational complexity within a tolerant level. Since the search in GBI is greedy and no backtracking is made, which patterns are extracted by GBI depends on which pairs are selected for chunking. This means that patterns that overlap each other have no longer been extracted and that there can be many patterns that are not extracted by GBI. B-GBI can help alleviate this problem but cannot solve it completely because the chunking process is still involved.

On the other hand, a majority of methods widely used for data mining are for data that do not have structure and that are represented by attribute-value pairs. Decision trees [21, 22] and induction rules [4, 18] relate attribute values to target classes. Association rules often used in data mining also uses this attribute-value pair representation. These methods can induce rules such that they are easy to understand. However, the attribute-value pair representation is not suitable to represent a more general data structure such as graph-structured data. This means that most of useful methods in data mining are not directly applicable to graph-structured data.

In the domain of inductive logic programming (ILP) [19], there are two systems that can construct a decision tree from structured data: Top-down Induction of Logical Decision trees (TILDE) [1] and Structural Classification and Regression Trees (S-CART) [14]. Although they were developed independently, they share the same theoretical framework and can construct a first-order logical decision tree, which is a binary tree where each node in the tree is associated with either a literal or a conjunction of literals. Namely, each node can represent a relational or structured data. They can utilize a substructure represented by one or more literals to generate a new node in a decision tree, but available structures are limited to those that are predefined.

In this chapter, for the purpose of constructing a decision tree for graph-structured data, we propose two algorithms: One is an algorithm to extract typical patterns from graph-structured data, called chunkingless graph-based induction (Cl-GBI), and the other is an algorithm to construct a decision tree from graph-structured data using Cl-GBI. Although Cl-GBI is an improved version of B-GBI, it does not employ the pairwise chunking strategy. Instead, the most frequent pairs are regarded as new nodes and given new node labels in the subsequent steps but none of them are chunked. In other words, they are used as pseudonodes, thus allowing extraction of overlapping subgraphs. We evaluate Cl-GBI on two datasets, the PTE dataset from the Predictive Toxicology Evaluation Challenge [20, 23] and the hepatitis dataset provided by Chiba University, and show that Cl-GBI can extract more typical substructures than B-GBI.

The decision tree construction algorithm, called decision tree chunkingless graph-based induction (DT-ClGBI), is a revised version of our previous algorithm called decision tree graph-based induction (DT-GBI) [7, 9] and can construct a decision tree for graph-structured data while simultaneously constructing substructures used as attributes for classification task by means of Cl-GBI instead of B-GBI adopted in DT-GBI. In this context, substructures means subgraphs or patterns that appear in a given graph database. Patterns extracted by Cl-GBI are regarded as attributes of graphs, and their existence/nonexistence is used as attribute values. Namely, in contrast to TILDE and S-CART, DT-ClGBI does not require the user to define available substructures in advance. Since attributes (features) are constructed while a classifier is being constructed, DT-ClGBI can be conceived as a method for feature construction. Using both synthetic and real-world datasets, we experimentally show DT-ClGBI can construct decision trees from graph-structured data that achieve reasonably good predictive accuracy.

This chapter is organized as follows: Section 9.2 briefly describes the framework of GBI. Section 9.3 points out the problem caused by the nature of chunking in GBI. Section 9.4 describes the details of Cl-GBI and reports the results of experimental evaluation. Section 9.5 explains DT-ClGBI and its working mechanism of how a decision tree is constructed using a simple example, and reports experimental results for both synthetic and real-world datasets. Section 9.6 concludes the chapter.

9.2 GRAPH-BASED INDUCTION REVISITED

9.2.1 Principle of GBI

Graph-based induction employs the idea of extracting typical patterns by stepwise pair expansion as shown in Figure 9.1. In the original GBI, an assumption is made that typical patterns represent some concepts/substructures, and "typicality" is characterized by the pattern's frequency or the value of some evaluation function of its frequency. We can use statistical indices as an evaluation function, such as frequency itself, information gain [21], gain ratio [22] and Gini index [3], all of which are based on frequency. In Figure 9.1 the shaded pattern consisting of nodes 1, 2, and 3 is thought typical because it occurs three times in the graph. GBI first finds

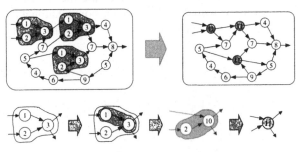

Figure 9.1. Principle of GBI.

the 1 → 3 pairs based on its frequency, chunks them into a new node 10, then in the next iteration finds the 2 → 10 pairs, chunking them into a new node 11. The resulting node represents the shaded pattern.

It is possible to extract typical patterns of various sizes by repeating the above three steps. Note that the search is greedy and no backtracking is made. This means that in enumerating pairs no pattern that has been chunked into one node is restored to the original pattern. Because of this, all the "typical patterns" that exist in the input graph are not necessarily extracted, and patterns that partially overlap are never generated, that is, any two patterns are either disjoint or perfect inclusion. The problem of extracting all the isomorphic subgraphs is known to be NP-complete. Thus, GBI aims at extracting only meaningful typical patterns of a certain size. Its objective is not finding all the typical patterns nor finding all the frequent patterns.

As described earlier, GBI can use any criterion that is based on the frequency of paired nodes. However, for finding a pattern that is of interest, any of its subpatterns must be of interest because of the nature of repeated chunking. In Figure 9.1 the pattern 1 → 3 must be typical for the pattern 2 → 10 to be typical. Said differently, unless pattern 1 → 3 is chunked, there is no way of finding the pattern 2 → 10. The frequency measure satisfies this monotonicity property. However, if the criterion chosen does not satisfy this monotonicity property, repeated chunking may not find good patterns even though the best pair based on the criterion is selected at each iteration. To resolve this issue GBI was improved to use two criteria, one based on frequency measures for chunking and the other for finding discriminative patterns after chunking. The latter criterion does not necessarily exhibit the monotonicity property. Any function that is discriminative can be used, such as information gain [21], gain ratio [22] and Gini index [3], and some others.

9.2.2 Beamwise Graph-Based Induction (B-GBI)

Since the search in GBI is greedy and no backtracking is made, which patterns (subgraphs) are extracted by GBI depends on which pair is selected for chunking. There can be many patterns that are not extracted by GBI. In Figure 9.2, if the pair B−C is selected for chunking beforehand, there is no way to extract the substructure A−B−D even if it is a typical pattern.

A beam search is incorporated into GBI in B-GBI [17] within the framework of greedy search in order to relax this problem, increase the search space, and extract more discriminative patterns while still keeping the computational complexity within a tolerant level. A certain number of pairs ranked from the top are selected to be

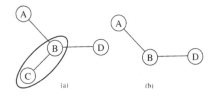

Figure 9.2.   Missing patterns due to chunking order.

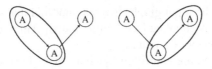

Figure 9.3. Two different pairs representing identical patterns.

chunked individually. To prevent each branch from growing exponentially, the total number of pairs to be chunked (the beam width) is fixed at every time of chunking. Thus, at any iteration step, there are always a fixed number of chunking steps performed in parallel.

Another improvement made in conjunction with B-GBI is canonical labeling. GBI assigns a new label to each newly chunked pair. Because it recursively chunks pairs, it happens that the new pairs that have different labels are the same pattern. A simple example is shown in Figure 9.3. They represent the same pattern but the ways they are constructed are different. To identify if two pairs represent the same pattern, each pair is represented by its canonical label [6], and they are considered to be identical only when the labels are the same.

9.3 PROBLEM CAUSED BY CHUNKING IN B-GBI

As described in Section 9.2.2, B-GBI increases the search space by running GBI in parallel. As a result, B-GBI can help alleviate the problem of overlapping subgraphs but cannot solve it completely because the chunking process is still involved. It happens that some of the overlapping patterns are not discovered by B-GBI. For example, suppose in Figure 9.2 the pair B–C is most frequent, followed by the pair A–B. When $b = 1$, there is no way that the pattern A–B–D is discovered because the pair B–C is chunked first, but by setting $b = 2$, the pair A–B can be chunked in the second beam and if the substructure A–B–D is frequent enough, there is a chance that the pair (A–B)–D is chunked at next iteration. However, setting b very large is prohibitive from a computational point of view.

Any subgraph that B-GBI can find is along the way in the chunking process. Thus, it happens that a pattern found in one input graph is unable to be found in the other input graph even if it does exist in the graph. An example is shown in Figure 9.4, where even if the pair A–B is selected for chunking and the substructure

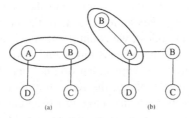

Figure 9.4. Pattern is found in one input graph but not in the other.

D–A–B–C exists in the input graphs, we may not find that substructure because an unexpected pair A–B is chunked [see Figure 9.4(b)]. This causes a serious problem in counting the frequency of a pattern.

Complete graph mining algorithms such as AGM [13], AcGM [12], FSG ([15] and Chapter 6 of this book), gSpan ([27] and Chapter 5 of this book), FFSM [10], and the like do not face the problem of overlapping subgraphs since they can find all frequent patterns in a graph database. However, these methods are designed to find existence or nonexistence of a certain pattern in one transaction and not to count how many times a certain pattern appears in one transaction. They also cannot give information on the positions of each pattern in any transaction of the graph database, which is required by domain experts.

Heuristic algorithms for graph mining such as SUBDUE ([5] and Chapter 7 of this book), GBI [28] and GREW [16], on the other hand, are designed for the purpose of enumerating typical patterns in a single large graph. Specially, B-GBI [17] can find (not all) typical patterns in either a single large graph or a graph database. However, all of them are not designed to detect the positions of patterns in any graph transaction. In Section 9.4, we will introduce a novel algorithm that can overcome the problem of overlapping subgraphs incurred by both GBI and B-GBI. The proposed algorithm, called Cl-GBI (chunkingless graph-based induction), employs a "chunkingless chunking" strategy, where frequent pairs are never chunked but used as pseudonodes in the subsequent steps, thus allowing extraction of overlapping subgraphs. It can also give the positions of patterns present in each graph transaction as well as find frequent patterns in either a single large graph or a graph database.

9.4 CHUNKINGLESS GRAPH-BASED INDUCTION (CL-GBI)

9.4.1 Approach

The basic ideas of Cl-GBI are described as follows. Those pairs that connect two adjoining nodes in the graphs are counted for their frequencies and a certain fixed number of pairs (the beam width) ranked from the top are selected. In B-GBI, each of the selected pairs is registered as one node and this node is assigned a new label. Then, the graphs in the respective state are rewritten by replacing all the occurrences of the selected pair with a node with the newly assigned label (pairwise chunking).

In Cl-GBI, we also register the selected pairs as new nodes and assign new labels to them. But those pairs are never chunked and the graphs are not "compressed" nor copied into respective states as in B-GBI. In the presence of the pseudonodes (i.e., newly assigned label nodes), we count the frequencies of pairs consisting of at least one new pseudonode. The other is either one of pseudonodes including those already created in the previous steps or an original one. In other words, the other is one of the existing nodes. Among the remaining pairs (after selecting the most frequent pairs) and the new pairs that have just been counted for their frequencies, we select the most frequent pairs with the number equal to the beam width specified in advance.

These steps are repeated for a predetermined number of times, each of which is referred to as a level. Those pairs that satisfy a typicality criterion (e.g., pairs whose information gain exceeds a given threshold) among all the extracted pairs are the output of the algorithm.

A frequency threshold is used to reduce the number of pairs being considered to be typical patterns. Another possible method to reduce the number of pairs is to eliminate those pairs whose typicality measure is low even if their frequency count is above the frequency threshold. The two parameters, beam width and number of levels, control the search space. The frequency threshold is another important parameter.

As in B-GBI, the Cl-GBI approach can handle both directed and undirected graphs as well as both general and induced subgraphs. It can also extract typical patterns in either a single large graph or a graph database and can use both document and total frequency optionally. Document frequency of a pattern is the number of graphs that contain the pattern or its ratio to the total number of graphs in the graph database, while total frequency is the total number of occurrences of the pattern in the graph database. Usually, we use document frequency for extracting typical patterns from a graph database, and total frequency for enumerating typical patterns in a single graph.

9.4.2 Algorithm of Cl-GBI

Given a graph database, two natural numbers b (beam width) and N_e (number of levels), and a frequency threshold θ, the new "chunkingless chunking" strategy repeats the following three steps.

Step 1   Extract all the pairs consisting of two connected nodes in the graphs, register their positions using node id (identifier) sets, and count their frequencies. From the second level on, extract all the pairs consisting of two connected nodes with at least one node being a new pseudonode.

Step 2   Select the b most frequent pairs from among the pairs extracted at step 1 (from the second level on, from among the unselected pairs in the previous levels and the newly extracted pairs). Each of the b selected pairs is registered as a new node. If either or both nodes of the selected pair are not original but pseudonodes, they are restored to the original patterns before registration.

Step 3   Assign a new label to each pair selected at step 2 but do not rewrite the graphs. Go back to step 1.

These steps are repeated N_e times (N_e levels). All the pairs extracted at step 1 in all the levels (i.e., level 1 to level N_e), including those that are not used as pseudonodes, are ranked based on a typicality criterion using a discriminative function such as information gain, gain ratio, or Gini index. Note that those pairs that have frequency count below a frequency threshold θ are eliminated, which means that there are three parameters b, N_e, and θ to control the search in Cl-GBI.

The output of Cl-GBI algorithm is a set of ranked typical patterns, each of which comes together with the positions of every occurrence of the pattern in each transaction of the graph database (given by the node id sets) as well as the number of occurrences.

9.4.3   Implementation Issues of Cl-GBI

The first issue concerns frequency counting. To count the number of occurrences of a pattern in a graph transaction, the canonical labeling employed in [17] is adopted. However, canonical labeling alone cannot solve the problem completely as shown in Figure 9.5. Suppose that the pair A → B is registered as a pseudonode N in the graph shown in Figure 9.5(a). How many times should the pair N → B be counted here? If only the canonical label is considered, the answer is 2 because there are two pseudonodes N_1 and N_2 as shown in Figure 9.5(b), and both N_1 → B and N_2 → B are counted separately. However, the pair N → B should be counted once. Our solution is to incorporate the canonical label with the node id set. If both the canonical label and the node id set are identical for two subgraphs, we regard that they are the same and count once.

The second issue regards the relations between a pseudonode and those nodes that are embedded inside. Think of the pseudonode C and the two embedded nodes A, B in Figure 9.6(a). What are the relations between C and A or C and B? In the case of enumerating frequent induced subgraphs, there is not any relation between C and A nor C and B. This is because a pair in this case must consist of two nodes and all links between them. However, in the case of extracting frequent general subgraphs, there is still a link between C and A as well as a link between C and B. To differentiate between the graphs shown in Figures 9.6(a) and 9.6(b), a flag indicating whether it is a self-loop or not is required. In summary, a pair consists of six elements: labels of two nodes and label of the link between them, information of which two nodes inside the pair the link is connected to, and a self-loop flag. In the case of enumerating frequent induced subgraphs, all links between two nodes should be considered.

The third issue involves the problem of how to reduce the computation cost. To avoid multiple extractions of the same pattern, we only consider those pairs that connect two disjoint nodes/pseudonodes. This is because for any pattern consisting of two overlapping nodes/pseudonodes, there exists a pair of disjoint nodes/pseudonodes that can construct that pattern. This simple strategy helps reduce the number

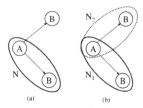

Figure 9.5.   Example of frequency counting.

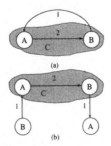

Figure 9.6. Relations between a pseudonode and its embedded nodes.

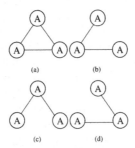

Figure 9.7. Three occurrences but counted once.

of pairs to be extracted significantly, and the computation cost is thus reduced accordingly.

9.4.4 Unsolved Problem of Cl-GBI

We found that there is still a problem in frequency counting that the use of both the canonical label and the node id set cannot solve. Think of the graph in Figure 9.7(a). The three subgraphs A−A−A illustrated in Figures 9.7(b), 9.7(c), and 9.7(d) share the same canonical label and the same node id set. Our current Cl-GBI cannot distinguish between these three. However, this problem arises only when extracting general subgraphs. It causes no problem in the case of enumerating frequent induced subgraphs.

9.4.5 Experimental Evaluation of Cl-GBI

To assess the performance of the Cl-GBI approach, we conducted some experiments on both synthetic and real-world graph-structured datasets. The proposed Cl-GBI algorithm was implemented in C^{++}. It should be noted that all graphs/subgraphs reported here are connected ones.

In the first experiment, we verify that Cl-GBI is capable of finding all frequent patterns in a single graph that other graph mining algorithms cannot. An example of such a single graph is shown in Figure 9.8(a). Suppose that the problem here is to find frequent induced subgraphs that occur at least three times in the graph.

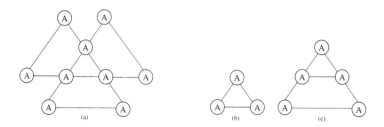

Figure 9.8.   Example of finding frequent patterns in a single graph.

Figure 9.8(c) shows an example of frequent induced subgraph which occurs three times in the graph (support = 3).

The current algorithms that are designed for extracting frequent patterns in a single graph such as GBI [28], B-GBI [17], SUBDUE ([5] and Chapter 7 of this book), or GREW [16] cannot discover the pattern shown in Figure 9.8(c) because three occurrences of this pattern are not disjoint but overlapping. The complete graph mining algorithms like AcGM [12], FSG ([15] and Chapter 6 of this book), gSpan ([27] and Chapter 5 of this book), FFSM [10], and so forth in the case that they are adapted to find frequent patterns in a single graph, also cannot find that pattern because of the antimonotonicity property they use to prune the search. Since the pattern shown in Figure 9.8(b) occurs only once in the graph and thus cannot be extracted, the pattern shown in Figure 9.8(c), which is one of its supergraphs, is also unable to be found. The proposed Cl-GBI algorithm, on the other hand, can find all 36 frequent induced subgraphs, including the one shown in Figure 9.8(c), with $b = 3$, $N_e = 5$.

In the second experiment, we show that Cl-GBI can extract all the frequent patterns of a graph database. This experiment was conducted on the PTE dataset [20], which contains the molecular structure data of carcinogenic compounds. This dataset was originally provided for the Predictive Toxicology Evaluation Challenge [23], and it contains information on 340 chemical compounds. There are 4 different types of bonds (1, 2, 3 and 7) and 24 different atoms (As, Ba, Br, C, Ca, Cl, Cu, F, H, Hg, I, K, Mn, N, Na, O, P, Pb, S, Se, Sn, Te, Ti and Zn). In addition, the atoms take some different states, and thus the total number of atom types is 66.

Each compound was converted to an undirected graph as in [15]. Thus, there are 340 graph transactions in total. Each node corresponds to an atom, whose label is made of a pair of the atom element and the atom type. Each link is placed for every bond. Link label directly corresponds to the bond type. By the conversion, there are 66 node labels and 4 link labels produced in total. The average size of this graph database is 27.0 in terms of the number of nodes, and 27.4 in terms of the number of links. The largest graph transaction contains 214 nodes and 214 links.

Tables 9.1 and 9.2 show the number of levels N_e needed to enumerate all frequent general and induced subgraphs, respectively, of the PTE dataset given various values of θ and the same beam width of 10. The number of frequent patterns extracted from the PTE dataset, for both cases, is verified by the AcGM [12] algorithm, which can find all frequent patterns in a graph database. As is easily predicted,

TABLE 9.1   Extracting Frequent General Subgraphs from PTE Dataset Using Cl-GBI

Frequency Threshold (θ)	30%	20%	10%
Number of frequent patterns	68	190	844
Beam width (b)	10	10	10
Number of levels (N_e) needed	12	18	84

TABLE 9.2   Extracting Frequent Induced Subgraphs from PTE Dataset Using Cl-GBI

Frequency Threshold (θ)	30%	20%	10%
Number of frequent patterns	49	139	537
Beam width (b)	10	10	10
Number of levels (N_e) needed	4	7	18

this Cl-GBI algorithm can find all the frequent patterns by setting b and N_e large enough.

In the third experiment we compare the performance of the Cl-GBI and B-GBI [17] algorithms with respect to the number of frequent induced subgraphs extracted from the hepatitis dataset. This dataset, which was provided by Chiba University, contains long time-series data (from 1982 to 2001) on laboratory examination of 771 patients of hepatitis B and C. In this experiment, we used the information on those patients as to whether the interferon therapy was effective or not in case they underwent the therapy. Information on each patient was converted to a directed graph after some preprocesses such as cleansing, averaging, and discretizing resulting values of examinations, and the obtained graph database has 142 graph transactions. The average size of this graph database is 111.6 in terms of the number of nodes, and 116.5 in terms of the number of links. The details of this data preparation can be found in [8].

We set $\theta = 20\%$ and $b = 5$. Table 9.3 shows some experimental results obtained from the hepatitis dataset. It is shown that Cl-GBI can find much more frequent patterns than B-GBI given the same beam width. It should be noted that one of the nice aspects of B-GBI is that the size of the input graph keeps reducing progressively as the chunking proceeds, and thus the number of pairs to be considered also progressively decreases accordingly. In the case of Cl-GBI, the number of pairs to be considered keeps increasing because the number of pseudonodes keeps increasing as the search proceeds. Thus, it is important to select appropriate values for b and N_e.

TABLE 9.3   Number of Frequent Induced
Subgraphs Extracted from Hepatitis Dataset Using
Cl-GBI by Setting $\theta = 20\%$ and $b = 5$

Algorithm	Number of Levels N_e	Number of Frequent Patterns
B-GBI	N/A	2186
Cl-GBI	5	3147
Cl-GBI	10	12489

9.5   DECISION TREE CHUNKINGLESS GRAPH-BASED INDUCTION (DT-CLGBI)

9.5.1   Decision Tree for Graph-Structured Data

As mentioned in Section 9.1, the attribute–value pair representation is not suitable
for graph-structured data, although both attributes and their values are essential for
a classification or prediction task because a class is related to some attribute values
in most cases. In a decision tree, each node and a branch connecting the node to its
child node correspond to an attribute and one of its attribute values, respectively.
Thus, to formulate the construction of a decision tree for graph-structured data, we
define attributes and their values as follows:

- Attribute: a pattern/subgraph in graph-structured data
- Value of an attribute: existence/nonexistence of the pattern in each graph

Since the value of an attribute is either yes (the pattern corresponding to the attribute
exists in the graph) or no (the pattern does not exist), the resulting decision tree
is represented as a binary tree. Namely, data (graphs) are divided into two groups:
one consists of graphs with the pattern, and the other consists of graphs without it. Figure 9.9 illustrates the decision tree constructed based on this approach.
One remaining question is how to determine patterns that are used as attributes

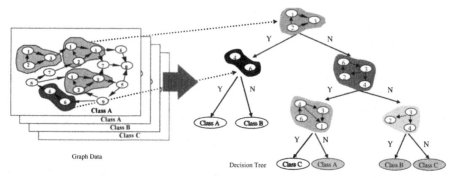

Figure 9.9.   Decision tree for classifying graph-structured data.

for graph-structured data. Our approach to this question is described in the next subsection.

9.5.2   Feature Construction by Cl-GBI

The algorithm we propose here, called decision tree chunkingless graph-based induction (DT-ClGBI), utilizes Cl-GBI to extract patterns from graph-structured data and uses them as attributes for a classification task, whereas our previous algorithm, decision tree graph-based induction (DT-GBI), adopted B-GBI to extract patterns. Namely, DT-ClGBI invokes Cl-GBI at each node of a decision tree and selects the most discriminative pattern from those which were extracted by Cl-GBI. Then the data (graphs) are divided into two groups, that is, one with the pattern and the other without the pattern as described above. For each group, the same process is recursively applied until the group contains graphs of a single class like the ordinary decision tree construction method such as C4.5 [22]. The algorithm of DT-ClGBI is summarized in Figure 9.10

In DT-ClGBI, each of the parameters of Cl-GBI, b, N_e, and θ, can be set to different values at different nodes in a decision tree. All patterns extracted at a node are inherited to its descendant nodes to prevent a pattern that has already been extracted in the node from being extracted again in its descendants. This means that, as the construction of a decision tree progresses, the number of patterns to be considered at a node progressively increases, and the size of a pattern newly extracted can be larger than existing patterns. Thus, although initial patterns at the

```
DT-ClGBI(D)
    INPUT
    D: a graph database
    begin
        Create a node DT for D
        if termination condition reached
            return DT
        else
            P := Cl-GBI(D)    (with b, Ne, and θ specified)
            Select the most discriminative pattern p from P
            Divide D into Dy (with p) and Dn (without p)
            for Di := Dy, Dn
                DTi := DT-ClGBI(Di)
                Augment DT by attaching DTi as its child
                along yes (no) branch
        return DT
    end
```

Figure 9.10. Algorithm of DT-ClGBI.

start of search consist of two nodes and the link between them, attributes useful for the classification task can be gradually grown up into larger patterns (subgraphs) by applying Cl-GBI recursively. In this sense, DT-ClGBI can be conceived as a method for feature construction, since features, that is, attributes (patterns) useful for the classification task, are constructed during the application of DT-ClGBI. Note that it is possible to extract all the patterns at the root node using all the data by setting N_e large enough and inherit them to the lower nodes. Patterns that are to be used at lower nodes should give smaller discriminative power at the root node, but Cl-GBI retains them and passes down to the lower node. The question is how to find this pattern where it is needed without running Cl-GBI using all the dataset. Recursive calling of Cl-GBI with inheritance facilitates this.

However, recursive partitioning of data until each subset in the partition contains data of a single class often results in overfitting to the training data and thus degrades the predictive accuracy of resulting decision trees. To avoid overfitting, and improve predictive accuracy, DT-ClGBI incorporates "pessimistic pruning" used in C4.5 [22] that prunes an overfitted tree based on the confidence interval for binomial distribution. This pruning is a postprocess that follows the algorithm in Figure 9.10. DT-ClGBI is also implemented in C^{++}.

Note that the criterion for selecting a pair that becomes a pseudonode in Cl-GBI and the criterion for selecting a discriminative pattern in DT-ClGBI can be different. In the following, document frequency of a pair is used as the former criterion, and information gain of a pattern is used as the latter criterion.[1]

9.5.3 Working Example of DT-ClGBI

Suppose DT-ClGBI receives a set of 4 graphs in the upper left-hand side of Figure 9.11. Both the beam width b and the number of levels N_e of Cl-GBI are set to 1 at every node of a decision tree to simplify the working of DT-ClGBI in this example, and the frequency threshold θ is set to 0%. Cl-GBI called inside of DT-ClGBI enumerates all the pairs in these graphs and extracts 11 kinds of pairs from the data. These pairs are: a→a, a→b, a→c, a→d, b→a, b→b, b→c, b→d, d→a, d→b, d→c. The existence/nonexistence of the pairs in each graph is converted into the ordinary table representation of attribute–value pairs, as shown in Figure 9.12. For instance, graph 1, graph 2, and graph 3 have the pair a→a but graph 4 does not have it. This is shown in the first column in Figure 9.12.

Then Cl-GBI selects the most frequent pair "a→a," assigns new label "e" to it to generate a pseudonode as shown in the upper right-hand side of Figure 9.11, and terminates. Next, DT-ClGBI selects the most discriminative pattern, that is, the pattern (pair) with the highest evaluation for classification (i.e., information gain) from the enumerated pairs, and uses it to divide the data into two groups at the root node. In this example, the pair "a→a" is selected. As a result, the input data is divided into two groups: one consisting of graph 1, graph 2, and graph 3 and the other consisting of only graph 4.

---

[1]We did not use information gain ratio because DT-ClGBI constructs a binary tree.

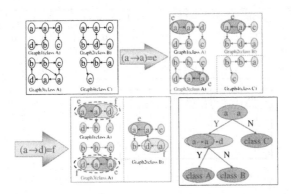

Figure 9.11. Example of decision tree construction by DT-ClGBI.

Graph	a→a	a→b	a→c	a→d	b→a	b→b	b→c	b→d	d→a	d→b	d→c
1 (class A)	1	1	0	1	0	0	0	1	0	0	1
2 (class B)	1	1	1	1	0	0	0	0	1	1	0
3 (class A)	1	0	1	1	1	1	1	0	0	1	0
4 (class C)	0	1	0	0	0	1	1	0	0	0	0

Figure 9.12. Attribute–value pairs at the first step.

The above process is applied recursively at each node to grow up the decision tree while constructing attributes (patterns) useful for the classification task at the same time. In this example, since the former group consists of graphs belonging to different classes, again Cl-GBI is applied to it, while the latter group is no longer divided because it contains a single graph of class C. For the former group, pairs in graph 1, graph 2, and graph 3 are enumerated and the attribute–value table is updated as shown in Figure 9.13. Note that the pairs included in the table for the parent node are inherited. In this case, the pair "a→d" is selected by Cl-GBI as the most frequent pair to be a pseudonode "f," while the pair "e→d" is selected as the most discriminative pattern by DT-ClGBI. Consequently, the graphs are separated into two partitions, each of which contains graphs of a single class as shown in the lower left-hand side of Figure 9.11. The constructed decision tree is shown in the lower right-hand side of Figure 9.11.

9.5.4 Classification Using the Constructed Decision Tree

Unseen new graph data must be classified once the decision tree has been constructed. Here again, the problem of subgraph isomorphism arises to test if the input graph contains the pattern (subgraph) specified in the test node of the tree. To alleviate this problem, we utilize Cl-GBI again. Theoretically, if the test pattern actually exists in the input graph, Cl-GBI can find it by setting the beam width b and the number of levels N_e large enough and by setting the frequency threshold to

Graph	a→b	a→c	a→d	b→a	b→b	b→c	b→d	d→a	d→b	d→c
1 (class A)	1	0	1	0	0	0	1	0	0	1
2 (class B)	1	1	1	0	0	0	0	1	1	0
3 (class A)	0	1	1	1	1	1	0	0	1	0

e→b	e→c	e→d	b→e	d→e
1	0	1	0	0
1	1	0	0	1
0	1	1	1	0

Figure 9.13. Attribute–value pairs at the second step.

0. However, note that nodes and links that never appear in the test pattern are never used to form the test pattern in Cl-GBI. Therefore, we can remove such nodes and links from the input graph before applying Cl-GBI to reduce its running time. This approach is summarized as follows:

Step 1   Remove nodes and links that never appear in the test pattern from the input graph.

Step 2   Apply Cl-GBI to the resulting input graph setting the parameters b and N_e large enough, while setting the parameter θ to 0.

Step 3   Test if one of the canonical labels of extracted patterns with the same size as the test pattern is equal to the canonical label of the test pattern.

In general, step 1 results in a small graph and Cl-GBI can run very quickly without any constraints on N_e and b. However, if we need to set these constraints, we may not be able to obtain the correct answer because we do not know how large these parameters should be. In that sense, this procedure can be regarded as an approximate solution to the subgraph isomorphism problem.

9.5.5   Experimental Evaluation of DT-ClGBI

To evaluate the performance of DT-ClGBI, we conducted some experiments on both synthetic and real-world datasets consisting of connected graphs. As for the synthetic datasets, we randomly generated three kinds of datasets having the average size of 30, 40, and 50, respectively, in terms of the number of nodes in a graph. We denote them by T30, T40, and T50, respectively. Every dataset has 300 connected graphs as transactions. The links were attached randomly with the probability of 20%. The number of node and link labels are 5 and 10, respectively, and each label was randomly assigned with equal probability. Then we equally divided each dataset into two classes, namely "active" and "inactive," and randomly embedded similarly generated 4 kinds of basic patterns in each transaction of the class "active," which are connected subgraphs having the average size of 4. The number of basic patterns to be embedded in a transaction of the class "active" was randomly selected in the

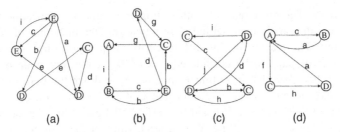

Figure 9.14.   Basic patterns.

range between 1 and 4, each of which was in turn chosen from the set of 4 basic patterns by equal probability. Thus each transaction of the class "active" includes from 1 to 4 basic patterns, and some of them may happen to be the same. The basic patterns embedded are shown in Figure 9.14.

We also check if there is any basic pattern included in a graph of the class "inactive" by Cl-GBI as described in Section 9.5.4. If there is, the corresponding node and link labels are changed in a way that the basic pattern no longer exists in the transaction. In other words, basic patterns are those that help discriminate the two classes.

In the first experiment, to confirm that DT-ClGBI can achieve good predictive accuracy on the synthetic datasets, we applied DT-ClGBI to the three datasets with two different settings, limiting the total number of levels of Cl-GBI to 8 to save computation time. This limit was determined based on our estimation that running Cl-GBI with $N_e = 8$ could be enough to extract the basic patterns. In the first setting, that is, setting 1, Cl-GBI was invoked only at the root node with $N_e = 8$. At the other nodes, we simply recalculated information gain for those patterns that have already been extracted at the root node. In the second setting, that is, setting 2, Cl-GBI was invoked at the root node with $N_e = 4$ and at every other node with the half value of N_e at its parent node. The minimal value of N_e retains 1 until the limitation of the total number of levels is violated, which was always the case in this experiment. In all the experiments for the synthetic datasets, the parameters b and θ are fixed to 5 and 10%, respectively.

Table 9.4 summarizes the error rate and computation time for both settings. Each value in this table is the average obtained by 10-fold cross validation, and the computation times are wall-clock runtimes on PC with Athlon MP 1900+ and

TABLE 9.4   Comparisons of Different Settings for DT-ClGBI in Average Error Rate for Test Sets and Computation Time of 10-fold Cross Validation

	Setting 1		Setting 2	
Datasets	Test Error (%)	Computation Time (s)	Test Error (%)	Computation Time (s)
T30	0	2,063	0	913
T40	0.67	12,262	0.67	5,003
T50	0.67	44,500	1.00	19,957

3 GB memory. The error rates of both settings are very low and almost the same, but the computation time of setting 2 is much smaller than that of setting 1 on every dataset. The reason for this difference in computation time is that in setting 2 the number of pairs to be considered in Cl-GBI at the root node of a decision tree is much less than that in setting 1, and as we go down the tree, some branches terminate and the total number of data that Cl-GBI has to work on becomes less and less. From these results, it is said that DT-ClGBI can achieve good accuracy on the synthetic datasets, but setting N_e to a large value at the root node of a decision tree is not a good approach from the viewpoint of computation time.

Figures 9.15 and 9.16 show the decision trees constructed at the final cycle of the 10-fold cross validation on the dataset T50 with setting 1 and 2, respectively. In this cycle, the tree constructed with setting 1 was more accurate than the tree constructed with setting 2, but the difference between them was not significant. From this figure, it is found that most test patterns of the decision trees are subgraphs of the basic patterns. In addition, we observed that they were extracted at the root node even in the case of setting 2. For example, the fourth test pattern in Figure 9.16, that is, pattern 530485 is a subgraph of the basic pattern (a) and was extracted at

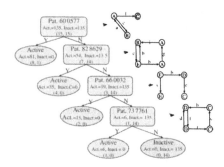

Figure 9.15. Example of decision trees in the case of the setting 1.

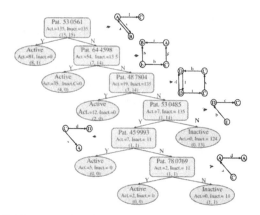

Figure 9.16. Example of decision trees in the case of the setting 2.

TABLE 9.5   Comparisons of Different Settings for DT-ClGBI in Average Error Rate for Test Sets and Average Size of Resulting Decision Trees of 10-fold Cross Validation

	Setting 3		Setting 4	
Datasets	Test Error (%)	Tree Size	Test Error	Tree Size
T30	1.33	17.2	0	9.2
T40	5	18.0	3.33	12.8
T50	9.67	26.8	7.67	22.8

the root node of the tree. This means that such subgraphs of the basic patterns are discriminative enough to classify the synthetic datasets we generated and $N_e = 4$ is enough to extract them at the root node. However, it is noted that generally it is impossible to know an appropriate value for N_e at the root node in advance.

The results in the above experiment may raise a question: Can DT-ClGBI really find discriminative patterns at intermediate nodes even if it cannot find them at the root node of a decision tree? To answer this question, in the next experiment, we applied DT-ClGBI to the same synthetic datasets with two other different settings, limiting N_e to 2 at the root node of a decision tree, and compared resulting decision trees in the average error rate and size obtained by 10-fold cross validation. The limit of N_e prohibits many discriminative subgraphs of the basic patterns from being extracted at the root node. We took this method for this experiment because it is very difficult to embed patterns in a transaction ensuring that their subgraphs are not discriminative enough. In the first setting, that is, setting 3, Cl-GBI was only invoked at the root node with $N_e = 2$ similarly to setting 1. In the second setting, that is, setting 4, Cl-GBI was invoked at the root node with $N_e = 2$ and at every other node with $N_e = 1$. In addition, the total number of levels of Cl-GBI in setting 4 was limited to 6.

Table 9.5 shows the average error rate and the average size of decision trees constructed by 10-fold cross validation of DT-ClGBI with both settings. We can find that for every dataset the error rate with setting 4 is better than that with setting 3, and the resulting decision trees with setting 4 are much smaller than those with setting 3. This means that patterns extracted at intermediate nodes of a decision tree with setting 4 are very effective for improving the predictive accuracy and for simplifying the resulting decision tree. Thus, it is said that in DT-ClGBI calling Cl-GBI at intermediate nodes is effective for extracting discriminative patterns that are useful for classification. This also contributes to reducing computation time as is the case for settings 1 and 2 (see Table 9.4).

Finally, we verified DT-ClGBI can construct decision trees that achieve reasonably good predictive accuracy also on a real-world dataset. For that purpose, we used the hepatitis dataset used in Section 9.4.5 again. The classification task here is to classify patients into two classes, LC (liver cirrhosis) and non-LC (non-liver cirrhosis) based on their fibrosis stages, which are categorized into five stages in the dataset: F0 (normal), F1, F2, F3, and F4 (severe = liver cirrhosis). All 43 patients at F4 stage were used as the class LC, while all 4 patients at F0 stage and 61 patients at F1 stage were used as the class non-LC. This ratio of LC to non-LC was determined based on [26]. The records for each patient were converted into a directed graph as mentioned in Section 9.4.5. The resulting graph database has 108 graph transactions. The average size of a graph transaction in this database is 316.2 in terms of the number of nodes and 386.4 in terms of the number of links.

Through some preliminary experiments on this database using DT-ClGBI, we found that existence of some graphs often makes the resulting decision tree too complicated and worsen the predictive accuracy. This has led us to adopt a two-step approach, first to divide the patients into "typical" and "nontypical," and second to construct a decision tree for each group of the patients. To divide the patients in the first step, we ran 10-fold cross validation of DT-ClGBI on this database, varying its parameters b and N_e in the ranges of {5, 6, 8, 10} and {6, 8, 10, 12}, respectively. The frequency threshold θ was fixed to 10%. Namely, we conducted 10-fold cross validation 16 times with different combinations of these parameters and obtained totally 160 decision trees in this step. Note that we ran Cl-GBI only at the root node of a decision tree and used those patterns that have been extracted at the root node for the succeeding nodes by recalculating their information gain to save computation time. Then we classified graphs whose average error rate is 0% into typical and the others into nontypical. As a result, for the class LC, 28 graphs are classified into the subset typical and the other 15 graphs into nontypical, while for the class non-LC, 48 graphs are classified into typical and 17 graphs into nontypical.

In the second step, we applied DT-ClGBI to each subset again adopting the best parameter setting in the first step with respect to the predictive accuracy, where $b = 8$ and $N_e = 10$. Also in this step, we ran Cl-GBI only at the root node. The contingency tables for test sets of the 10-fold cross validation are reported in Table 9.6, and examples of decision trees constructed for each subset in this step and patterns that appear in them are shown in Figures 9.17 and 9.18.

The predictive accuracy (average of 10-fold cross validation) for the subset "typical" is 97.4%, and that for "nontypical" is 78.1%. And then the overall accuracy is 91.7%, which is much better than the accuracy obtained by applying DT-ClGBI to the original dataset with $b = 8$ and $N_e = 10$, that is, 83.4%. The resulting decision trees in Figures 9.17 and 9.18 are very simple and thus have good interpretability. We can find typical features for a patient with liver cirrhosis in the extracted patterns such as "GOT is High" or "PLT is Low." From these results, we can say that DT-ClGBI can achieve reasonably good predictive accuracy on a real-world dataset and extract discriminative features embedded in the dataset as a subpattern.

TABLE 9.6   Contingency Table with the Number of Instances

Actual Class	Predicted Class (LC = F4, non-LC = {F0 + F1})					
	for "Typical"		for "Nontypical"		Overall	
	LC	non-LC	LC	non-LC	LC	non-LC
LC	27	1	12	3	39	4
non-LC	1	47	4	13	5	60

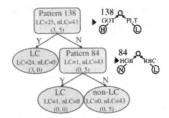

Figure 9.17. Example of decision trees constructed in the second step for the class "typical."

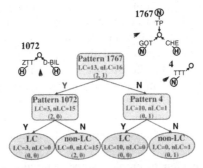

Figure 9.18. Example of decision trees constructed in the second step for the class "nontypical."

The underlying assumption of our problem setting of graph data classification is the existence of discriminative substructures in the graph. There are other attempts to construct features without using the graph mining approach. One successful approach is the TFS (topological fragment spectra) method [24] that is used in the domain of drug molecule classification. The TFS method represents a molecule (graph transaction) by a multidimensional vector called TFS that is obtained by (1) enumerating all the possible fragments (paths including branches) smaller than a specified size and (2) quantifying each fragment by a numerical score such as the total mass numbers of atoms contained in the fragment. As a result, a TFS forms a spectrum of frequencies of these scores for a molecule. Thus, it is expected that if the spectrums of two molecules are similar, these molecules are similar. We can regard that each numeric score is an attribute and its frequency is its value. It is reported that with this representation a support vector machine (SVM) achieved good classification accuracy [25]. However, the problem of the TFS approach is its interpretability. Different fragments can result in the same numeric score. Further, SVM does not give insight into the discriminative substructures. Thus, we did not attempt to compare DT-ClGBI with the TFS method. Another successful approach is to enumerate linear fragments of a molecule and use them as its attributes and their existence/nonexistence as their values (see Chapter 14 of this book). This may not work for general graphs. Since a molecule can be regarded as a single connected graph and a linear fragment as one of its subgraphs, it is obvious that this feature construction method can be covered by DT-ClGBI.

9.6 CONCLUSIONS

In this chapter, we proposed a novel algorithm called Cl-GBI, which can discover typical patterns in either a single large graph or a graph database, and also an algorithm called DT-ClGBI, which can construct a decision tree for graph-structured data using Cl-GBI. Cl-GBI employs a "chunkingless chunking" strategy that helps overcome the problem of overlapping subgraphs. Experiments conducted on both synthetic and real-world graph-structured data confirmed its effectiveness. Also, Cl-GBI can give the correct number of occurrences of a pattern as well as their positions in each transaction of the graph database, which are very useful for algorithms such as DT-ClGBI that need correct counting. In DT-ClGBI, substructures, or patterns useful for a classification task are constructed on the fly by means of Cl-GBI during the construction process of a decision tree. The experimental results using synthetic and real-world datasets showed DT-ClGBI can construct decision trees that achieve good predictive accuracy for graph-structured data.

As future work, we plan to employ some heuristics to speed up the Cl-GBI algorithm in order to extract larger typical subgraphs at an early stage in the search process. This could also improve the performance of DT-ClGBI. Construction of a decision tree to divide a graph database into two groups in the two-step approach we took for the hepatitis dataset should also be considered. Moreover, it is necessary to experimentally compare DT-ClGBI with other methods, especially with ILP-based ones such as TILDE and S-CART.

REFERENCES

1. H. Blockeel and L. De Raedt. Top-down induction of first-order logical decision tree. *Artificial Intelligence*, 101:285–297, 1998.
2. C. Borgelt and M. R. Berthold. Mining molecular fragments: Finding relevant substructures of molecules. In Proceedings of the 2nd the IEEE International Conference on Data Mining (ICDM 2002), Maebashi City, Japan, pp. 51–58, 2002.
3. L. Breiman, J. H. Friedman, R. A. Olshen, and C. J. Stone. *Classification and Regression Trees*. Wadsworth & Brooks/Cole Advanced Books & Software, Pacific Grove, California, 1984.
4. P. Clark and T. Niblett. The CN2 Induction Algorithm. *Machine Learning*, 3(4):261–283, 1989.
5. D. J. Cook and L. B. Holder. Substructure discovery using minimum description length and background knowledge, *Artificial Intelligence Research*, 1:231–255, 1994.
6. S. Fortin. The graph isomorphism problem, Technical Report TR96-20, Department of Computer Science, University of Alberta, Edmonton, Canada, 1996.

7. W. Geamsakul, T. Matsuda, T. Yoshida, M. Motoda, and T. Washio. Classifier construction by graph-based induction for graph-structured data, In Proceedings of the 7th Pacific-Asia Conference on Knowledge Discovery and Data Mining (PAKDD 2003), pp. 52–62, 2003.

8. W. Geamsakul, T. Yoshida, K. Ohara, H. Motoda, H. Yokoi, and K. Takabayashi. Constructing a decision tree for graph-structured data and its applications, *Journal of Fundamenta Informatiae, Special issue on Advances in Mining Graphs, Trees and Sequence*, 66(1–2):131–160, 2005.

9. W. Geamsakul, T. Matsuda, T. Yoshida, H. Motoda, and T. Washio. Performance evaluation of decision tree graph-based induction, Proceedings of the International Conference on Discovery Science, pp. 128–140, 2003.

10. J. Huan, W. Wang, and J. Prins. Efficient Mining of Frequent Subgraphs in the Presence of Isomorphism. In Proceedings of the 3rd IEEE International Conference on Data Mining (ICDM 2003), pp. 549–552, 2003.

11. J. Huan, W. Wang, A. Washington, J. Prins, R. Shah, and A. Tropsha. Accurate classification of protein structural families using coherent subgraph analysis. In Proceedings of the 9th Pacific Symposium on Biocomputing, pp. 411–422, 2004.

12. A. Inokuchi, T. Washio, K. Nishimura, and H. Motoda. A fast algorithm for mining frequent connected subgraphs, IBM Research Report RT0448, Tokyo Research Laboratory, IBM, Japan, 2002.

13. A. Inokuchi, T. Washio, and H. Motoda. Complete mining of frequent patterns from graphs: Mining graph data. *Machine Learning*, 50(3):321–354, 2003.

14. S. Kramer and G. Windmer. Inducing classification and regression trees in first order logic. In *Relational Data Mining* (Sašo Džeroski, Nada Lavrač Eds.), Chapter 6, Springer, Berlin, 140–159, 2001.

15. M. Kuramochi and G. Karypis. An efficient algorithm for discovering frequent subgraphs. *IEEE Trans. Knowledge and Data Engineering*, 16(9):1038–1051.

16. M. Kuramochi and G. Karypis. GREW–A scalable frequent subgraph discovery algorithm. In Proceedings of the 4th IEEE International Conference on Data Mining (ICDM 2004), pp. 439–442, 2004.

17. T. Matsuda, H. Motoda, T. Yoshida, and T. Washio. Mining patterns from structured data by beam-wise graph-based induction. In Proceedings of the 5th International Conference on Discovery Science (DS 2002), pp. 422–429, 2002.

18. R. S. Michalski. Learning flexible concepts: Fundamental ideas and a method based on two-tiered representation. *Machine Learning: An Artificial Intelligence Approach*, 3:63–102, 1990.

19. S. Muggleton, L. de Raedt. Inductive logic programming: Theory and methods. *Journal of Logic Programming*, 19(20):629–679, 1994.

20. http://web.comlab.ox.ac.uk/oucl/research/areas/machlearn/PTE/.

21. J. R. Quinlan. Induction of decision trees. *Machine Learning*, 1:81–106, 1986.

22. J. R. Quinlan. *C4.5: Programs for Machine Learning*, Morgan Kaufmann, San Mateo, California, 1993.

23. A. Srinivasan, R. D. King, S. H. Muggleton, and M. Sternberg. The predictive toxicology evaluation challenge. In Proceedings of the 15th International Joint Conference on Artificial Intelligence (IJCAI-97), pp. 1–6, 1997.

24. Y. Takahashi, H. Ohoka, and Y. Ishiyama. Structural similarity analysis based on topological fragment spectra. *Advances in Molecular Similarity*, 2:93–104, 1998.

25. Y. Takahashi, S. Fujishima, K. Nishikoori, H. Kato, and T. Okada. Identification of dopamine D1 receptor agonists and antagonists under existing noise compounds by

TFS-based ANN and SVM. *Journal of Computational Chemistry, Japan*, 4(2):43–48, 2005.

26. Y. Yamada, E. Suzuki, H. Yokoi, and K. Takabayashi. Decision-tree induction from time-series data based on a standard-example split test. In Proceedings of the 12th International Conference on Machine Learning, pp. 840–847, 2003.

27. X. Yan and J. Han. gSpan: Graph-based structure pattern mining. In Proceedings of the 2nd IEEE International Conference on Data Mining (ICDM 2002), pp. 721–724, 2002.

28. K. Yoshida and M. Motoda. CLIP: Concept learning from inference patterns. *Artificial Intelligence*, 75(1): 63–92, 1995.

10

SOME LINKS BETWEEN FORMAL CONCEPT ANALYSIS AND GRAPH MINING

MICHEL LIQUIÈRE

LIRMM, Montpellier, France

10.1 PRESENTATION

This chapter presents a formal model to learning from examples represented by labeled graphs. This formal model is based upon lattice theory and in particular Galois lattices. We widen the domain of formal concept analysis by the use of the Galois lattices model with structural descriptions of examples and concepts. The operational implementation of our model, called "Graal" (for GRAph And Learning), constructs a Galois lattice for any description language provided that the operations of comparison and generalization are determined for that language. These operations exist in the case of labeled graphs satisfying a partial order relation (homomorphism).

This chapter is concerned as well with the known problems regarding propositionalization (i.e., the transformation of a structural description into a propositional description). Using classical lattice results, we have a formal model for the transformation of a structural machine learning problem into a propositional one.

Mining Graph Data, Edited by Diane J. Cook and Lawrence B. Holder
Copyright © 2007 John Wiley & Sons, Inc.

10.2   BASIC CONCEPTS AND NOTATION

10.2.1   Labeled Graphs

Let L denote set of *labels*. A labeled digraph (for labeled directed graph) $G = (V, E, \alpha)$ is a 3-tuple where V is the finite set of *vertices*, $E \subseteq V \times V$ is the set of *edges*, $\alpha: V \to L$ is a function assigning labels to each vertex. Edge $(v, v') \in E$ originates at node $v \in V$ and terminates at node $v' \in V$. For a vertex $v \in V$, we note $N^+(v) = \{v'|(v, v') \in E\}$ and $N^-(v) = \{v'|(v', v) \in E\}$.

For two labeled graphs $G_1 = (V_1, E_1, \alpha_1)$, $G_2 = (V_2, E_2, \alpha_2)$, a *homomorphism* h is a mapping $h : V_1 \to V_2$ for which $\forall v \in V_1$, $\alpha_1(v) = \alpha_2(h(v))$, and $\forall (v_1, v_2) \in E_1 \Rightarrow (h(v_1), h(v_2)) \in E_2$. A *subgraph isomorphism* is a homomorphism with an injective mapping (see Fig. 10.1). An *isomorphism* is a homomorphism with a bijective mapping.

10.2.2   Order

A *preorder* on a set X is a binary relation $\leq$ on X that is reflexive and transitive. If $x \leq y$ and $y \leq x$, then we shall write $x \cong y$ and say that x and y are equivalent elements.

A *partial order* (or poset) on a set X is an anty-symmetric preorder. If $x \leq y$ and $y \leq x$, then $x = y$ (isomorphic element). Given a preorder on X and an equivalence relation $\cong$ on X, the preorder induces a partial order in the quotient space $X/\cong$.

A *lattice* is a partial order $(L, \leq, \vee, \wedge)$ where every pair of elements (x,y) has a unique meet (also named inf for infimum) $x \wedge y$ and a unique join (also named sup for supremum) $x \vee y$ [5] and satisfying the following properties: $\forall x, y, z \in L$.

$x \wedge (y \wedge z) = (x \wedge y) \wedge z$	$x \vee (y \vee z) = (x \vee y) \vee z$	(associativity)
$x \wedge y = y \wedge x$	$x \vee y = y \vee x$	(commutativity)
$x \wedge x = x$	$x \vee x = x$	(idempotence)
$x \wedge (y \vee z) = x$	$x \vee (y \wedge z) = x$	(absortion).

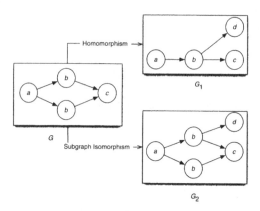

Figure 10.1.   Homomorphism and subgraph isomorphism.

10.3 FORMAL CONCEPT ANALYSIS

"Formal concept analysis (FCA) has been introduced by Wille [32] and applied in many quite different realms like psychology, sociology, anthropology ... " [11]. An important contribution of FCA is a mathematical definition of a concept. From a philosophical point of view, a concept is a unit of thought consisting of two parts, the extension and the intension. The extension covers all objects belonging to the concept and the intension comprises all attributes valid for each of those objects. Hence objects and attributes play a prominent role together with several relations such as the incidence relation, the hierarchical "subconcept–superconcept" relation between concepts, and the implication between attributes (or objects). We know [32] that, from a *context* $(\mathcal{E}, \mathcal{A}, \mathcal{R})$ where $\mathcal{E}$ is a set of objects (examples), $\mathcal{A}$ a set of properties (attributes), and $\mathcal{R}$ a binary relation between $\mathcal{E}$ and $\mathcal{A}$, we can construct a minimal ordered set named a concept lattice [32], or Galois lattice [10].

We have the following formal definitions: With $e \in \mathcal{E}$, $a \in \mathcal{A}$, $e \mathcal{R} a$ is asserting that "the object e verifies the attribute a" equivalent to "the attribute a is true for the object e".

In a context $(\mathcal{E}, \mathcal{A}, \mathcal{R})$, each *concept* (E, A), with $E \subseteq \mathcal{E}$ and $A \subseteq \mathcal{A}$, follows the properties:

- E contains each of the objects $e \in \mathcal{E}$ verifying all the attributes in A.
  With $A \subseteq \mathcal{A}$, $e(A) = \{e \in \mathcal{E} | e \mathcal{R} a, \forall a \in A\}$ represents the *extension part* of A.
- A contains each of the attributes $a \in \mathcal{A}$ true for the objects in E.
  With $E \subseteq \mathcal{E}$, $d(E) = \{a \in \mathcal{A} | e \mathcal{R} a, \forall e \in E\}$ represents the *intention part* of E.
- A pair (E, A) with $E \subseteq \mathcal{E}$ and $A \subseteq \mathcal{A}$, is a *concept* iff $d(E) = A$ and $e(A) = E$.

For a context $(\mathcal{E}, \mathcal{A}, \mathcal{R})$, the structure $L(\mathcal{E}, \mathcal{A}, \mathcal{R}) = \{(E, A) | (E, A)$ is a concept of $(\mathcal{E}, \mathcal{A}, \mathcal{R})\}$ ordered by $(E_1, A_1) \geq (E_2, A_2) \Leftrightarrow A_1 \subseteq A_2$ (formally equivalent to $E_2 \subseteq E_1$) is a concept lattice [32]. This classical partial order relation between concepts is the relation "to be more general then" used in the machine learning community.

For a context $(\mathcal{E}, \mathcal{A}, \mathcal{R})$ and two concepts $(E_1, A_1), (E_2, A_2)$, the operations $\wedge$ and $\vee$ are:

$(E_1, A_1) \vee (E_2, A_2) = (E, A)$ with $E = e(A_1 \cap A_2)$ and $A = A_1 \cap A_2$
$(E_1, A_1) \wedge (E_2, A_2) = (E, A)$ with $E = E_1 \cap E_2$ and $A = d(E_1 \cap E_2)$.

The $\wedge$ operation (going down in the lattice) is defined from the intersection of the extensions while the $\vee$ operation (going up in the lattice) corresponds to the intersection of the intensions (see Fig. 10.2). Using machine learning terminology, the construction of the Galois lattice using the $\vee$ operation is a *generalization search*. From two concepts the $\vee$ operation builds a more general concept. If we base our search on the $\wedge$ operation, we have a *specialization search*.

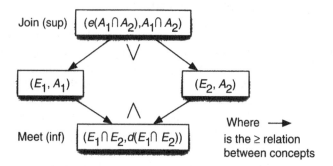

Figure 10.2.   Classical Galois lattice operations.

10.3.1   Galois Lattice Construction from a Binary Relation

There are many algorithms for the construction of the Galois lattice for a con-
text $(\mathcal{E}, \mathcal{A}, \mathcal{R})$ [11, 17]. In this chapter we present the algorithm [25] because it
is simple, it uses general lattice operations, and it is incremental (online algo-
rithm): It computes the lattice at step $n + 1$ from the lattice at step n and the
description of a new example (at the first step we begin with an empty lattice).
In the next section we generalize this algorithm for a general description of the
examples.[1]

The presentation of Norris's algorithm uses the $\vee$ operation. In this case, it is
a generalization algorithm: We begin from some specific descriptions and construct
some less specific description. Inversely, if we want to construct the lattice from
the general descriptions to specific descriptions, we have to replace the $\mathcal{R}$ relation
by the inverse $\mathcal{R}^{-1}$ relation, and we have to give to the algorithm the couples
$(\{a\}, E)$ where $a \in \mathcal{A}$, and $E \subseteq \mathcal{E}$. This is named "the duality principle for concept
lattice" [11].

Algorithm 10.1   Norris's Algorithm: Concept Lattice Construction Algorithm

Procedure: maximal

Data: A couple (E, A)

Data: A list L_n of concepts

Result: A Boolean

if *(there is a* $(E_j, A_j) \in L_n$ *with* $A_j = A$) **then**
   return false;

else
   return true;

end

---

[1]In fact, for this purpose, we may use directly any Galois lattice construction algorithms that uses general
operations $\vee$ between descriptions. This is the case for [13, 25, 26] but not for [6].

Procedure: AddExample

Data: A new Example: a couple $(\{e\}, A)$, with e an identifier of an example, and $A \subseteq \mathcal{A}$ the description of the example e.

Data: A list L_n of concepts

Result: A list L_{n+1} of concepts

$L_{n+1} \leftarrow L_n$;

for *each concept* $(E_i, A_i) \in L_n$ **do**

    if $(A_i \subseteq A)$ **then**

        Replace (E_i, A_i) in L_n by $(E_i \cup \{e\}, A_i)$;

    else

        // $\vee$ operation;

        compute $(E', A') \leftarrow (e(A_i \cap A), (A_i \cap A))$;

        if $maximal((E', A'), L_{n+1})$ **then**

            add (E', A') to L_{n+1} ;

        end

    end

end

if $maximal((\{e\}, A), L_{n+1})$ **then**

    add $(\{e\}, A)$ to L_{n+1} ;

end

return L_{n+1};

Algorithm: Norris's Algorithm:

Data: A context $(\mathcal{E}, \mathcal{A}, \mathcal{R})$

Result: The list L of all concepts in $(\mathcal{E}, \mathcal{A}, \mathcal{R})$

$L \leftarrow \emptyset$;

for *each example* $e \in \mathcal{E}$ **do**

    $L \leftarrow$ AddExample$((\{e\}, d(\{e\})), L)$;

end

return L;

Let us consider the relation $\mathcal{R}$ between $\mathcal{E} = \{e_0, e_1, e_2, e_3, e_4\}$ and $\mathcal{A} = \{a_0, a_1, a_2, a_3, a_4, a_5, a_6\}$ defined by the table in Figure 10.3.

This binary relation gives the Galois lattice of Figure 10.4.

10.4 EXTENSION LATTICE AND DESCRIPTION LATTICE GIVE CONCEPT LATTICE

In this paragraph, we give another formalization of a Galois lattice based on a Galois connection between a lattice $\mathcal{X}$ and a lattice $\mathcal{D}$. The lattice $\mathcal{X}$ is defined from a set of examples $\mathcal{E}$, with $\mathcal{X} = \prod(\mathcal{E})$, the power set of the finite set $\mathcal{E}$. We name this

$$\begin{bmatrix} & a_0 & a_1 & a_2 & a_3 & a_4 & a_5 & a_6 \\ e_0 & 1 & 0 & 0 & 0 & 1 & 1 & 1 \\ e_1 & 0 & 0 & 1 & 0 & 1 & 1 & 1 \\ e_2 & 0 & 0 & 0 & 1 & 1 & 1 & 0 \\ e_3 & 0 & 1 & 0 & 0 & 1 & 0 & 1 \\ e_4 & 0 & 0 & 1 & 1 & 0 & 1 & 1 \end{bmatrix}$$

Figure 10.3. Relation $\mathcal{R}$.

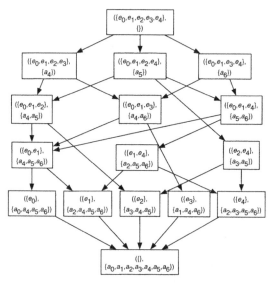

Figure 10.4. Galois lattice of $\mathcal{R}$, on the form of its Hasse diagram.

lattice the *extension lattice*. The lattice $\mathcal{D}$ named *description lattice* comes from a *description language* $\mathcal{L}$, which is a set of well-formed expressions with:

$\geq_{\mathcal{L}}$: A partial order between the expressions in $\mathcal{L}$. We note $=_{\mathcal{L}}$ the equality between two expressions of $\mathcal{L}$. For example, for graphs this equality can be an isomorphism relation.

$\otimes_{\mathcal{L}}$: $\mathcal{L} \times \mathcal{L} \to \mathcal{L}$ a product operator with for $D_1 \in \mathcal{L}$ and $D_2 \in \mathcal{L}$, $D = D_1 \otimes_{\mathcal{L}} D_2$ of $\mathcal{L}$ verify: $D \geq_{\mathcal{L}} D_1$, $D \geq_{\mathcal{L}} D_2$ and $\forall D' | (D' \geq_{\mathcal{L}} D_1)$ and $(D' \geq_{\mathcal{L}} D_2) \Rightarrow D' \geq_{\mathcal{L}} D$.

This is a classical least general generalization operator [27].

The relation between these two lattices is given by the two mappings $d: \mathcal{X} \to \mathcal{D}$ and $e: \mathcal{D} \to \mathcal{X}$ (see Fig. 10.5). We now generalize the notion of context in FCA into the notion of structural context [21] also named general context in [7].

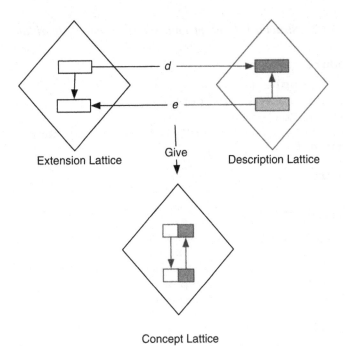

Concept Lattice

Figure 10.5. Relations between the two spaces.

A *structural context* C is a triplet $(\mathcal{E}, \mathcal{L}, d)$ where $\mathcal{E}$ is a set of examples, $\mathcal{L}$ is a description language, and $d: \prod(\mathcal{E}) \to \mathcal{L}$ a mapping. This mapping gives also the intension (also named description) $d(\{e\})$ for an example $e \in \mathcal{E}$. We can easily transform each context $(\mathcal{E}, \mathcal{A}, \mathcal{R})$ into a structural context $(\mathcal{E}, \mathcal{L}, d)$, where $\mathcal{L}$ is the power set of $\mathcal{A}$.

For a structural context $(\mathcal{E}, \mathcal{L}, d)$ a concept is a pair (E, D) with:

- $D \in \mathcal{L} | D = \bigotimes_{\mathcal{L}} d(\{e\})$ for $e \in E$.
- $E \in \prod(\mathcal{E}) | E = \{e_i \in \mathcal{E} | D \geq_{\mathcal{L}} d(e_i)\}$.

For two concepts $(E_1, D_1), (E_2, D_2)$ we have the partial order relation: $(E_1, D_1) \geq_c (E_2, D_2) \Leftrightarrow E_1 \supseteq E_2$ and $D_1 \geq_{\mathcal{L}} D_2$.

Proposition 10.1 For a structural context $(\mathcal{E}, \mathcal{L}, d)$ the set of concepts, ordered by the relation $\geq_c$, is a Galois lattice [21].

From the previous definitions, we can define the lattice operations $\wedge$ (meet) and $\vee$ (join). $(E_1, D_1) \vee (E_2, D_2) = (E, D)$ with $D = (D_1 \otimes_{\mathcal{L}} D_2)$ and $E = \{e \in \mathcal{E} | d(\{e\}) \leq_{\mathcal{L}} D\}$. $(E_1, D_1) \wedge (E_2, D_2) = (E, D)$ with $E = E_1 \cap E_2$ and $D = \otimes_{\mathcal{L}} d(e_i)$, for all $e_i \in E$. We denote this *structural Galois lattice* $\mathcal{G}(\mathcal{E}, \mathcal{L}, d)$. With this result we can generalize the Norris's algorithm and use it a for the generation of a Galois lattice from a structural context:

Algorithm 10.2 *Structural Concept Lattice Construction Algorithm*

Procedure: maximal

Data: A couple (E, D)
Data: A list L_n of concepts
Result: A Boolean
if *(there is a $(E_j, D_j) \in L_n$ with $D_j =_{\mathcal{L}} D$)* **then**
    return false;
else
    return true;
end

Procedure: AddExample

Data: A new Example: a couple $(\{e\}, D)$, with e an
        identifier of an example, and $D \in \mathcal{L}$ a
        description of the example e
Data: A list L_n of concepts
Result: A list L_{n+1} of concepts
$L_{n+1} \leftarrow L_n$;
for *each concept $(E_i, D_i) \in L_n$* **do**
    if *$(D_i \geq_{\mathcal{L}} D)$* **then**
        Replace (E_i, D_i) in L_n by $(E_i \cup \{e\}, D_i)$;
    else
        // $\vee$ operation;
        compute $(E', D') \leftarrow (e(D_i \otimes_{\mathcal{L}} D), (D_i \otimes_{\mathcal{L}} D))$;
        if *maximal$((E', D'), L_{n+1})$* **then**
            add (E', D') to L_{n+1} ;
        end
    end
end
if *maximal$((\{e\}, D), L_{n+1})$* **then**
    add $(\{e\}, D)$ to L_{n+1} ;
end
return L_{n+1};

Algorithm: Norris's algorithm for structural
language:

Data: A structural context $(\mathcal{E}, \mathcal{L}, d)$
Result: The list L of all concepts in $(\mathcal{E}, \mathcal{L}, d)$
$L \leftarrow \emptyset$;
for *each example $e \in \mathcal{E}$* **do**
    $L \leftarrow$ AddExample$((\{e\}, d(e)), L)$;
end
return L;

This algorithm uses the operators $\geq_{\mathcal{L}}, =_{\mathcal{L}},^2$ and $\otimes_{\mathcal{L}}$. This is a generalization method; if we want to use it on a specialization search, we need an initial set of graphs. This remark is further developed in Section 10.6.

The complexity of the construction of a Galois lattice with this algorithm for a description language $\mathcal{L}$ is a function of:

1. The time and space complexity of the operations $(\geq_{\mathcal{L}}, \otimes_{\mathcal{L}})$.
2. The number of nodes in the lattice.
3. The algorithm used: In our case, for a description language $\mathcal{L}$ and a lattice L, the complexity of the AddExample function is $O(|L|^2 \times P + |L| \times T)$ where P represents the complexity of the $\geq_{\mathcal{L}}$ and T represents the complexity of the $\otimes_{\mathcal{L}}$ operation.

We will see in the next section an application of this formalization to a graph description of the examples, and we will evaluate the complexity of the method in this case.

10.5 GRAPH DESCRIPTION AND GALOIS LATTICE

For a description language we want to use labeled digraphs. This representation is interesting because it is used in many applications [8]; furthermore, conceptual graphs [29] are a specific case of labeled digraphs.

Now we have to select the partial order we consider between labeled directed graphs (digraphs). A classical partial order relation is the subgraph isomorphism relation. The interpretation of such a relation between graphs is natural but there are two drawbacks: the complexity of the search of such a relation between two graphs and the definition of a product operator for this relation. There is another classical preorder relation between graphs based on the homomorphism projection. This preorder is also used in inductive logic programming (ILP) under the name θ-subsumption [27] and in conceptual graph models with the term "projection" [29]. Now, our goal is to choose the operators $\geq_g$ and $\otimes_g$ for our labeled digraphs.

10.5.1 $\geq_g$ for Labeled Graphs

For two labeled digraphs G_1 and G_2, we note $G_1 \geq_g G_2 \Leftrightarrow$ there is a homomorphism from G_1 *into* G_2. For a set of labeled digraphs, the homomorphism relation is only a preorder because the antisymmetric property is not fulfilled [34]. In order to use Algorithm 10.2, a partial order between elements of the description language is not necessary, but it is better for the interpretation of the results. For the homomorphism relation we have the following equivalence relation so we can construct

---

^{2}With a preorder we can replace $=_{\mathcal{L}}$ by another equivalence relation $\cong_{\mathcal{L}}$.

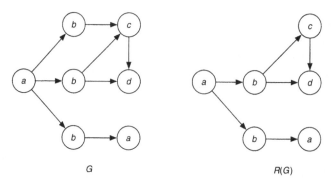

Figure 10.6. Labeled graph G and its core graph R(G).

a quotient space with this relation. For this purpose, we use the class of core labeled graphs [34].

Equivalence relation $\cong_g$: For two labeled digraphs G_1 and G_2, we note $G_1 \cong_g G_2$ if both $G_1 \geq_g G_2$ and $G_2 \geq_g G_1$.

Definition 10.1 (Core)   A labeled graph G is called a core iff it has no strict labeled subgraph $G' \subset G$ with $G' \cong G$.

Proposition 10.2   Concerning the equivalence relation defined above ($\cong_g$), an equivalence class of labeled graphs contains one and only one core labeled graph, which is the (unique) labeled graph with the smallest vertex number [34].

We can construct the core graph of a graph as proved by [23]. This operation is called reduction (notation R), and we have the following property:

Proposition 10.3   If, for a class of labeled graphs, the homomorphism is polynomial, then the reduction operation R is polynomial [23]. We use $R(G)$ to represent the core labeled graph of a labeled graph G (Fig. 10.6).

A labeled graph description of an example can be converted to an equivalent ($\cong$) core labeled graph, using the R operation.

Proposition 10.4   The restriction of $\geq_g$ to the set of core labeled graphs is a lattice [34].

For this lattice, the $\wedge$ operation is based on the disjoint sum $\oplus$ of two graphs,[3] $G_1 \wedge G_2 = R(G_1 \oplus G_2).$[4] The $\vee$ operation is more complex and is defined in the following discussion.

---

[3]We construct $G : (V, E, \alpha)$ from $G_1 : (V_1, E_1, \alpha_1)$ $G_2 : (V_2, E_2, \alpha_2)$ with $V = V_1 \cup V_2$, $E = E_1 \cup E_2$.
[4]We do not use this operation in the algorithm.

10.5.2 Homomorphism and Product Operator $\otimes_g$

For a description language $\mathcal{L}$ we need a product operator ($\otimes_{\mathcal{L}}$), and there is a classical product operator for the homomorphism relation between labeled digraphs.

Definition 10.2 ($\otimes_g$ operation) For two labeled digraphs $G_1 = (V_1, E_1, \alpha_1)$, $G_2 = (V_2, E_2, \alpha_2)$ we define $G : G_1 \otimes_g G_2$ by $G = (V, E, \alpha)$ with: $V = V_1 \times V_2$ with $\alpha_1(V_1) = \alpha_2(V_2)$ and $((x_1, x_2),(y_1, y_2)) \in E \Leftrightarrow (x_i, y_i) \in V_i (i = 1, 2)$.

Proposition 10.5 For the homomorphism relation, $G_1 \otimes_g G_2$ is the product of G_1 and G_2.

Proof. A proof of this proposition can be found in [4].

Given the core labeled digraph G_2, G_2 the product is $\otimes_{gc}$ with $G_1 \otimes_{gc} G_2 = R(G_1 \otimes_g G_2)$, and we have a partial order [21].

10.5.3 Complexity

Using the generalization relation $\geq_g$, the product $\otimes_g$ and Algorithm 10.2 we can build the Galois lattice for a set of examples described by labeled digraphs. But the complexity of the operation $\geq_g$ is NP-hard for this kind of graphs.

Size of Galois Lattice. Given a context $(\mathcal{E}, \mathcal{A}, \mathcal{R})$, the number of nodes in the lattice is less than or equal to $2^{\min(|\mathcal{E}|,|\mathcal{A}|)}$. This property comes from the relation between the two spaces (partition space and description space).

In the case of a general description of the example, the size of the Galois lattice is $\min(2^{|\mathcal{E}|}, \text{SD})$ (where SD is the size of the description lattice). In general, for a structural description of the example we have $\text{SD} \gg 2^{|\mathcal{E}|}$ [15].

Complexity of the $\otimes_g$ operation. For two labeled graphs $G_1 = (V_1, E_1, \alpha_1)$ and $G_2 = (V_2, E_2, \alpha_2)$, a time complexity of the product $\otimes_g$ is $O(|V_1| \times |V_2| \times (|E_1| + |E_2|))$ the size can be $O(|E_1| \times |E_2|)$.

So for a set of labeled digraphs Γ the size of $\otimes_g G_i$ with $G_i \in \Gamma$ can be exponential. For core graphs reduction operation R, in the $\otimes_{gc}$ operator, is NP-hard and the size can be also $O(|E_1| \times |E_2|)$.

Complexity of the $\geq_g$ Operation. For general labeled digraphs the complexity of the homomorphism operation is NP-hard. However, the homomorphism is polynomial for paths, trees, and locally injective digraphs (see Definition 10.3). Then the product operator is also polynomial for these type of graphs.

10.5.4   Locally Injective Digraph

We define a large class of labeled graphs, named LIG graphs, where the operations $\geq_{\text{lig}}$ and $\otimes_{\text{lig}}$ have a polynomial complexity.

Definition 10.3 [Locally Injective Digraph (LIG Digraph)]   A labeled digraph $G : (V, E, \alpha)$ is *locally injective* $\Leftrightarrow$ for each vertex $v \in V$, $\forall v_1, v_2 \in N^+(v)$, $v_1 \neq v_2 \Rightarrow \alpha(v_1) \neq \alpha(v_2)$ and $\forall v_1, v_2 \in N^-(v)$, $v_1 \neq v_2 \Rightarrow \alpha(v_1) \neq \alpha(v_2)$.

In Figure 10.7, G_1 is a LIG digraph, G_2 is not a LIG digraph because, for the node labeled with c, there are two neighbors labeled with the same label b.

Proposition 10.6   The homomorphism operation is polynomial for locally injective digraphs.

In fact, the subgraph isomorphism relation is also polynomial for this class of digraphs [21].

Proposition 10.7   For two LIG digraphs $G_1, G_2 : G = G_1 \otimes_g G_2$ is a LIG digraph.

These properties are interesting because for LIG we can use the operations $\geq_g$ and $\otimes_g$ with a polynomial complexity.
     We can use the previous algorithm directly on LIG digraphs, but the product operator is redundant. In many cases we obtain a digraph where some part is equivalent to or more general than other parts. To avoid this drawback we can use the core LIG graph and the $\otimes_{gc}$ operator, which is polynomial for LIG graphs. This kind of graph is used in the examples in Figure 10.8. In this figure the nodes with gray background are $\wedge$-irreducible elements (nodes with one and only one predecessor). We use this element in Section 10.6.

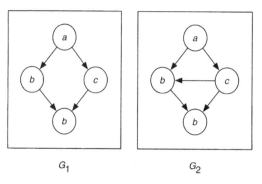

G_1 G_2

Figure 10.7.   Example of LIG digraphs and not LIG digraphs.

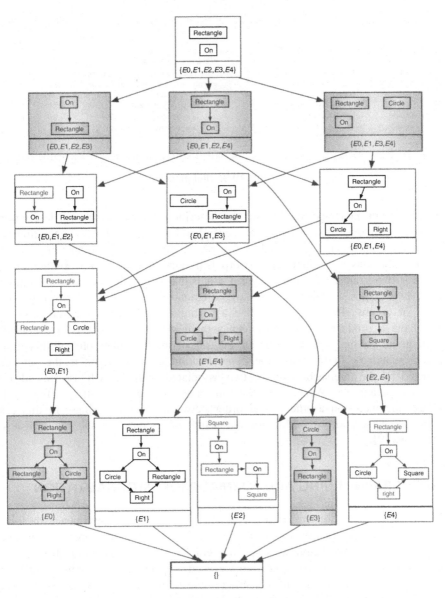

Figure 10.8. Structural Galois lattice in the form of its Hasse diagram.

10.5.5 Complexity for Some Description languages

In this section we give some complexity results for the construction of a structural Galois lattice when examples are described by an expression of a description language (see Fig. 10.9). There are many description languages $\mathcal{L}$; we present here a small subset and give the time and size complexity of the operations $\geq_{\mathcal{L}}, \otimes_{\mathcal{L}}$. In

	≥	$O(\geq)$	⊗	$O(\otimes)$	Size
Set	⊆	P	∩	P	P
String	Lexical order	P	maximal common prefix	P	P
Rooted Tree	Rooted tree Inclusion	P	maximal common prefix tree	P	P
LIG Graph	Homomorphism	P	graph product and Reduction	P	?
Automata	Homomorphism	P	automata product and Minimization	P	E
Graph	Homomorphism	NP	graph product and Reduction	NP	E
Clause	θ-subsumption	NP	rlgg operator [27]	NP	E
Graph	Subgraph isomorphism	NP	no product operator		

where P = polynomial, NP = NP-hard, and E = Exponential

Figure 10.9. Relation $\mathcal{R}$.

fact, for some language, the size of the description generalizing two descriptions increases just by a factor. However, several utilizations of the $\otimes_{\mathcal{L}}$ operation can produce a description exponential in size, so we give also this complexity.

For the classical subgraph isomorphism relation there is no product operator. For two graphs we cannot have a graph that is the unique generalization of these two graphs. In this case we have to extend the description from a graph to a set of graphs. Then we can define a partial order between the set of graphs and have a product operator between sets of graphs. To do this, we consider each graph as the representation of the set of all the subgraphs of this graph (itself included). In this case the generalization of two graphs is the set constructed by the intersection of the two associated sets. We will use this idea in the next paragraphs. We can use the same methodology with strings if we want to use the inclusion as the partial order between a set of strings. In this case there is a very good algorithm for the search of all repeated substrings in a set of strings [9].

10.6 GRAPH MINING AND FORMAL PROPOSITIONALIZATION

In the previous section we used a product operator. As a consequence the algorithm, for the construction of the lattice, uses a generalization operator (generalization method). We begin with specific graphs and build generalizations of the graphs. In many practical application we need a specialization method. We begin with some general graphs and construct some more specific graphs using a specialization operator. Using this top-down search, we can select concepts by the cardinality of their extension part. This defines a semilattice [22] named iceberg lattice [30].

For a propositional description of the examples, and a context $(\mathcal{E}, \mathcal{A}, \mathcal{R})$, $(\mathcal{A}, \mathcal{E}, \mathcal{R}^{-1})$ is also a context and we can use the classical concept lattice algorithm on this dual context. "In other words: if we exchange the roles of objects and attributes, we obtain the dual concept lattice" [11]. For a structural context, we have not the duality principle directly; however, using some fundamental lattice properties we can overcome this problem.

Many machine learning problems have a solution in a propositional description of the examples, for example, the Q-learning [28]. When we want to deal with a structural description, many new complexities appear (time and size complexity of the generalization and matching operations). For this problem, a natural idea is to convert the structural problem into a propositional one. This transformation does not change the complexity of the problem, but classical methods can be used with this new description.

For our problem: Structural FCA and specialization search, this methodology seems really promising. To do this transformation we need a set SP of structural patterns (features). Then, we can use this set to recode each example with a Boolean vector. Each boolean in this vector gives the information about the presence or absence of a feature of SP in the description of the example. For a description language $\mathcal{L}$, the test of presence or absence uses a $\geq_{\mathcal{L}}$ operator. Such a transformation from a structural machine learning problem to a propositional problem is called propositionalization [16]. There are many recent works in this domain [2, 16, 18]. These methods search the "best" features. Most of the authors define this problem as being the selection of a subset of m attributes from a set of n attributes. This gives a theoretical complexity of C_n^m that renders the model unusable in the general case. Then the propositionalization methods use heuristics for the selection of the attributes, like the stochastic selection [2] or syntactic selection [19].

Our purpose has a great proximity but proposes to consider the problem from another point of view. The idea is to find a propositional language for a problem of classification equivalent to a description language. This equivalence is not an expressiveness equivalence of the languages. This has already been studied in first-order logic with the universal relationship notion.

In this chapter, using a fundamental theorem (Theorem 10.1 thereafter), we define a new equivalence between languages. This equivalence uses the structure of our classifications spaces. We consider that, for a set of examples, two description languages are equivalent if they give the same concept lattice.

To show our solution, we have to define how two languages will be considered as equivalent for a set of examples. Then, it will be possible to search for a propositional language equivalent, in the specified senses, to the structural language.

10.6.1   Some Properties of Lattices and Galois Lattices

The elements $\vee$-irreducible (resp. $\wedge$-irreducible) in a lattice are the elements with exactly a lower neighbor (resp. upper neighbor). We note $J(L)$ [resp. $M(L)$] the set of all $\vee$-irreducible ($\wedge$-irreducible) elements of a lattice L.

For a lattice L, we can construct a binary relation $B(L)$ between $J(L)$ and $M(L)$ with $B(j_i, m_j) = 1 \Leftrightarrow j_i \leq m_j$ (for $j_i \in J(L)$ and $m_j \in M(L)$).

The $\wedge$-irreducible (resp. $\vee$-irreducible) elements are the element that cannot be derived from a $\wedge$ (resp. $\vee$) operation between elements of the lattice. So other elements are not essential because we can reconstruct them.

Proposition 10.8   In a finite lattice each element is the meet (resp. join) of $\wedge$-irreducible (resp. $\vee$-irreducible) elements.

THEOREM 10.1 (Lattice and Galois Lattice)   All finite lattice L is isomorph with the Galois lattice build from the binary relation $B(L) = (J(L), M(L), \leq)$.

This result has been known since 1970 [3] and is also a fundamental theorem in FCA [11].

For a context $(\mathcal{E}, \mathcal{A}, \mathcal{R})$ the set $\mathcal{A}$ (resp. $\mathcal{E}$) is a superset of the set of $\wedge$-irreducible (resp. $\vee$-irreducible) elements. So from a given context we can remove all the reducible elements. The new context (named reduced context [11]) gives a Galois lattice, which is isomorphic with $L(\mathcal{E}, \mathcal{A}, \mathcal{R})$. This property is also true if we remove only a subset of the reducible elements and, in particular, a subset of the reducible attributes. We develop this point of view in the next section and give a methodology for formal propositionalization.

10.6.2 Language Equivalence for a Set of Examples

In the real world, the choice of a good description language has been one of the most important tasks of human knowledge acquisition. In fact, the search of a simpler compact language, allowing to express the differences and resemblances between objects, is the foundation of many domains. For example, in the area of chess, a good player (computer or human) needs a set of concepts (like open column, center domination, etc.) to evaluate a position.

One of the simplest languages is the propositional language. Therefore, we are interested in the problem to know if there exists, for a set of examples, a propositional language equivalent to a given structural language. And more precisely what are the conditions for the emergence of this propositional language.

In the framework of our model, we obtain the following definition:

Definition 10.4 (Context Equivalence)   Two structural contexts $(\mathcal{E}, \mathcal{L}_1, d_1)$ and $(\mathcal{E}, \mathcal{L}_2, d_2)$ are called equivalent iff $\mathcal{G}(\mathcal{E}, \mathcal{L}_1, d_1) \equiv \mathcal{G}(\mathcal{E}, \mathcal{L}_2, d_2)$ where $\equiv$ is a lattice isomorphism that conserve the extension part.

So, for the same set of examples, all classifications of the examples (extension part of the concept) obtained with the language $\mathcal{L}_1$ and with the language $\mathcal{L}_2$ are the same. We have the same definition between a structural context $(\mathcal{E}, \mathcal{L}, d)$ and a propositional context $(\mathcal{E}, \mathcal{A}, \mathcal{R})$. They are called equivalent iff $\mathcal{G}(\mathcal{E}, \mathcal{L}, d) \equiv L(\mathcal{E}, \mathcal{A}, \mathcal{R})$.

Using this definition and Theorem 10.1, we can now give a property on the existence of a propositional language equivalent to a structural description language for a set of examples.

Proposition 10.9   For a structural context $(\mathcal{E}, \mathcal{L}, d)$ there is a propositional context $(\mathcal{E}, \mathcal{A}, \mathcal{R})$ equivalent.

Partial proof: Proposition 10.8 and Theorem 10.1 show that each of the elements in the lattice can be characterized from the $\wedge$-irreducible elements. This is therefore also true in the case of structural Galois lattice. Thus the concepts of $(\mathcal{E}, \mathcal{L}, d)$ are in one-to-one correspondence with the concepts of $(\mathcal{E}, \mathcal{A}, \mathcal{R})$ where $\mathcal{A}$ is the set $M(L)$ of $\wedge$-irreducible elements of $L : \mathcal{G}(\mathcal{E}, \mathcal{L}, d)$.

10.6.3   Search for $\wedge$-Irreducible Elements

Definition 10.8 gives a formal definition of propositionalization and Proposition 10.9 proves the existence of a propositional context equivalent to a structural context. Now the problem is the search for the set of $\wedge$-irreducible elements (or a subset for an approximation). A theorical method can build the complete structural Galois lattice for a structural context, then we can easily find the $\wedge$-irreducible elements (all element with only one predecessor in the Hasse diagram). In Figure 10.8, we show a distribution of the $\wedge$-irreducible elements in the lattice (nodes with gray background).

With this method, we can find the set of irreducible elements. However, even for simple structural description languages $\mathcal{L}$, the complexity of the $\otimes_{\mathcal{L}}$ and $\geq_{\mathcal{L}}$ operations and the size of the lattice limit this method to a small set of examples.

In fact three other considerations induce the use of a specialization method:

1. Since all element in a complete lattice are the meet ($\wedge$) of $\wedge$-irreducible element, the $\wedge$-irreducible elements are (statistically) more "on the top" of the lattice.
2. We can limit the exploration to $\wedge$-irreducible elements seen on "enough" examples (classical support data mining heuristic).
3. We can use classical pattern mining for the construction, from general to specific, of set of patterns.

In [11] there are some important properties for the search of $\wedge$-irreducible (or $\vee$-irreducible) elements in the case of a propositional context.

Proposition 10.10 ($\wedge$-irreducible)   For a context $(\mathcal{E}, \mathcal{A}, \mathcal{R})$ where $\forall a_1, a_2 \; {}_{(a_1 \neq a_2)}$ $e(a_1) \neq e(a_2)$, $a \in \mathcal{A}$ is a $\wedge$-irreducible attribute $\Longleftrightarrow \{e \in \mathcal{E} | \neg(e \mathcal{R} a)$ and $\forall a' \neq a \in d(e(a))^5 \; e \mathcal{R} a\} \neq \emptyset$. We note irre($a$) this set.

---

[5]Closure of a, noted a'' by Wille [32].

We can see that a $\wedge$-irreducible attribute a is characterized by this set of examples (irre(a)). The example in irre(a) does not verify the attribute a but is true for all attribute b with $e(a) \subset e(b)$. In fact the a attribute is fundamental for the separation of the examples of irre(a) and the set of other examples.

Proposition 10.10 gives a polynomial algorithm for seeking the $\wedge$-irreducible element of a binary relation $\mathcal{R}$. But for a propositionalisation problem, the number of attributes we have to test can be exponential. Then the relation $\mathcal{R}$ can be really big. Furthermore, in some application, like reinforcement learning [28], the pattern occurs step after step. So an incremental algorithm is better for this case.

Definition 10.5   A context $(\mathcal{E}, \mathcal{A}, \mathcal{R})$ is $\wedge$-reduce iff all the attributes $a \in \mathcal{A}$ are $\wedge$-irreducible.

Proposition 10.11   For a $\wedge$-reduce context $(\mathcal{E},\mathcal{A}_n,\mathcal{R}_n)$, the context $(\mathcal{E},\mathcal{A}_{n+1},\mathcal{R}_{n+1})$, built by the addition of a new attribute a_{n+1} to the context $(\mathcal{E},\mathcal{A}_n,\mathcal{R}_n)$ with $\mathcal{A}_{n+1} = \mathcal{A}_n \cup a_{n+1}$ and $\mathcal{R}_{n+1}$ is the new binary relation between $\mathcal{E}$ and $\mathcal{A}_{n+1}$.

The context $(\mathcal{E},\mathcal{A}_{n+1},\mathcal{R}_{n+1})$ is $\wedge$-reduce iff the three following properties are verified:

    i. $\nexists a_k \in \mathcal{A}_n | e(a_k) = e(a_{n+1})$.
    ii. a_{n+1} is a $\wedge$-irreducible for the context $(\mathcal{E},\mathcal{A}_{n+1},\mathcal{R}_{n+1})$.
    iii. $\forall a_k \in \mathcal{A}_n | e(a_k) \subset e(a_{n+1}) \Rightarrow irre(a_k) \cap e(a_{n+1}) \neq \emptyset$.

Property 10.11 gives Algorithm 10.3.

Algorithm 10.3   Incremental addFeature method

Procedure: update
Data: $C_n : (\mathcal{E}, \mathcal{A}_n, \mathcal{R}_n)$
Data: attribute a
Result: C_n modified
$F \leftarrow \{a_k \in \mathcal{A}_n | e(a_k) \subseteq e(a)\}$;
for a_k in F **do**
   irre(a_k) $\leftarrow$ irre(a_k) $\cap$ e(a);
   if irre(a_k) $= \emptyset$ **then**
      $C_n \leftarrow C_n - a_k$;
   end
end
return C_n;

Procedure: is-Irreducible

Data: $C_n : (\mathcal{E}, \mathcal{A}_n, \mathcal{R}_n)$
Data: attribute a
Result: True if a is $\wedge$-irreducible else False;
Result: We compute also irre(a) if a is
       $\wedge$-irreducible;
$F \leftarrow \{a_k \in \mathcal{A}_n | e(a) \subseteq e(a_k)\}$;
if $F = \emptyset$ **then**
   irre$(a) \leftarrow \mathcal{E} - e(a)$
else
   //Initialization of irre(a);
   irre$(a) \leftarrow e(a_j) - e(a)$ // With a_j one element in
   F;
   $F \leftarrow F - a_j$;
   for a_k *in* F **do**
      if $e(a_k) = e(a)$ **then**
         return False;
      else
         irre$(a) \leftarrow$ irre$(a) \cap e(a_k)$;
         if $irre(a) = \emptyset$ **then**
            return *False*;
         end
      end
   end
end
return $irre(a) \neq \emptyset$;

Algorithm: addFeature: (Incremental Algorithm)

Data: A structural context $(\mathcal{E}, \mathcal{L}, d)$
Data: $C_n : (\mathcal{E}, \mathcal{A}_n, \mathcal{R}_n)$
Data: feature $a \in \mathcal{L}$
Result: $C_{n+1} : (\mathcal{E}, \mathcal{A}_{n+1}, \mathcal{R}_{n+1})$
if $is\text{-}Irreducible(C_n, a)$ **then**
   $C_{n+1} \leftarrow$ update(C_n, a) ;
else
   $C_{n+1} \leftarrow C_n$;
end
return C_{n+1};

10.6.4 ∧-Irreducibles and Pattern Mining

The idea is to use the algorithm addFeature on features extracted from graph descriptions of examples by a graph mining method. In fact, each graph is replaced by the set (or a subset) of its subgraphs. The graph mining algorithms are optimized for a complete specialization search, they build the graphs at step n from graphs of step $n - 1$. These methods use different parameters and principally a support parameter that is, for a given pattern, the minimum number of examples where this pattern is present. In our algorithm Apropos (given thereafter), we will use the pattern mining method as a feature generator. We name it PatternMiningMethod() and consider that the result of this method is a set of expressions of the description language $\mathcal{L}$.

Algorithm **10.4** *APropos Algorithm*

Procedure: Apropos
Data: A structural context $(\mathcal{E}, \mathcal{L}, d)$
Data: s: support parameter
Data: m: maximal depth
Result: A context $(\mathcal{E}, \mathcal{A}, \mathcal{R})$ (then a set of
        features $\mathcal{A}$)
$n \leftarrow 0$;
$C_0 \leftarrow (\mathcal{E}, \emptyset, \mathcal{R}_0)$;
while $(n \leq m)$ **do**
  $F \leftarrow$ PatternMiningMethod$((\mathcal{E}, \mathcal{L}, d)$, s, n, .. and
  other specific parameters)
  for f in F **do**
    $C_{n+1} \leftarrow$ addFeature$((\mathcal{E}, \mathcal{L}, d), C_n, f)$
  end
  $n \leftarrow n + 1$;
end
return C_{n+1};

In the algorithm Apropos, we can employ any pattern mining method that uses the expression of a description language $\mathcal{L}$. For the description of the example by graphs, there is now a large set of graph mining methods [14, 24, 31]. To reduce the number of features, we are tempted to replace the expressions with less specific even if that results in some loss of information. For FCA, this transformation is formalized in [12].

In 1989 we have proposed a machine learning method for the search of repeated elementary paths present in a set of graphs [20]. This method is a levelwise method (Apriori method [1]) that used the classical support criteria and with no limitation on the structure of the graph. Some more recent methods on the same problem are found in [24, 33]. We use this graph mining method in the algorithm APropos for the set of examples in Figure 10.8. If we limit the search to path with 3 nodes maximum, seen 3 times or more on the set of example in Figure 10.8, we obtain the set of features shown in Figure 10.10.

$$\begin{bmatrix} & E_0 & E_1 & E_2 & E_3 & E_4 \\ \texttt{Rectangle} & 1 & 1 & 1 & 1 & 1 \\ \texttt{on} & 1 & 1 & 1 & 1 & 1 \\ \texttt{right} & 1 & 1 & 0 & 0 & 1 \\ \texttt{Circle} & 1 & 1 & 0 & 1 & 1 \\ \texttt{Rectangle} \to \texttt{on} & 1 & 1 & 1 & 0 & 1 \\ \texttt{on} \to \texttt{Rectangle} & 1 & 1 & 1 & 1 & 0 \\ \texttt{on} \to \texttt{Circle} & 1 & 1 & 0 & 0 & 1 \\ \texttt{Rectangle} \to \texttt{on} \to \texttt{Circle} & 1 & 1 & 0 & 0 & 1 \end{bmatrix}$$

Figure 10.10. Paths with maximal length 2 with support 3.

$$\begin{bmatrix} & E_0 & E_1 & E_2 & E_3 & E_4 \\ \texttt{Circle} & 1 & 1 & 0 & 1 & 1 \\ \texttt{Rectangle} \to \texttt{on} & 1 & 1 & 1 & 0 & 1 \\ \texttt{on} \to \texttt{Rectangle} & 1 & 1 & 1 & 1 & 0 \\ \texttt{right, on} \to \texttt{Circle, Rectangle} \to \texttt{on} \to \texttt{Circle} & 1 & 1 & 0 & 0 & 1 \end{bmatrix}$$

Figure 10.11. $\wedge$-Irreducible path elements.

Using the Apropos algorithm, on this set of paths we obtain the following $\wedge$-reduced context (Fig. 10.11):

For this example, we have limited our search to paths where we have only a subset of all $\wedge$-irreducible elements since:

1. All the graphs are not explored (path length 2).
2. The support parameters reduce the set of paths.

With a support of 1, we have the set of $\wedge$-irreducible path elements (length < 2) shown in Figure 10.12.

The Galois lattice construct from this binary relation is the Galois lattice of Figure 10.4 because the binary relation is the binary relation of Figure 10.3. For this example, the Galois lattice constructed will be isomorph to the structural Galois lattice of Figure 10.8 since the set of $\wedge$-irreducible elements are the same. So in this case, the description of the examples with labeled graph or by path of length 2 are equivalent.

10.6.5 Experimentation

We have made another experimentation on real data. These data are the classical mutagenesis problem described with graphs [8]. For this experimentation we do not use the example/counterexample knowledge. To reduce the complexity, for the

	E_0	E_1	E_2	E_3	E_4
Rectangle $\rightarrow$ on	1	1	1	0	1
on $\rightarrow$ Rectangle	1	1	1	1	0
Circle	1	1	0	1	1
Square, on $\rightarrow$ Square	0	0	1	0	1
Circle $\rightarrow$ right	0	1	0	0	1
right $\rightarrow$ circle, Rectangle $\rightarrow$ right	1	0	0	0	0
Circle $\rightarrow$ on	0	0	1	0	0

Figure 10.12. $\wedge$-Irreducible path elements (lenght < 2, support 1).

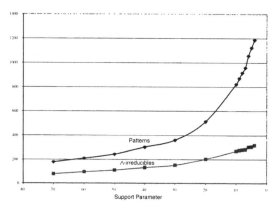

Figure 10.13. Experimentation on mutagenesis data: modification of the support parameter.

search of patterns, we use, here also, two parameters: the support number s [this would mean that the corresponding pattern (path or tree) occurs in s graphs in the original database] and the maximal size of the patterns m (specify that the largest pattern to be examined has m nodes). We have made one experimentation where s varies from 70 to 4 where m is 6 (see Fig. 10.13). In the second experimentation (Fig. 10.14) s is 20 and m changes from 2 to 8.

From the set of patterns extracted from the graphs by the algorithm Gaston [24], we have merged in one pattern all the patterns with exactly the same extension. This gives a more precise idea on the set of $\wedge$-irreducibles. For example, with parameters $s = 20$, $m = 8$, we obtain 63,487 patterns from Gaston, but there are only 1121 patterns with different extensions.

The vertical axis gives the number of patterns. The horizontal axis is the support parameter. We have 230 examples and the support varies from 70 (30.4%) to 4 (1.7%). The upper curve (Patterns) is the number of patterns [for each couple of patterns $p_1 \neq p_2, e(p_1) \neq e(p_2)$] extracted by Gaston. The lower curve ($\wedge$-irreducibles) is the number of $\wedge$-irreducible elements for the corresponding set of

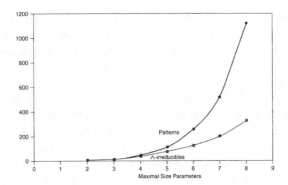

Figure 10.14. Experimentation on mutagenesis data: modification of the maximal size parameter.

patterns. The memory space needed is less then 600k and the time for the search of all the ∧-irreducibles elements varies from 2 to 6 s on an IMac G5 1.8 MHz computer.

For the first experimentation, the reduction factor is 3.5 for 1027 patterns and 5.2 for 3781 patterns. For the second experimentation, the reduction factor is 1.1 for 10 patterns (with 1 or 2 nodes) to a reduction of 3, 4 for 1121 patterns (with a maximum of 8 nodes).

From these experimentations, we find an augmentation of the reduction factor that reduces the exponential augmentation of the number of patterns. We think that this property comes from the position of the ∧-irreducible elements in a lattice.

We have for an element x of (a lattice) L denote by J_x the subset of ∧-irreducibles below x and let M_x, dually, be the set of ∨-irreducibles above x. A fundamental if obvious remark is that the higher is x, the larger is J_x while M_x is smaller. From a technical viewpoint, the structure of a lattice L is obviously encoded into the order relation on irreducibles by the bijection $x \rightarrow (J_x, M_x)$ (all $x \in L$, since x is less than or equal to y iff M_x is a superset of M_y and $M_{x \vee y} = M_x \cap M_y$ [10].

Now, if we consider Property 10.10, we can say (with only an empirical proof here) that the ∧-irreducible elements are "statically more" near the top of the lattice. So the specialization search is a good heuristics for the research of a subset of the ∧-irreducible elements.

10.7 CONCLUSION

In this chapter we have given some links between formal concept analysis and graph mining. The generalization of the formal concept analysis to general description of the examples, and in particular graph description, gives a formal definition of the search space. We prove that, with specific graph description of the examples, we can have a partial order and a product operator. In this case the structure of the classification space is a lattice. The structure of this space allows the use of classical algorithms for the construction of all the general concepts. But the construction of this lattice is made by a generalization operator. In many real-world applications, we need a specialization operator. In the second part of the chapter we proposed an answer to this problem with the definition of an equivalence between description language. This equivalence uses some particular elements of the lattice: the irreducible elements. These elements allow a re-description of the example into a propositional description, but the research of these elements is a new problem. We propose an online algorithm for the search of a subset of the set of ∧-irreducible elements. So we can remove from the set of patterns found with a graph mining algorithm all the patterns that are redundant for the classification task. An experimentation follows that prove the validity of the approach for the reduction of a large number of patterns to a more practical size.

ACKNOWLEDGMENTS

We greatly thank Stephano Cerri and Diane Cook for their helpful comments and suggestions on an earlier version of this manuscript.

REFERENCES

1. R. Agrawal, T. Imelinsky, and A. Swami, Mining association rules between sets of items in large database. In Proceedings of ACM SIGMOD '93 Conference, pp. 207–216, May 26–28, Washington, DC, 1993.
2. E. Alphonse and C. Rouveirol. Lazy propositionalization for relational learning in horn W. *14th European Conference on Artificial Intelligence, (ECAI'00)*, pp. 256–260, IOS Press, Berlin, 2000.
3. M. Barbut and B. Monjardet, Ordre et classification. In *Hachette*, Paris, Hachette, 1970, p. 174,.
4. C. Berge, Graphes and Hypergraphes, In *Bordas*, North-Holland Pub. Co., Paris, pp. xiv, 528, 1973.
5. G. Birkhoof. *Lattice theory*, 3rd ed. American Mathematical Society, Providence, RI, 1967.
6. J. P. Bordat. Calcul pratique du treillis de Galois d'une correspondance. *Math. Sci. Humaines*. 24° année, 96:31–7, 1986.
7. L. Chaudron and N. Maille, Generalized formal concept analysis, ICCS'2000, Lecture Notes in Artificial Intelligence, Vol. No. 1867, pp. 357–370, 2000.
8. D. J. Cook and L. B. Holder. Graph-based data mining. *IEEE Intelligent Systems*, 15(2):32–41, 2000.
9. M. Crochemore and W. Rytter. Usefulness of the Karp-Miller-Rosenberg algorithm in parallel computations on strings and arrays. *Theoretical Computer Science*, 88:(1) 59–82, 1991.
10. V. Duquenne. Latticial structure in data analysis. *Theoritical Computer Science*, 217:407–436, 1999.
11. B. Ganter and R. Wille. *Formal Concept Analysis, Mathematical Foundation*. Springer, Heidelberg, 1998.

12. B. Ganter and S. Kuznetsov. Pattern structures and their projections. *ICCS'2001* Lecture Notes in Computer Science, Vol. 2120, Proceedings of the 9th International Conference on Conceptual Structures, pp. 129–142, 2001.

13. R. Godin and R. Missaoui. An incremental concept formation approach for learning from databases. *Theoritical Computer Science.* 133:387–419, 1994.

14. A. Inokuchi, T. Washio, and H. Motoda. Complete mining of frequent patterns from graphs: Mining graph data. *Journal of Machine Learning*, 50:321–354, 2003.

15. J. Kearns. *The Computational Complexity of Machine Learning.* MIT Press, Cambridge, Massachusetts, 1994.

16. S. Kramer, N. Lavrac, P. A. Flach, S. Kramer, N. Lavrac, and P. A. Flach. Propositionalization approaches to relational data mining. In N. Lavrac and S. Dzeroski, eds. *Relational Data Mining.* Springer, Berlin Heidelberg New York, 2001.

17. S. Kuznetsov and S. Obiedkov. Algorithms for the construction of concept lattices and their diagram graphs. Principles of Data Mining and Knowledge Discovery (PKDD 2001), Freiburg, Germany, 2001.

18. N. Lavrac, F. Zelezny, and P. A. Flach. RSD: Relational subgroup discovery through first-order feature construction. *Inductive Logic Programming Conference.* Springer, Sydney, 2002.

19. N. Lavrac and P. Flach. An extended transformation approach to inductive logic programming. *ACM Trans. Comput. Log.* 2(4):458–494, 2001.

20. M. Liquière and J. Sallantin. INNE: A structural learning algorithm for noisy Data. *First Conference Tools for AI, IEEE International Workshop*, Fairfax,VA, pp. 70–76, 1989.

21. M. Liquiere and J. Sallantin. Structural machine learning with Galois lattice and graphs. *Machine Learning: Proceedings of the 1998 International Conferences.* Margan Kaufmann, (ICML 98), pp. 305–313, San Francisco, CA, USA, 1998.

22. E. Mephu Nguifo. Galois lattice: A framework for concept learning—design, evaluation and refinement. pp. 461–467, 6° Tool with AI, New Orleans TAI, 1994.

23. M. L. Mugnier. Knowledge representation and reasoning based on graph homomorphisms. Proceedings 8th International Conference on Conceptual Structures, ICCS'2000, Lecture Notes in Artificial Intelligence, Vol. 1867, pp. 172–192, 2000.

24. S. Nijssen and J. Kok. A quickstart in frequent structure mining can make a difference. Proceedings of the SIGKDD. http://www.liacs.nl/home/snijssen/gaston/, 2004.

25. E. M. Norris. An algorithm for computing the maximal rectangles of a binary relation. *Journal of ACM,* 21:356–366, 1974.

26. L. Nourine and O. Raynaud. A fast algorithm for building lattices. *Information Processing Letters.* 71:199–204, 1999.

27. G. D. Plotkin. A further note on inductive generalization. In B. Meltzer and D. Michie, eds. *Machine Intelligence*, Vol. 6, Edinburgh University Press, Edinburgh, pp. 101–124, 1971.

28. M. Ricordeau. Q-concept-learning: Generalization with concept lattice representation in reinforcement learning. Proceedings of the 15th IEEE International Conference on Tools with Artificial Intelligence, IEEE Computer Society, 2003.

29. J. F. Sowa. *Conceptual Structures: Information Processing in Mind and Machine.* Addison-Wesley, Reading, MA, 1984.

30. G. Stumme, R. Taouil, Y. Bastide, and L. Lakhal. Computing iceberg concept lattices with TITANIC. *Data Knowl. Eng.* 42(2):189–222, 2002.

31. T. Washio and H. Motoda. State of the art of graph-based data mining. *ACM SIGKDD Explorations Newsletter* 5(1):59–68, 2003.

32. R. Wille. Knowledge acquisition by methods of formal concept analysis. In E. Diday, ed. *Data Analysis, Learning Symbolic and Numeric Knowledge*. ISBN:0-941743-64-0. Nova Science Publishers, pp. 365–380, New York, 1989.
33. M. Zaki. SPADE: An efficient algorithm of mining frequent sequences. *Machine Learning Journal*. 42(1/2):31–60, 2001.
34. H. Zhou. Multiplicativity. Part I. Variations, multiplicative graphs and digraphs. *Journal of graph Theory*. 15(5):469–488, 1991.

11

KERNEL METHODS FOR GRAPHS

THOMAS GÄRTNER,[1] TAMÁS HORVÁTH,[1] QUOC V. LE,[2] ALEX
J. SMOLA,[2] AND STEFAN WROBEL[1,3]

[1] *Fraunhofer AIS, Schloß Birlinghoven, Sankt Augustin, Germany*
[2] *Statistical Machine Learning Program, NICTA and ANU Canberra, Australia*
[3] *Department of Computer Science III, University of Bonn, Bonn Germany*

11.1 INTRODUCTION

Supervised learning is one of the most commonly considered data mining scenarios. The *supervised learning problem*—which we will concentrate on in this chapter—is to find a function that estimates a fixed but unknown functional or conditional dependence between objects and one of their properties—given some exemplary objects for which this property has been observed. The objects with observed property are called *training instances* and those for which the property has to be estimated are *test instances*. In the most common setting, known as *induction*, a good model of the dependence has to be found without knowing the test instances. A less common but nevertheless important problem is—given training and test instances—to find a model that has good predictive performance on the test data. This setting is known as *transduction*. In both cases, whenever the property takes one of a finite set of possible values, we speak of *classification*; whenever it takes real values, we speak of *regression*.

Mining Graph Data, Edited by Diane J. Cook and Lawrence B. Holder

Traditionally, machine learning research has focused on data that can directly be represented in a Euclidean space. Today, such methods are frequently applied in the analysis of business and scientific data. However, it has turned out that more and more of the collected data has a structure that hinders direct representation in a Euclidean space. Two powerful languages that can be used to represent such data are first- or higher-order logic [27] and graphs. Although graphs are special relational structures, for complexity reasons it is useful to consider them separately. In this chapter, we focus on graphs as our representation language only.

Two of the most frequently occurring machine learning problems involving graph structures are classification of vertices in a graph and classification of graphs in a graph database, that is, a set of disjoint graphs. A simple example of the first problem is the classification of Web pages on the World Wide Web, given the links between the pages. A simple example of the second problem is the classification of chemical compounds, given their atom bond structure. Both of these learning problems will be tackled in later parts of this chapter.

Our approach is to adapt kernel methods [32], a class of well-established learning algorithms, to graphs, rather than developing a new learning algorithm for graph-structured data from scratch. Today kernel methods are one of the most widely—and most successfully—applied classes of learning algorithms. Interestingly, one can look at two parts of kernel methods almost independently: a *kernel function* and a kernel-based *learning algorithm*. The kernel function is a positive definite function on the instances and "isolates" the learning algorithm from the instances, that is, the learning algorithm does not need to access any particular aspect of an instance—it relies on kernels between instances only. Hence, kernel methods can be applied to any data structure by defining a suitable kernel function. The main purpose of this chapter is to review some recent advances in the rapidly developing research area of *kernel methods for graphs*. This chapter is based on the work previously reported in [11–14, 18, 19] and is organized as follows: Section 11.2 presents recent results on making the classification of graphs tractable. Section 11.3 presents recent work on making transductive multiclass classification of vertices in a large graph tractable. Section 11.4 concludes.

11.2 GRAPH CLASSIFICATION

In this section we consider the classification of instances that have a natural representation as labeled graphs. That is, each instance is described by a graph $g = (V, E, \alpha, \beta)$, where the elements of V and E are labeled by the symbols of some appropriately chosen alphabets L_V and L_E, respectively (i.e., $\alpha : V \to L_V$ and $\beta : E \to L_E$). Depending on the set E of edges, g is either directed (i.e., when $E \subseteq V \times V$) or undirected (i.e., when $E \subseteq \{X \in 2^V : |X| = 2\}$). For a set G of graphs, we assume without loss of generality that the elements of G are all defined over the same vertex and edge alphabets, and that the graphs in G are either all directed or all undirected. For further basic notions and notations on graphs, the reader is referred to the introductory chapter of this book.

To classify graph-structured instances with kernel methods, for example, support vector machines [6], it suffices to define a valid (positive definite) kernel function $k : G \times G \to \mathbb{R}$, also referred to as *graph kernel*. In the design of practically useful graph kernels, one would require them to be computable in polynomial time in the size of the graphs and to distinguish between nonisomorphic graphs, that is, the underlying embedding function $\phi : g \mapsto k(g, \cdot)$ to be injective modulo isomorphism. Such graph kernels are called *complete* graph kernels. For the computation of complete graph kernels, the following result holds [13].

Proposition 11.1   Computing any complete graph kernel is at least as hard as deciding whether two graphs are isomorphic.

Since the graph isomorphism problem is believed to be not in P (and thus most likely not efficiently solvable, even though it is also probably not NP-complete), one has to resort to graph kernels where the underlying embedding functions may map some nonisomorphic graphs to the same point. Although this may seem to be a severe relaxation, empirical results indicate that several incomplete graph kernels have excellent predictive performance on real-world datasets. In the subsequent sections, for example, we also present some noncomplete graph kernels that have good predictive performance on a large benchmark chemical graph dataset.

The design of most graph kernels is based on the following general idea:

1. Map each graph to some distinguished, not necessarily finite, (multi)set of patterns occurring in the graph.
2. Define the graph kernel for each pair g_1, g_2 of graphs as a kernel on the (multi)sets assigned to g_1 and g_2.

As an example of this approach, we mention the family of graph kernels based on mapping each graph to a certain set of subgraphs occurring frequently in the set of graphs representing the instances. Encouraging empirical results have recently been reported for this approach (see [8] and Chapter 7 of this book). Such frequent pattern-based graph kernels, however, involve the problem of choosing a trade-off between predictive power and runtime, as both these conflicting requirements depend on the choice of the frequency threshold. As an alternative to graph kernels based on frequent subgraphs, in this section we describe two graphs kernels, the *walk kernel* (WK) [13] and the *cyclic pattern kernel* (CPK) [18, 19], which do not depend on such a frequency threshold.

11.2.1 Walk-Based Graph Kernels

Following the general methodology sketched above, in this section we define two kernels for labeled *directed* graphs. Both kernel definitions are based on mapping graphs to multisets of label sequences corresponding to *walks* in the graphs. Such graph kernels are called *walk kernels*.

To go into the technical details, we need some basic definitions. Let G be a set of labeled directed graphs (i.e., each graph in G is defined over the same alphabets of vertex and edge labels) and let $g = (V, E, \alpha, \beta)$ be a graph in G. A *walk* w in g is a sequence of vertices $w = v_1, v_2, \ldots v_{\ell+1}$ such that $(v_i, v_{i+1}) \in E$ for every $i = 1, \ldots, \ell$. The *length* of the walk is equal to the number of edges in this sequence, that is, ℓ in the above case. To compute walk kernels in a compact way, we use the *adjacency matrix M_g* of g, defined by

$$[M_g]_{ij} = \begin{cases} 1 & \text{if } (v_i, v_j) \in E \\ 0 & \text{otherwise} \end{cases}$$

for every $v_i, v_j \in V$.

Another central concept for the definition of walk kernels is the notion of products of labeled directed graphs. Let $g_1 = (V_1, E_1, \alpha_1, \beta_1)$ and $g_2 = (V_2, E_2, \alpha_2, \beta_2)$ be directed graphs over the same alphabets L_V and L_E of vertex and edge labels, respectively. Then the *direct product* of g_1 and g_2 is the directed graph $g_1 \times g_2 = (V, E, \alpha, \beta)$ over the alphabets L_V and L_E, where

$$V = \{(v_1, v_2) \in V_1 \times V_2 : \alpha_1(v_1) = \alpha_2(v_2)\}$$
$$E = \{((u_1, u_2), (v_1, v_2)) \in V \times V :$$
$$(u_1, v_1) \in E_1, (u_2, v_2) \in E_2, \text{ and } \beta_1(u_1, v_1) = \beta_2(u_2, v_2)\}$$

and α (resp. β) maps each vertex (resp. each edge) to the common label of its components.

The Direct Product Kernel. The walk kernel described first in this section is based on defining one feature for every possible label sequence and then counting how many walks in a graph match this label sequence. The inner product in this feature space can be computed with the following closed form: Let G be a set of directed graphs and let $g_1 = (V_1, E_1, \alpha_1, \beta_1)$ and $g_2 = (V_2, E_2, \alpha_2, \beta_2)$ be graphs in G. Let $M_{g_1 \times g_2}$ and $V_{g_1 \times g_2}$ denote the adjacency matrix and the vertex set of the direct product $g_1 \times g_2$, respectively. With a sequence of weights $\lambda_0, \lambda_1, \ldots$ ($\lambda_i \in \mathbb{R}; \lambda_i \geq 0$ for all $i \in \mathbb{N}$) the *direct product kernel* is defined as

$$k_\times(g_1, g_2) = \sum_{i,j=1}^{|V_{g_1 \times g_2}|} \left[\sum_{\ell=0}^{\infty} \lambda_\ell M_{g_1 \times g_2}^\ell \right]_{ij}$$

if the limit exists. Note that every entry $\left[M_{g_1 \times g_2}^\ell \right]_{ij}$ of the ℓth power of $M_{g_1 \times g_2}$ is the number of walks of length ℓ from $v_i = (v_{i,1}, v_{i,2})$ to $v_j = (v_{j,1}, v_{j,2})$ in the product graph $g_1 \times g_2$. This is in turn equal to the number of all possible pairs of walks of length ℓ from $v_{i,1}$ to $v_{j,1}$ in g_1 and from $v_{i,2}$ to $v_{j,2}$ in g_2 with the same label sequence.

To compute this kernel function one can make use of polynomial time-computable closed forms or resort to approximations by short walks on the graphs (see [13] for further details).

The Noncontiguous Sequence Kernel. A variant of the above kernel function that can be used whenever the label sequences are unlikely to match exactly is the following kernel. Let g_1 and g_2 be the graphs as defined above and let $g_1 \circ g_2$ be their direct product when ignoring the labels in g_1 and g_2. With a sequence of weights $\lambda_0, \lambda_1, \ldots$ ($\lambda_i \in \mathbb{R}$; $\lambda_i \geq 0$ for all $i \in \mathbb{N}$) and a factor $0 \leq \mu \leq 1$ penalizing gaps, the *noncontiguous sequence kernel* is defined as

$$k_*(g_1, g_2) = \sum_{i,j=1}^{|V_{g_1 \times g_2}|} \left[\sum_{\ell=0}^{\infty} \lambda_\ell \left((1-\mu) M_{g_1 \times g_2} + \mu M_{g_1 \circ g_2} \right)^\ell \right]_{ij}$$

if the limit exists.

This kernel is very similar to the direct product kernel. The only difference is that instead of $M_{g_1 \times g_2}$, the matrix $(1 - \mu) M_{g_1 \times g_2} + \mu M_{g_1 \circ g_2}$ is used. The relationship can be seen by adding—parallel to each edge—a new edge labeled # with weight $\sqrt{\mu}$ in both factor graphs.

11.2.2 Cycle-Based Graph Kernels

In this section we present another graph kernel, called the cyclic pattern kernel (CPK), which is based on mapping graphs into a distinguished set of cycles and trees. Using the labels of vertices and edges, these cycles and trees are then mapped to strings called *cyclic* and *tree patterns*. For two graphs, CPK is defined as the cardinality of the intersection of their cyclic and tree patterns. Although below we define CPK for undirected graphs, we note that the approach can easily be adapted to directed graphs as well.

To simplify the description, we first define a function that will be used many times in what follows. Let U be a set and $k_\cap : 2^U \times 2^U \to \mathbb{R}$ be the function defined by

$$k_\cap : (S_1, S_2) \mapsto |S_1 \cap S_2| \tag{11.1}$$

for every $S_1, S_2 \subseteq U$. The proof of the next proposition follows directly from the definition.

Proposition 11.2 $k_\cap$ is a kernel.

In order to define CPK, we recall some basic notions related to graphs. Let $g = (V, E, \alpha, \beta)$ be an undirected graph. Two vertices of g are *adjacent* if they are connected by an edge. The *degree* of a vertex $v \in V$ is the number of vertices adjacent to v. A graph $g' = (V', E', \alpha', \beta')$ is a *subgraph* of g, if $V' \subseteq V$, $E' \subseteq E$,

$\alpha'(v) = \alpha(v)$ for every $v \in V'$, and $\beta'(e) = \beta(e)$ for every $e \in E'$. A walk of g is a sequence $w = v_1, v_2, \ldots, v_{\ell+1}$ of vertices of g such that $\{v_i, v_{i+1}\} \in E$ for every $i = 1, \ldots, \ell$. If there is a walk between any pair of its vertices, g is *connected*. A *connected component* of g is a maximal subgraph of g that is connected. A vertex $v \in V$ is an *articulation* vertex if its removal increases the number of connected components of g. If it contains no articulation vertex, g is *biconnected*. A *biconnected component* of g is a maximal subgraph that is biconnected. A subgraph c of g forms a *simple cycle* if it is connected and each of its vertices has degree 2. We denote by $\mathcal{S}(g)$ the set of simple cycles of g. We note that the number of simple cycles can grow *faster* than $2^{|V|}$.

It holds that the biconnected components of a graph g are pairwise edge disjoint and thus form a partition on the set of edges of g. This partition, in turn, corresponds to the following equivalence relation on the set of edges: Two edges are equivalent if and only if they belong to a common simple cycle. This property of biconnected components implies that an edge of a graph belongs to a simple cycle if and only if its biconnected component contains more than one edge. Edges not belonging to simple cycles are called *bridges*. The subgraph of a graph g formed by its bridges is denoted by $\mathcal{B}g$. Clearly, each bridge of a graph is a singleton biconnected component, and $\mathcal{B}(g)$ is a forest.

11.2.2.1 CPK Defined by the Set of All Simple Cycles.
Let G be a set of undirected graphs over the vertex and edge alphabets L_V and L_E, respectively. We assume without loss of generality that L_V and L_E are disjoint. Let $\Sigma = L_V \cup L_E$, Γ be an alphabet, and let π be a mapping from the set of simple cycles and trees labeled by Σ to Γ^* such that

  (i) π is injective modulo isomorphism on the set of cycles and trees,
  (ii) π can be computed in polynomial time.

We note that such a π always exists and can easily be constructed (see, e.g., [19, 39]). Using π, the set of *cyclic* and *tree patterns* of a graph $g \in G$ is defined by

$$C(g) = \{\pi(C) : C \in \mathcal{S}(g)\} \qquad (11.2)$$

$$T(g) = \{\pi(T) : T \text{ is a connected component of } \mathcal{B}(g)\} \qquad (11.3)$$

respectively. The *cyclic pattern kernel* for G is then defined by

$$k_S(g_1, g_2) = k_\cap(C(g_1), C(g_2)) + k_\cap(T(g_1), T(g_2)) \qquad (11.4)$$

for every $g_1, g_2 \in G$. Since the sets $C(g)$ and $T(g)$ are disjoint for every $g \in G$, k_S is a kernel by Proposition 11.2.

As mentioned previously, kernels can often be computed efficiently using some closed form, that is, without explicitly performing the embedding into the feature space. Unfortunately, unless $P = NP$, k_S cannot be computed in polynomial time. In fact, one can show the following stronger negative result.

Algorithm **11.1** *Computing CPK*

Require: labeled undirected graphs g_1 and g_2
Ensure: $k_S(g_1, g_2)$

```
1: for i = 1, 2 do begin
2:     compute B(gᵢ)
3:     compute T(gᵢ) from B(gᵢ)
4:     compute S(gᵢ)
5:     compute C(gᵢ) from S(gᵢ)
6: end
7: return |C(g₁) ∩ C(g₂)| + |T(g₁) ∩ T(g₂)|
```

THEOREM 11.1   Computing k_S is at least as hard as counting simple cycles of length k in a graph.

This problem is, however, unlikely to be fixed-parameter tractable [10]. That is, for any graph with n vertices, most likely there is no algorithm counting simple cycles of length k in time $f(k) \cdot n^c$, where $f : \mathbb{N} \to \mathbb{N}$ is some computable function and c is some constant.

Based on the above negative complexity result, one can consider Algorithm 11.1, which calculates CPK by explicitly performing the embedding of graphs into the feature space. Since the biconnected components of a graph can be computed efficiently [34] and their number is bounded by the size (i.e., number of vertices) of the graph, steps 2 and 3 of Algorithm 11.1 can be computed in polynomial time. Since the set of simple cycles of a graph can be enumerated with polynomial delay [29], steps 4 and 5 can be performed in time polynomial in the sizes of g_1 and g_2, and in their number of simple cycles, which in turn can be exponential in the size of these graphs. Thus Algorithm 11.1 provides an efficient way of computing k_S for well-behaved graph databases, that is, for sets of disjoint graphs containing graphs with a small number of simple cycles.

The restriction to such well-behaved graph databases is necessary as while $S(g)$ can be listed in time polynomial in $|S(g)|$ [29], for the enumeration of the set $C(g)$ of cyclic patterns the following negative result holds [19].

Proposition 11.3   If P ≠ NP, the set of cyclic patterns of a labeled undirected graph cannot be enumerated in output-polynomial time.

Hence, although k_S depends only on *cyclic patterns* defined in (11.2), in steps 3 and 4 of Algorithm 11.1 we compute the set of all *simple cycles*. Algorithm 11.1 is thus polynomial in $|S(g_1)|$ and $|S(g_2)|$ rather than in $|C(g_1)|$ and $|C(g_2)|$.

Restricting CPK to well-behaved graph databases is a theoretically rather severe restriction because graphs containing exponentially many simple cycles may only have polynomially or even constantly many cyclic patterns. As an example, let g be the biconnected graph defined in Figure 11.1 such that g's vertices and edges

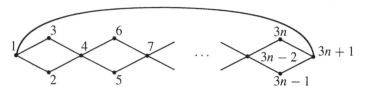

Figure 11.1. Graph of order $3n + 1$ such that edges and vertices are labeled by the same symbol (not shown in the figure). It contains $2^n + n$ simple cycles but only 2 cyclic patterns.

are labeled by the same symbol. Also g is made up of $3n + 1$ vertices and contains $2^n + n$ simple cycles, which in turn form, however, only two different cyclic patterns. Despite the worst-case exponential number of simple cycles, it turns out in practice that there are large-scale real-world graph databases that are well-behaved, for example, the NCI-HIV dataset consisting of more than 42,000 molecules, that we will use to evaluate our graph kernels.

In the next sections we present two approaches relaxing the above computational limitation of CPK.

11.2.2.2 CPK for Graphs with Bounded Treewidth.
The proof of Proposition 11.3 is based on a polynomial-time reduction from the NP-complete Hamiltonian cycle problem. This and many other NP-hard computational problems dealing with graphs become, however, polynomially solvable when restricted to graphs with *bounded treewidth* (see, e.g., [3] for an overview). Treewidth [31] is a measure of tree-likeness of graphs. More precisely, let $g = (V, E, \alpha, \beta)$ be a graph. A *tree decomposition* of g is a pair $(T, \mathcal{V})$, where $T = (I, F)$ is a tree and $\mathcal{V} = \{V_i \subseteq V : i \in I\}$ is a family of subsets of V such that

- (i) $\bigcup_{i \in I} V_i = V$,
- (ii) For every $\{u, v\} \in E$ there is an $i \in I$ such that $\{u, v\} \subseteq V_i$
- (iii) For every $i, j, k \in I$ it holds that if j is on the path from i to k in T, then $V_i \cap V_k \subseteq V_j$.

The *width* of a tree decomposition $(T, \mathcal{V})$ of g is $\max_{i \in I} |V_i| - 1$, and the *treewidth* of g is the width of a tree decomposition of g with the smallest width. Clearly, the treewidth of a tree is 1, and the treewidth of a simple cycle is 2.

The class of bounded treewidth graphs includes many practically relevant graph classes (see, e.g., [4] for an overview). We also note that graphs with small treewidth may have exponentially many simple cycles. As an example, consider again the graph given in Figure 11.1. This graph has treewidth 2 for every $n > 0$ and one of its tree decompositions with minimum treewidth is shown in Figure 11.2.

Using the positive result on the regular-language-constrained simple path problem for bounded treewidth graphs given in [1], the following positive result can be shown.

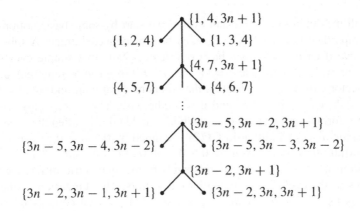

Figure 11.2. Tree decomposition of width 2 for the graph given in Figure 11.1.

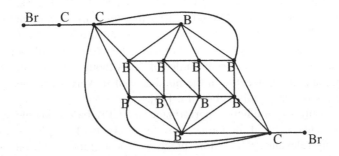

Figure 11.3. Chemical compound $C_3H_2B_{10}Br_2$ from the NCI-HIV dataset (NSC Id: 676529).

THEOREM 11.1 ([18])   For $i = 1, 2$, let $g_i = (V_i, E_i, \alpha_i, \beta_i)$ be graphs with constant bounded treewidth. Then $k_S(g_1, g_2)$ can be computed in time polynomial in

$$\max\{|V_1|, |V_2|, |\mathcal{C}(g_1)|, |\mathcal{C}(g_2)|\}$$

We omit the proof of this theorem and refer the reader to [18] for further details.

11.2.2.3   CPK Based on Relevant Cycles. Consider the chemical structure graph depicted in Figure 11.3. This graph has a single biconnected component with 12 vertices, containing 12,878 simple cycles. To avoid the computation of such a big set of simple cycles, in this section we empirically investigate whether another, possibly smaller set of cyclic structures can be employed without significant loss of predictive performance. For that we consider cyclic patterns based on the *relevant* cycles [28] of a graph. For instance, the molecular graph in Figure 11.3 has only 19 relevant cycles, each with length 3. (Note that this graph contains altogether 20 simple cycles of length 3.)

To recall the definition of relevant cycles, we start by some basic notions from algebraic graph theory. Let $g = (V, E, \alpha, \beta)$ be a biconnected graph. A subgraph c of g is a *cycle* if each vertex of c has even degree. Note that simple cycles are a special case of cycles. Each cycle $c = (V', E', \alpha', \beta')$ of g can be represented by the incidence vector $\mathbf{c}$ of its edges. That is, the components of $\mathbf{c}$ are indexed by E, and for every $e \in E$, $\mathbf{c}_e = 1$ if $e \in E'$ and it is 0 otherwise. The vectors corresponding to cycles in g form a vector space over the binary field GF(2), called the *cycle space* (or *cycle vector space*). Addition in GF(2) is defined by $0 \oplus 0 = 0$, $0 \oplus 1 = 1$, and $1 \oplus 1 = 0$. Multiplication is given by $0 \otimes 0 = 0$, $0 \otimes 1 = 0$, and $1 \otimes 1 = 1$. Thus, vector addition in the cycle space corresponds to the symmetric difference of the sets of edges of the cycles represented by the vectors. The dimension of the cycle space of g is its *cyclomatic number* $\nu(g) = |E| - |V| + 1$ (g consists of a single connected component). To represent the cycle space of g, one can consider any of its bases with minimum length, where the length of a basis is the sum of the number of edges of the cycles represented by the vectors belonging to the basis. Since the minimum basis of a graph's cycle space is not necessarily unique, the cyclic structure of a graph is described by the union of all minimum bases of its cycle space [28]. This canonical set of cycles is called the set of *relevant cycles*. In [37], it is shown that *relevant cycles* can be enumerated with polynomial delay and counted in time polynomial in the order of G. Although the set of relevant cycles of a graph can be exponential in the number n of its vertices in the worst case, its cardinality is typically only cubic in n [15].

To measure the predictive performance of CPK based on relevant cycles, in our experiments we used monotone increasing subsets of simple cycles that can be generated by relevant cycles. More precisely, for a graph g and integer $k \geq 1$, let $\mathcal{R}_k(g)$ denote the set

$$\mathcal{R}_k(g) = \begin{cases} \text{set of relevant cycles of } g & \text{if } k = 1 \\ \{c \oplus c' \in \mathcal{S}(g) : c \in \mathcal{R}_{k-1}(g) \text{ and } c' \in \mathcal{R}_1(g)\} & \text{otherwise} \end{cases}$$

Since $\mathcal{R}_1(g) \subseteq \mathcal{S}(g)$, it holds that $\mathcal{R}_1(g) \subseteq \mathcal{R}_2(g) \subseteq \cdots \subseteq \mathcal{R}_{\nu(g)}(g) = \mathcal{S}(g)$. We note that the set of relevant cycles of a graph is the union of the relevant cycles of its biconnected components. Since biconnected components of a graph are enumerable in linear time [34], and relevant cycles of a biconnected graph are enumerable with polynomial delay [37], $\mathcal{R}_k(g)$ can be computed in time polynomial in $|\mathcal{R}_k(g)|$ for any arbitrary graph g.

For a graph database G and integer $k \geq 1$, the CPK based on $\mathcal{R}_k(g)$, denoted $k_{\mathcal{R}_k}$, is then defined by

$$k_{\mathcal{R}_k}(g_1, g_2) = k_\cap(P_{\mathcal{R}_k}(g_1), P_{\mathcal{R}_k}(g_2)) + k_\cap(T(g_1), T(g_2))$$

for every $g_1, g_2 \in G$, where $P_{\mathcal{R}_k}(g) = \{\pi(c) : c \in \mathcal{R}_k(g)\}$ and $T(g)$ is defined by (11.3) for every $g \in G$. The remarks above along with Proposition 11.2 imply that $k_{\mathcal{R}_k}$ is a kernel that can be computed in time polynomial in $\max\{n_1, |\mathcal{R}_k(g_1)|, n_2, |\mathcal{R}_k(g_2)|\}$, where n_1 and n_2 denote the number of vertices of g_1 and g_2, respectively.

11.2.3 Empirical Evaluation

In this section, we evaluate the predictive power of walk-based and cyclic pattern kernels on the NCI-HIV dataset[1] of chemical compounds that has frequently been used in the empirical evaluation of graph mining approaches (see, e.g., [5, 7, 8, 25]). Each compound in this dataset is described by its chemical structure and classified into one of three categories based on its capability to inhibit the HIV virus: confirmed inactive (CI), moderately active (CM), or active (CA). The dataset contains 42,689 molecules, 423 of which are active, 1081 are moderately active, and 41,185 are inactive. Since more than 99% of the corresponding 42,689 chemical graphs contain less than 1000 cycles, the cyclic and tree patterns for the whole dataset can be computed in about 10 min.

Practical Considerations for the Walk Kernel. For molecule classification the number of vertex labels is limited by the number of elements occurring in natural compounds. It is therefore reasonable to not just use the element of the atom as its label. Instead, we use the pair consisting of the atom's element and the multiset of all neighbors' elements as the label.[2] In the HIV dataset this increases the number of different labels from 62 to 1391.

The size of this dataset, in particular the size of the graphs in this dataset, hinders the computation of walk-based graph kernels by means of eigendecompositions on the product graphs. The largest graph contains 214 atoms (not counting hydrogen atoms). If all had the same label, the product graph would have 45,796 vertices. As different elements occur in this molecule, the product graph has less vertices. However, it turns out that the largest product graph (without the vertex coloring step) still has 34,645 vertices. The vertex coloring above changes the number of vertices with the same label, thus the product graph is reduced to 12,293 vertices. For each kernel computation, either eigendecomposition or inversion of the adjacency matrix of a product graph has to be performed. With cubic time complexity, such operations on matrices of this size are not feasible.

The only chance to compute graph kernels in this application is to approximate them. There are two choices. First we consider counting the number of walks in the product graph up to a certain depth. In our experiments it turned out that counting walks with 13 or less vertices is still feasible. An alternative is to explicitly construct the image of each graph in the feature space. In the original dataset 62 different labels occur and after the vertex coloring 1391 different labels occur. The size of the feature space of label sequences of length 13 is then $62^{13} > 10^{23}$ for the original dataset and $1391^{13} > 10^{40}$ with the vertex coloring. We would also have to take into account walks with less than 13 vertices, but at the same time not all walks will occur in at least one graph. The size of this feature space hinders explicit computation. We thus resorted to counting walks with 13 or less vertices in the product graph.

---

[1] http://cactus.nci.nih.gov/ncidb/download.html.
[2] We note that more sophisticated vertex coloring algorithms are frequently employed in isomorphism tests.

Experimental Methodology and Results. We compare our approach to the results presented in [7] and [8]. The classification problems considered there were: (1) distinguish CA from CM, (2) distinguish CA and CM from CI, and (3) distinguish CA from CI. Additionally, we will consider (4) distinguish CA from CM and CI. For each problem, the area under the ROC curve (AUC), averaged over a five-fold cross validation, is given for different misclassification cost settings.

To choose the parameters of the walk-based graph kernel (to be precise, we use the direct product kernel) we proceeded as follows. We split the smallest problem (1) into 10% for parameter tuning and 90% for evaluation. First we tried different parameters for the exponential weight (10^{-3}, 10^{-2}, 10^{-1}, 1, 10) in a single nearest-neighbor algorithm (leading to an average AUC of 66, 66, 67.4, 75.9, and 33.8%) and decided to use 1 from now. Next we needed to choose the complexity (regularization) parameter of the SVM. Here we tried different parameters (10^{-3}, 10^{-2}, 10^{-1} leading to an average AUC of 69.4, 71.6, and 70.8%) and found the parameter 10^{-2} to work best. Evaluating with an SVM and these parameters on the remaining 90% of the data, we achieved an average AUC of 82.0% and standard deviation 0.024.

For cyclic pattern kernels, only the complexity constant of the support vector machine has to be chosen. Here, the heuristic as implemented in SVM light [20] is used. Also, we did not use any vertex coloring with cyclic pattern kernels.

To compare our results to those achieved in previous work, we fixed these parameters and reran the experiments on the full data of all four problems. Table 11.1 summarizes these results and the results with FSG reported in [7]. In [8] the authors of [7] describe improved results (FSG*). There, the authors report results obtained with an optimized threshold on the frequency of patterns.[3] Clearly, the graph kernels proposed here outperform FSG and FSG* over all problems and misclassification cost settings.

To evaluate the significance of our results we proceeded as follows: As we did not know the variance of the area under the ROC curve for FSG, we assumed the same variance as obtained with graph kernels. Thus, to test the hypothesis that graph kernels significantly outperform FSG, we used a pooled sample variance equal to the variance exhibited by graph kernels. As FSG and graph kernels were applied in a five-fold cross validation, the estimated standard error of the average difference is the pooled sample variance times $\sqrt{\frac{2}{5}}$. The test statistic is then the average difference divided by its estimated standard error. This statistic follows a t distribution. The null hypothesis—graph kernels perform no better than FSG—can be rejected at the significance level α if the test statistic is greater than the corresponding percentile of the t distribution.

Table 11.1 shows the detailed results of this comparison. Walk-based graph kernels perform always better or at least not significantly worse than any other kernel. Cyclic pattern kernels are sometimes outperformed by walk-based graph kernels but can be computed much more efficiently. For example, in the classification

---

[3]In [8] also including a description of the three-dimensional shape of each molecule is considered. We do not compare our results to those obtained using the three-dimensional information. We are considering to also include three-dimensional information in our future work and expect similar improvements.

TABLE 11.1   Area under the ROC Curve (AUC) in Percentage for Different Costs and Problems[a]

Task	Cost	Walk-Based Kernels (AUC in %)	Cyclic Pattern Kernels (AUC in %)	FSG (AUC in %)	FSG* (AUC in %)
(1)	1.0	81.8($\pm$2.4)	81.3($\pm$1.4)	77.4 $\ll_w$ $\ll_c$	81.0
(1)	2.5	82.5($\pm$3.2)	82.7($\pm$1.3)	78.2 $<_w$ $\ll_c$	79.2 $<_w \ll_c$
(2)	1.0	81.5($\pm$1.5)	77.5($\pm$1.7) $\ll_w$	74.2 $\ll_w$ $\ll_c$	76.5 $\ll_w$
(2)	35.0	79.9($\pm$1.1)	80.1($\pm$1.7)	77.8 $\ll_w$ $<_c$	79.4
(3)	1.0	94.2($\pm$1.5)	91.9($\pm$1.1) $<_w$	86.8 $\ll_w$ $\ll_c$	83.9 $\ll_w \ll_c$
(3)	100.0	94.4($\pm$1.5)	92.9($\pm$1.0) $<_w$	91.4 $\ll_w$ $<_c$	90.8 $\ll_w$ $\ll_c$
(4)	1.0	92.6($\pm$1.5)	90.8($\pm$2.4) $<_w$	—	—
(4)	100.0	92.8($\pm$1.3)	92.1($\pm$2.6)	—	—

[a] $<_w$, significant loss against walk-based kernels at 10%; $\ll_w$, significant loss against walk-based kernels at 1%; $<_c$, significant loss against cyclic pattern kernels at 10%; $\ll_c$, significant loss against cyclic pattern kernels at 1%.

TABLE 11.2   Area under the ROC Curve (AUC) in Percentage for Different Costs and Problems [$\gg_x$ (resp. $>_x$) Significant Win at 5% (resp. 10%) Level wrt $k_{\mathcal{R}_x}$]

Task	Cost	$k_{\mathcal{R}_1}$ (AUC in %)	$k_{\mathcal{R}_2}$ (AUC in %)	$k_{\mathcal{R}_3}$ (AUC in %	$k_{\mathcal{R}_\infty} = k_S$ (AUC in %)
(1)	1.0	80.1($\pm$4.5)	81.5($\pm$3.1) $\gg_1$	81.4($\pm$3.2)	81.3($\pm$3.3)
(1)	2.5	82.1($\pm$4.6)	83.0($\pm$4.1) $\gg_1$	82.9($\pm$3.9)	82.7($\pm$4.2)
(2)	1.0	75.4($\pm$2.2)	77.1($\pm$2.3) $\gg_1$	77.8($\pm$1.8) $\gg_1$	77.8($\pm$1.9) $\gg_1$
(2)	35.0	79.5($\pm$2.8)	80.0($\pm$2.7)	80.4($\pm$2.4) $\gg_1$	80.5($\pm$2.5) $\gg_{1,2}$
(3)	1.0	91.1($\pm$3.4)	92.5($\pm$2.9) $\gg_1$	92.5($\pm$2.7) $\gg_1$	92.6($\pm$2.8) $\gg_1$
(3)	100.0	93.7($\pm$2.4)	93.9($\pm$1.9)	93.6($\pm$1.8)	93.4($\pm$2.0)
(4)	1.0	89.2($\pm$3.2)	90.7($\pm$2.5) $\gg_1$	90.8($\pm$2.7) $\gg_1$	90.8($\pm$2.7) $\gg_1$
(4)	100.0	92.9($\pm$2.6)	92.9($\pm$3.2) $>_\infty$	92.6($\pm$2.9) $\gg_\infty$	92.2($\pm$3.0)

problem (4) fivefold cross validation with walk-based graph kernels finished in about 8 h while cyclic pattern kernels needed only 20 min.

Empirical Evaluation of $k_{\mathcal{R}_k}$. To evaluate CPK based on relevant cycles, we looked at the predictive performance of $k_{\mathcal{R}_1}, k_{\mathcal{R}_2}, k_{\mathcal{R}_3}$, and k_S. We used the above-described dataset and compared the different kernels with a paired t test on the same folds.[4] The results are given in Table 11.2. They indicate that $k_{\mathcal{R}_2}$ can be used in most of the cases without a significant loss of predictive performance wrt (with respect to) k_S and that $k_{\mathcal{R}_3}$ was never outperformed by k_S. Hence, the alternative definition of CPK allows one to apply it to graph databases containing graphs even with exponentially many simple cycles.

[4] However, note that the folds were different from the ones used in the above experiments.

11.3   VERTEX CLASSIFICATION

In this section we consider the classification vertices in a graph. That is, the vertices represent the instances of the learning problem, and we want to make use of our knowledge about common properties or relations between instances. While kernel functions between graphs (Section 11.2) can be used pretty directly with most available kernel methods, kernels for vertices raise somewhat different computational challenges. In this section we first give an overview of several kernels for vertices that have been defined in the literature. It turns out that for some of these more efficient learning is possible in a transductive setting. We thus develop an algorithm for *transductive Gaussian processes classification* of vertices that exploits this and perform an empirical evaluation on several datasets.

11.3.1   Kernels for Vertices

To apply Bayesian and other kernel methods to instances represented as vertices in a graph, we (only) need to define a covariance kernel on these instances. Hence, in this section we describe kernels that are based on knowledge about common properties of instances or relations between instances. We begin with a review of several kernels defined in the literature and discuss some efficiency issues later.

Kernel Definitions.   The best known kernel in this class is the *diffusion kernel* [24]. The motivation behind this and similar kernel functions is that it is often easier to describe the local neighborhood of an instance than to describe the structure of the whole instance space. For example, the neighborhood of an instance could be all instances that differ from this one only by the presence or absence of one property. When working with molecules such properties might be bonds, functional groups, or particular chemical properties. The neighborhood relation obviously induces global information about the make up of the instance space. The approach taken in the kernels described in this section is to try to capture this global information in a kernel function merely based on the neighborhood description. To model the structure of instance spaces, undirected graphs or hypergraphs can be used. While the use of hypergraphs is less common in the literature, it appears more natural.

A hypergraph is defined by a set of vertices V—the instances—and a set of edges E, where each edge is a subset of vertices. Each edge of the hypergraph can be interpreted as some property that all vertices of the edge have in common. For documents, for example, the edges could correspond to words or citations that they have in common; in a metric space the hyperedge could include all vertices with distance less than a given threshold from some point. By a walk in a hypergraph we understand a sequence of vertices and edges $v_1, e_1, v_2, \ldots v_{\ell+1}$ where $v_i, v_{i+1} \in e_i \in E$ for all $1 \leq i \leq \ell$.

For a hypergraph we define the $|V| \times |E|$ matrix B by $B_{ij} = 1$ if and only if $v_i \in e_j$ and $B_{ij} = 0$, otherwise. Let then the $|V| \times |V|$ matrix D be defined by $D_{ii} = \sum_j \left[B^\top B \right]_{ij} = \sum_j \left[B^\top B \right]_{ji}$. The matrices $B^\top B$ and $L = D - B^\top B$ are positive definite by construction. The matrix L is known as the graph Laplacian. Often the normalized Laplacian is used.

Conceptually, kernel matrices are then defined as the limits of matrix power series of the form

$$K = \sum_{i=0}^{\infty} \lambda_i \left(B^\top B\right)^i \quad \text{or} \quad K = \sum_{i=0}^{\infty} \lambda_i \left(-L\right)^i$$

with suitable parameters λ_i. These power series can be interpreted as measuring the number of walks of different lengths between given vertices. Sometimes the limit is approximated by a finite sum.

Limits of such power series can be computed by means of an eigenvalue decomposition of $B^\top B$ or $-L$, and a "recomposition" with modified eigenvalues. The modification of the eigenvalues is usually such that the order of eigenvalues is kept, while all eigenvalues are forced to become positive.

Examples for such kernel functions are the *diffusion kernel* [24]

$$K = \sum_{i=0}^{\infty} \frac{\beta^i}{i!}(-L)^i$$

the *von Neumann kernel* [23]

$$K = \sum_{i=1}^{\infty} \gamma^{i-1} \left(B^\top B\right)^i$$

and the *regularized Laplacian kernel* [33]

$$K = \sum_{i=1}^{\infty} \gamma^i (-L)^i$$

For exponential power series like the diffusion kernel, the limit can be computed by exponentiating the eigenvalues, while for geometrical power series, the limit can be computed by the formula $1/(1 - \gamma e)$, where e is an eigenvalue of $B^\top B$ or $-L$, respectively. Instead of computing K by manipulating the eigenvalues, one can alternatively compute kernels like the regularized Laplacian by inverting the matrix $1 + \gamma L$ (**1** denotes the identity matrix). A general framework and analysis of these kernels can be found in [33].

Efficiency Issues. While kernel functions between graphs (Section 11.2) can be used pretty directly with most available kernel methods, kernels for vertices raise somewhat different computational challenges. If the instance space is big, the computation of the kernels as defined above might be too expensive. Most kernel methods rely on computing Kv for some vector v several times in the course of the algorithm. Obtaining K by matrix inversion or eigenvalue decomposition is expensive and even if L is sparse, K hardly is, making Kv expensive as well.

However, there is another issue: Assume that we are given the inverse kernel matrix K^{-1} on training and test set and we wish to perform induction only. In this case we need to compute the kernel matrix (or its inverse) restricted to the training set. Let

$$K^{-1} = \begin{bmatrix} A & B \\ B^\top & C \end{bmatrix}$$

Then the upper left-hand corner (representing the training set part only) of K is given by the Schur complement $\left(A - B^\top C^{-1} B\right)^{-1}$. Computing the latter is costly as it involves the inversion of C and we usually assume a very large amount of unlabeled data. Moreover, neither the Schur complement nor its inverse are typically sparse.

Here we have a nice connection between graphical models and graph kernels. Assume that the target property t is a normal random variable with conditional independence properties. In this case the inverse covariance matrix has nonzero entries only for variables with a direct dependency structure. This follows directly from an application of the Clifford–Hammersley theorem to Gaussian random variables [26]. In other words, if we are given a graphical model of normal random variables, their conditional independence structure is reflected by K^{-1}.

In the same way as in graphical models marginalization may induce dependencies, computing the kernel matrix on the training set only may lead to dense matrices, even when the inverse kernel on training and test data combined is sparse. The bottom line is there are cases where it is computationally cheaper to take both training and test set into account and optimize over a larger set of variables rather than dealing with a smaller dense matrix (which is expensive to obtain).

Later in this chapter we develop a transductive algorithm that can use the inverse kernel matrix instead of the kernel matrix—thus avoiding expensive matrix inversion or eigenvalue decompositions with graph kernels—and exploit the sparsity of the graph.

11.3.2 Bayesian Kernel Methods

Before we describe our transductive Gaussian processes classifier in detail, in this section we give a brief introduction to Bayesian kernel methods.

Exponential Family Distributions. We begin with a brief review of exponential family distributions. For the purpose of learning algorithms we are usually interested in the joint density $p(x, y|\theta)$ or the conditional density $p(y|\theta, x)$ of random variables x, y with parameters θ. A density $p(x, y|\theta)$ with $(x, y) \in \mathcal{X} \times \mathcal{Y}$ is in the exponential family whenever it can be expressed as

$$p(x, y|\theta) = \exp\left[\langle \phi(x, y), \theta \rangle - g(\theta)\right]$$

where

$$g(\theta) = \log \int_{\mathcal{X} \times \mathcal{Y}} \exp\left[\langle \phi(x, y), \theta \rangle\right] dx \, dy$$

is called the *log-partition function*, $\phi(x, y)$ are *sufficient statistics* of (x, y), and $\langle \cdot, \cdot \rangle$ denotes the inner product. It holds then that

$$\frac{\partial}{\partial \theta} g(\theta) = \mathbf{E}_{p(x, y|\theta)}\left[\phi(x, y)\right]$$

$$\frac{\partial^2}{\partial \theta \, \partial \theta'} g(\theta) = \mathbf{E}_{p(x, y|\theta)}\left[\phi(x, y)\phi(x, y)'\right] - \mathbf{E}_{p(x, y|\theta)}\left[\phi(x, y)\right] \mathbf{E}_{p(x, y|\theta)}\left[\phi(x, y)'\right]$$

$$= \mathbf{Cov}_{p(x, y|\theta)}\left[\phi(x, y)\right]$$

and it can directly be seen that $p(x, y|\theta)$ is convex in θ.

From the joint exponential densities above, we can derive the conditional exponential densities as

$$p(y|x, \theta) = \exp\left[\langle \phi(x, y), \theta \rangle - g(\theta|x)\right]$$

where

$$g(\theta|x) = \log \int_{\mathcal{Y}} \exp\left(\langle \phi(x, y), \theta \rangle\right) \, dy$$

is the *conditional log-partition function*. It holds then that

$$\frac{\partial}{\partial \theta} g(\theta|x) = \mathbf{E}_{p(y|x, \theta)}\left[\phi(x, y)\right]$$

$$\frac{\partial^2}{\partial \theta \, \partial \theta'} g(\theta|x) = \mathbf{Cov}_{p(y|x, \theta)}\left[\phi(x, y)\right]$$

and it can directly be seen that $p(y|x, \theta)$ is convex in θ.

Bayesian Parameter Estimation. To estimate the label y of a new test point x from data $(X, Y) = \{(x_1, y_1), \dots, (x_{|X|}, y_{|X|})\}$ under the assumption of a parameterized family of distributions, like the exponential, we need to compute

$$p(y|x, X, Y) = \int p(y|x, \theta) p(\theta|X, Y) \, d\theta$$

To avoid the integral over θ, we can alternatively use

$$p(y|x, \theta^*)$$

where

$$\theta^* = \arg\max_{\theta} p(\theta|X, Y)$$

The quantity $p(\theta|X, Y)$ is known as the *posterior* of the parameters and is related to the *likelihood* of the parameters $p(X, Y|\theta)$ as follows:

$$p(\theta|X, Y) = p(X, Y|\theta)\frac{p(\theta)}{p(X, Y)}$$

As $p(X, Y)$ is independent of θ, we can maximize the posterior $p(\theta|X, Y)$ by maximizing the likelihood $p(X, Y|\theta)$ times the prior $p(\theta)$ or—equivalently—minimize the negative log-posterior:

$$-\log p(\theta|X, Y) = -\log \prod_{i=1}^{n} \exp\left[\langle\phi(x_i, y_i), \theta\rangle - g(\theta|x_i)\right] - \log p(\theta) + c'$$

$$= \sum_{i=1}^{n} g(\theta|x_i) - \sum_{i=1}^{n} \langle\phi(x_i, y_i), \theta\rangle - \log p(\theta) + c'$$

where $c' = \log p(X, Y) - \log(X|\theta) = \log p(X, Y) - \log(X) = p(Y|X)$ is independent of θ. Hence, it is a constant in the optimization problem and can be ignored.

Bayesian Kernel Methods. It turns out that under certain—rather general—conditions on the prior of the above minimization problem, it can be shown that the optimal parameters lie in the span of the sufficient statistics of the training data. This observation is known as the representer theorem. Thus we can replace θ by

$$\theta = \sum_j \int_{\mathcal{Y}} \alpha_{jy}\phi(x_j, y)\, dy$$

and minimize

$$\sum_{i=1}^{n} g(\theta|x_i) - \sum_{i,j=1}^{n} \int_{\mathcal{Y}} \alpha_{jy} \langle\phi(x_i, y_i), \phi(x_j, y)\rangle\, dy - \log p(\theta)$$

Assuming, for example, an uncorrelated Gaussian prior $p(\theta) \sim \mathcal{N}(0, \sigma^2 \mathbf{1})$ implies a Gaussian process on the random variable $t(x, y) := \langle\phi(x, y), \theta\rangle$ with covariance kernel k

$$\sigma^2 k((x, y), (x', y')) := \sigma^2 \langle\phi(x, y), \phi(x', y')\rangle = \mathrm{Cov}\left[t(x, y), t(x', y')\right] \qquad (11.5)$$

We thus obtain the minimization problem

$$\sum_{i=1}^{n} g(\theta|x_i) - \sum_{i,j=1}^{n} \int_{\mathcal{Y}} \alpha_{jy} k\left((x_i, y_i), (x_j, y)\right)\, dy$$

$$- \sum_{i,j=1}^{n} \int_{\mathcal{Y}\times\mathcal{Y}} \alpha_{iy'}\alpha_{jy} k\left((x_i, y'), (x_j, y)\right)\, dy\, dy'$$

with

$$g(\theta|x) = \log \int_{\mathcal{Y}} \exp \left[\sum_{j} \int_{\mathcal{Y}} \alpha_{jy} k\left((x_i, y'), (x_j, y)\right) \, dy \right] dy'$$

It remains to define a suitable joint covariance kernel $k : (\mathcal{X} \times \mathcal{Y}) \times (\mathcal{X} \times \mathcal{Y}) \to \mathbb{R}$. Usually this problem is simplified to the problem of defining the covariance of the instances $k_{\mathcal{X}} : \mathcal{X} \times \mathcal{X} \to \mathbb{R}$ based on the domain and the problem of defining the covariance of the labels based on the learning task $k_{\mathcal{Y}} : \mathcal{Y} \times \mathcal{Y} \to \mathbb{R}$. For regression often $k_{\mathcal{Y}}(y, y') = yy'$ is used; for classification often $k_{\mathcal{Y}}(y, y') = \delta_{yy'}$ is used. The joint covariance kernel is then simply the product $k((x, y), (x', y')) = k_{\mathcal{X}}(x, x') k_{\mathcal{Y}}(y, y')$. In the remainder of this section we will focus on $k_{\mathcal{X}}(\cdot, \cdot) = \langle \phi_{\mathcal{X}}(\cdot), \phi_{\mathcal{X}}(\cdot) \rangle$ and usually refer to it as k and ϕ.

11.3.3 Multiclass Transduction

A common problem of many transductive approaches is that they scale badly with the amount of unlabeled data, which prohibits the use of massive sets of unlabeled data. In this section, we thus present a transductive Gaussian process classifier for multiclass estimation problems. It performs particularly effective on graphs and other data structures for which the kernel matrix or its inverse have special numerical properties that allow fast matrix vector multiplication. An empirical evaluation of this algorithm shows improved predictive performance compared to both standard Gaussian process classification as well as transductive algorithms. We perform classification on a dataset consisting of a massive digraph with 75,888 vertices and 508,960 edges. To the best of our knowledge it has so far not been possible to perform transduction on graphs of this size in reasonable time (with standard hardware). On standard data our method shows competitive or better performance.

11.3.3.1 Gaussian Processes. We begin with a brief overview over Gaussian process multiclass classification [38] recast in terms of exponential families. Denote by $\mathcal{X} \times \mathcal{Y}$ with $\mathcal{Y} = \{1, \ldots, |\mathcal{Y}|\}$ the domain of observations and labels. Moreover let $X := \{x_1, \ldots, x_{|\mathcal{X}|}\}$ and $Y := \{y_1, \ldots, y_{|\mathcal{X}|}\}$ be the set of observations. It is our goal to estimate $y|x$ via

$$p(y|x, \theta) = \exp\left(\langle \phi(x, y), \theta \rangle - g(\theta|x)\right)$$

where

$$g(\theta|x) = \log \sum_{y \in \mathcal{Y}} \exp\left(\langle \phi(x, y), \theta \rangle\right)$$

$\phi(x, y)$ are the joint sufficient statistics of x and y and $g(\theta|x)$ is the conditional log-partition function that takes care of the normalization. We impose a normal prior on θ, leading to the following negative joint likelihood in θ and Y:

$$\mathcal{P} := -\log p(\theta, Y|X) = \sum_{i=1}^{|\mathcal{X}|} \left[g(\theta|x_i) - \langle \phi(x_i, y_i), \theta \rangle \right] + \frac{1}{2\sigma^2} \|\theta\|^2 + c \quad (11.6)$$

for some constant c independent of θ and Y. For transduction purposes, $p(\theta, Y|X)$ will prove more useful than $p(\theta|Y, X)$.

Parametric Optimization Problem. In the following we assume isotropy[5] among the class labels, that is, $\langle \phi(x, y), \phi(x', y') \rangle = \delta_{y,y'} \langle \phi(x), \phi(x') \rangle$ where δ is the Kronecker δ. This allows us to decompose θ into $\theta_1, \ldots, \theta_{|\mathcal{Y}|}$ such that

$$\langle \phi(x, y), \theta \rangle = \langle \phi(x), \theta_y \rangle \text{ and } \|\theta\|^2 = \sum_{y=1}^{|\mathcal{Y}|} \|\theta_y\|^2 \quad (11.7)$$

Applying the representer theorem in conjunction with (11.7) gives

$$\theta_y = \sum_{i=1}^{|\mathcal{X}|} \alpha_{iy} \phi(x_i) \quad (11.8)$$

where $\alpha \in \mathbb{R}^{|\mathcal{X}| \times |\mathcal{Y}|}$.

Let $\mu \in \mathbb{R}^{|\mathcal{X}| \times |\mathcal{Y}|}$ with $\mu_{ij} = 1$ if $y_i = j$ and $\mu_{ij} = 0$ otherwise, and $K \in \mathbb{R}^{|\mathcal{X}| \times |\mathcal{X}|}$ with $K_{ij} = \langle \phi(x_i), \phi(x_j) \rangle$. The joint log-likelihood (11.6) can then be expressed in terms of α and K, yielding

$$\mathcal{P} := \sum_{i=1}^{|\mathcal{X}|} \log \sum_{y=1}^{|\mathcal{Y}|} \exp \left([K\alpha]_{iy} \right) - \operatorname{tr} \mu^\top K\alpha + \frac{1}{2\sigma^2} \operatorname{tr} \alpha^\top K\alpha + c \quad (11.9)$$

where tr denotes the trace. Equivalently we could expand (11.6) in terms of $t := K\alpha$:

$$-\log p(\theta|X, Y) = \sum_{i=1}^{|\mathcal{X}|} \log \sum_{y=1}^{|\mathcal{Y}|} \exp \left([t]_{iy} \right) - \operatorname{tr} \mu^\top t + \frac{1}{2\sigma^2} \operatorname{tr} t^\top K^{-1} t + c \quad (11.10)$$

This is commonly done in Gaussian process literature, and we will use both formulations, depending on the problem we need to solve: If $K\alpha$ can be computed effectively, as is the case with string kernels, we use the α parameterization. Conversely, if $K^{-1}\alpha$ is cheap, as, for example, with graph kernels, we use the t parameterization.

---

[5]This is not a necessary requirement for the efficiency of our algorithm, however, it greatly simplifies the presentation.

DERIVATIVES. Second-order methods such as conjugate gradient require the computation of derivatives of $-\log p(\theta, Y|X)$ with respect to θ in terms of α or t. Using the shorthand $\pi \in \mathbb{R}^{m \times n}$ with $\pi_{ij} := p(y = j|x_i, \theta)$ we have

$$\partial_\alpha P = K(\pi - \mu + \sigma^{-2}\alpha) \tag{11.11a}$$

$$\partial_t P = \pi - \mu + \sigma^{-2} K^{-1} t \tag{11.11b}$$

To avoid spelling out tensors of fourth order for the second derivatives (since $\alpha \in \mathbb{R}^{|\mathcal{X}| \times |\mathcal{Y}|}$), we state the action of the latter as bilinear forms on vectors $\beta, \gamma, u, v \in \mathbb{R}^{|\mathcal{X}| \times |\mathcal{Y}|}$:

$$\partial_\alpha^2 P[\beta, \gamma] = \mathrm{tr}(K\gamma)^\top (\pi. * (K\beta)) - \mathrm{tr}(\pi. * K\gamma)^\top (\pi. * (K\beta))$$
$$+ \sigma^{-2} \mathrm{tr}\, \gamma^\top K\beta \tag{11.12a}$$

$$\partial_t^2 P[u, v] = \mathrm{tr}\, u^\top (\pi. * v) - \mathrm{tr}(\pi. * u)^\top (\pi. * v) + \sigma^{-2} \mathrm{tr}\, u^\top K^{-1} v. \tag{11.12b}$$

We used the Matlab notation of .* to denote element-wise multiplication of matrices.

Let $L \cdot |\mathcal{Y}|$ be the computation time required to compute $K\alpha$ and $K^{-1}t$, respectively. One may check that $L = O(|\mathcal{X}|)$ implies that each conjugate gradient (CG) descent step can be performed in $O(|\mathcal{X}|)$ time. Combining this with rates of convergence for Newton-type or nonlinear CG solver strategies yields overall time costs in the order of $O(|\mathcal{X}| \log |\mathcal{X}|)$ to $O(|\mathcal{X}|^2)$ worst case, a significant improvement over conventional $O(|\mathcal{X}|^3)$ methods.

11.3.3.2 *Transductive Inference by Variational Methods.*

As we are interested in transduction, the labels Y (and analogously the data X) decompose as $Y = Y_{\text{train}} \cup Y_{\text{test}}$. To directly estimate $p(Y_{\text{test}}|X, Y_{\text{train}})$ we would need to integrate out θ, which is usually intractable. Instead, we now aim at estimating the mode of $p(\theta|X, Y_{\text{train}})$ by variational means. With the KL divergence D and an arbitrary distribution q the well-known bound (see, e.g., [22])

$$-\log p(\theta|X, Y_{\text{train}}) \leq -\log p(\theta|X, Y_{\text{train}}) + D(q(Y_{\text{test}}) \| p(Y_{\text{test}}|X, Y_{\text{train}}, \theta)) \tag{11.13}$$

$$= -\sum_{Y_{\text{test}}} (\log p(Y_{\text{test}}, \theta|X, Y_{\text{train}}) - \log q(Y_{\text{test}})) \, q(Y_{\text{test}}) \tag{11.14}$$

holds. This bound (11.14) can be minimized with respect to θ and q in an iterative fashion. The key trick is that while using a factorizing approximation for q we restrict the latter to distributions that satisfy balancing constraints. That is, we require them to yield marginals on the unlabeled data, which are comparable with the labeled observations.

DECOMPOSING THE VARIATIONAL BOUND. To simplify (11.14) observe that

$$p(Y_{\text{test}}, \theta | X, Y_{\text{train}}) = p(Y_{\text{train}}, Y_{\text{test}}, \theta | X) / p(Y_{\text{train}} | X) \qquad (11.15)$$

In other words, the first term in (11.14) equals (11.9) up to a constant independent of θ or Y_{test}. With $q_{ij} := q(y_i = j)$, we define $\mu_{ij}(q) = q_{ij}$ for all $i > |\mathcal{X}_{\text{train}}|$ and $\mu_{ij}(q) = 1$ if $y_i = 1$ and 0 otherwise for all $i \leq |\mathcal{X}_{\text{train}}|$. In other words, we are taking the expectation in μ over all unobserved labels Y_{test} with respect to the distribution $q(Y_{\text{test}})$. We have

$$\sum_{Y_{\text{test}}} q(Y_{\text{test}}) \log p(Y_{\text{test}}, \theta | X, Y_{\text{train}}) = \sum_{i=1}^{|\mathcal{X}|} \log \sum_{j=1}^{|\mathcal{Y}|} \exp\left([K\alpha]_{ij}\right)$$

$$- \operatorname{tr}\mu(q)^\top K\alpha + \frac{1}{2\sigma^2} \operatorname{tr}\alpha^\top K\alpha + c \qquad (11.16)$$

which we can of course also expand in $t = K\alpha$ as in (11.10). The second term in (11.14) is the entropy of q. Since q factorizes, we have

$$\sum_{Y_{\text{test}}} q(Y_{\text{test}}) \log q(Y_{\text{test}}) = \sum_{i=|\mathcal{X}_{\text{train}}|+1}^{|\mathcal{X}|} q_{ij} \log q_{ij} \qquad (11.17)$$

For fixed q the optimization over θ proceeds as in the previous section for fully observed models. We now discuss optimization over q.

MARGINAL CONSTRAINTS ON q. It is unreasonable to assume that q may be chosen freely from all factorizing distributions (the latter would lead to a straightforward EM algorithm for transductive inference): If we observe a certain distribution of labels on the training set, for example, for binary classification we see 45% positive and 55% negative labels, then it is very unlikely that the label distribution on the test set deviates significantly. Hence we should make use of this information.

If $|\mathcal{X}| \gg |\mathcal{X}_{\text{train}}|$, a naive application of the variational bound can lead to cases where q is concentrated on one class—the increase in likelihood for a resulting very simple classifier completely outweighs any balancing constraints implicit in the data. This is confirmed by experimental results. It is, incidentally, also the reason why SVM transduction optimization codes [21] impose a balancing constraint on the assignment of test labels. We impose the following conditions:

$$r_j^- \leq \sum_{i=|\mathcal{X}_{\text{train}}|+1}^{|\mathcal{X}|} q_{ij} \leq r_j^+ \qquad \text{for all } j \in \mathcal{Y}$$

and

$$\sum_{j=1}^{|\mathcal{Y}|} q_{ij} = 1 \qquad \text{for all } i \in \left\{|\mathcal{X}|_r, \ldots, |\mathcal{X}|\right\}$$

Here the constraints $r_j^- = p_{emp}(y = j) - \epsilon$ and $r_j^+ = p_{emp}(y = j) + \epsilon$ are chosen such as to correspond to confidence intervals given by finite sample size tail bounds. In other words we set $p_{emp}(y = j) = |\mathcal{X}_{train}|^{-1}|\{i : 1 \leq i \leq |\mathcal{X}_{train}| \wedge y_i = j\}|$ and ϵ such as to satisfy

$$\Pr \left\{ \left| |\mathcal{X}_{train}|^{-1} \sum_{i=1}^{|\mathcal{X}_{train}|} \xi_i - |\mathcal{X}_{test}|^{-1} \sum_{i=1}^{|\mathcal{X}_{test}|} \xi_i' \right| > \epsilon \right\} \leq \delta \qquad (11.18)$$

for iid $\{0, 1\}$ random variables ξ_i and ξ_i' with mean p. This is a standard ghost-sample inequality. It follows directly from [17, Eq. (2.7)] after application of a union bound over the class labels that

$$\epsilon \leq \sqrt{\log(2|\mathcal{Y}|/\delta)|\mathcal{X}|/ (2|\mathcal{X}_{train}||\mathcal{X}_{test}|)} \qquad (11.19)$$

11.3.3.3 Optimization. *Optimization in α and t:* $\mathcal{P}$ is convex in α (and in t since $t = K\alpha$). This means that Newton–conjugate gradient (NCG) and Newton–Raphson (NR) can be used for optimization. We use the following procedure [9]:

- Compute updates $\alpha \longleftarrow \alpha - \eta \partial_\alpha^2 \mathcal{P}^{-1} \partial_\alpha \mathcal{P}$ via
  - Solve the linear system approximately by conjugate-gradient iterations.
  - Find optimal η by line search.
- Repeat until the norm of the gradient is sufficiently small.

Key is the fact that the arising linear system is only solved approximately, which can be done using very few CG iterations. Since each of them is $O(|\mathcal{X}|)$ for fast kernel-vector computations the overall cost is a subquadratic function of $|\mathcal{X}|$.

OPTIMIZATION IN q. Is somewhat less straightforward: We need to find the optimal q in terms of KL divergence subject to the marginal constraint. Denote by τ the part of $K\alpha$ pertaining to test data, or more formally $\tau \in \mathbb{R}^{|\mathcal{X}_{test}| \times |\mathcal{Y}|}$ with $\tau_{ij} = [K\alpha]_{i+|\mathcal{X}_{train}|,j}$. We have

$$\underset{q}{\text{Minimize}} \quad \text{tr}\, q^\top \tau + \sum_{i,j} q_{ij} \log q_{ij} \qquad (11.20a)$$

$$\text{Subject to} \quad q_j^- \leq \sum_i q_{ij} \leq q_j^+ \text{ for all } j \in \mathcal{Y} \qquad (11.20b)$$

$$\sum_j q_{ij} = 1 \text{ for all } i \in \{1, \ldots, |\mathcal{X}|\} \text{ and } q_{ij} \geq 0 \qquad (11.20c)$$

Using Lagrange multipliers, one can show that q needs to satisfy $q_{ij} = \exp(-\tau_{ij})$ $b_i c_j$ where $b_i, c_j \geq 0$. Solving for $\sum_j^{|\mathcal{Y}|} q_{ij} = 1$ yields

$$q_{ij} = \frac{\exp(-\tau_{ij})c_j}{\sum_{l=1}^{|\mathcal{Y}|} \exp(-\tau_{il})c_l}$$

This means that instead of an optimization problem in $|\mathcal{X}_{\text{test}}| \times |\mathcal{Y}|$ variables, we only need to optimize over $|\mathcal{Y}|$ variables subject to $2|\mathcal{Y}|$ constraints. Using $\gamma_j = \log c_j$, it is

$$\underset{c}{\text{Minimize}} \quad \sum_{i,j} \frac{\exp(\gamma_j - \tau_{ij})}{\sum_{l=1}^{|\mathcal{Y}|} \exp(\gamma_l - \tau_{il})} \left(\gamma_j - \log \sum_{l=1}^{|\mathcal{Y}|} \exp(\gamma_l - \tau_{il}) \right) \qquad (11.21a)$$

$$\text{Subject to} \quad q_j^- \le \sum_i \frac{\exp(\gamma_j - \tau_{ij})}{\sum_{l=1}^{|\mathcal{Y}|} \exp(\gamma_l - \tau_{il})} \le q_j^+ \qquad (11.21b)$$

This problem now only depends on $|\mathcal{Y}|$ variables. It can be solved by standard second-order methods. As initialization we choose γ_i such that the per class averages match the marginal constraint while ignoring the per sample balance. After that a small number of Newton steps suffices for optimization.

11.3.4   Related Work

String Kernels:. Efficient computation of string kernels using suffix trees was described in [36]. In particular, it was observed that expansions of the form $\sum_{i=1}^{|\mathcal{X}|} \alpha_i k(x_i, x)$ can be evaluated in linear time in the length of x, provided some preprocessing for the coefficients α and observations x_i is performed. This preprocessing is independent of x and can be computed in $O(\sum_i |x_i|)$ time. The efficient computation scheme covers all kernels of type

$$k(x, x') = \sum_s w_s \#_s(x) \#_s(x') \qquad (11.22)$$

for arbitrary $w_s \ge 0$. Here, $\#_s(x)$ denotes the number of occurrences of s in x and the sum is carried out over all substrings of x. This means that computation time for evaluating $K\alpha$ is again $O(\sum_i |x_i|)$ as we need to evaluate the kernel expansion for all $x \in X$. Since the average string length is independent of m, this yields an $O(m)$ algorithm for $K\alpha$.

VECTORS. If $k(x, x') = \phi(x)^\top \phi(x')$ and $\phi(x) \in \mathbb{R}^d$ for $d \ll |\mathcal{X}|$, it is possible to carry out matrix vector multiplications in $O(|\mathcal{X}|d)$ time. This is useful for cases where we have a sparse matrix with a small number of low-rank updates (e.g., from low-rank dense fill-ins).

EXISTING TRANSDUCTIVE APPROACHES. For SVMs use nonlinear programming [2] or EM-style iterations for binary classification [21]. Moreover, on graphs various methods for unsupervised learning have been proposed [40, 41], all of which are mainly concerned with computing the kernel matrix on training and test set jointly. Other formulations impose that the label assignment on the test set be consistent with the assumption of confident classification [35]. Others again exploit the fact that training and test set have similar marginal distributions [21].

The approach described in this chapter takes advantage of all three properties. Our formulation is particularly efficient whenever $K\alpha$ or $K^{-1}\alpha$ can be computed in linear time, where K is the kernel matrix and α is a coefficient vector. We approach the problem as follows:

- We require consistency of training and test marginals. This avoids problems with overly large majority classes and small training sets.
- Kernels (or their inverses) are computed on training and test set simultaneously. On graphs this can lead to considerable computational savings.
- Self-consistency of the estimates is achieved by a variational approach. This allows us to make use of Gaussian process multiclass formulations.

11.3.5 Empirical Evaluation

Unfortunately, we are not aware of other multiclass transductive learning algorithms. To still be able to compare our approach to other transductive learning algorithms, we performed experiments on some benchmark datasets. To investigate the performance of our algorithm in classifying vertices of a graph, we choose the WebKB dataset.

BENCHMARK DATASETS. Table 11.3 reports results on some benchmark datasets. To be able to compare the error rates of the transductive multiclass Gaussian process classifier proposed in this chapter, we also report error rates from [2] and compare to a simple Gaussian process classifier. The reported error rates are for 10-fold cross validations. Parameters were chosen on a single split inside each training fold.

GRAPH MINING. To illustrate the effectiveness of our approach on graphs we performed experiments on the well-known WebKB dataset. This dataset consists of 8275 Web pages classified into 7 classes. Each Web page contains textual content and/or links to other Web pages. As we are using this dataset to evaluate our graph mining algorithm, we ignore the text on each Web page and consider the dataset as a labeled directed graph. To have the data set as large as possible, in contrast to most other work, we did not remove any Web pages.

Table 11.4 reports the results of our algorithm on different subsets of the WebKB data as well as on the full data. We use the co-linkage graph and report results for "inverse" 10-fold stratified cross validations, that is, we use 1 fold as training data and 9 folds as test data. Parameters are the same for all reported experiments and were found by experimenting with a few parameter sets on the "Cornell" subset only. It turned out that the class membership probabilities are not well-calibrated on this dataset. To overcome this, we predict on the test set as follows: For each class the instances that are most likely to be in this class are picked (if they have not been picked for a class with lower index) such that the fraction of instances assigned to this class is the same on the training and test set. We will investigate the reason for this in future work.

The setting most similar to ours is probably the one described in [40]. Although a directed graph approach outperforms an undirected approach, we resorted to kernels

TABLE 11.3   Error Rates on Some Benchmark Datasets (Mostly from UCI)

Dataset	#Inst.	#Attr	Ind. GP (%)	Transd. GP (%)	S^{3}VMmip (%)[a]
cancer	699	9	3.4 ± 4.1	2.1 ± 4.7	3.4
cancer (progn.)	569	30	6.1 ± 3.7	6.0 ± 3.7	3.3
heart (cleave.)	297	13	15.0 ± 5.6	13.0 ± 6.3	16.0
housing	506	13	7.0 ± 1.0	6.8 ± 0.9	15.1
ionosphere	351	34	8.6 ± 6.3	6.1 ± 3.4	10.6
pima	769	8	19.6 ± 8.1	17.6 ± 8.0	22.2
sonar	208	60	10.5 ± 5.1	8.6 ± 3.4	21.9
glass	214	10	20.5 ± 1.6	17.3 ± 4.5	—
wine	178	13	19.4 ± 5.7	15.6 ± 4.2	—
tictactoe	958	9	3.9 ± 0.7	3.3 ± 0.6	—
cmc	1473	10	32.5 ± 7.1	28.9 ± 7.5	—
USPS	9298	256	5.9	4.8	—[b]

[a]The last column is the error rates reported in [2].
[b]In [2] only subsets of USPS were considered due to the size of this problem.

TABLE 11.4   Results on WebKB for "inverse" 10-fold Cross Validation

Dataset	$|V|$	$|E|$	Error (%)	Dataset	$|V|$	$|E|$	Error (%)
Cornell	867	1793	10	Misc	4113	4462	66
Texas	827	1683	8	All	8275	14370	53
Washington	1205	2368	10	Universities	4162	9591	12
Wisconsin	1263	3678	15				

for undirected graphs, as those are computationally more attractive. We will investigate computationally attractive digraph kernels in future work and expect similar benefits as reported by [40]. Though we are using more training data than [40], we are also considering a more difficult learning problem (multiclass without removing various instances). To investigate the behavior of our algorithm with less training data, we performed a 20-fold inverse cross validation on the "wisconsin" subset and observed an error rate of 17% there.

To show that the runtime performance of our algorithm is sufficient for classifying the vertices of massive graphs, we also performed initial experiments on the Epinions dataset [30]. The dataset is a social network consisting of 75,888 people connected by 508,960 "trust" edges. Additionally the dataset comes with a list of 185 "to previewers" for 25 topic areas. We tried to predict these but only got 12% of the top reviewers correct. As we are not aware of any predictive results on this task, we suppose this low accuracy is inherent to this task. However, the main point is that the experiments prove that the algorithm can be run on very large graph datasets.

11.4 CONCLUSIONS AND FUTURE WORK

In this chapter we reviewed some recent advances in the area of *kernel methods for graphs*. On the one hand we described kernels that can be used for graphs in a graph database. On the other hand we reviewed kernels that can be used for vertices in a single graph and introduced a transductive algorithm that allows efficient classification of vertices in a graph.

11.4.1 Graph Classification

The obvious approach to define kernels on objects that have a natural representation as a graph is to map each graph to a set of subgraphs and measure the intersection of the two sets. As mentioned above, such graph kernels cannot be computed efficiently if the mapping is required to be unique up to isomorphism.

In the literature different approaches have been tried to overcome this problem. Graepel [16] restricted the image of the graph to paths up to a given size, and [7] only considers the set of connected graphs that occur frequently as subgraphs in the graph database.

In this work we presented two effectively computable kernels for graphs. Although the underlying decompositions are not unique up to isomorphism, our experiments on a large chemical dataset indicate that the above complexity limitation does not hinder successful classification of molecules.

In future work we consider investigating more specialized kernels for molecules: Often more important than the chemical structure graph for the activity of a molecule is the three-dimensional (3D) structure of the molecule. Depending on the energy level, graphs can take different conformations and thus different 3D structures. Developing kernels that take 3D information into account might facilitate better predictive performance in such domains.

11.4.2 Vertex Classification

Current kernel methods for graphs do not scale very well with the available amount of unlabeled data. It turned out that for certain graph kernels it is more efficient to perform transduction than induction. We presented a transductive Gaussian process classifier for multiclass estimation problems. It performs particularly effectively on graphs and other data structures for which the kernel matrix or its inverse has special numerical properties that allow fast matrix vector multiplication. That said, also on standard dense problems we observed very large improvements (typically a 10% reduction of the training error) over standard induction.

Structured Labels and Conditional Random Fields are a clear area where one could extend the transductive setting. The key obstacle to overcome in this context is to find a suitable marginal distribution: With increasing structure of the labels the confidence bounds per subclass decrease dramatically. A promising strategy is to use only partial marginals on maximal cliques and enforce them directly, similarly to an unconditional Markov network.

Other Marginal Constraints than matching marginals are also worth exploring. In particular, constraints derived from exchangeable distributions such as those used by latent dirichlet allocation are a promising area to consider. This may also lead to connections between GP classification and clustering.

Sparse $O(m^{1.3})$ *Solvers for Graphs* have recently been proposed by the theoretical computer science community. It is worth exploring their use for inference on graphs.

Acknowledgments

The authors thank Mathew Richardson and Pedro Domingos for collecting the Epinions data and Deepayan Chakrabarti and Christos Faloutsos for providing a preprocessed version. Parts of this work were carried out when TG was visiting NICTA. National ICT Australia is funded through the Australian government's *Backing Australia's Ability* initiative, in part through the Australian Research Council. This work was partially supported by grants of the ARC, the Pascal Network of Excellence, and by the DFG project (WR 40/2-1) *Hybride Methoden und Systemarchitekturen für heterogene Informationsräume.*

REFERENCES

1. C. Barrett, R. Jacob, and M. Marathe. Formal-language-constrained path problems. *SIAM Journal on Computing*, 30(3):809–837, 2000.
2. K. Bennett. Combining support vector and mathematical programming methods for classification. In *Advances in Kernel Methods—Support Vector Learning*, pp. 307–326, MIT Press, Cambridge, 1998.
3. H. L. Bodlaender. A tourist guide through treewidth. *Acta Cybern.*, 11(1–2):1–22, 1993.
4. H. L. Bodlaender. A partial *k*-arboretum of graphs with bounded treewidth. *Theoretical Computer Science*, 209(1–2):1–45, 1998.
5. C. Borgelt and M. R. Berthold. Mining molecular fragments: Finding relevant substructures of molecules. In *Proceedings of the 2002 IEEE International Conference on Data Mining*. IEEE Computer Society, 2002.
6. B. E. Boser, I. M. Guyon, and V. N. Vapnik. A training algorithm for optimal margin classifiers. In D. Haussler, ed. *Proceedings of the Annual Conference on Computational Learning Theory*, pp. 144–152, ACM Press, New York, 1992.
7. M. Deshpande, M. Kuramochi, and G. Karypis. Automated approaches for classifying structures. In Proceedings of the 2nd ACM SIGKDD Workshop on Data Mining and Bioinformatics, 2002.
8. M. Deshpande, M. Kuramochi, and G. Karypis. Frequent sub-structure based approaches for classifying chemical compounds. In *Proceedings of the 3rd IEEE International Conference on Data Mining*, pp. 35–42. IEEE Computer Society, 2003.
9. R. Fletcher. *Practical Methods of Optimization*. Wiley, New York, 1989.
10. J. Flum and M. Grohe. The parameterized complexity of counting problems. *SIAM Journal on Computing*, 33(4):892–922, 2004.
11. T. Gärtner. A survey of kernels for structured data. *SIGKDD Explorations*, 5(1):49–58, July 2003.
12. T. Gärtner. Predictive graph mining with kernel methods. In *Advanced Methods for Knowledge Discovery from Complex Data*. Springer, Berlin, 2005.

13. T. Gärtner, P. A. Flach, and S. Wrobel. On graph kernels: Hardness results and efficient alternatives. In Proceedings of the 16th Annual Conference on Computational Learning Theory and the 7th Kernel Workshop, 2003.

14. T. Gärtner, Q. V. Le, S. Burton, A. J. Smola, and S. V. N. Vishwanathan. Large-scale multiclass transduction. *Advances in Neural Information Processing Systems, 18*: 411–418, 2006.

15. P. M. Gleiss and P. F. Stadler. Relevant cycles in biopolymers and random graph. In Proceedings of the 4th Slovene International Conference in Graph Theory, 1999.

16. T. Graepel. PAC-Bayesian pattern classification with kernels. Ph.D. thesis, TU Berlin, 2002.

17. W. Hoeffding. Probability inequalities for sums of bounded random variables. *Journal of the American Statistical Association*, 58:13–30, 1963.

18. T. Horváth. Cyclic pattern kernels revisited. In *Proceedings of Advances in Knowledge Discovery and Data Mining, 9th Pacific-Asia Conference, PAKDD 2005*, Vol. 3518 of *LNAI*, pp. 791–801. Springer, Berlin, 2005.

19. T. Horváth, T. Gärtner, and S. Wrobel. Cyclic pattern kernels for predictive graph mining. In Proceedings of the 10th ACM SIGKDD International Conference on Knowledge Discovery and Data Mining, pp. 158–167, 2004.

20. T. Joachims. Making large-scale SVM learning practical. In B. Schölkopf, C. J. C. Burges, and A. J. Smola, eds. *Advances in Kernel Methods—Support Vector Learning*, pp. 169–184, MIT Press, Cambridge, MA, 1999.

21. T. Joachims. *Learning to Classify Text Using Support Vector Machines: Methods, Theory, and Algorithms*. The Kluwer International Series in Engineering and Computer Science. Kluwer Academic, Boston, 2002.

22. M. I. Jordan, Z. Ghahramani, Tommi S. Jaakkola, and L. K. Saul. An introduction to variational methods for graphical models. *Machine Learning*, 37(2):183–233, 1999.

23. J. Kandola, J. Shawe-Taylor, and N. Cristianini. Learning semantic similarity. In *Advances in Neural Information Processing Systems*, Vol. 15. MIT Press, Cambridge, MA, 2003.

24. I. R. Kondor and J. D. Lafferty. Diffusion kernels on graphs and other discrete structures. In Proceedings of the 19th International Conference on Machine Learning, pp. 315–312, 2002.

25. S. Kramer, L. De Raedt, and C. Helma. Molecular feature mining in HIV data. In Proceedings and the Seventh ACM SIGKDD International Conference on Knowledge Discovery and Data Mining, pp. 136–143, 2002.

26. S. L. Lauritzen. *Graphical Models*. Oxford University Press, Oxford, 1996.

27. J. W. Lloyd. *Logic for Learning*. Springer, Berlin, 2003.

28. M. Plotkin. Mathematical basis of ring-finding algorithms at CIDS. *J. Chem. Doc.*, 11:60–63, 1971.

29. R. C. Read and R. E. Tarjan. Bounds on backtrack algorithms for listing cycles, paths, and spanning trees. *Networks*, 5(3):237–252, 1975.

30. M. Richardson and P. Domingos. Mining knowledge-sharing sites for viral marketing. In Proceedings of the Eighth ACM SIGKDD International Conference on Knowledge Discovery and Data Mining, 2002.

31. N. Robertson and P. D. Seymour. Graph minors. II. Algorithmic aspects of tree-width. *Journal of Algorithms*, 7(3):309–322, 1986.

32. B. Schölkopf and A. J. Smola. *Learning with Kernels*. MIT Press, Cambridge, MA, 2002.

33. A. J. Smola and I. R. Kondor. Kernels and regularization on graphs. In B. Schölkopf and M. K. Warmuth, eds. *Proceedings of the Annual Conference on Computational Learning Theory*, Lecture Notes in Computer Science. Springer, Berlin, 2003.
34. R. Tarjan. Depth-first search and linear graph algorithms. *SIAM Journal on Computing*, 1(2):146–160, 1972.
35. V. Vapnik. *Statistical Learning Theory*. Wiley, New York, 1998.
36. S. V. N. Vishwanathan and A. J. Smola. Fast kernels for string and tree matching. In K. Tsuda, B. Schölkopf, and J.P. Vert, eds. *Kernels and Bioinformatics*, MIT Press, Cambridge, MA, 2004.
37. P. Vismara. Union of all the minimum cycle bases of a graph. *Electronic Journal of Combinatorics*, 4(1):73–87, 1997.
38. C. K. I. Williams and D. Barber. Bayesian classification with Gaussian processes. *IEEE Transactions on Pattern Analysis and Machine Intelligence PAMI*, 20(12):1342–1351, 1998.
39. M. Zaki. Efficiently mining frequent trees in a forest. In *Proceedings of the 8th ACM SIGKDD International Conference on Knowledge Discovery and Data Mining*, pp. 71–80, ACM Press, New York, 2002.
40. D. Zhou, J. Huang, and B. Schölkopf. Learning from labeled and unlabeled data on a directed graph. In International Conference on Machine Learning, 2005.
41. X. Zhu, J. Lafferty, and Z. Ghahramani. Semi-supervised learning using gaussian fields and harmonic functions. In International Conference on Machine Learning ICML'03, 2003.

12

KERNELS AS LINK ANALYSIS MEASURES

MASASHI SHIMBO AND TAKAHIKO ITO

Nara Institute of Science and Technology
Ikoma, Nara 630-0192, Japan

12.1 INTRODUCTION

Citations have been a major source of information for analyzing the relationship between scientific papers, authors, and journals from the early days of bibliometric studies. With the recent proliferation of hypertext documents on the Web, many attempts have been made to apply the methods proposed in bibliometrics to the structural analysis of the Web and also to develop new methods. Two notable methods, PageRank [5] and HITS (Hypertext-Induced Topic Search) [15], have emerged from the research and are used extensively to evaluate the *importance*, or popularity, of Web pages.

Another type of link analysis measure has been studied in bibliometrics, with an intention to quantify the *relatedness*, or similarity, of two given documents. Co-citation coupling [22] is one such classical measure of relatedness. This method is still widely used, for example, by the well-known scientific literature search system CiteSeer [4] to recommend related papers to users.

These two lines of link analysis measures, namely, importance and relatedness, have been discussed independently in the literature. One of the objectives of this chapter is to present a unified framework that accounts for both importance and relatedness and to make it possible to define measures intermediate between the two.

Mining Graph Data, Edited by Diane J. Cook and Lawrence B. Holder

Our approach is based on a parametric family of positive semidefinite kernels [21] defining an inner product of nodes in a graph. In particular, we show that the parameterization of Kandola et al.'s Neumann kernels [12] allows us to identify the co-citation and bibliographic coupling [14] relatedness as an extreme of the parameter range and the HITS importance as the other extreme. Between these established relatedness and importance measures lies a spectrum of intermediate link analysis measures, obtained by tuning a single parameter.

Another topic addressed in this chapter is the limitation of the standard relatedness measures of co-citation and bibliographic coupling. These methods are not capable of computing relatedness if the documents have no common citations (in the case of bibliographic coupling) or if they are not jointly cited by any document (in co-citation). Even in these situations, the Neumann kernels can give nonzero scores to the pair of documents, unless they belong to different connected components. However, these scores are biased toward importance at the same time, making them unsuitable for evaluating relatedness. This limitation is overcome with different kernels based on the graph Laplacian (the regularized Laplacian kernels [23] and diffusion kernels [16]), which give a relatedness measure consistently over their parameter range. By introducing a new parameter to the Laplacian-based kernels, we obtain link analysis measures that are intermediate between the HITS importance and the relatedness measure given by the Laplacian-based kernels.

While several kernels on graph nodes have been proposed, their utility were discussed mainly in the context of machine learning, and their usage and characteristics as link analysis measures have not been fully investigated. We analyze some properties of these kernels that are essential for them to be used as link analysis measures. We also discuss the practical issues that may be encountered in the application of these kernels, including parameter tuning and approximation methods.

12.2 PRELIMINARIES

Most of the existing link analysis measures can be classified into two types: relatedness and importance. Relatedness measures quantify the similarity of two nodes in a graph or the relevance of one node to another. Importance, on the other hand, is the measure for ranking a given group of nodes in order of their significance, impact, or popularity within the group. In this section, we review several link analysis methods of relatedness and importance that are relevant to subsequent discussions. For more comprehensive surveys on link analysis, see [8] and [2, Chapter 5].

We assume that nodes in the graph are indexed by natural numbers and identify a node with its index, but when it is necessary the ith node is denoted by n_i. Throughout this chapter, uppercase letters denote matrices, and boldface letters denote column vectors. For a matrix A and a vector $\mathbf{v}$, A^T and $\mathbf{v}^\mathrm{T}$ respectively, denote the transpose of A and $\mathbf{v}$. All matrices are square unless noted otherwise. For integers i and j, $A(i, j)$ represents the (i, j)-element of a matrix A, and $\mathbf{v}(i)$ represents the ith component of vector $\mathbf{v}$. We denote by $\mathbf{1}$ the vector of all 1's. For any matrix A, let $\rho(A)$ be the spectral radius of A.

12.2.1 Relatedness: Co-citations and Bibliographic Couplings

A general assumption underpinning link analysis is that in graphs modeling a network structure such as bibliographic citations and the Web, an edge (a citation or hyperlink) between a pair of nodes (papers or Web pages) signifies the nodes being in some sense related. Hence the degree of relatedness can be inferred from the node proximity induced by the existence of edges.

Co-citation [22] and bibliographic coupling [14] are the standard methods of computing relatedness between documents from citation network, or, more generally, between nodes in a graph.

Co-citation coupling defines relatedness between documents as the number of other documents citing them both. And bibliographic coupling defines relatedness between two documents as the number of common references cited by the two. These measures can be formally defined in terms of the adjacency matrix of a citation graph. Given an adjacency matrix A, the number of co-citations between nodes i and j is given by the (i, j)-element of the *co-citation matrix* $A^{T}A$. Similarly, *bibliographic coupling matrix* AA^{T} gives the values of bibliographic coupling. Because these matrices are symmetric, their graph counterparts, the *co-citation graph* and *bibliographic coupling graph*, are undirected. Figure 12.1 illustrates a citation graph and its induced co-citation and bibliographic coupling graphs.

12.2.2 Importance: Kleinberg's HITS

Because of the difficulty in computing the importance of documents from their contents, citation counts have long been used as the index of document importance. A support for this approach was provided by several researchers; even though citations are made for various reasons, a positive correlation was observed between the number of citations and the significance or impact of the cited work. See [17] for the list of literature.

Kleinberg's HITS [15], along with PageRank [5], is a more recent and sophisticated method for evaluating document importance. HITS assigns two scores to each document (node), called the authority and hub scores. The assumption behind HITS is the existence of the *mutual reinforcement* relation between authorities and hubs. That is, authoritative documents are the ones that are cited by many hub documents, and hub documents are the ones that cite many authorities. Let A be the adjacency matrix of a citation graph. The HITS algorithm computes the following recursion over $n = 0, 1, \ldots$ starting from $\mathbf{a}_{(0)} = \mathbf{h}_{(0)} = \mathbf{1}$.

$$\mathbf{a}_{(n+1)} = \frac{A^{T}\mathbf{h}_{(n)}}{|A^{T}\mathbf{h}_{(n)}|} \qquad \mathbf{h}_{(n+1)} = \frac{A\mathbf{a}_{(n+1)}}{|A\mathbf{a}_{(n+1)}|} \tag{12.1}$$

The ith component of the *authority vector* $\lim_{n \to \infty} \mathbf{a}_{(n)}$ represents the *authority score* of node i. Similarly, the *hub vector* $\lim_{n \to \infty} \mathbf{h}_{(n)}$ gives the *hub scores*. It is well known that if the multiplicity of the dominant eigenvalue of $A^{T}A$ and AA^{T} is 1 (i.e., if it is a simple eigenvalue), the authority and hub vectors exist and equal the dominant eigenvectors of $A^{T}A$ and AA^{T}, respectively. Moreover, it can be shown that these scores are all nonnegative.

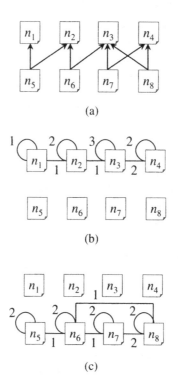

Figure 12.1. (a) Citation graph, and (b) the induced co-citation and (c) bibliographic coupling graphs.

12.3 KERNEL-BASED UNIFIED FRAMEWORK FOR IMPORTANCE AND RELATEDNESS

In this section and the next, we present some formulations of link analysis measures that are intermediate between importance of nodes and their relatedness. These formulations are based on the family of symmetric positive semidefinite kernels [21], which define an inner product of nodes in a graph.

The concept of an intermediate between importance and relatedness might sound ill-defined since importance is a measure defined on individual nodes, whereas relatedness is defined between them. However, given an importance score vector $\mathbf{v}$ such as the HITS authority vector, $\mathbf{v}\mathbf{v}^T$ defines a matrix in which every row (and column) i gives a ranking of nodes identical to the one given by $\mathbf{v}$ except for[1] an i such that $\mathbf{v}(i) \neq 0$. Importance can thus be treated as a function over a pair of nodes, or a matrix, as well.

---

[1] The assumption of $\mathbf{v}(i) \neq 0$ is weaker than it appears. For example, HITS assigns an authority score $\mathbf{v}(i) \neq 0$ to every node i if the co-citation graph is connected.

12.3.1   Neumann Kernels

Kandola et al. [12] proposed the *Neumann kernels* for computing document similarity from terms occurring in documents analogously to latent semantic analysis [7]. We discuss the interpretation of these kernels in the context of link analysis.

The Neumann kernel in its original form is defined in terms of the term-by-document (rectangular) matrix X whose (i, j)-element is the frequency of the ith term occurring in document j. To compute the kernel matrices, document correlation matrix $K = X^T X$ and term correlation matrix $M = X X^T$ are first constructed. The Neumann kernel matrices are then derived from K and M as follows.

Definition 12.1   Let X be a term-by-document matrix, and let $K = X^T X$ and $M = X X^T$. The *Neumann kernel* matrices with *diffusion factor* $\gamma (\geq 0)$, denoted by $\hat{K}_\gamma$ and $\hat{M}_\gamma$, are defined as the solution to the following system of equations.

$$\hat{K}_\gamma = \gamma X^T \hat{M}_\gamma X + K \qquad \hat{M}_\gamma = \gamma X^T \hat{K}_\gamma X + M \qquad (12.2)$$

The similarity between documents i and j is given by the (i, j)-element of $\hat{K}_\gamma$, and the term similarity is given by $\hat{M}_\gamma$.

Equation (12.2) implies an alternative representation based on the Neumann series:

$$\hat{K}_\gamma = K \sum_{n=0}^{\infty} \gamma^n K^n \qquad \hat{M}_\gamma = M \sum_{n=0}^{\infty} \gamma^n M^n \qquad (12.3)$$

Hence, when γ is less than the reciprocal of the spectral radius of K and M, the solution exists and is given by

$$\hat{K}_\gamma = K (I - \gamma K)^{-1} \qquad \hat{M}_\gamma = M (I - \gamma M)^{-1} \qquad (12.4)$$

12.3.2   Link Analysis with Neumann Kernels

The recurrence over $\hat{K}$ and $\hat{M}$ in Eq. (12.2) implies that the Neumann kernels evaluate similarity between documents from term similarity, and vice versa. This complementary relation is reminiscent of the recursion in Eq. (11.1) between the authorities and hubs in HITS. We apply the Neumann kernels to link analysis on the basis of this particular similarity to HITS. Specifically, we use the adjacency matrix A of a citation graph in place of the term-by-document matrix X. Hence we have $K = A^T A$ and $M = A A^T$, which coincide with the co-citation and bibliographic coupling matrices, respectively. Plugging them into Eq. (12.3) yields the Neumann kernels based solely on citation information. For convenience, we introduce the shorthand

$$N_\gamma(B) = B \sum_{n=0}^{\infty} (\gamma B)^n \qquad (12.5)$$

and write

$$\hat{K}_\gamma = N_\gamma(A^\mathsf{T}A) = A^\mathsf{T}A \sum_{n=0}^{\infty} \gamma^n (A^\mathsf{T}A)^n \qquad (12.6)$$

$$\hat{M}_\gamma = N_\gamma(AA^\mathsf{T}) = AA^\mathsf{T} \sum_{n=0}^{\infty} \gamma^n (AA^\mathsf{T})^n \qquad (12.7)$$

As seen from Eqs. (12.6) and (12.7), $\hat{M}_\gamma$ can be obtained simply by transposing A in Eq. (12.6). For this reason, our discussion below focuses on $\hat{K} = N_\gamma(A^\mathsf{T}A)$.

The relationship between HITS and the Neumann kernels thus obtained from citation graph turns out much deeper than just the superficial resemblance of their recursive forms. We will show that when viewed as a ranking method, the Neumann kernels subsume the HITS importance ranking as a special case.

12.3.3 Interpretation

Equation (12.5) shows that the Neumann kernel matrix $N_\gamma(A^\mathsf{T}A)$ is a weighted sum of $(A^\mathsf{T}A)^n$ over every $n = 1, 2, \ldots$. Given that the (i, j)-element of the term $(A^\mathsf{T}A)^n$ represents the number of paths of length n between nodes i and j in the co-citation graph, we see that each element of the kernel matrix equals the weighted sum of the number of paths between nodes.

Each term $(A^\mathsf{T}A)^n$ can be further interpreted as follows. When length n is small, the number $(A^\mathsf{T}A)^n(i, j)$ of paths between i and j is nonzero only if their distance is less than n. In other words, $(A^\mathsf{T}A)^n$ with small n captures the degree of proximity between any two nodes. Since relatedness is a measure based on proximity, $(A^\mathsf{T}A)^n(i, j)$ can be interpreted as indicating relatedness between nodes i and j. In particular, $(A^\mathsf{T}A)^1$ is exactly the co-citation matrix.

In contrast, it is not clear what this quantity indicates when n is sufficiently large. The following theorem and corollary state that the number of paths of length n between i and j is an indicator of the importance of these nodes.

THEOREM 12.1   Let λ be the dominant eigenvalue of a nonnegative symmetric matrix $A^\mathsf{T}A$. If λ is a simple eigenvalue (i.e., λ has a multiplicity of one), there exists a unit eigenvector $\mathbf{v}$ corresponding to λ such that $(A^\mathsf{T}A/\lambda)^n \to \mathbf{v}\mathbf{v}^\mathsf{T}$ as $n \to \infty$.

Proof. Let $A^\mathsf{T}A = \sum_{i=1}^{m} \lambda_i \mathbf{v}_i \mathbf{v}_i^\mathsf{T}$ be a spectral decomposition of $A^\mathsf{T}A$, with $\lambda = \lambda_1$, $\mathbf{v} = \mathbf{v}_1$, and $\mathbf{v}_i^\mathsf{T}\mathbf{v}_j = 0$ for every $i \neq j$, $1 \leq i, j \leq m$. By raising both sides to the power of n, we have for each $n = 1, 2, \ldots$,

$$\left(A^\mathsf{T}A\right)^n = \sum_{i=1}^{m} \lambda_i^n \mathbf{v}_i \mathbf{v}_i^\mathsf{T}$$

Dividing both sides of the equation by λ_1^n yields

$$\left(\frac{A^{\mathrm{T}}A}{\lambda_1}\right)^n = \mathbf{v}_1\mathbf{v}_1^{\mathrm{T}} + \sum_{i=2}^{m}\left(\frac{\lambda_i}{\lambda_1}\right)^n \mathbf{v}_i\mathbf{v}_i^{\mathrm{T}}$$

Because $\lambda = \lambda_1 > \lambda_i$ for every $i = 2, \ldots, m$, the second term on the right-hand side vanishes as $n \to \infty$.   □

Corollary 12.1   Let the dominant eigenvalue of nonnegative symmetric matrix $A^{\mathrm{T}}A$ be simple, and let $\mathbf{v}$ be an eigenvector corresponding to the dominant eigenvalue. For any node i with $\mathbf{v}(i) \neq 0$ and any node pair (j, k) such that $\mathbf{v}(j) > \mathbf{v}(k)$, there exists an integer m satisfying $(A^{\mathrm{T}}A)^n(i, j) > (A^{\mathrm{T}}A)^n(i, k)$ for all $n \geq m$.

When the co-citation graph is connected (or equivalently, $A^{\mathrm{T}}A$ is irreducible), Corollary 12.1 tells us that for every row (and column) vector of $(A^{\mathrm{T}}A)^n$, the node ranking induced by the magnitude of its components tends toward the HITS authority ranking (determined by the dominant eigenvector $\mathbf{v}$ of $A^{\mathrm{T}}A$). The similar argument is possible for bibliographic coupling and the Neumann kernels $N_\gamma(AA^{\mathrm{T}})^n$.

Putting the above discussions together, we see that summing $(A^{\mathrm{T}}A)^n$ and $(AA^{\mathrm{T}})^n$ over $n = 1, 2, \ldots$ as in Eqs. (12.6) and (12.7) can be interpreted as the mixture of relatedness (when n is small) and importance (when n is large). As a special case, the Neumann kernels subsume co-citation and bibliographic coupling at $\gamma = 0$. Near the opposite extreme of the parameter range, that is, $\gamma \simeq 1/\rho(A^{\mathrm{T}}A)$, the rankings induced by the Neumann kernels are also identical to the HITS importance, as stated by the following theorem.

THEOREM 12.2   Let λ be the dominant eigenvalue of a nonnegative symmetric matrix $A^{\mathrm{T}}A$. If λ is a simple eigenvalue, there exists a unit eigenvector $\mathbf{v}$ corresponding to λ such that

$$\left(\lambda^{-1} - \gamma\right) N_\gamma(A^{\mathrm{T}}A) \to \mathbf{v}\mathbf{v}^{\mathrm{T}} \quad as \quad \gamma \to \lambda^{-1} - 0$$

Proof. Let $A^{\mathrm{T}}A = \sum_{i=1}^{m} \lambda_i \mathbf{v}_i \mathbf{v}_i^{\mathrm{T}}$ be a spectral decomposition of $A^{\mathrm{T}}A$, with $\lambda = \lambda_1$ the dominant eigenvector and $\mathbf{v} = \mathbf{v}_1$. By the infinite series representation of the Neumann kernels [Eq. (12.6)], we have

$$N_\gamma(A^{\mathrm{T}}A) = \sum_{i=1}^{m}\left(\sum_{n=1}^{\infty}\gamma^{n-1}\lambda_i^n\right)\mathbf{v}_i\mathbf{v}_i^{\mathrm{T}} = \sum_{i=1}^{m}\frac{\lambda_i}{1 - \gamma\lambda_i}\mathbf{v}_i\mathbf{v}_i^{\mathrm{T}}$$

Multiplying both sides by $(1/\lambda_1 - \gamma)$ and rearranging terms yields

$$\left(\frac{1}{\lambda_1} - \gamma\right) N_\gamma(A^{\mathrm{T}}A) = \mathbf{v}_1\mathbf{v}_1^{\mathrm{T}} + \sum_{i=2}^{m}\frac{\lambda_i(1 - \gamma\lambda_1)}{\lambda_1(1 - \gamma\lambda_i)}\mathbf{v}_i\mathbf{v}_i^{\mathrm{T}}$$

where the second term on the right-hand side vanishes as $\gamma \to (1/\lambda_1) - 0$.   □

12.4 LAPLACIAN KERNELS AS A RELATEDNESS MEASURE

12.4.1 Limitations of Co-citation Relatedness

In Section 12.3, we argued that the Neumann kernels provide a unifying framework that subsumes both the HITS importance and the co-citation and bibliographic coupling relatedness. Unfortunately, this means the measures induced by the Neumann kernels inherit not only the advantages of co-citation and bibliographic coupling but also their limitations. The two limitations we address are as follows.

Limitation 12.1   Co-citation coupling assigns a nonzero relatedness score to a pair of documents only if they are commonly referenced by a document.

In the citation graph of Figure 12.1(a), documents n_1 and n_3 are not jointly cited by any document, and this results in the absence of an edge between n_1 and n_3 in the co-citation graph of Figure 12.1(b). They are hence unrelated to each other in terms of co-citation coupling. Because real-world graphs are typically sparse, it is often desirable to capture even weak relationship between nodes, such as n_1 and n_3 in this case. The relationship between these nodes might not be as strong as n_1 and n_2, or n_2 and n_3, but the fact that they are both co-cited with n_2 by other nodes still conveys valuable information.

Limitation 12.2   Co-citation coupling determines the relatedness between nodes i and j only on the basis of the number of nodes commonly citing the two. Nodes citing only one of i and j are neglected, and in effect, the number of citations from those nodes does not affect the relatedness between i and j in any way.

The following example illustrates why Limitation 12.2 is an issue.

EXAMPLE 12.1   In the graph depicted in Figure 12.2, n_3 represents a frequently linked web page such as Google and Yahoo. By contrast, n_1 is a much less popular page cited only by page n_4. Nodes n_1 and n_2 are co-cited by n_4, and nodes n_2 and n_3 by n_5. Our intuition dictates that n_2 is more related to n_1 than to n_3, as one would not conclude a page is related (or similar) to Google or Yahoo (n_3) just because it is jointly cited with them. However, because each of the page pairs (n_1, n_2) and (n_2, n_3) has a co-citation count of one (owing to n_4 and n_5, respectively), n_1 and n_3 are estimated as equally related to n_2 in terms of co-citation relatedness.

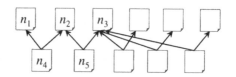

Figure 12.2.   Citation graph illustrating a limitation of co-citation relatedness.

The Neumann kernels, when viewed as relatedness measures, do not provide a solution to the above problems. They count the paths of any length between two nodes (see Section 12.3.3), so at first sight they appear to alleviate Limitation 12.1 if parameter $\gamma > 0$. However, increased γ biases the induced measures toward importance at the same time, making them unsuitable for evaluating relatedness. This bias also incurs Limitation 12.2. When applied to the graph of Figure 12.2, the Neumann kernels with nonzero γ assign a greater score to n_3 than to n_1 with respect to node n_2, which contradicts our intuition even further than co-citation. For example, the submatrix of a Neumann kernel for nodes n_1 through n_3 is shown below. Here, $\gamma = 0.18 \simeq 0.9\lambda^{-1}$ where λ is the spectral radius of the co-citation coupling matrix:

$$N_{0.18} = \begin{pmatrix} 1.89 & 3.05 & 3.40 \\ 3.05 & 8.34 & 15.49 \\ 3.40 & 15.49 & 46.12 \end{pmatrix} \tag{12.8}$$

In this submatrix, $N_{0.18}(2, 1) = 3.05 < 15.49 = N_{0.18}(2, 3)$. The same holds for smaller γ; at $\gamma = 0.02 \simeq 0.1\lambda^{-1}$, for instance, $N_{0.02}(2, 1) = 1.06 < 1.13 = N_{0.02}(2, 3)$.

12.4.2 Regularized Laplacian Kernels

We argue that the kernels based on the *graph Laplacian* [6] overcomes the limitations of Section 12.4.1.

Definition 12.2   Let B be the adjacency matrix of an undirected graph G with positive edge weights. The (combinatorial) *Laplacian* of G is defined as $L(B) = D(B) - B$, where $D(B)$ is a diagonal matrix with diagonals

$$D(B)(i, i) = \sum_j B(i, j) \tag{12.9}$$

Smola and Kondor define the regularized Laplacian kernels [23]) as follows.

Definition 12.3   Let B be a nonnegative symmetric matrix and let $\gamma \geq 0$. The matrix

$$R_\gamma(B) = (I + \gamma L(B))^{-1} \tag{12.10}$$

is called the *regularized Laplacian kernel* on the graph induced by B with diffusion factor γ.

To apply this kernel to link analysis, we take as B the co-citation matrix $A^T A$ or the bibliographic coupling matrix AA^T to ensure the symmetry of B.

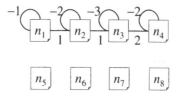

Figure 12.3. Graph induced by the negative Laplacian of the co-citation graph of Figure 12.1(b).

When $\gamma\lambda < 1$, where $\lambda = \rho(L(B))$, the right-hand side of (12.10) is the closed-form solution to the infinite series

$$R_\gamma(B) = \sum_{n=0}^{\infty} \gamma^n \, (-L(B))^n \qquad (12.11)$$

It is also possible to obtain this formula by using $-L(B)$ in place of the adjacency matrix B and dropping the first factor B from Eq. (12.5).

Note that the series in Eq. (12.11) converges only if $\gamma\lambda < 1$, where $\lambda = \rho(L(B))$, or the spectral radius of $L(B)$; the matrix (12.10) may exist even when (12.11) does not converge. From the practical standpoint, however, restricting the parameter range to $\gamma\lambda < 1$ has some merits in that an approximate computation method based on infinite series representation is applicable (see Section 12.5.2).

As a bonus, the infinite series representation admits the interpretation of kernel computation as path counting, paralleling the discussion of Section 12.3.3. In the case of the regularized Laplacian kernels, counting takes place not in a co-citation or a bibliographic coupling graph but in the graph induced by taking the negative of their Laplacian as the adjacency matrix. The difference is that self-loop edges in the latter graph have negative weights. Figure 12.3 depicts an example of the graph induced by $-L(B)$, where B is the adjacency matrix of the co-citation graph of Figure 12.1(b).

12.4.3 Relatedness Measure Induced by the Regularized Laplacian Kernels

Unlike the Neumann kernels, the regularized Laplacian kernels do not suffer from the limitations mentioned Section 12.4.1.

Both kernels count all paths regardless of length, and thus assign a nonzero value to any pair of nodes as long as they are connected. The regularized Laplacian kernels remain a relatedness measure even if diffusion factor γ is increased, by virtue of negative weights assigned to self-loops. During path counting, paths through these self-loops are also taken into account. This effectively reduces the score of authoritative nodes because the self-loops at these nodes typically have heavier weights; as seen from Eq. (12.9), the weight of a self-loop at a node is the negated sum of the edge weights incident to the node.

For the Neumann kernels, paths are counted in the co-citation graph. In this case, self-loops have positive weights, and hence in the Neumann kernels, counting paths through self loops amounts to amplifying the inner products (scores) involving authoritative nodes.

EXAMPLE 12.2 Recall the graph depicted in Figure 12.2 As argued in Example 12.1, it is more natural to regard n_2 as more related to n_1 than to n_3. The regularized Laplacian kernel matches this intuition by assigning a greater relatedness score to n_1 than to n_3 relative to n_2. Below is the regularized Laplacian kernel with $\gamma = 0.18 \simeq 0.9\lambda^{-1}$, where λ is the spectral radius of the negative Laplacian of the co-citation matrix.

$$R_{0.18} = \begin{pmatrix} 0.87 & 0.12 & 0.01 \\ 0.12 & 0.76 & 0.08 \\ 0.01 & 0.08 & 0.62 \end{pmatrix}$$

Again, only the submatrix of the kernel for nodes n_1 through n_3 is shown. Here, we have $R_{0.18}(2, 1) > R_{0.18}(2, 3)$.

If discounting high-degree nodes is all that is needed to overcome Limitation 12.2, one may argue that there should be simpler ways. However, straightforward discounting methods often fail to provide a relatedness measure that matches our intuition consistently over the parameter range; these methods may work at a relatively small γ, but as γ gets larger, they are either biased toward importance or give a measure inconsistent with our intuition on relatedness.

For example, it is easily verified that simply normalizing the Neumann kernel matrices (12.8) by $\overline{N}(i, j) = N(i, j)/\sqrt{N(i, i)N(j, j)}$ does not solve the problem, for $\overline{N}_{0.18}(2, 1) = 0.77 < 0.79 = \overline{N}_{0.18}(2, 3)$. Using the distance of nodes in the kernel feature spaces does not work either.

Another alternative is to use the column transition matrix $\overline{A}$, obtained by normalizing the adjacency matrix A so that its column sums equal to one[2] and applying the Neumann kernels to the co-citation matrix $\overline{A}^T\overline{A}$. This scheme appears to work at first sight. At $\gamma = 0.68 \simeq 0.9\lambda^{-1}$, it gives

$$N_{0.68}(\overline{A}^T\overline{A}) = \begin{pmatrix} 9.25 & 5.67 & 0.87 \\ 5.67 & 3.80 & 0.81 \\ 0.87 & 0.81 & 1.31 \end{pmatrix}$$

and this matrix evaluates n_2 as more related to n_1 than to n_3 as desired. Note however that n_3 is more related to n_1 than to n_2, that is, $N_{0.68}(\overline{A}^T\overline{A})(3, 1) > N_{0.68}(\overline{A}^T\overline{A})(3, 2)$. This is against intuition since the relatedness between n_3 and n_1 must be deduced from the paths from n_3 to n_1 in the co-citation graph, all of which pass through n_2 on the way. Indeed, using the same proof technique as

---

[2]The column transition matrix $\overline{A}$ is used in [11] to derive weighted co-citation matrix $\overline{A}^T\overline{A}$.

Theorem 12.2, we can show that $N_\gamma(\overline{A^\mathrm{T} A})$ in the limit $\gamma \to 1/\lambda$ gives identical ranking $n_1 > n_2 > n_3$ to all of the nodes n_1, n_2, and n_3.

This anomaly suggests that although using column transition matrices may be a handy heuristic, it discounts the score of high-degree nodes too severely. In consequence, it does not provide a relatedness measure consistently over its parameter range, nor a framework for providing intermediate measures between relatedness and importance that match our intuition.

By contrast, such anomaly does not occur in the regularized Laplacian kernels even in the limit of γ. The following theorem states that these kernels in the limit assign a uniform score to all the nodes in the same connected component of the co-citation graph $B = A^\mathrm{T} A$.

THEOREM 12.3   Let $B \in \mathbb{R}^{m \times m}$ be a nonnegative symmetric matrix whose graph counterpart is connected. The regularized Laplacian kernel $R_\gamma(B)$ converges to $(1/m)\mathbf{1}\mathbf{1}^\mathrm{T}$ as $\gamma \to \infty$.

Proof. (Sketch) First note that $(0, 1/m^{1/2})$ is an eigenpair of $L(B)$. By combining this fact with the technique similar to Theorem 12.2, we have the theorem.   □

Another question regarding the regularized Laplacian kernels is whether they yield a nonnegative score vector. This is obvious for the Neumann kernels, which are based on the powers of the (nonnegative) adjacency matrix, but the regularized Laplacian kernels are defined as the power series of (the negative of) the Laplacian, which contains negative elements. However, for the cases in which $\gamma\lambda < 1$, the proof of their nonnegativity can be obtained as follows.

THEOREM 12.4   The regularized Laplacian kernel $R_{\gamma,\alpha}(B)$ is nonnegative if $\gamma\rho(L(B)) < 1$.

Proof. It suffices to show that the matrix $P = I + \gamma L(B)$ is an M-matrix [3, 19]; that is, $P(i, j) \leq 0$ for every $i \neq j$ and $P^{-1} \geq 0$. Furthermore, a standard theorem in matrix analysis tells us that P is an M-matrix iff there exists a nonnegative matrix $Q = sI - P$ such that $s > \rho(P)$. Below, we show that such Q (and s) exists. Let $D(B)$ be the diagonal matrix as defined in Definition 12.2, and let $d = \max_i D(B)(i, i)$. If $s \geq 1 + \gamma d$, then $(s - 1)I - \gamma D(B)$ is nonnegative, and so is Q, because

$$Q = sI - P$$
$$= sI - (I + \gamma L(B))$$
$$= (s - 1)I - \gamma(D(B) - B)$$
$$= \big[(s - 1)I - \gamma D(B)\big] + \gamma B$$

We also have $s > \rho(Q)$ since

$$
\begin{aligned}
\rho(Q) &= \rho((s-1)I - \gamma L(B)) \\
&= \|(s-1)I - \gamma L(B)\|_2 \\
&\leq \|(s-1)I\|_2 + \|\gamma L(B)\|_2 \\
&= s - 1 + \gamma\|L(B)\|_2 \\
&= s - 1 + \gamma\rho(L(B)) \\
&< s
\end{aligned}
$$

Here, $\|A\|_2$ denotes the 2-norm of a matrix A, which equals to the spectral radius $\rho(A)$ when A is symmetric. $\qquad\square$

12.4.4 Controlling Bias in the Regularized Laplacian Kernels

While parameter γ controls the bias between importance and relatedness in the Neumann kernels, it does not serve the same purpose in the regularized Laplacian kernels. We will verify this claim in the experiment of Section 12.7.3. As an alternative way to control bias in the regularized Laplacian kernels, we introduce a new parameterization scheme.

Definition 12.4   Let B be the adjacency matrix of an undirected graph G with positive weights, and let $0 \leq \alpha \leq 1$. The *modified Laplacian* $L_\alpha(B)$ of G is defined as $L_\alpha(B) = \alpha D(B) - B$, where $D(B)(i,i) = \sum_j B(i,j)$ as in Definition 12.2.

Definition 12.5   Let B be a nonnegative symmetric matrix, and G be its induced graph. For $\gamma \geq 0$ and $0 \leq \alpha \leq 1$, the following infinite series, if convergent, is called the *modified regularized Laplacian kernel* on G:

$$
R_{\gamma,\alpha}(B) = \sum_{n=0}^{\infty} \gamma^n \, (-L_\alpha(B))^n \tag{12.12}
$$

where $L_\alpha(B)$ is the modified Laplacian of G.

Note that we defined the modified Laplacian kernels as a series, so $R_{\gamma,\alpha}(B)$ is defined only when the series converges, that is, $\gamma\rho(L_\alpha(B)) < 1$.

When $\alpha = 1$, $R_{\gamma,\alpha}(B)$ reduces to the (original) regularized Laplacian kernel $R_\gamma(B)$, and hence their elements represent relatedness between nodes. As α decreases toward 0, each row (column) vector of the kernel matrix bears more and more the character of an importance vector, provided that γ is sufficiently large. In particular, at $\alpha = 0$, $R_{\gamma,\alpha}(B)$ reduces to $I + \gamma N_\gamma(B)$, where $N_\gamma(B)$ is the Neumann kernel given by Eq. (12.5).

We can also show that the modified regularized Laplacian kernels are positive semidefinite, meaning that they indeed define an inner product and hence are compatible with support vector machines and other state-of-the-art kernel-based machine learning tools. To show their positive semidefiniteness, we first need a lemma.

LEMMA 12.1   Let $A \in \mathbb{R}^{m \times m}$ be a symmetric matrix, and let $f(\cdot)$ be a function such that $f(A)$ is convergent. If $f(\lambda)$ is nonnegative for all eigenvalues λ of A, then $f(A)$ is symmetric positive semidefinite.

Proof. Let $A = U^{\mathrm{T}} \Lambda U$ be an eigendecomposition of A, where U is an orthogonal matrix and Λ is a diagonal matrix with all its diagonals being eigenvalues of A. Hence by assumption, $f(\Lambda(i, i)) \geq 0$ for all $i = 1, \ldots, m$. We also have $f(A) = U^{\mathrm{T}} f(\Lambda) U$, where $f(\Lambda)$ is a diagonal matrix with $f(\Lambda)(i, i) = f(\Lambda(i, i))$. Therefore, $f(\Lambda)$ is a nonnegative diagonal matrix, and $C = f(\Lambda)^{1/2}$ exists. Substituting $f(\Lambda) = C^{\mathrm{T}} C$ in $f(A)$, we have $f(A) = (CU)^{\mathrm{T}} (CU)$.   □

THEOREM 12.5   For any nonnegative symmetric matrix B, the modified regularized Laplacian kernel $R_{\gamma,\alpha}(B)$, defined as a convergent series in Eq. (12.12), is symmetric positive semidefinite.

Proof. Let $f(x) = \sum_{n=0}^{\infty} (-\gamma x)^n$. The convergence of $R_{\gamma,\alpha}(B)$ implies $|\gamma \lambda| < 1$ for any eigenvalue λ of $L_\alpha(B)$. Hence we have $f(\lambda) = (1 + \gamma \lambda)^{-1} \geq 0$ for any eigenvalue λ of $L_\alpha(B)$. Now that $f(L_\alpha(B)) = R_{\gamma,\alpha}(B)$ by Definition 12.5 and $L_\alpha(B)$ is symmetric, the symmetric positive semidefiniteness of $R_{\gamma,\alpha}(B)$ follows from Lemma 12.1.   □

We can also prove the modified regularized Laplacian kernels are nonnegative.

THEOREM 12.6   The modified regularized Laplacian kernel $R_{\gamma,\alpha}(B)$ is nonnegative.

Proof. Omitted. The proof mostly parallels that of Theorem 12.4, using $L_\alpha(B)$ in place of $L(B)$.   □

From this theorem, it follows that each column vector of these kernel matrices can be viewed as a nonnegative score vector, just like the HITS importance vectors.

12.4.5 Diffusion Kernels

The kernels presented above are based on the Neumann series,[3] but other series can be used. Using the matrix exponential in place of the Neumann series yields the so-called *diffusion (heat) kernels*, originally developed in the context of spectral graph theory [6]. It was first introduced to machine learning community by Kondor and Lafferty [16].

---

[3]The reason we presented the kernels based on the Neumann series first is mainly for the ease of exposition; that is, it is easier to show the connection of the Neumann series based kernels to the existing measures of relatedness and importance, namely, co-citation coupling and HITS.

Definition 12.6 Let G be an undirected graph with positive weights, and B be its adjacency matrix. The *diffusion kernel* matrix H_γ on G with diffusion factor $\gamma \geq 0$ is given by

$$H_\gamma(B) = \exp(-\gamma L(B)) = \sum_{n=0}^{\infty} \frac{\gamma^n (-L(B))^n}{n!} \tag{12.13}$$

It can be shown that $H_\gamma(B)$ also converges to a uniform matrix as $\gamma \to \infty$. This property suggests that diffusion kernels can also be considered as a relatedness measure.

It is conceivable to use the modified Laplacian $L_\alpha(B) = \alpha D(B) - B$ in place of the Laplacian $L(B)$ with diffusion kernels. The resulting kernel $H_{\gamma,\alpha}(B)$ allows for controlling the bias between relatedness and importance just like the modified regularized Laplacian kernels.

THEOREM 12.7 For any nonnegative symmetric matrix B, $H_{\gamma,\alpha}(B) = \exp(-\gamma L_\alpha(B))$ is symmetric positive semidefinite.

Proof. First observe that $\exp(A)$ is positive semidefinite for any symmetric matrix A, which follows from Lemma 12.1 by setting $f(A) = \exp(A)$ and the fact that $f(x) = \exp(x) \geq 0$ for any $x \in \mathbb{R}$. Since $-\gamma L_{\gamma,\alpha}(B)$ is symmetric, $H_{\gamma,\alpha}(B) = f(-\gamma L_{\gamma,\alpha}(B))$ is symmetric positive semidefinite. $\square$

Using a technique similar to the one used for Theorem 12.2, we can also show that when $\mathbf{v}$ is the HITS authority vector of a citation graph A, and λ is the dominant eigenvector of $A^\mathsf{T} A$, the matrix $H_{0,\gamma}(A^\mathsf{T} A)/\exp(\gamma\lambda)$ converges to $\mathbf{v}\mathbf{v}^\mathsf{T}$ as $\gamma \to \infty$.

12.5 PRACTICAL ISSUES

In this section, we discuss some issues that may be encountered in the practical application of the kernel-based link analysis.

12.5.1 Parameter Tuning

The suitable values of the bias parameters, γ and α, must be chosen according to the objective of the task. As White and Smyth [24] put it (in regard to similar parameters in their proposed link analysis methods), the choices for the parameters are inherently subjective, and there appears to be no general methodology for determining parameter values.

For instance, consider a recommendation system for scientific literature. Given a small list of the "root" papers a user found interesting, the system should recommend other papers that may be of interest to the user. In this case, the degree of the user's acquaintance with the field of the given papers is likely to affect how much the system should bias (through parameter tuning) its decision toward authoritative papers in the field, but such knowledge on the user is often outside the scope of link analysis.

Given that choosing the "right" parameter requires user modeling that itself constitutes a difficult problem, a practical alternative should be to present to the user (or an external user-modeling module) a set of ranking lists (kernel matrices) under various parameter values, and let her choose the one that best fits their needs.

An issue with this approach is how we determine sample points efficiently. As we will see in Section 12.7, the character of the link analysis measures induced by these kernels turns out to be far from linear in the parameters. To this end, the derivatives of the kernel matrices with respect to the bias parameter can serve this purpose. For some kernels, the derivatives can be analytically computed from the kernel matrices at a given point γ as follows. We regard B as fixed and hence write $d/d\gamma$ instead of $\partial/\partial\gamma$.

$$\frac{dN_\gamma(B)}{d\gamma} = \left(N_\gamma(B)\right)^2 \tag{12.14}$$

$$\frac{dR_\gamma(B)}{d\gamma} = -L(B)\left(R_\gamma(B)\right)^2 \tag{12.15}$$

$$\frac{dH_\gamma(B)}{d\gamma} = -L(B)H_\gamma(B) \tag{12.16}$$

Let us take the Neumann kernel N_γ as an example. When we have N_γ for some γ at hand, we can compute the first-order approximation $\tilde{N}_{\gamma+\Delta\gamma}$ of the matrix $N_{\gamma+\Delta\gamma}$ by

$$N_{\gamma+\Delta\gamma}(B) \simeq \tilde{N}_{\gamma+\Delta\gamma}(B) = N_\gamma(B) + \Delta\gamma\frac{dN_\gamma(B)}{d\gamma} \tag{12.17}$$

where $dN_\gamma(B)/d\gamma$ is given by Eq. (12.14). By comparing the rankings induced by $\tilde{N}_{\gamma+\Delta\gamma}(B)$ and $N_\gamma(B)$, we can estimate how likely a change may occur in a given range $[\gamma, \gamma+\Delta\gamma]$.

The cost of this estimation is that of multiplication and summation in Eqs. (12.14) and (12.17). There is no need to compute $N_{\gamma+\Delta\gamma}(B)$ from scratch, until a suitable sampling interval is fixed. The feasibility of this approximation method will be empirically examined in Section 12.7.5.

12.5.2   Computational Issues

Computing the entire kernel matrix requires matrix inversion or exponentiation. Hence the computational complexity is roughly $O(|V|^3)$ where $|V|$ is the number of nodes in the graph, and this may be a computational burden when the graph is large. However, the standard techniques for matrix computation [10, Section 11.2] allow the approximation of kernel computation with the sum of the first k terms of the infinite series in Eqs. (12.6), (12.10), and (12.12). In Section 12.7.6, we evaluate the quality of these approximations using real data.

Furthermore, if one is concerned with the importance of nodes relative to a single node i rather than the entire kernel matrix, or if the entire kernel matrix cannot be kept on memory, we can reduce the space requirement by summing $\left(\gamma A^{\mathrm{T}} A\right)^n \mathbf{u}_i$ over $n = 1, \ldots, k$, where $\mathbf{u}_i$ is a unit vector with only 1 at the ith component; the computation now reduces to that of vector sums and the matrix–vector multiplication similar to HITS.

12.6 RELATED WORK

12.6.1 Relative Importance

"Relative importance" is a new link analysis measure recently proposed by White and Smyth [24]. This measure is defined as the "importance of nodes in a graph relative to one or more root nodes." In this view, HITS and PageRank are "global" importance algorithms. White and Smyth made a convincing argument that simply applying global importance algorithms to the subgraph surrounding the root nodes does not yield a precise estimate of relative importance because the root nodes are not given any special preference during importance computation.

The Neumann and modified Laplacian kernels fit naturally as relative importance, and as a bonus clarify the relationship between relative importance and relatedness (namely, the co-citation and bibliographic coupling relatedness).

12.6.2 Kernel-Based Link Analysis

Smola and Kondor [23] pointed out the connection between kernels on graph nodes and some importance computation methods including HITS. Their formulation appears quite different from the formulation presented in this chapter. They state that for a given node in a regular graph, its HITS score is given by the length of its corresponding vector in the feature space induced by Laplacian-based kernels. In the formulation presented in this chapter, by contrast, the HITS importance scores are obtained as an extremum of the Neumann kernels, which relies on an adjacency matrix instead of the Laplacian. In addition, the formulation takes the components of the kernel matrices directly as the scores indicating relatedness or importance, instead of the distance of nodes in the kernel-induced feature space.

Saerens and his colleagues [20] have recently proposed to use the *average commute time* as a distance measure between nodes in a graph. This quantity is based on the Markov-chain model of random walks and is shown to be computable from the pseudoinverse of the Laplacian. Since the graph Laplacian is positive semidefinite, so is its pseudoinverse.

12.6.3   Katz Status Index and Other Link Analysis Measures

The Katz status index[4] [13] is defined as the column sum of the infinite series

$$Z_\gamma(A) = \sum_{n=1}^{\infty} (\gamma A)^n = (I - \gamma A)^{-1} - I$$

where A is a (possibly asymmetric) adjacency matrix and $0 \le \gamma \le 1$ is called the attenuation parameter. Since $Z_\gamma(A) = \gamma N_\gamma(A)$, $Z_\gamma(A)$ is also a symmetric positive semidefinite kernel provided that A is symmetric. Although the matrix $Z_\gamma(A)$ is sometimes used as the "similarity" matrix, our analysis of the Neumann kernels in Section 12.4 implies that $Z_\gamma(A)$ is also biased toward importance as n gets greater.

Acharyya and Ghosh [1] discuss computing node importance similar to the Katz status index. They use the (row) transition matrix[5] $\hat{A}$ obtained by normalizing the rows of the given adjacency matrix A, instead of A itself.

Unlike the kernel-based formulation developed in this chapter, Acharyya and Ghosh [1] do not use the components of the matrix as indicating the relative merit of nodes with respect to a root node, but they use the dominant eigenvector of the matrix as the importance vector.

In the same paper, Acharyya and Ghosh [1] also pursued an automatic parameter tuning method, and they succeeded in detecting a stable parameter range using the maximum-entropy principle. Although a principled approach to parameter tuning is quite attractive, stability is only one of the many criteria for optimality, and we believe there is much room for further research.

12.7   EVALUATION WITH BIBLIOGRAPHIC CITATION DATA

To evaluate the characteristics of the kernel-based link analysis measures introduced in the previous sections, we applied them to a co-citation network of papers on natural language processing (NLP), which is a connected graph consisting of 2280 nodes (papers).

The reason for using this rather small graph is to make it possible to compute the exact kernel matrices. We evaluate the quality of the approximated computation method in Section 12.5.2, but this requires the exact kernel matrices as the baseline, whose computation would be infeasible with larger graphs.

To build this co-citation graph, we first collected the bibliographic references in the papers from major NLP journals and conferences, which resulted in a citation digraph consisting of 2867 nodes (papers). This citation digraph is then converted to a co-citation graph, and finally we extracted its largest connected component with 2280 nodes.

---

[4]We thank Prof. Marco Saerens for pointing out the similarity of the Neumann kernels and the Katz status index to us.

[5]Note the row transition matrix $\hat{A}$ is different from the column transition matrix $\overline{A}$ discussed in Section 12.4.3.

Note that we employed only the connected component because of the limitation in HITS. If there were multiple connected components, only nodes in one component would have nonzero HITS authority scores, since the scores are based on the dominant eigenvector; the rest of the nodes in the other components would have zero scores. In contrast, the kernel methods, which compute not a vector but a matrix, have no problem in assigning nontrivial scores between any node pairs in each connected components, regardless of the number of connected components.

After the co-citation graph is built, kernel matrices were computed with MATLAB from the closed-form representation of individual kernels; for example, $A^{T}A(I - \gamma A^{T}A)^{-1}$ for the Neumann kernels. Each kernel matrix was treated as a ranking method by taking the ith row vector of the matrix as the score vector for the ith node (paper). Given the ith score vector, or the ranking induced thereof, we call the ith node as the *root node* of this ranking, borrowing the terminology from [24].

We examine the results as follows. To gain intuition about the characteristics of individual kernels, we first present the ranking lists with respect to a certain root node. Note, however, that we are not trying to draw any conclusion about the general properties of the kernels from this specific example; when we observed some trends in this example, they were subsequently verified using the entire data.

12.7.1 An Illustrating Example

We first present the ranking lists for a root paper "Empirical Studies in Discourse" by M. A. Walker and J. D. Moore, *Computational Linguistics* 23(1):1–12, 1997 in Table 12.1. This table lists all 24 papers that are ranked as top 10 in at least one of the 11 ranking lists shown in the right-hand side of the table. The "Topic" field shows the topic category of the papers, parsing (P), word-sense disambiguation (W), machine translation (T), standard corpus (data set) (C), and discourse (D). An—in column CC (co-citation coupling) indicates that the paper was not co-cited with the root paper. The diffusion factor γ for kernels is shown as a normalized factor relative to the reciprocal λ^{-1} of the spectral radius $\lambda = \rho(B)$, where B is the co-citation matrix for the Neumann kernel (NK), or the negative of its Laplacian for the regularized Laplacian (RLK) and diffusion (DK) kernels. Hence the admissible range of parameter γ is $0 \leq \gamma < 1/\lambda$ for Neumann kernels. For the regularized Laplacian kernels to have infinite series representation (12.11), the same parameter range is required (but this time λ is the spectral radius of the Laplacian). In this experiment, we used this parameter range for the regularized Laplacian to save space, for the character of the regularized Laplacian for larger γ is mostly similar to the diffusion kernels. And $0 \leq \gamma\lambda < 1$ is also the range in which the approximated computation of Section 12.5.2 is applicable.

Before analyzing the results, let us clarify the reason we selected this root paper, "Empirical Studies in Discourse," as an illustrating example. It is a paper on discourse processing, and presumably, the papers most related to it must be those on discourse processing as well. In contrast, papers on discourse are not ranked among the top 10 HITS authority list (column H in Table 12.1). Hence we can expect

TABLE 12.1 Ranking List Relative to the Paper "Empirical Studies in Discourse"[a]

Paper Title	Topic	H	CC	NK ($\gamma\lambda$) 0.1	0.9	0.99	RLK ($\gamma\lambda$) 0.1	0.999	0.1	DK ($\gamma\lambda$) 0.1	1	10	100
Building a large annotated corpus of English: The Penn Treebank	C	**1**	2	**2**	**1**	**1**	**7**	**7**	**7**	**7**	**7**	**7**	50
A stochastic parts program and noun phrase parser for unrestricted text	P	**2**	—	12	7	2	12	23	12	12	16	133	419
Statistical decision-tree models for parsing	P	**3**	—	**8**	**5**	**3**	**8**	**8**	**8**	**8**	**9**	46	341
A new statistical parser based on bigram lexical dependencies	P	**4**	—	**9**	**6**	**4**	**10**	**10**	**9**	**9**	**10**	47	355
Unsupervised word sense disambiguation rivaling supervised methods	W	**5**	—	62	15	**5**	75	134	75	75	92	293	527
Word-sense disambiguation using statistical models of Roget's categories trained	W	**6**	—	365	12	**6**	32	54	32	32	44	150	377
The mathematics of statistical machine translation	T	**7**	—	374	26	**9**	551	553	544	544	552	623	981
Three generative, lexicalised models for statistical parsing	P	**8**	—	11	11	**7**	11	14	11	11	11	61	381
Transformation-based error-driven learning and natural language processing	P	**9**	—	21	14	**8**	24	47	24	24	28	201	594
Integrating multiple knowledge sources to disambiguate word sense	W	**10**	—	63	18	**10**	74	113	74	74	87	234	442
Attention, intentions, and the structure of discourse	D	50	2	**3**	**3**	25	**6**	**6**	**6**	**6**	**6**	**6**	11

Assessing agreement on classification tasks: The kappa statistic	D	76	**2**	**4**	**4**	41	**5**	**5**	**5**	**5**	**5**	**5**
Centering: A framework for modeling the local coherence of discourse	D	96	—	**10**	33	90	**9**	**9**	**10**	**8**	**9**	80
Combining multiple knowledge sources for discourse segmentation	D	115	—	14	62	110	13	13	13	12	**10**	75
A prosodic analysis of discourse segments in ···	D	198	—	15	96	182	15	11	15	13	**8**	14
The reliability of a dialogue structure coding scheme	D	201	**2**	**5**	**8**	94	**4**	**4**	**4**	**4**	**4**	**4**
Message Understanding Conference tests of discourse processing	D	604	**2**	**6**	**9**	156	**3**	**3**	**3**	**3**	**3**	**3**
Discourse segmentation by human and automated means	D	685	—	62	336	691	52	31	52	42	12	**8**
Empirical studies in discourse	D	771	**1**	**1**	**2**	95	**1**	**1**	**1**	**1**	**1**	**1**
A proposal for incremental dialogue evaluation	D	780	—	142	514	845	101	79	101	96	30	**9**
Human-machine problem solving using spoken language systems	D	1026	—	203	786	1171	146	133	146	146	77	**10**
Experiments in evaluating interactive spoken language systems	D	1046	—	204	805	1199	144	131	144	144	69	**7**
Preventing false inferences	D	1054	—	205	812	1213	145	132	145	145	70	**6**
Effects of variable initiative on linguistic behavior in human-computer spoken natural language dialogue	D	1061	**2**	**7**	**10**	205	**2**	**2**	**2**	**2**	**2**	**2**

[a]HITS (H), co-citation (CC), and the Neumann (NK), regularized Laplacian (RLK) and diffusion (DK) kernels. Top-10 ranks are shown in boldface.

the set of globally authoritative papers to be quite distinct from the set of papers most related to this specific paper. Consequently, we can observe whether a given ranking is inclined toward importance or relatedness, by comparing the ranking list with those of HITS and co-citation. Indeed, except for one paper, most of the papers with nonzero co-citation scores deal with discourse, as seen from their titles and the topic column. The only exception is the most authoritative paper, the one on a standard NLP dataset ("Penn Treebank").

The root node having a co-citation with the authoritative paper unrelated to discourse processing also helps to grasp the characteristics of the ranking methods. By examining where this authoritative paper is ranked relative to the other (presumably more related) co-cited papers on discourse, we can see whether the ranking list is biased toward importance or not. Moreover, we can regard these co-cited papers as a real-world example of the graph structure depicted in Figure 12.2 (in Section 12.4.1); the root paper is n_2, the Penn Treebank paper corresponds to n_3, and each of the other co-cited papers on discourse corresponds to n_1.

12.7.2 Neumann Kernels

Looking at the columns for the Neumann kernels (NK) in Table 12.1, we see that at $\gamma\lambda = 0.1$, the top-ranked paper is the root paper itself, followed by six papers that are co-cited with the root paper. As γ is increased, the ranking is more inclined toward that of HITS. At $\gamma\lambda = 0.99$, the top-10 list of papers is almost identical to HITS, while the ranking of the root paper drops to 95th.

The trend observed above is consistent with the role of γ we claimed in Section 12.3.3; as γ tends toward λ^{-1}, the measures induced by the Neumann kernels deviate away from the co-citation relatedness toward the HITS importance.

To confirm that this trend is not specific to this example, we evaluated the correlation between the top-10 paper lists produced by HITS and the Neumann kernels, for all 2280 root papers in the graph. Following White and Smyth [24], we use the minimizing Kendall (K-min) distance [9] to evaluate the correlation. Table 12.2 shows the K-min distance between the top-10 lists induced by the Neumann kernels and HITS, averaged over all 2280 root nodes. A small K-min distance means the two rankings are similar. It is equal to 0 if all top-10 items are identical, and takes the maximum value of 100 if there are no common items in the top-10 lists.

Table 12.2 indicates that the rankings induced by the Neumann kernels are indeed biased toward the HITS ranking as γ is increased.

TABLE 12.2 K-min Distance Between the Neumann Kernels and HITS, Averaged over 2280 Root Nodes

$\gamma\lambda$	0.1	0.5	0.9	0.99	0.999	0.9999	0.99999
Average K-min	87.3	86.3	72.0	26.4	5.5	1.1	0.0

12.7.3 Regularized Laplacian and Diffusion Kernels

Consider the column labeled 0.1 below "RLK" in Table 12.1, that is, the ranking list induced by the regularized Laplacian kernel with $\gamma = 0.1\lambda^{-1}$. We will compare this ranking with that of the Neumann kernel with $\gamma = 0.1\lambda^{-1}$. In these ranking lists, the set of the top seven papers is identical, which also matches those with nonzero co-citation scores. However, even though both methods rank the root paper itself at the top, the ordering of the other six is completely reversed. The regularized Laplacian kernel ranks the most authoritative "Penn Treebank" paper at the seventh, while the Neumann kernel ranks it as the second.

This result seems to suggest the property of the Neumann kernels that their ranking is biased toward importance even when diffusion rate γ is relatively small $(0.1\lambda^{-1})$. By contrast, the regularized Laplacian kernels seem less prone to the HITS ranking, even when γ is as large as $0.999\lambda^{-1}$.

We again verify whether this trend is generic to the entire data. Table 12.3 lists the average K-min distance among the rankings of the regularized Laplacian kernels, diffusion kernels, and HITS, with the average taken over 2280 root nodes. The rankings of the regularized Laplacian kernels seem to be extremely stable; the difference in K-min between $\gamma = 0.01\lambda^{-1}$ and $0.999\lambda^{-1}$ is only 0.5. All over this parameter range, these rankings do not resemble that of HITS; the K-min distances are consistently above 95 over this parameter range, while the distance between the Neumann kernel at $\gamma = 0.1$ is 87.3 (see Table 12.2).

To see if the regularized Laplacian kernels really is a relatedness measure, we need to measure the correlation between this kernel and co-citation. However, this time we cannot use the K-min distance for top-10 lists, as the number of co-citations can be less than 10. Instead, we have verified that for every root paper in the dataset, all the papers that are co-cited with the root paper are ranked topmost by the regularized Laplacian kernels with $\gamma = 0.1\lambda^{-1}$ and $0.01\lambda^{-1}$.

TABLE 12.3 K-min Distance Between the Regularized Laplacian Kernels and HITS, Averaged over 2280 Root Nodes

		RLK ($\gamma\lambda$)				DK ($\gamma\lambda$)				
	$\gamma\lambda$	0.01	0.1	0.5	0.999	0.01	0.1	1	10	100
RLK	0.01	0.0	0.1	0.1	0.5	3.5	0.6	0.2	5.5	24.0
	0.1		0.0	0.1	0.4	3.5	0.7	0.2	5.4	23.9
	0.5			0.0	0.4	3.7	0.7	0.2	5.4	23.9
	0.999				0.0	4.0	1.0	0.5	5.1	23.7
DK	0.01					0.0	3.5	3.6	8.9	27.4
	0.1						0.0	0.6	5.9	24.5
	1							0.0	5.4	24.0
	10								0.0	20.3
	100									0.0
HITS		95.7	95.7	95.8	96.1	95.7	95.7	95.8	99.3	99.8

TABLE 12.4   Average K-min Distance Between the Modified Regularized Laplacian Kernels and Other Link Analysis Measures: HITS (as a Baseline Measure of Importance), and the Original Regularized Laplacian Kernel with $\gamma = 0.1\lambda^{-1}$ (as a Measure of Relatedness)

α	0.01	0.05	0.1	0.15	0.2	0.3	0.5	0.75
HITS	0.0	12.9	35.0	51.8	89.6	94.7	96.0	96.1
Original RLK	95.0	93.3	95.3	94.3	33.5	17.3	8.2	4.1

The trends observed in the regularized Laplacian kernels are evident in the diffusion kernels as well. The ranking of the diffusion kernels with $\gamma = 0.1\lambda^{-1}$ for "Empirical Studies in Discourse" is mostly identical to that of the regularized Laplacian kernel with the same γ, shown in Table 12.1. By increasing γ to $100\lambda^{-1}$, we see that the diffusion kernels undervalue authoritative papers such as "The Penn Treebank" more severely. Although this paper is co-cited with the root paper, it is left out of the top-10 list in this case. This explains the trend that K-min distance between the diffusion kernels and HITS increases as γ is increased (see Table 12.3). Although it is not shown in the table, this trend continues for $\gamma > 100\lambda^{-1}$ because increased γ makes a "uniform" relatedness measure as we mentioned earlier, which does not resemble the relatedness measures given by co-citation or the regularized Laplacian kernels at small γ.

12.7.4   Modified Regularized Laplacian Kernels

We now examine the effect of parameter α on the modified regularized Laplacian kernels. To see if this parameter controls the trade-off between relatedness and importance as claimed, we compare the rankings induced by these kernels with those of HITS and the regularized Laplacian kernel with $\gamma = 0.1\lambda^{-1}$. The latter two are used as the benchmark of the importance and relatedness measures, respectively.[6] We again use the average K-min distance over all 2280 root nodes to assess the correlation. The diffusion rate γ is set to $0.99999\lambda^{-1}$, where λ is the spectral radius of the modified Laplacian of the co-citation graph.

The result is shown in Table 12.4. The modified regularized Laplacian kernels tend to be more similar to the HITS ranking as α is decreased. We see that the top-10 ranking induced by the modified regularized Laplacian kernel with $\alpha = 0.01$ is identical to that of HITS. When α is large, the rankings of the modified regularized Laplacian kernels are similar to the relatedness measure induced by the (unmodified) regularized Laplacian kernel.

---

[6]We did not use co-citation as the baseline for relatedness because of the same reason stated in Section 12.7.3; that is, the number of co-citations are often less than 10 and K-min distance cannot handle this case. The regularized Laplacian kernel was used instead, on the basis of the argument in Section 12.7.3.

TABLE 12.5   Average K-min Distance Between the Top-10 Rankings Induced by the Neumann Kernels at γ and $\gamma + \Delta\gamma$, Using the Exact Value of $N_{\gamma+\delta\gamma}$ and Estimated Values of $\tilde{N}_{\gamma+\delta\gamma}$ by Eq. (12.17)

(a) $\Delta\gamma = 0.099\lambda^{-1}$

$\gamma\lambda$	0.1	0.2	0.3	0.4	0.5	0.6	0.7	0.8	0.9
Exact	0.27	0.20	0.30	0.47	0.88	1.32	2.32	7.11	70.26
Approx.	0.07	0.13	0.22	0.34	0.65	0.99	1.56	3.31	9.74

(b) $\Delta\gamma = 0.009\lambda^{-1}$

$\gamma\lambda$	0.9	0.91	0.92	0.93	0.94	0.95	0.96	0.97	0.98	0.99
Exact	1.38	1.49	1.77	2.21	2.81	3.38	4.51	6.45	9.93	26.27
Approx.	1.23	1.32	1.61	1.89	2.33	3.14	3.42	4.85	5.49	7.73

12.7.5   Estimating Parameter Sensitivity

This subsection verifies the feasibility of the method described in Section 12.5 for estimating the parameter range in which kernel matrices (or, to be precise, the rankings they induce) are sensitive to parameter change.

The row labeled "Exact" in Table 12.5(a) shows the average K-min distance between the top-10 rankings induced by the Neumann kernel N_γ and $N_{\gamma+\Delta\gamma}$ for each γ (shown as the factor of $1/\lambda$), where $\Delta\gamma = 0.099\lambda^{-1}$. This quantity measures how much the top-10 rankings change as the diffusion factor changes from γ to $\gamma + \Delta\gamma$. Similarly, the row labeled "Approx." shows the average K-min distance between N_γ and the first-order approximation $\tilde{N}_{\gamma+\Delta\gamma}$ of $N_{\gamma+\Delta\gamma}$ by formula (12.17). As we see from the table, the K-min for the approximation is larger in regions (i.e., $\gamma\lambda = 0.9$) where actual values of K-min are larger, relative to the regions where they are not (cf. Table 12.2). Note that the large discrepancy in the value of K-min at this point (70.26 vs. 9.74) is only because the step size $\Delta\gamma = 0.009\lambda^{-1}$ is too large to capture the change of kernel matrices in this region. When one finds a surge in the value of estimated K-min in a region as in this case, the region must be sampled with a finer-grained step size. The discrepancy in the K-min distance between the actual and estimated kernels in the parameter range $\gamma\lambda = 0.9$ to 0.99, with step size $\Delta\gamma = 0.009\lambda^{-1}$ are shown in Table 12.5(b).

12.7.6   Approximation

In the final experiment, we examine the feasibility of approximating kernel matrices with the sum of the first k terms of their power series representations. These representations are given by Eq. (12.6) for the Neumann kernels, and Eq. (12.11) for the regularized Laplacian kernels. The average K-min distances between the kernel matrix obtained from the closed form and the approximations with various step size k are shown in Table 12.6. Although not present in the table, the standard deviation at $k = 1000$ is 5 for the Neumann kernel with $\gamma\lambda = 0.999$ and is less than 1 for the

TABLE 12.6 Average K-min Distance Between the Exact and Approximate Solutions: Neumann Kernels (NK), the Regularized Laplacian Kernels (RLK) and Diffusion Kernels (DK)[a]

k	NK ($\gamma\lambda$)				RLK ($\gamma\lambda$)				DK ($\gamma\lambda$)		
	0.1	0.9	0.99	0.999	0.1	0.9	0.99	0.999	0.1	1	10
5	0.0	10.0	50.7	75.1	0.0	0.4	1.1	1.3	0.5	0.1	14.4
10	0.0	6.0	42.5	66.5	0.0	0.4	0.6	0.6	0.5	0.1	21.1
50	0.0	0.1	14.9	31.6	0.0	0.0	0.5	0.5	0.5	0.1	0.1
100	0.0	0.0	6.5	21.2	0.0	0.0	0.4	0.5	0.5	0.1	0.1
500	0.0	0.0	0.1	4.8	0.0	0.0	0.1	0.4	0.5	0.1	0.1
1000	0.0	0.0	0.0	1.6	0.0	0.0	0.0	0.3	0.5	0.1	0.1

[a]The approximations are given by the sum of the first k terms of their power series representation.

other kernel/parameter pairs. The values for diffusion kernels (DK) not converging to 0 is due to round-off errors.

12.8 SUMMARY

Although several kernels on graph nodes have been proposed in the context of machine learning, their property as link analysis measures was not previously discussed in depth. This chapter has demonstrated the merit of viewing these kernels as link analysis measures.

The Neumann kernels, when applied to citation graphs, yield a link analysis measure intermediate between co-citation/bibliographic coupling relatedness and the HITS importance, providing a new account of the relationship between these well-known measures. The parameter of the Neumann kernels allows us to tune the bias of the induced measure between relatedness and importance depending on the application.

The kernels based on the graph Laplacian, including the regularized Laplacian and diffusion kernels, have some favorable properties as a relatedness measure that are missing not only from the co-citation and bibliographic coupling but also from the Neumann kernels.

With these frameworks connecting the link analysis methods and graph kernels, it might be possible to transfer the expertise gained on the link analysis methods to the graph kernels, and vice versa. For example, it has been pointed out that HITS suffers from the "tightly knit community effect" [18], and now that the Neumann kernels are shown to subsume HITS as a special case, it would be interesting to see whether and how the limitation carries over to the Neumann kernels.

REFERENCES

1. S. Acharyya and J. Ghosh. A maximum entropy framework for higher order link analysis on discrete graphs. In Workshop on Link Analysis for Detecting Complex Behavior (LinkKDD 2003), Washington, DC, August 27, 2003.

2. P. Baldi, P. Frasconi, and P. Smyth. *Modeling the Internet and the Web: Probabilistic Methods and Algorithms.* Wiley, Hoboken, NJ, 2003.

3. A. Berman and R. J. Plemmons. *Nonnegative Matrices in the Mathematical Sciences.* Classics in Applied Mathematics 9. SIAM, Philadelphia, PA, 1994.

4. K. D. Bollacker, S. Lawrence, and C. L. Giles. CiteSeer: An autonomous web agent for automatic retrieval and identification of interesting publications. In Proceedings of the 2nd International ACM Conference on Autonomous Agents, pp. 116–123, 1998.

5. S. Brin and L. Page. The anatomy of a large-scale hypertextual (web) search engine. *Computer Network and ISDN Systems*, 30(1–7):107–117, 1998.

6. F. R. K. Chung. *Spectral Graph Theory.* American Mathematical Society, Providence, RI, 1997.

7. S. C. Deerwester, S. T. Dumais, T. K. Landauer, G. W. Furnas, and R. A. Harshman. Indexing by latent semantic analysis. *Journal of the American Society for Information Science*, 41(6):391–407, 1990.

8. D. Dhyani, W. K. Ng, and S. S. Bhowmick. A survey of Web metrics. *ACM Computing Surveys*, 34(4):469–503, 2002.

9. R. Fagin, R. Kumar, and D. Sivakumar. Comparing top k lists. *SIAM Journal on Discrete Mathematics*, 17(1):134–160, 2003.

10. G. H. Golub and C. F. Van Loan. *Matrix Computation*, 3rd ed. Johns Hopkins University Press, Baltimore, MD, 1996.

11. T. Joachims. Text categorization with support vector machines: Learning with many relevant features. Technical Report LS-8 Report 23, Computer Science Department, University of Dortmund, Dortmund, Germany, 1997.

12. J. Kandola, J. Shawe-Taylor, and N. Cristianini. Learning semantic similarity. In *NIPS 15*, 673–680. MIT Press, Cambridge, MA, 2003.

13. L. Katz. A new status index derived from sociometric analysis. *Psychmetrika*, 18(1): 39–43, 1953.

14. M. M. Kessler. Bibliographic coupling between scientific papers. *American Documentation*, 14(1):10–25, 1963.

15. J. M. Kleinberg. Authoritative sources in a hyperlinked environment. *Journal of the ACM*, 46:604–632, 1999.

16. R. Kondor and J. Lafferty. Diffusion kernels on graphs and other discrete input spaces. In Proceedings of the 18th International Conference on Machine Learning, pp. 21–24, 2001.

17. C. le Pair. The citation gap of applicable science. In A. F. J. van Raan, ed., *Handbook of Quantitative Studies of Science and Technology*, Chapter 17, pp. 537–553, North-Holland, Amsterdam, 1988.

18. R. Lempel and S. Moran. The stochastic approach for link-structure analysis. *ACM Transactions on Information Systems*, 19(2):131–160, 2001.

19. C. D. Meyer. *Matrix Analysis and Applied Linear Algebra.* Society for Industrial and Applied Mathematics, Philadelphia, PA, 2001.

20. M. Saerens, F. Fouss, L. Yen, and P. Dupont. The principal component analysis of a graph, and its relationship to spectral clustering. In *Proceedings of ECML*, pp. 371–383, Springer, Berlin, Germany, 2004.

21. B. Schölkopf and A. J. Smola. *Learning with Kernels: Support Vector Machines, Regularization, Optimization, and Beyond.* MIT Press, Cambridge, MA, 2002.

22. H. Small. Co-citation in the scientific literature: A new measure of the relationship between two documents. *Journal of the American Society for Information Science*, 24:265–269, 1973.

23. A. J. Smola and R. Kondor. Kernels and regularization of graphs. In Proceedings of COLT'03, pp. 144–158, 2003.
24. S. White and P. Smyth. Algorithms for estimating relative importance in networks. In Proceedings of the KDD'03, pp. 266–275, 2003.

13

ENTITY RESOLUTION IN GRAPHS

INDRAJIT BHATTACHARYA AND LISE GETOOR

University of Maryland, College Park, Maryland

13.1 INTRODUCTION

In many applications, there are a variety of ways of referring to the same underlying real-world entity. For example, J. Doe, Jonathan Doe, and Jon Doe may all refer to the same person. In addition, entity references may be linked or grouped together. For example, Jonathan Doe may be married to Jeanette Doe and may have dependents James Doe, Jason Doe, and Jacqueline Doe, and Jon Doe may be married to Jean Doe and J. Doe may have dependents Jim Doe, Jason Doe, and Jackie Doe. Given such data, we can build a graph from the entity references, where the nodes are the entity references and edges (or often hyperedges) in the graph indicate links among the references.

However, the problem is that for any real-world entity there may well be more than one node in the graph that refers to that entity. In the example above, we may have three nodes all referring to the individual Jonathan Doe, two nodes referring to Jeanette Doe, two nodes referring to each of James Doe, Jason Doe, and Jacqueline Doe. Further, because the edges are defined over entity references, rather than entities themselves, the graph does not accurately reflect the relationships between entities. For example, until we realize that Jon Doe refers to the same person as Jonathan Doe, we may not think that Jon Doe has any children, and until we realize that J. Doe refers to the same person as Jonathan Doe, we will not realize that he is married.

Mining Graph Data, Edited by Diane J. Cook and Lawrence B. Holder
Copyright © 2007 John Wiley & Sons, Inc.

Thus an important first step in any graph mining algorithm is transforming such a *reference graph*, where nodes are entity references and edges are among entity references, into an *entity graph*, where nodes are the entities themselves and edges are among entities. Given a collection of references to entities, we would like to (a) determine the collection of "true" underlying entities, (b) correctly map the entity references in the collection to these entities, and (c) correctly map the entity reference relationships (edges in the reference graph) to entity relationships (edges in the entity graph).

This problem comes up in many guises throughout computer science. Examples include computer vision, where we need to figure out when regions in two different images refer to the same underlying object (the correspondence problem); natural language processing, when we would like to determine which noun phrases refer to the same underlying entity (co-reference resolution); and databases, where, when merging two databases or cleaning a database, we would like to determine when two tuple records are referring to the same real-world object (deduplication and/or record linkage).

Why do these ambiguities in entity references occur? Oftentimes data may have data entry errors, such as typographical errors. Or multiple representations are possible, such as abbreviations, alternate representations, and so on. Or in a database, we may have different keys—one person's database may use social security numbers while another uses name and address. Regardless, an exact comparison does not suffice for resolving entity references in such cases.

In data cleaning, deduplication [18, 27] is important for both accurate analysis, for example, determining the number of customers, and for cost effectiveness, for example, removing duplicates from mailing lists. In information integration, determining approximate joins [9] is important for consolidating information from multiple sources; most often there will not be a unique key that can be used to join tables in distributed databases, and we must infer when two records from different databases, possibly with different structures, refer to the same entity.

Traditional approaches to entity resolution and deduplication are based on approximate string matching criteria. These work well for correcting typographical errors and other types of noisy references, but do not make use of domain knowledge, such as common abbreviations, synonyms, or nicknames, and do not learn mappings between values. More sophisticated approaches can make use of domain-specific attribute similarity functions and mapping functions and consider the reference not just as a string but as a more structured object, such as a person entity that has first name, middle name, last name, address, and so on.

More recent approaches make use of attribute similarity measures but, in addition, take graph (i.e., relational) similarity into account. For example, if we are comparing two census records for Jon Doe and Jonathan Doe, we should be more likely to match them if they are both married to Jeannette Doe and they both have dependents James Doe, Jason Doe and June Doe. In other words, the string similarity of the attributes is taken into account, but so too is the similarity of the people to whom the person is related.

The problem becomes even more interesting when we do not assume that the related entities have already been resolved. In fact, when the relations are among entities of the same type, determining that two references refer to the same individual may in turn allow us to make additional inferences. In other words, the resolution process becomes *iterative*.

As mentioned earlier, the problem may be viewed in terms of a graph where the nodes represent the entity references that need to be resolved and the edges correspond to the observed relations among them. We will call this the *reference graph* capturing relations among the entity references. Our census example is shown in Figure 13.1(a). Before this graph can be mined for potential patterns or features, it needs to be "cleaned." Many nodes in this graph are duplicates in that they refer to the same underlying entity. The task is to identify which references correspond to the same entity and then consolidate them to create the *entity graph*. Figure 13.1(b) shows the entity graph after the references in the reference graph have been resolved. First, note that even in this simple graph, the entity graph is much smaller than the reference graph. In addition to the reduction in graph size that comes with resolving references, resolution is necessary for discovering the true patterns in the entity graph as well. Reference graphs are often collections of disconnected subgraphs. Unless they are resolved, the edges involving the same entity will be dispersed over its many references. Models built from such a graph will be inaccurate. In the example, the connections from the Jeanette Doe entity to the Jacqueline Doe entity can only be seen in the resolved entity graph.

Graph-based approaches for entity resolution take the edges into account as well for resolving references. In the above example, we may decide that the two Jason Doe references are the same, based on the fact that there is an exact string match and their fathers' have similar last names (though in general, we would not want to always do this; certainly two J. Does do not necessarily refer to the same person). Having done this, we may make use of the fact that both J. Doe and Jonathan Doe have a common dependent, to merge them.

Using the relational evidence provided in the graph has the potential benefit that we may produce more accurate results than if we use only attribute similarity measures. In particular, we may be able to decrease the false-positive rate because we can set our string match threshold more conservatively. But it has the downside

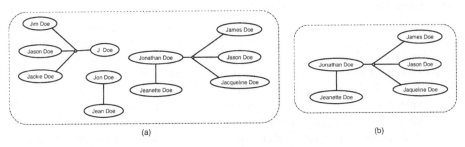

Figure 13.1. Example of (a) a reference graph for simple example given in the text and (b) the resolved entity graph.

that the process is more expensive computationally; first, as we go beyond simply comparing attributes to comparing edges and subgraphs, the similarity computation becomes more expensive. Second, as we iterate, we must continue to update the similarities as new resolutions are made.

In the next section, we review related work on entity resolution; most of the work that we describe does not take a graph-based approach. In Section 13.3, we introduce another more realistic motivating example for graph-based entity resolution. In Section 13.4, we formalize the graph-based entity resolution problem in Section 13.5, we define several similarity measures appropriate for entity resolution in graphs, and in Section 13.6 we describe a clustering algorithm that uses them to perform entity resolution. In Section 13.7, we describe some experimental results using the different similarity measures on two real-world datasets.

13.2 RELATED WORK

There has been a large body of work on deduplication, record linkage, and co-reference resolution. Here we review some of the main work, but the review is not exhaustive. For a nice summary report, see [40].

String Similarity. The traditional approach to entity resolution looks at textual similarity in the descriptions of the entities. For example, whether or not two citations refer to the same paper depends on the similarity measure such as edit distance between the two citation strings. There has been extensive work on defining approximate string similarity measures [8, 11, 27, 29] that may be used for unsupervised entity resolution. The other approach is to use adaptive supervised algorithms that learn string similarity measures from labeled data [6, 12, 36, 39]. One of the difficulties in using a supervised method for resolution is constructing a good training set that includes a representative collection of positive and negative examples. One approach that avoids the problem of training set construction is active learning [37, 39], where the user is asked to label ambiguous examples by the learner.

Theoretical Bounds. Cohen et al. [10] studies the theoretical problem of "hardening a soft database" that has many co-referent entries. Hardening refers to the task of figuring out which pairs of soft identifiers refer to the same real-world object. Given the likelihood of being co-referent for each soft pair and a probability distribution of possible hard databases, hardening is defined as the optimization problem of finding the most likely hard model given the soft facts. A cost is associated with each hard tuple that is added to the database and for each co-reference decision made. They show that this optimization problem is NP-hard even under strong restrictions. They propose a greedy agglomerative clustering approach for an approximate solution. This algorithm's complexity is linear in the number of entries in the soft database.

Efficiency. Given that solving the entity resolution problem optimally is computationally expensive, an important focus is on efficiency issues in data cleaning, where the goal is to come up with inexpensive algorithms for finding approximate solutions to the problem. The key mechanisms for doing this involve computing the matches efficiently and employing techniques commonly called "blocking" to quickly find potential duplicates and eliminate nonduplicates from consideration [18, 19, 25, 28]. The merge/purge problem was posed by Hernández and Stolfo [18] with efficient schemes to retrieve potential duplicates without resorting to quadratic complexity. They use a "sorted neighborhood method" where an appropriate key is chosen for matching. Records are then sorted or grouped according to that key and potential matches are identified using a sliding window technique. However, some keys may be badly distorted so that their matches cannot be spanned by the window and such cases will not be retrieved. The solution proposed is a multipass method over different keys and then merging the results using transitive closure. Monge and Elkan [28] combine the union find algorithm with a priority queue lookup to find connected components in an undirected graph. McCallum et al. [25] propose the use of canopies to first partition the data into overlapping clusters using a cheap distance metric and then use a more accurate and expensive distance metric for those data pairs that lie within the same canopy. Gravano et al. [17] propose a sampling approach to quickly compute cosine similarity between tuples for fast text-joins within an SQL framework. Chaudhuri et al. [8] use an error-tolerant index for data warehousing applications for probabilistically looking up a small set of candidate reference tuples for matching against an incoming tuple. This is considered "probabilistically safe" since the closest tuples in the database will be retrieved with high probability. This is also efficient since only a small number of matches needs to be performed.

Probabilistic Modeling. The groundwork for posing entity resolution as a probabilistic classification problem was done by Fellegi and Sunter [14], who extend the ideas of Newcombe et al. [31] for labeling pairs of records from two different files to be merged as "match" or "nonmatch" on the basis of agreement among their different fields. They estimate the conditional probabilities of these field agreement values given that the pair is really from the match class or the nonmatch class. They show that if the agreement values for the different fields are conditionally independent given the class, then these probabilities can be estimated in an unsupervised fashion. Winkler [41] builds upon this work for cases when the conditional independence assumption cannot be made and uses a generalized expectation maximization algorithm for estimating parameters to separate matches and nonmatches. More recently, hierarchical graphical models have been proposed [35] that use a separate match variable for each attribute and an overall match variable that depends on all of these lower level matches.

Graph-Based Approaches. Approaches that take into account relational structure of the entities for data integration have been proposed [1, 4, 13, 20, 21, 30]. Ananthakrishna et al. [1] introduce relational deduplication in data warehouse applications where there is a dimensional hierarchy over the relations. They augment the string similarity measure between two tuples with the similarity between their foreign key relations across the hierarchy which they call children sets. To avoid comparison between all pairs of tuples in a relation, they propose a grouping strategy that makes uses of the relational hierarchy as well.

Kalashnikov et al. [21] enhance feature-based similarity between an ambiguous reference and the many entity choices for it with relationship analysis between the entities, like affiliation and co-authorship. They propose a "content attraction principle" hypothesizing that an ambiguous reference will be more strongly connected via such relationships to its true entity compared to other entity choices for it. They translate this principle to a set of nonlinear equations that relate all the connection strengths in the entity graph and those between a reference and its choice entities. A solution to this nonlinear optimization problem yields the connection strengths and the strongest connection determines the entity choice for each reference.

Neville et al. [30] explore different graph partition schemes for clustering in graphs where the edge weights reflect attribute similarity between nodes. By varying the edge weights and edge existence probabilities conditioned on the cluster labels, they compare algorithms that consider only attributes and those that combine attribute and relational evidence. They report that spectral techniques for partitioning [38] work better than other min-cut and k-clustering approaches, but combining attribute and relational information proves detrimental for clustering.

The Subdue system proposed by Jonyer et al. (see [20] and Chapter 8 of this book) is a scheme for conceptual clustering of structured data. In addition to partitioning the data, conceptual clustering also summarizes the clusters with conceptual descriptions of objects contained in them. Subdue generates a hierarchical conceptual clustering by discovering substructures in the data using the minimum description length principle. This helps to compress the graph and represent conceptual structure as well.

Doan et al. [13] explore a profiler-based approach for tying up disjoint attributes for sanity checks using domain knowledge. For example, on merging two objects (9, John Smith) and (John Smith, 120k) from two tables with schemas (age, name) and (name, salary), we get a person whose age is 9 years and whose salary is 120K. This would be deemed an unlikely match by a profiler.

In earlier work of our own [4], we propose different measures for relational similarity in graphs and show how this can be combined with attribute similarity for improved entity resolution in collaboration graphs. We also relate the problem of graph-based entity resolution to discovering groups of collaborating entities in graph [3] and suggest that the two tasks may be performed jointly so that a better solution for one of these tasks leads to improvements in the other as well.

Probabilistic Inference in Graphs. Probabilistic models that take into account interaction between different entity resolution decisions have been proposed for named entity recognition in natural language processing and for citation matching. McCallum and Wellner [24] use conditional random fields for noun co-reference and use clique templates with tied parameters to capture repeated relational structure. They do not directly model explicit links among entities.

Li et al. [23] address the problem of disambiguating "entity mentions," potentially of multiple types, in the context of unstructured textual documents. They propose a probabilistic generative model that captures a joint distribution over pairs of entities in terms of co-mentions in documents. In addition, they include an appearance model that transforms mentions from the original form. They evaluate a discriminative pairwise classifier for the same task that is shown to perform well. However, they show both empirically and theoretically that direct clustering over the pairwise decisions can hurt performance for the mention matching task when the number of entity clusters is more than two.

Parag and Domingos [32] use the idea of merging evidence to allow the flow of reasoning between different pairwise decisions over multiple entity types. They are able to achieve significant benefit from generalizing the mapping of attribute matches to multiple references, for example, being able to generalize from one match of the venue "Proc. of SIGMOD" with "Proceedings of the International Conference on Management of Data" to other instances.

Pasula et al. [33] propose a generic probabilistic relational model framework for the citation matching problem. Because of the intractability of performing exact probabilistic inference, they propose sampling algorithms for reasoning over the unknown set of entities. Milch et al. [26] propose a more general approach to the identity uncertainty problem. They present a formal generative language for defining probability distribution over worlds with unknown objects and identity uncertainty. This can be seen as a probability distribution over first-order model structures with varying number of objects. They show that the inference problem is decidable for a large class of these models and propose a rejection sampling algorithm for estimating probabilities.

In other work of our own [5], we have adapted the latent dirichlet allocation model for documents and topics and extended it to propose a generative group model for joint entity resolution. Instead of performing a pairwise comparison task, we use a latent group variable for each reference, which is inferred from observed collaborative patterns among references in addition to attribute similarity to predict the entity label for each reference.

Tools. A number of frameworks and tools have been developed. Galhardas et al. [15] propose a framework for declarative data cleaning by extending SQL with specialized operators for matching, clustering, and merging. The WHIRL system [9] integrates a logical query language for doing "soft" text joins in databases with efficient query processing. Potter's Wheel [34], Active Atlas [39], and D-Dupe [7] are some other data cleaning frameworks that involve user interaction.

Application Domains. Data cleaning and reference disambiguation approaches have been applied and evaluated in a number of domains. The earliest application is on medical data [31]. Census data is an area where detection of duplicates poses a significant challenge, and Winkler [41] has successfully applied his research and other baselines to this domain. A great deal of work has been done making use of bibliographic data [4, 19, 22, 25, 33, 37]. Almost without exception, the focus has been on the matching of citations. Work in co-reference resolution and disambiguating entity mentions in natural language processing [23, 24] has been applied to text corpora and newswire articles such as the TREC corpus. For detailed evaluation of algorithm performance, researchers have also resorted to synthetic [4, 30] and semisynthetic [1, 8] datasets where various features of the data can be varied in a controlled fashion.

Evaluation Metrics. As has been pointed out by Sarawagi and Bhamidipaty [37], the choice of a good evaluation metric is an issue for entity resolution tasks. Mostly, resolution has been evaluated as a pairwise classification problem. Accuracy may not be the best metric to use since datasets tend to be highly skewed in their distribution over duplicate and nonduplicate pairs; often less than 1% of all pairs are duplicates. In such a scenario, a trivial classifier that labels all pairs as nonduplicates would have 99% accuracy. Though accuracy has been used by some researchers [8, 27, 30, 41], most have used precision over the duplicate prediction and recall over the entire set of duplicates. Observe that a classifier that indiscriminately labels all pairs as nonduplicates will have high precision but zero recall. The two measures are usually combined into one number by taking their harmonic mean. This is the so-called F1 measure. Another option that has been explored is weighted accuracy, but this may report high accuracy values even when precision is poor. Cohen et al. [11] rank all candidate pairs by distance and evaluate the ranking. In addition to the maximum F1 measure of the ranking, they consider the noninterpolated average precision and interpolated precision at specific recall levels.

Some other approaches to this problem have posed it as a clustering task, where references that correspond to the same entity are associated with the same cluster. Performance measures that evaluate the qualities of the clusters generated compared to the true clusters are more relevant in such cases. Monge and Elkan [28] use a notion of cluster purity for evaluation. Each of the generated clusters may either match a true cluster, be a subset of a true cluster, or include references from more than one cluster. They treat the first two cases as pure clusters while the third category of clusters is deemed impure. They use the number of pure and impure clusters generated as the evaluation metric. We have proposed an alternative evaluation metric for this clustering task where we measure the diversity of each constructed cluster of entity references in terms of the number of references to different real entities that it contains [4] and the dispersion of each entity over the number of different clusters. We show that dispersion-diversity plots capture the quality of the clusters directly and can be used to evaluate the trade-off in a fashion similar to precision recall curves.

13.3 MOTIVATING EXAMPLE FOR GRAPH-BASED ENTITY RESOLUTION

Throughout the rest of this chapter, we will motivate the problem of entity resolution in graphs using an illustrative entity resolution task from the bibliographic domain.

Consider the problem of trying to construct a database of papers, authors, and citations from a collection of paper references, perhaps collected by crawling the Web. A well-known example of such a system is CiteSeer [16], an autonomous citation indexing engine. CiteSeer is an important resource for CS researchers and makes searching for electronic versions of papers easier. However, as anyone who has used CiteSeer can attest, there are often multiple references to the same paper, citations are not always resolved, and authors are not always correctly identified [33, 37].

An Example. The most commonly studied bibliographic entity resolution task is resolving paper citations. Consider the following example from [37]:

- R. Agrawal, R. Srikant. *Fast algorithms for mining association rules in large databases.* In VLDB-94, 1994.
- Rakesh Agrawal and Ramakrishnan Srikant. *Fast Algorithms for Mining Association Rules.* In Proc. of the 20th Int'l Conference on Very Large Databases, Santiago, Chile, September 1994.

These very different strings are citations of the same paper and clearly string edit distance alone will not work. However, if we extract the different fields or attributes of the paper, we may have better luck. Sometimes, paper resolution can be done based simply on the title. We can use one of the many existing methods for string matching, perhaps even tuned to the task of title matching. However, there is additional relational information, in terms of the venue, the authors of the paper, and the citations made by the paper; this additional information may provide further evidence to the fact that the two references are the same.

13.3.1   Issues in Graph-Based Entity Resolution

Multitype Entity Resolution. In the above example and in the earlier census example, we had one type of entity (e.g., papers, people) and we were trying to resolve them. We refer to this as *single-type entity* resolution. But note that while the above two citation strings are used to motivate the paper resolution problem, it is more interesting to see that they present an illustrative example of multitype entity resolution, the problem of resolving entity references when there are different types of entities to be resolved. The citations refer to papers, but in addition to the paper title, they contain references to other types of entities, which themselves may be ambiguous. In particular, the strings refer to *author* and *venue* entities in addition to *paper* entities. This brings up a scenario where multiple types of entities need to be resolved simultaneously. Assuming that the strings have been correctly parsed to separate out the different fields—which is a difficult problem by itself—the first citation string mentions R. Agrawal and R. Srikant as the authors and VLDB-94, 1994 as the venue, while the second has Rakesh Agrawal and Ramakrishnan Srikant as the authors and Proc. of the 20th Int'l Conference on Very Large Databases, Santiago, Chile, September 1994 as the venue reference. Not all of these pairs are easy to disambiguate individually. While it may not be too difficult to resolve R. Srikant and Ramakrishnan Srikant, Agrawal is an extremely common Indian last name, and it is certainly not obvious that R. Agrawal refers to the same author entity as Rakesh Agrawal. As for the venue references, unless one knows that for two resolved papers their venues must be the same, it is very difficult, if not impossible, to resolve the two venue references without specialized domain knowledge.

Collective Entity Resolution. For our example, it is not hard to observe the dependence among the different resolution decisions across multiple classes. We can make the resolution decisions *depend* on each other in cases where they are *related*. When the resolution decisions are made *collectively*, they can be much easier to make. In the citation domain the relevant entities are *papers, authors,* and *venues* and the relevant relationships are *write* and *published in,* that is, authors *write* papers that get *published in* venues. An author is a *co-author* of another author if they write the same paper. We can use these relations to make one resolution decision lead to another. We may begin by resolving R. Srikant and Ramakrishnan Srikant, of which we are most confident. This may lead us to believe R. Agrawal and Rakesh Agrawal are the same author, since they are *co-authors* with the resolved author entity Srikant. Now we can go back to the paper citations that, in addition to having very similar titles, have now been determined to be *written by* the same author entities. This makes us more confident that the two paper references map to the same paper entity. Following the same thread, two identical papers citations must have identical venue citations. So we may resolve the apparently disparate venue references.

Graph-Based Evidence for Entity Resolution. It is now easier to see how graphs help in representing these dependencies. The reference graph has nodes for each entity reference. If the references correspond to multiple entity classes, then we have multiple node types. Edges represent the relations that hold between the references. We may also have more than one type of edge to represent the different types of relations that exist. The co-author relation illustrates the need for hyperedges that may involve more than two references. Since all observed authors of a paper are co-authors, this relation is naturally captured as a hyperedge that spans all the author references from a paper. Note that this may alternatively be captured using a quadratic number of binary edges, but this leads to a dramatic increase in the graph size.

Local Versus Global Resolution. When we resolve two references, such as the resolution of two venue strings Proc. of the 20th Int'l Conference on Very Large Databases, Santiago, Chile, September 1994 and VLDB-94, 1994, we should probably resolve all other occurrences of these venue references. In other words, once we have decided the two references are the same, we should also resolve any other venue references that exactly match these two references. We call this type of resolution *global resolution*. For certain entity references it makes the most sense and can speed things up significantly. However, it may not always be appropriate. In the case of names, for example, R. Agrawal may refer to the Rakesh Agrawal entity in one case, while some other instance of R. Agrawal might refer to a different entity Rajeev Agrawal. We refer to this latter resolution strategy as *local resolution*.

Additional Sources of Relational Evidence. Other bibliographic relations can also potentially be used when available. Papers are *cited by* other papers and two papers are *co-cited* if they are cited by the same paper entity. If two similar paper references are co-cited by the same paper entity, that may serve as additional evidence that they are the same. However, graph-based evidence may also be *negative* in some cases. For example, two paper references that are *cited by* the same paper entity are unlikely to be duplicates, as are two author references that are *co-authors* of the same paper entity.

13.3.2 Author Resolution in Graphs

Within the context of entity resolution in bibliographic data, we first look at the problem of resolving author references in graphs leveraging co-author relationships. Suppose that we have two different papers, and we are trying to determine if there are any authors in common between them. We can do a string similarity match between the author names, but often references to the same person vary significantly. The most common difference is the variety of ways in which the first name and middle name are specified. For an author entity Jeffrey David Ullman, we may see references J. D. Ullman, Jeff Ullman, Ullman, J. D. and so on. For the most part, these types of transformations can be handled by specialized code that checks for common name presentation transforms. However, we are still presented with the dilemma of determining whether a first name or middle name is the same as some initial; while the case of matching J. D. Ullman and Jeffrey D. Ullman seems quite obvious, for common names such as J. Smith or X. Wang the problem is more difficult. Existing systems take name frequency into account and will give unusual names higher matching scores. But this still leaves the problem of determining whether a reference to J. Smith refers to James Smith or Jessica Smith.

As mentioned before, additional context information can be used in the form of co-author relationships. If the co-authors of J. Smith for these two papers are the same, then we should take this into account and give the two references a higher matching score. But in order to do this we must have already determined that the other two author references refer to the same individual; thus it becomes a chicken and egg problem.

Consider the example shown in Figure 13.2(a), where we have four paper references, each with a title and author references. In order to resolve these references, we may begin by examining the author references to see which ones we consider to be the same. In the first step, we might decide that all of the Aho references refer to the same author entity because Aho is an unusual last name. This corresponds to resolving all of the Aho references into a single entity. However, suppose based on name information alone, we are not quite sure that the Ullman references are to the same author, and we are certainly not sure about the Johnson references, since

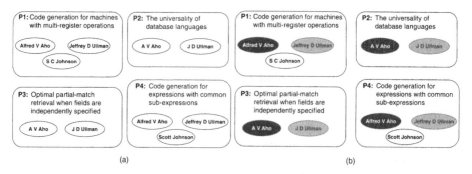

Figure 13.2. (a) An example author/paper resolution problem. Each box represents a paper reference (in this case unique) and each oval represents an author reference. (b) The resolved entities corresponding to the example author/paper resolution problem.

Johnson is a very common name. But, deciding that the Aho references correspond to the same entity gives us additional information for the Ullman references. We know that the references share a common co-author. Now with higher confidence we can consolidate the Ullman references. Based solely on the Aho entity consolidation, we do not have enough evidence to consolidate the Johnson references. However, after making the Ullman consolidations, we may decide that having two co-authors in common is enough evidence to tip the balance, and that all the Johnson references correspond to the same Johnson entity. Figure 13.2(b) shows the final result after all the author references have been correctly resolved, where references to the same entity are shaded accordingly.

As illustrated in the above example, the problem of graph-based author resolution is likely to be an iterative process: As we identify co-author/collaborator relationships among author entities, this will allow us to identify additional potential co-references. We can continue in this fashion until all of the entities have been resolved. Figure 13.3(a) shows the reference graph for this example and Figure 13.3(b) shows the resulting entity graph.

13.4 GRAPH-BASED ENTITY RESOLUTION: PROBLEM FORMULATION

In the graph-based entity resolution problem, we have a reference graph—a graph over some collection of references to entities—and from this graph we would like to identify the (unique, minimal) collection of individuals or entities to which they should be mapped, and the induced entity graph. In other words, we would like to find a many-to-one mapping from references to entities. Figure 13.3(a) shows the reference graph for the author resolution example and Figure 13.3(b) shows the resulting entity graph.

In what follows, lowercase characters e and r denote entities and references and qualified uppercase letters such as $e.A$ and $r.E$ denote attributes of the variables, and we will use $e.a$ and $r.e$ to denote particular values of variables (short-hand for

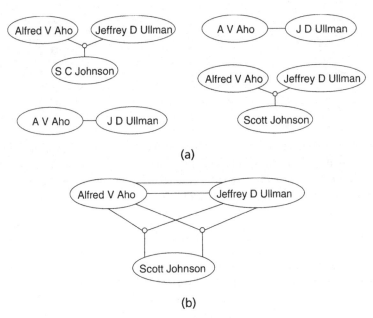

Figure 13.3. (a) Reference graph and (b) entity graph for the author resolution example.

$e.A = a$). In the single-entity resolution problem, we are given a set of references $\mathcal{R} = \{r_i\}$, where each reference r has its attributes $r.A$. The references correspond to entities $\mathcal{E} = \{e_i\}$ so that each reference r has a hidden entity label $r.E$. Each entity e also has its own attributes $e.A$, but the entities are not directly observed. What we observe are the attributes $r.A$ of individual references. We can imagine $r.A$ to be generated by some distortion process from the attributes of the corresponding entity $r.E$. Obviously, the entity labels of the references are not observed. The problem is to recover the hidden set of entities $\mathcal{E} = \{e_i\}$ and the entity labels $r.E$ of individual references given the observed attributes of the references.

We use relational information among references to help us in collective entity resolution. We will assume that the references are observed, not individually, but as members of hyperedges. We are given a set of hyperedges $\mathcal{H} = \{h_i\}$, and the membership of a reference in a hyperedge is captured by its hyperedge label $r.H$. In general, it is possible for a reference to belong to multiple hyperedges, and all of our approaches can be extended to handle this. However, in this chapter, we will consider a simpler scenario where each reference occurs in a single hyperedge. If reference r occurs in hyperedge h, then $r.H = h$. Note that unlike the entity labels, we *know* the association of hyperedges and references. The hyperedges can help us make better predictions if we assume that they are indicative of associative patterns among the entities. In other words, the entity labels of references that occur in the same hyperedge are related to each other. Now the resolution decisions are not independent. Instead of finding the entity labels for each reference individually, our task is to predict the entity labels of the references collectively, where the entity label $r.E$ of any reference r is directly influenced by the choice of entity label $r'.E$

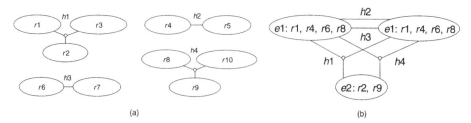

(a)   (b)

Figure 13.4. (a) More abstract representation of the reference graph for the author resolution example; the r's are references and the h's are hyperedges. (b) Abstract representation
for the entity graph for the author resolution example; the e's are entities, the references they
correspond to are listed, and the h's are hyperedges.

for another reference r' if they are associated with the same hyperedge, that is,
$r.H = r'.H$.

To make this more concrete, consider our earlier example. Figure 13.4(a) shows
the references and hyperedges. Each observed author name corresponds to a reference, so there are 10 references r_1 through r_{10}. In this case, the attributes are
the names themselves, so, for example, $r_1.A$ is Alfred Aho, $r_2.A$ is S.C. Johnson, and $r_3.A$ is Jeffrey Ullman. The set of true entities ε is $\{e_1, e_2, e_3\}$ as shown
in Figure 13.4(b), where $e_1.A =$ Alfred V. Aho, $e_2.A =$ S.C. Johnson, and $e_3.A =$
Jeffrey D. Ullman. Clearly, r_1, r_4, r_6, and r_8 correspond to Alfred V. Aho, so that
$r_1.E = r_4.E = r_6.E = r_8.E = e_1$. Similarly, $r_3.E = r_5.E = r_7.E = r_{10}.E = e_2$ and
$r_2.E = r_9.E = e_3$. There are also the hyperedges, which correspond to each set of
authors for a paper. There are four papers, so that $\mathcal{H} = \{h_1, h_2, h_3, h_4\}$. The references r_1 through r_3 are associated with hyperedge h_1 since they are the observed
author references in the first paper. This is represented as $r_1.H = r_2.H = r_3.H =$
h_1. We may similarly capture the hyperedge associations of the other references.
The problem here is to figure out from the attributes $r.A$ and the hyperedge labels
$r.h$ of the references that there are three distinct entities such that r_1, r_4, r_6, and r_8
correspond to one entity, r_3, r_5, r_7, and r_{10} correspond to the second, and r_2 and
r_9 correspond to the third. Here we have introduced the notation assuming that all
references correspond to the same class of entities. It may be extended to handle
multiple entity classes and multiple types of hyperedges between references. In the
rest of this chapter, we will use the term edge as a substitute for hyperedge. It
will be understood that an edge may involve more than two nodes unless explicitly
stated otherwise.

Entity Resolution as a Clustering Problem. Alternatively, the task of collective entity resolution may be viewed as a graph-based clustering problem where
the goal is to cluster the references so that those that have identical entity labels
are in the same cluster. One approach to finding this mapping is to use a greedy
agglomerative clustering approach. At any stage of the clustering algorithm, the set
of *entity clusters* reflect our current beliefs about the underlying entities. In other
words, each constructed entity cluster should correspond to one underlying entity,

and all references in that cluster should be to that entity. To start off, each reference belongs to a separate cluster, and at each step the pair of clusters (or entities) that are most likely to refer to the same entity are merged. The key to the success of a clustering algorithm is the similarity measure that is employed. In graph-based clustering for entity resolution, we want to use a similarity measure that takes into account the similarity in the attributes of the references in the two clusters as well as the relational similarity. In addition, the measure should take into account the *related* resolution decisions that have been made previously. Accordingly, similarity measures are extended to consider both reference attributes and edge-based patterns.

The *attribute similarity component* of the similarity measure takes into account the similarity of the attributes $r.A$ of the references in the two clusters. In the author resolution case, it measures the similarity between two observed reference names. In addition, the *graph-based similarity component* takes into account the similarity of the relations that the two entities or clusters participate in. Each cluster is associated with a set of references, and through these references the cluster is associated with a set of edges to other references. But what is it about these other references that we want to consider? We want to look not at their attributes but at the resolution decisions that have been taken on them. Specifically, we want to look at the entity cluster labels of these references, or which clusters they currently belong to. To illustrate this using our example from Figure 13.2, suppose we have already resolved the Aho and Ullman references in clusters c_1 and c_2, respectively, and we are looking at the current similarity of the two Johnsons, which are yet to be resolved and still belong to separate entity clusters, say c_{3a} having r_2 and c_{3b} having r_9. The two clusters are associated with one edge each, h_1 and h_4, respectively. We want to factor into the similarity measure *not the names* of the other references in the two edges but the fact that both of them are associated with the same resolved entity clusters c_1 and c_2, which is what makes them similar.

An issue that is brought out by the above discussion is the dynamic nature of the graph-based similarity component. Initially, when all references belong to distinct entity clusters, the two Johnson clusters will not be considered similar enough. But their graph-based similarity goes up in stages as first the Aho references and then the Ullman references are resolved. In the following section, we describe an iterative clustering algorithm that leverages this dynamic nature of graph-based similarity.

13.5   SIMILARITY MEASURES FOR ENTITY RESOLUTION

In this section, we define the similarity measure between two entity clusters as a weighted combination of the attribute similarity and graph-based similarity between them and highlight the computational and other algorithmic issues that are involved.

For two entity clusters c_i and c_j, their similarity may be defined as

$$\text{sim}(c_i, c_j) = (1 - \alpha) \times \text{sim}_{\text{attr}}(c_i, c_j) + \alpha \times \text{sim}_{\text{graph}}(c_i, c_j) \quad 0 \le \alpha \le 1$$

where $\text{sim}_{\text{attr}}(\ )$ is the similarity of the attributes and $\text{sim}_{\text{graph}}(\ )$ is the graph-based similarity between the two entity clusters, and they are linearly combined with weights α and $1 - \alpha$. In the following two subsections, we discuss the two similarity components in detail.

13.5.1 Attribute Similarity

We assume the existence of some basic similarity measure that takes two reference attributes and returns a value between 0 and 1 that indicates the degree of similarity between them. A higher value indicates greater similarity between the attributes. We are not making any other assumptions about the attribute similarity measure. Any measure that satisfies these assumptions can be used. This is particularly helpful since such similarity measures are often tuned for specific domains and may be available as library functions. Depending on the domain, it is possible to adapt a different attribute measure and tie it in with our algorithm seamlessly.

However, we need $\mathrm{sim}_{attr}(\ )$ to define the similarity of attributes between two entity clusters. Each entity cluster is a collection of references with their own attributes. So we need to use the similarity measure that takes two attributes to define the similarity between two clusters of attributes. This is similar to finding the aggregate distance between two clusters given a pairwise distance measure. Many approaches like *single link, average link*, and *complete link* have been proposed [2] where some aggregation operation is used to combine the pairwise distances between the two clusters. The duplicate relation is typically transitive: If references r_i are r_j are duplicates, then all other duplicates of r_i will also be duplicates of r_j. So the single link measure that takes the minimum pairwise distance (or the maximum pairwise similarity) between two clusters is the most relevant for our purposes.

The first issue with a cluster linkage metric is that it is computationally intensive. The number of attribute similarity computations involved in finding the similarity between two clusters is quadratic in the average number of references in each cluster. So pairwise comparison is a problem for larger clusters. Further, as clusters merge and new references are added to a cluster, similarities need to be recomputed. This repeated similarity computation adds to the complexity. There are a number of ways this problem may be addressed. First, updating the cluster similarities is not computationally challenging when using the single link metric. They may be incrementally modified as clusters merge. Specifically, when two clusters c_i and c_j are merged to create a new cluster c_{ij}, the similarity to a third cluster c_k may be updated as $\mathrm{sim}_{attr}(c_{ij}, c_k) = \max(\mathrm{sim}_{attr}(c_i, c_k), \mathrm{sim}_{attr}(c_j, c_k))$. Second, when we are dealing with attributes such as names, though a cluster may contain hundreds of references, there will be very few distinct names. So for two clusters c_i and c_j, computing attribute similarity involves $|\mathrm{distinct}(c_i)| \times |\mathrm{distinct}(c_j|$ similarity computations, rather than $|c_i| \times |c_j|$ computations, where $\mathrm{distinct}(c)$ is the number of unique reference attribute values in cluster c. Finally, if we have some way of finding an "average" attribute value for each of the clusters, the problem is much simpler.

This last point also touches on the issue of the semantics of the attribute similarity in the case where we are clustering in order to perform entity resolution. By design, each cluster corresponds to one resolved entity e. This entity may have been referred to using many different attributes, as is evident from the attributes $r.A$ of the references belonging to that cluster. But if we assume this attribute to be single valued, then there really is one true value of the entity attribute $e.A$. For instance, the reference attributes may have been A. Aho, Alfred V. Aho, A.V Aho, and so on, but the true entity attribute is Alfred V. Aho. We may define this *representative cluster attribute* as the most likely one given the reference attributes in that cluster. So, when computing the attribute similarity of the two entity clusters, it makes more sense to look at the similarity of the two representative attributes for the clusters, rather than considering all of the attributes in each cluster. This makes a difference in situations where we have detected differences that may be due to typographical errors in reference attributes within an entity cluster but have still been able to resolve them correctly. When computing similarity values for this entity cluster, we possibly do not want these noisy reference attributes to play a role.

13.5.2 Graph-Based Similarity

Next, we address the graph-based similarity measure between two entity clusters considering the entities that they are related to via the observed edges. There are many possible ways to define this similarity. We explore some possibilities here and focus on relevant issues.

13.5.2.1 Edge Detail Similarity.
Just as in the case of the attributes, each entity cluster is associated with a set of edges to which the references contained in it belong. Recall that each reference r is associated with an observed hyperedge $r.H$. Then the edge set $c.H$ for an entity cluster c may be defined as $c.H = \{h|r.H = h, r \in c\}$. To ground this in terms of our example in Figure 13.4, after we have resolved all the entities so that cluster c_1 refers to the Aho entity, cluster c_2 to the Ullman entity, and cluster c_3 to the Johnson entity, then the edge set for the Aho cluster is $c_1.H = \{h_1, h_2, h_3, h_4\}$ having the edges corresponding to the four papers written by Aho. The edge set $c_2.H$ for Ullman is identical while that for Johnson is $c_3.H = \{h_1, h_4\}$.

We define a similarity measure for a pair of edges so that given this pairwise similarity measure, we can again use a linkage metric like single link to measure the relational similarity between two clusters. First, we define a pairwise similarity measure between two hyperedges. We have already noted that what we want to consider for relational similarity are the cluster labels of the references in each edge, and not their attributes. So for an edge, we consider the multiset of entity labels, one for each reference associated with it.[1] We will employ the commonly used Jaccard similarity measure over these multisets of entity labels. Specifically, let label(h_i) be the set of entity labels for an edge h. Then for a pair of edges h_i and h_j, we define their similarity as

$$\text{sim}(h_i, h_j) = \frac{|\text{label}(h_i) \bigcap \text{label}(h_j)|}{|\text{label}(h_i) \bigcup \text{label}(h_j)|}$$

[1]Though in general we can have a multiset of entity labels for an edge, all the labels are likely to be distinct when the edges represent co-authorship relations.

Given this pairwise similarity measure for edges, we can use an aggregation operation such as max to calculate the graph-based similarity between two entity clusters
as follows:

$$\text{sim}_{\text{graph}}(c_i, c_j) = \max_{(h_i, h_j)}\{\text{sim}(h_i, h_j)\} \qquad h_i \in c_i.H \qquad h_j \in c_j.H$$

This we will call the *edge detail similarity* between two clusters since it explicitly
considers every edge associated with each entity cluster. Observe that computing
the similarity between two edges is linear in the average number of references in
an edge, while computing the edge detail similarity of two clusters is quadratic in
the average number of edges associated with each cluster. An issue with using max
as the aggregation function is that a high similarity between a single pair of edges
may not always be a strong evidence for determining duplicates in domains where
the edges can be noisy as well. In the absence of noise, however, it is very likely
that two authors with similar names are duplicates if both are observed to write just
one paper with the same set of collaborating authors. Another advantage of using
max, as compared with other functions like avg, is that it allows cluster similarities
to be updated efficiently when related clusters merge.

13.5.2.2   *Neighborhood Similarity.*   Clearly, an issue with edge detail similarity
is the computational complexity. The solutions discussed in the context of attribute
similarity cannot be extended for edges. It is not trivial to incrementally update edge
detail similarity when clusters merge. Also, the reduction due to repeating entity
patterns in edges is not expected to be as significant as for attributes.

A second and more pertinent issue is whether the detailed pairwise similarity
computation for edges is really necessary for the task at hand. While it may make
sense for some applications, it may not be necessary to look at the structure of each
edge separately for the task of graph-based entity resolution. Using a more concrete
example, for two author entities e_1 and e_1' to be considered similar, it is not necessary
for both of them to co-author a paper with entities e_2, e_3, and e_4 together. Instead, if
e_1 participates in an edge $\{e_1, e_2, e_3, e_4\}$ and e_1' participates in three separate edges
$\{e_1', e_2\}$, $\{e_1', e_3\}$, and $\{e_1', e_4\}$, then that also should count as significant graph-based
evidence for their being the same author entity (if they have similar attributes as
well). In other words, whether or not two entity clusters with similar attributes have
the same edge structures, if they have the *same neighbor clusters* to which they have
edges, that is sufficient graph-based evidence for bibliographic entity resolution.

We will now formalize the notion of neighborhood clusters for an entity cluster.
Recall that we have defined label (l) as the multiset of entity labels for an edge h
and for an entity cluster c; and $c.H$ is the set of edges associated with it. Then we
can formally define the neighborhood multiset $c.N$ for c as follows:

$$c.N = \bigcup_m \text{label}(h_i) \qquad h_i \in c.H$$

where $\bigcup_m$ is the multiset union operator. Intuitively, we collapse the edge structure
and just look at how many times c has participated in the same edge with another

entity cluster. Note that we do not include c itself among its neighbors. Returning to our example from Figure 13.2, once the entities have been resolved and clusters c_1, c_2, and c_3 correspond to entities Aho, Ullman, and Johnson, respectively, the labels of the edges for cluster c_1 are $\{c_1, c_2, c_3\}$ for h_1, $\{c_1, c_2\}$ for h_2, $\{c_1, c_2\}$ for h_3, and $\{c_1, c_2, c_3\}$ for h_4. Then the neighborhood multiset $c_1.N$ is the collection of all cluster labels occurring in these four edges: $c_1.N = \{c_2, c_2, c_2, c_2, c_3, c_3\}$. Now, for the graph-based similarity measure between two entity clusters, we take the Jaccard similarity between their neighborhood multisets:

$$\text{sim}_{\text{graph}}(c_i, c_j) = \frac{|c_i.N \bigcap c_j.N|}{|c_i.N \bigcup c_j.N|}$$

We will call this the *neighborhood similarity* between the two entity clusters. Computing and updating neighborhood similarity is significantly cheaper computationally compared to the *edge detail similarity*. It is linear in the average number of neighbors per entity. Note that the neighborhood of an entity cluster has a natural interpretation for author entities and collaborator relationships. The neighborhood is the multiset of collaborators for an author entity. Also, by avoiding looking at individual links, neighborhood similarity is less susceptible to noisy edges than edge detail similarity.

Observe that when all edges are binary, the similarity measure between two edges becomes Boolean, as does the edge detail similarity between two clusters. Also, the edges can be mapped directly to the neighborhood set. Accordingly using the edge detail similarity is not appropriate any more and neighborhood similarity is sufficient.

13.5.3 Negative Evidence from Graphs

So far, we have considered graph structure as additional evidence for two author references actually referring to the same underlying author entity. However, graph-based evidence can be negative as well. A "soft" aspect of negative evidence is directly captured by the combined similarity measure. Imagine two references with identical names. If we only consider attributes, their similarity would be very high. However, if they do not have any similarity in their edge structures, then we are less inclined to believe that they correspond to the same entity. This is reflected by the drop in the their overall similarity when the graph-based similarity measure is factored in as well.

We may also imagine stronger graph-based constraints for clustering. In many relational domains, there is the constraint that no two references appearing in the same edge can be duplicates of each other. To take a real bibliographic example, if a paper is observed to be co-authored by M. Faloutsos, P. Faloutsos, and C. Faloutsos; then probably these references correspond to distinct author entities. We have such constraints for every edge that has more than one reference. This can be taken into account by the graph-based similarity measure. The similarity between two cluster pairs is zero if merging them violates any relational constraint.

13.6   GRAPH-BASED CLUSTERING FOR ENTITY RESOLUTION

Given the similarity measure for a pair of entity clusters, we can use a greedy
agglomerative clustering algorithm that finds the closest cluster pair at each step
and merges them. Here we discuss several implementation and performance issues
regarding graph-based clustering algorithms for entity resolution (GBC-ER).

13.6.1   Blocking to Find Potential Resolution Candidates

Initially, each reference belongs to a distinct entity cluster, and the algorithm iteratively looks to find the closest pair of clusters. Unless the datasets are small, it is
impractical to consider all possible pairs as potential candidates for merging. Apart
from the scaling issue, most pairs checked by an $O(n^2)$ approach will be rejected
since usually about 1% of all pairs are true duplicates. Blocking techniques are
usually employed to rule out pairs that are certain to be nonduplicates. Bucketing
algorithms may be used to create groups of similar reference attributes, and only
references within the same bucket are considered as potential duplicates.

The algorithm inserts all of these potential duplicate pairs into a priority queue
considering their similarities. Then it iteratively picks the pair with the highest
similarity and merges them. The algorithm terminates when the similarity for the
closest pair falls below a threshold.

13.6.2   Graph-Based Bootstrapping for Entity Clusters

The graph-based clustering algorithm begins with each reference assigned to a separate entity cluster. So to start with, the clusters are disconnected and there is
no graph-based evidence to make use of between clusters. As a result, all initial
merges occur based solely on attribute similarity, but we want to do this in a relatively conservative fashion. One option is to assign the same initial cluster to any
two references that have attributes v_1 and v_2, where either v_1 is identical to v_2, or
v_1 is an initialed form of v_2. For example, we may merge Alfred Aho references
with other Alfred Aho references or with A. Aho references. However, for domains
where last names repeat very frequently, like Chinese, Japanese, or Indian names,
this can affect precision quite adversely. For the case of such common last names,[2]
the same author label can be assigned to pairs only when they have document co-
authors with identical names as well. For example, two X. Wang references will be
merged when they are co-authors with Y. Li. (We will also merge the Y. Li references.) This should improve bootstrap precision significantly under the assumption
that while it may be common for different authors to have the same (initialed) name,
it is extremely unlikely that they will collaborate with the same author, or with two
other authors with identical names.

In addition to using a secondary source for determining common names, a data-
driven approach may also be employed. A [last name, first initial] combination in

---

[2]A list of common last names is available at http://en.wikipedia.org/wiki/List_of_most_popular_
family_names.

the data is ambiguous, if there exist multiple first names with that initial for the last name. For example, though Zabrinsky is not a common last name, K. Zabrinsky will be considered ambiguous if Ken Zabrinsky and Karen Zabrinsky occur as author references in the data. Ambiguous references or references with common last names are not bootstrapped in the absence of relational evidence in the form of co-authorships, as described above.

13.6.3 Finding the Closest Entity Clusters

Once potential duplicate entity clusters have been identified and clusters have been bootstrapped, the algorithm iterates over the following steps. At each step, it identifies the currently closest pair of clusters (c_i, c_j) from the candidate set and merges them to create a new cluster c_{ij}. It removes from the candidate set all pairs that involve either c_i or c_j and inserts relevant pairs for c_{ij}. It also updates the similarity measures for the "related" cluster pairs. All of these tasks need to be performed efficiently to make the algorithm scalable.

For efficient extraction of the closest pairs and updating of the similarities, an indexed max-heap data structure can be used. In addition to the actual similarity value, each entry in the max-heap maintains pointers to the two clusters whose similarity it corresponds to. Also, each cluster c_i indexes all the entries in the max-heap that stores similarities between c_i and some other cluster c_j. The maximum similarity entry can be extracted from the heap in $O(1)$ time. For each entry whose similarity changed because of the merge operation, the heap can be updated in $O(\log n)$ steps, where n is the number of entries in the heap. The requirement therefore is to be able to efficiently locate the entries affected by the merge. First, the set includes all entries that involve c_i or c_j. These are easily retrieved by the indexes maintained by c_i and c_j, and then one of the entries is replaced by c_{ij} and similarities are recomputed. The other set of entries whose similarities are affected are those that are *related* to c_i or c_j as neighbors. These may be retrieved efficiently as well if each cluster maintains indexes to its neighbors. For the initial set of clusters, the neighbors are those that correspond to references in the same edge. As clusters are merged, the new cluster inherits the neighbors from both of its parents. It also inherits the heap indexes from its two parents.

13.6.4 Evaluation Measures

Most approaches to entity resolution have viewed the problem as a pairwise classification task. However, the duplicate relation semantically defines an equivalence partitioning on the references and our approach is to find clusters of duplicate references. From this perspective, it would be preferable to evaluate the clusters directly instead of an evaluation of the implied binary duplicate predictions.

Given the author entity label $l_i \in L$ for each reference in the evaluation, if the resolution is perfect, then each generated cluster will exactly correspond to one author entity. When this is not the case, we may imagine two different scenarios—either a cluster includes references that have different entity labels or it fails to capture all references that have a particular entity label.

The first measure is defined over generated clusters $c_i \in C$. Ideally, each cluster should be homogeneous—all references in a cluster should have the same entity label. The greater the number of references to different entities in a cluster, the worse the quality of the cluster. Rather than naively counting the different entity labels in a cluster, we can measure the entropy of the distribution over entity labels in a cluster. This we call *cluster diversity*. As an example, suppose there are 10 references in a generated entity cluster; five of them correspond to one entity, three to a second entity, and the remaining two to a third entity. Then the probability distribution over entity labels for the references in this cluster is $\langle 0.5, 0.3, 0.2 \rangle$. In general, if this distribution is $\langle p_1, p_2, \ldots, p_k \rangle$, then the entropy is defined as $\sum_k -p_i \log p_i$. Note that the entropy is 0 when all references have the same real entity label and the label distribution is $\langle 1.0 \rangle$. The entropy is highest for a uniform distribution. Also, a uniform distribution over $k + 1$ labels has higher entropy than that over k labels. The weighted average of the diversity of clusters is considered as the first quality measure, where the entropy of each cluster is weighted by the number of references in it. To be more precise, the combined diversity over a clustering C is

$$\text{Div}(C) = \sum_{c_i \in C} \frac{|c_i|}{N} \text{div}(c_i)$$

where N is the total number of references, and div (c_i) is the entropy of the entity label distribution in cluster c_i.

However, just minimizing the cluster diversity is not enough since zero diversity can be achieved by assigning each reference to a distinct cluster. Obviously, this is not a good clustering since references that correspond to the same entity are dispersed over many different clusters. So a second measure of cluster quality is *entity dispersion* defined over each entity label l_i. Here for each entity, we consider the distribution over different cluster labels assigned to the references for that entity. As for cluster diversity, the entropy is measured for the distribution over cluster labels for each entity. We look at the weighted average of the entity dispersions, where the weight of an entity label is the number of references that have that label:

$$\text{Disp}(C) = \sum_{l_i \in L} \frac{|l_i|}{N} \text{disp}(l_i)$$

where disp (l_i) is entropy of the cluster label distribution for the references having entity label l_i.

Note that lower values are preferred for dispersion and diversity. A perfect clustering will have zero cluster diversity and zero entity dispersion. This is practically not achievable in most cases and a decrease in one will usually mean an increase in the other. We can plot the dispersion–diversity curve for each set of generated clusters. This shows the value of diversity achieved for any value of dispersion. Also, we can observe how the dispersion and diversity change as clusters are merged iteratively.

13.7 EXPERIMENTAL EVALUATION

To illustrate the power of graph-based entity resolution, we present evaluations on two citation datasets from different research areas and compare an implementation of the graph-based entity resolution algorithm (GBC-ER) with others based solely on attributes.

The first of the citation datasets is the CiteSeer dataset containing citations to papers from four different areas in machine learning, originally created by Giles et al. [16]. This has 2892 references to 1165 authors, contained in 1504 documents. The second dataset is significantly larger; arXiv (HEP) contains papers from high energy physics used in KDD Cup 2003.[3] This has 58,515 references to 9200 authors, contained in 29,555 papers. The authors for both datasets have been hand labeled.[4]

To evaluate the algorithms, we measure the performance of the algorithms for detecting duplicates in terms of the traditional precision, recall, and F1 on pairwise duplicate decisions in addition to our proposed dispersion–diversity measure for clusters. It is practically infeasible to consider all pairs, particularly for HEP, so a "blocking" approach is employed to extract the potential duplicates. This approach retains ~99% of the true duplicates for both datasets. The number of potential duplicate pairs of author references after blocking is 13,667 for CiteSeer and 1,534,661 for HEP.

As our baseline (ATTR), we compare with the hybrid *SoftTF-IDF* measure [11] that has been shown to outperform other unsupervised approaches for text-based entity resolution. Essentially, it augments the TF-IDF similarity for matching token sets with approximate token matching using a secondary string similarity measure. Jaro-Winkler is reported to be the best secondary similarity measure for *SoftTF-IDF*. We also experiment with the Jaro and the Scaled Levenstein measures. However, directly using an off-the-shelf string similarity measure for matching names results in very poor recall. From domain knowledge about names, we know that first and middle names may be initialed or dropped. A black-box string similarity measure would unfairly penalize such cases. To deal with this, ATTR uses string similarity only for last names and *retained* first and middle names. In addition, it uses drop probabilities p_{DropF} and p_{DropM} for dropped first and middle names, initial probabilities p_{FI} and p_{MI} for correct initials, and p_{FIr} and p_{MIr} for incorrect initials. The probabilities we used are 0.75, 0.001, and 0.001 for correctly initialing, incorrectly initialing, and dropping the first name, while the values for the middle name are 0.25, 0.7, and 0.002. We arrived at these values by observing the true values in the datasets and then hand-tuning them for performance. Our observation is that baseline resolution performance does not vary significantly as these values are varied over reasonable ranges.

ATTR only reports pairwise match decisions. Since the duplicate relation is transitive, we also evaluate ATTR*, which removes inconsistencies in the pairwise

[3]http://www.cs.cornell.edu/projects/kddcup/index.html.
[4]We would like to thank Aron Culotta and Andrew McCallum for providing the author labels for the CiteSeer dataset and David Jensen for providing the author labels for the HEP dataset. We performed additional cleaning for both.

match decisions in ATTR by taking a transitive closure. Note that this issue does not arise with GBC-ER; it does not make pairwise decisions. All of these unsupervised approaches ATTR, ATTR*, and GBC-ER need a similarity threshold for deciding duplicates. We consider the best F1 that can be achieved over all thresholds.

Table 13.1 records F1 achieved by the four algorithms with various string similarity measures coupled with SoftTF-IDF, while Table 13.2 shows the best F1 and the corresponding precision and recall for the four algorithms for each dataset over all secondary similarity measures. The recall includes blocking so that the highest recall achievable is 0.993 for CiteSeer and 0.991 for HEP.

The best baseline performance is with Jaro as secondary string similarity for CiteSeer and Scaled Levenstein for HEP. It is also worth noting that a baseline without initial and drop probabilities scores below 0.5 F1 using Jaro and Jaro-Winkler for both datasets. It is higher with Scaled Levenstein (0.7) but still significantly below the augmented baseline. Transitive closure affects the baseline differently in the two datasets. While it adversely affects precision for HEP, reducing the F1 measure as a result, it improves recall for CiteSeer and thereby improves F1 as well.

GBC-ER outperforms both forms of the baseline for both datasets. Also, for each secondary similarity measure GBC-ER with neighborhood similarity outperforms the baselines with that measure and is in turn outperformed by GBC-ER using edge detail similarity. For CiteSeer, GBC-ER gets close to the highest possible recall with very high accuracy. Improvement over the baseline is greater for HEP. While the improvement may not appear large in terms of F1, note that GBC-ER reduces

TABLE 13.1   Performance of ATTR, ATTR*, and GBC Using Neighborhood and Edge Detail Similarity in Terms of F1 Using Various Secondary Similarity Measures with SoftTF-IDF[a]

	CiteSeer			HEP		
	SL	JA	JW	SL	JA	JW
ATTR	0.980	**0.981**	0.980	**0.976**	0.976	0.972
ATTR*	0.989	**0.991**	0.990	**0.971**	0.968	0.965
GBC(Nbr)	**0.994**	**0.994**	**0.994**	0.979	**0.981**	0.981
GBC(Edge)	**0.995**	**0.995**	**0.995**	0.982	**0.983**	0.982

[a]The measures compared are Scaled Levenstein (SL), Jaro (JA), and Jaro Winkler (JW).

TABLE 13.2   Best F1 and Corresponding Precision and Recall for ATTR, ATTR*, and GBC-ER with Neighborhood and Edge Detail Similarity for CiteSeer and HEP Datasets

	CiteSeer			HEP		
	P	R	F1	P	R	F1
ATTR	0.990	0.971	0.981	0.987	0.965	0.976
ATTR*	0.992	0.988	0.991	0.976	0.965	0.971
GBC(Nbr)	0.998	0.991	**0.994**	0.990	0.972	**0.981**
GBC(Edge)	0.997	0.993	**0.995**	0.992	0.974	**0.983**

error rate over the baseline by 44% for CiteSeer (from 0.009 to 0.005) and by 29% for HEP (from 0.024 to 0.017). Also, HEP has more than 64, 6000 true duplicate pairs, so that a 1% improvement in F1 translates to more than 6400 correct pairs.

Looking more closely at the resolution decisions from CiteSeer, we were able to identify some interesting combinations of decisions by GBC-ER that would be difficult or impossible for an attribute-only model. There are instances in the dataset where reference pairs are very similar but correspond to different author entities. Examples include *(liu j, lu j)* and *(chang c, chiang c)*. GBC-ER correctly predicts that these are not duplicates. At the same time, there are other pairs that are not any more similar in terms of attributes than the examples above and yet are duplicates. These are also correctly predicted by GBC-ER using the same similarity threshold by leveraging common collaboration patterns. The following are examples: *(john m f, john m st), (reisbech c, reisbeck c k), (shortliffe e h, shortcliffe e h), (tawaratumida s, tawaratsumida sukoya), (elliott g, elliot g l), (mahedevan s, mahadevan sridhar), (livezey b, livezy b), (brajinik g, brajnik g), (kaelbing l p, kaelbling leslie pack), (littmann michael l, littman m), (sondergaard h, sndergaard h)*, and *(dubnick cezary, dubnicki c)*. An example of a particularly pathological case is *(minton s, minton andrew b)*, which is the result of a parse error. The attribute-only baselines cannot make the right prediction for both these sets of examples simultaneously, whatever the decision threshold, since they consider names alone.

Figures 13.5 and 13.6 show how performance varies for GBC-ER for the two datasets with varying combination weight α for attribute and graph-based similarity. Recall that when α is 0, the similarity is based only on attributes, and when α is 1 it is wholly graph-based. The plots show that GBC-ER with both neighborhood and edge detail similarity outperform the baselines over *all* values of α. Note that GBC-ER takes advantage of graph-based bootstrapping in these experiments, which explains why it is better than the baseline even when α is 0. The best performance for CiteSeer is around 0.5 while for HEP performance peaks around 0.1 and then trails off. It can also be observed that edge detail similarity is more stable in performance over varying α than neighborhood similarity. Significantly, once clusters have been bootstrapped using attribute and graph-based evidence, GBC-ER outperforms the baselines even when α is 1, which means that attributes are being overlooked altogether and clusters are merged using graph-based evidence alone.

The first row of plots in Figure 13.5 use single-link criterion for attribute similarity while those in the second measure similarity between representative cluster attributes. The differences in the two plots are observable for low values of α when attribute similarity plays a more dominant role. It can be seen that the single-link approach performs better than the cluster representative approach for all secondary similarity measures. The curves become identical for higher values of α when attributes matter less. A similar trend was observed for HEP as well.

Figure 13.7 shows performance of GBC-ER without using graph-based bootstrapping. When α is 0, GBC-ER is identical to ATTR*, which is verified by the results. As α increases from 0, performance improves over the baseline and then drops again. For HEP, performance falls sharply with higher α with neighborhood similarity. Edge detail similarity, however, still performs surprisingly well. Even

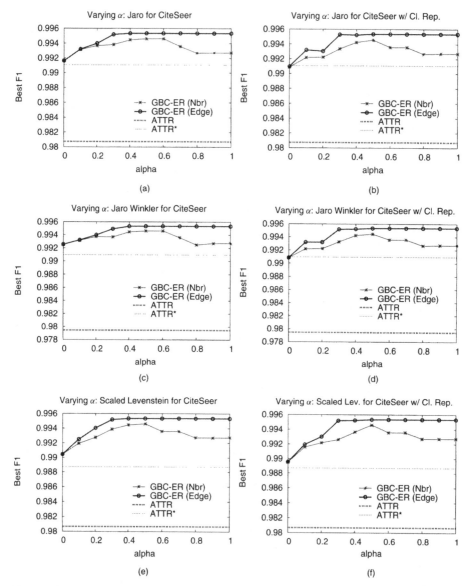

Figure 13.5. Best F1 measures achieved by GBC-ER with neighborhood and edge detail similarities over varying combination weight α for CiteSeer. Plots (a), (c), and (e) use single link for attribute similarity with Jaro, Jaro-Winkler, and Scaled Levenstein, respectively, as secondary similarity while plots (b), (d), and (f) use representative attributes for clusters with the same three secondary similarity measures.

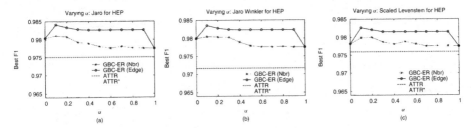

Figure 13.6. Best F1 measures achieved by GBC-ER with neighborhood and edge detail similarities over varying combination weight α for HEP. Plots (a)–(c) are with Jaro, Jaro-Winkler, and Scaled Levenstein, respectively, as secondary similarity for attributes.

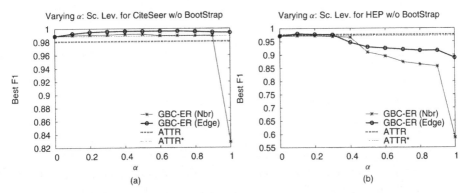

Figure 13.7. Effect of not using bootstrapping for initializing the entity clusters for (a) CiteSeer and (b) HEP using Scaled Levenstein as secondary similarity.

when α is 1, it does better than the baseline for CiteSeer and is able to achieve close to 0.9 F1 for HEP. This suggests that edge detail is a reliable indicator of identity even without considering attributes. It should, however, be noted that these results include blocking, which uses attributes to find potential duplicates. This suggests that given people with similar names, it is possible to identify duplicates with a high degree of reliability using edge detail similarity alone.

Figure 13.8 shows the precision–recall characteristics for the four algorithms with Jaro as secondary similarity. Plot 13.8(a) shows that all algorithms perform well for CiteSeer. Plot 13.8(b) concentrates on the region of difference between the algorithms. Still the curves for edge detail and neighborhood similarity are almost identical, as is the case for HEP in plot 13.8(c). But they consistently remain over the baselines for both datasets. Observe that ATTR* dominates ATTR for CiteSeer but the roles reverse for HEP. Note that GBC-ER starts above 90% recall. This is by virtue of graph-based bootstrapping. The corresponding precision is high as well validating the effectiveness of our graph-based bootstrapping scheme. The characteristics with other secondary similarity measures are similar.

Figure 13.9 illustrates how dispersion–diversity measures may be used. We show the results only for HEP with Jaro similarity. Other plots are similar. Recall that

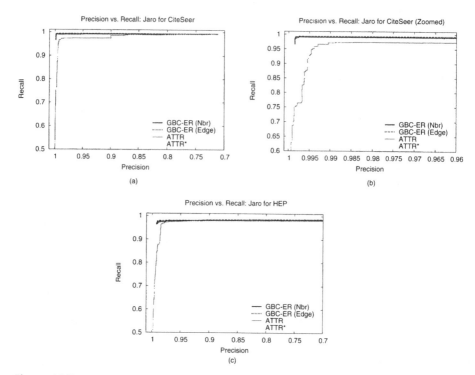

Figure 13.8. Precision–recall characteristics for (a)–(b) CiteSeer and (c) HEP using Jaro similarity. Plot (b) highlights the difference between the algorithms from plot (a).

while values closer to 1 are preferred for precision and recall, values closer to 0 are better for dispersion and diversity so that points on the dispersion–diversity curves in plot 13.9(a) that are closer to (0, 0) indicate better performance. Plot 13.9(a) shows that dispersion–diversity curves are very similar to precision–recall curves, but they evaluate the clusters directly. Note that since ATTR makes pairwise decisions and does not generate clusters of duplicate references, it cannot be evaluated using this approach. Figures 13.9(b)–(d) plot diversity and dispersion against number of clusters. Again, observe how graph-based bootstrapping significantly lowers the initial number of clusters for GBC-ER, thereby starting off with much lower dispersion. The initial number of clusters is lowered by 83% for HEP and by 58% for CiteSeer. That it is accurate can be inferred from the significantly lower diversity at the same number of clusters compared to ATTR*. The number of real author entities for HEP is 8967, which is shown using a vertical line in plots 13.9(b)–(d). Plot 13.9(b) highlights the difference in dispersion between the three algorithms, and we can see that GBC-ER improves dispersion over ATTR* at the correct number of entity clusters and that edge detail performs better than neighborhood similarity.

Finally, we look at the execution times of the algorithms. All experiments were run on a 1.1-GHz Dell PowerEdge 2500 Pentium III server. Table 13.3 records the execution times in CPU seconds of the baselines and different versions of GBC-ER

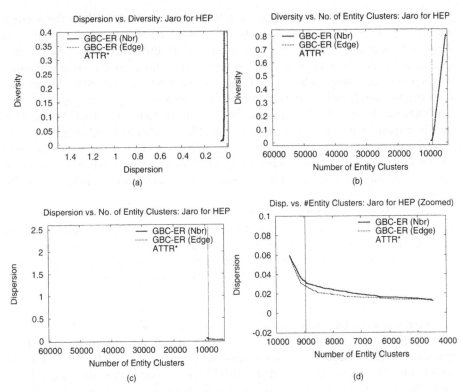

Figure 13.9. Performance measure using dispersion and diversity for HEP with Jaro similarity. The plots show (a) dispersion against diversity, (b) diversity and (c)–(d) dispersion over changing number of entity clusters.

TABLE 13.3   Execution Time of GBC-ER, ATTR, and
ATTR* in CPU Seconds for CiteSeer and HEP Datasets

	CiteSeer	HEP
ATTR	1.88	162.13
ATTR*	2.30	217.37
GBC(Nbr w/ Single Link)	3.14	543.83
GBC(Edge w/ Single Link)	3.18	690.44
GBC(Nbr w/ Cluster Rep.)	3.65	402.68
GBC(Edge w/ Cluster Rep.)	3.75	583.58

on the CiteSeer and HEP datasets. GBC-ER expectedly takes more time than the baselines. But it is quite fast for CiteSeer taking only twice as much time as ATTR*. It takes longer for HEP; about 7 times as long compared to the baseline. While using edge detail is more expensive than neighborhood similarity, it does not take significantly longer for either dataset. The complexity of edge detail depends on a number of factors. It grows quadratically with the average number of edges per entity and linearly with the average number of references in each edge. While the average edge size is the same for both datasets, the average number of edges per entity is 2.5 for CiteSeer and 6.36 for HEP, which explains the difference in execution times. In contrast, complexity of neighborhood similarity is linear in the average number of neighbors per entity, which is 2.15 for CiteSeer and 4.5 for HEP. Separately, we expected single-link clustering to be more expensive than using representative attributes for clusters. While this is true for HEP, the trend reverses for CiteSeer. This may be explained by taking into account the added overhead of updating the cluster representative every time new references are added to a cluster. Since CiteSeer only has an average of 2.5 references per entity, the cost of a naive updation scheme for cluster representatives overshadows the small gain in computing attribute similarity for clusters.

To see how the algorithms scale with increasing number of references in the dataset, we used a synthetic generator for ambiguous data. We preserved features of the real datasets wherever possible, like the average number of references and edges per entity, the degree of neighborhood for each entity and the average number of references per edge. The execution times of ATTR* and different versions of GBC-ER are plotted in Figure 13.10 against varying number of references in the dataset. We would like to stress that the execution time depends on other factors as well like the number of potential duplicate pairs in the data, which we were not able to control directly. So these numbers should not be compared with the execution times on the real datasets but instead should serve only for a comparative study of the different algorithms. The curves confirm that GBC-ER takes longer than the baseline, but they also show that the trend is roughly linear in the number of references for all

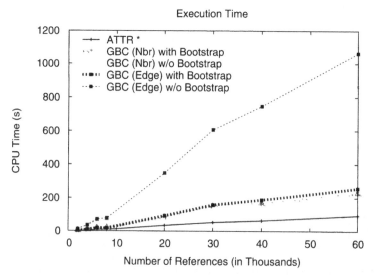

Figure 13.10. Execution time for ATTR* and GBC-ER with neighborhood and edge detail similarity over varying number of references in synthetic datasets.

versions of it. The plots also show the significant speedup that is achieved with graph-based bootstrapping in addition to the performance benefits that it provides.

13.8 CONCLUSION

In this chapter, we have seen how graph-based entity resolution may be posed as a clustering problem that augments a general category of attribute similarity measures with graph-based similarity among the entities to be resolved. We looked at two different similarity measures based on graphs and a graph-based clustering algorithm (GBC-ER) that shows performance improvements over attribute-based entity resolution in two real citation datasets. The combined similarity measure allows a smooth transition from an attribute-only measure at one extreme to one that is based just on graph-based similarity at the other. Depending on the reliability of the attributes or the relations in the application domain or the particular dataset, it is possible to attach higher weights to either of them.

The graph-based similarity measures are intended to capture the dependencies between resolution decisions. They do so by looking at the current entity labels of related entities. We presented two different graph-based similarity measures that consider relations between entities at different levels of detail. Edge-detail similarity explicitly considers each edge in which an entity participates. Neighborhood similarity reduces the computational complexity involved by collapsing the edge structure and just looking at the set of neighborhood entities for each entity cluster. The execution times for the similarity measures depends on the graph structure in the data. In general, edge detail takes more time than neighborhood similarity but is a more reliable indicator of identity. Even without considering attributes at all in the similarity measure, edge detail is able to achieve high accuracy in determining duplicates. In fact, in one of our datasets, even without using attributes at all, it does better than attribute-only baselines. This suggests that in domains where attributes are extremely unreliable or perhaps unavailable, it may still be possible to discover identity going by the graph patterns alone.

We consider two different options for measuring attribute similarity between entity clusters. The first is the single linkage criterion that looks at all pairwise attribute similarities between two clusters and chooses the highest one. This reduces to transitive closure over pairwise attribute decisions that is appropriate for entity resolution. This measure is expected to be computationally intensive, and, therefore, we propose an alternative measure that constructs the representative attribute for each entity cluster and then measures similarities between these representatives only. However, our experiments show that the single-linkage measure performs better in practice. Also, unless the representatives are recomputed quickly as clusters expand, this approach does not come with significant improvements in execution time either.

We evaluate a graph-based bootstrapping approach for initializing the entity clusters quickly and accurately. Experimental results show that bootstrapping reduces the initial number of clusters by 83% on our larger dataset without significantly compromising on precision. We show that graph-based bootstrapping also improves performance by uncovering patterns quickly that the graph-based similarity measures can leverage.

Since GBC-ER considers relational similarities that are expensive to compute and updates similarities iteratively, it is expectedly more costly than performing attribute similarity. However, our experiments on synthetic data show that both graph-based similarity measures scale gracefully over increasing number of references in the dataset. Also, the added cost clearly reaps bigger benefits, as shown by the performance plots. Database cleaning is not an operation that is likely to be performed very frequently and the increased computation time is not expected to be too critical. There may, of course, be situations where this approach is not likely to prove advantageous, for example, where distinctive cliques do not exist for the entities or if references for each edge appear randomly. There the user has the choice of falling back on traditional attribute similarity or setting α to assign lower weights for graph-based similarity.

In summary, entity resolution is an area that has been attracting growing attention to address the influx of structured and semistructured data from a multitude of heterogeneous sources. Accurate resolution is important for a variety of reasons ranging from cost effectiveness and reduction in data volume to accurate analysis for critical applications. In the case of graph data, it is especially important to look at entity resolution from a graph-based perspective. We have found graph-based entity resolution to be a powerful and promising approach that combines attribute similarity with relational evidence and shows improved performance over traditional approaches.

REFERENCES

1. R. Ananthakrishna, S. Chaudhuri, and V. Ganti. Eliminating fuzzy duplicates in data warehouses. In Proceedings of the 28th International Conference on Very Large Databases (VLDB-2002), Hong Kong, China, 2002.
2. P. Berkhin. Survey of clustering data mining techniques. Technical Report, Accrue Software, 2002.
3. I. Bhattacharya and L. Getoor. Deduplication and group detection using links. In Proceedings of the 10th ACM SIGKDD Workshop on Link Analysis and Group Detection (LinkKDD-04), August 2004.
4. I. Bhattacharya and L. Getoor. Iterative record linkage for cleaning and integration. In Proceedings of the SIGMOD 2004 Workshop on Research Issues on Data Mining and Knowledge Discovery, June 2004.
5. I. Bhattacharya and L. Getoor. A latent dirichlet model for entity resolution. Technical Report, University of Maryland, College Park, MD, 2005.
6. M. Bilenko and R. J. Mooney. Adaptive duplicate detection using learnable string similarity measures. In Proceedings of the Ninth ACM SIGKDD International Conference on Knowledge Discovery and Data Mining (KDD-2003), Washington, DC, 2003.
7. M. Bilgic, L. Licamele, L. Getoor, and B. Shneiderman. D-dupe: An interactive tool for entity resolution in social networks. In The 13th International Symposium on Graph Drawing (Poster), Limerick, Ireland, September 2005.

8. S. Chaudhuri, K. Ganjam, V. Ganti, and R. Motwani. Robust and efficient fuzzy match for online data cleaning. In Proceedings of the 2003 ACM SIGMOD International Conference on Management of Data, pp. 313–324, San Diego, CA, 2003.

9. W. Cohen. Data integration using similarity joins and a word-based information representation language. *ACM Transactions on Information Systems*, 18:288–321, 2000.

10. W. W. Cohen, H. Kautz, and D. McAllester. Hardening soft information sources. In Proceedings of the Sixth International Conference on Knowledge Discovery and Data Mining (KDD-2000) pp. 255–259, Boston, August 2000.

11. W. W. Cohen, P. Ravikumar, and S. E. Fienberg. A comparison of string distance metrics for name-matching tasks. In Proceedings of the IJCAI-2003 Workshop on Information Integration on the Web, pp. 73–78, Acapulco, Mexico, August 2003.

12. W. W. Cohen and J. Richman. Learning to match and cluster large high-dimensional data sets for data integration. In Proceedings of the Eighth ACM SIGKDD International Conference on Knowledge Discovery and Data Mining (KDD-2002), Edmonton, Alberta, 2002.

13. A. Doan, Y. Lu, Y. Lee, and J. Han. Object matching for data integration: A profile-based approach. In Proceedings of the IJCAI Workshop on Information Integration on the Web, Acapulco, Mexico, August 2003.

14. I. P. Fellegi and A. B. Sunter. A theory for record linkage. *Journal of the American Statistical Association*, 64:1183–1210, 1969.

15. H. Galhardas, D. Florescu, E. Simon, and D. Shasha. An extensible framework for data cleaning. In *ICDE '00: Proceedings of the 16th International Conference on Data Engineering*, p. 312. IEEE Computer Society, 2000.

16. C. Lee Giles, K. Bollacker, and S. Lawrence. CiteSeer: An automatic citation indexing system. In Proceedings of the Third ACM Conference on Digital Libraries, pp. 89–98, Pittsburgh, PA, June 23–26 1998.

17. L. Gravano, P. Ipeirotis, N. Koudas, and D. Srivastava. Text joins for data cleansing and integration in an rdbms. In 19th IEEE International Conference on Data Engineering, 2003.

18. M. A. Hernández and S. J. Stolfo. The merge/purge problem for large databases. In Proceedings of the 1995 ACM SIGMOD International Conference on Management of Data (SIGMOD-95) pp. 127–138, San Jose, CA, May 1995.

19. J. A. Hylton. Identifying and merging related bibliographic records. Master's thesis, Department of Electrical Engineering and Computer Science, MIT, Cambridge, MA, 1996.

20. I. Jonyer, L. B. Holder, and D. J. Cook. Graph-based hierarchical conceptual clustering. *Journal of Machine Learning Research*, 2(1–2):19–43, 2001.

21. D. V. Kalashnikov, S. Mehrotra, and Z. Chen. Exploiting relationships for domain-independent data cleaning. In SIAM International Conference on Data Mining (SIAM SDM), Newport Beach, CA, April 21–23 2005.

22. S. Lawrence, K. Bollacker, and C. L. Giles. Autonomous citation matching. In *Proceedings of the Third International Conference on Autonomous Agents*, New York, May 1999. ACM Press.

23. X. Li, P. Morie, and D. Roth. Semantic integration in text: From ambiguous names to identifiable entities. *AI Magazine. Special Issue on Semantic Integration*, 2005.

24. A. McCallum and B. Wellner. Conditional models of identity uncertainty with application to noun coreference. In Neural Information Processing Systems (NIPS), 2004.

25. A. K. McCallum, K. Nigam, and L. Ungar. Efficient clustering of high-dimensional data sets with application to reference matching. In Proceedings of the Sixth International

Conference on Knowledge Discovery and Data Mining (KDD-2000), pp. 169–178, Boston, MA, August 2000.

26. B. Milch, B. Marthi, D. Sontag, S. Russell, D. L. Ong, and A. Kolobov. Blog: Probabilistic models with unknown objects. In Proceedings IJCAI, 2005.

27. A. E. Monge and C. P. Elkan. The field matching problem: Algorithms and applications. In Proceedings of the Second International Conference on Knowledge Discovery and Data Mining (KDD-96), pp. 267–270, Portland, OR, August 1996.

28. A. E. Monge and C. P. Elkan. An efficient domain-independent algorithm for detecting approximately duplicate database records. In Proceedings of the SIGMOD 1997 Workshop on Research Issues on Data Mining and Knowledge Discovery, pp. 23–29, Tuscon, AZ, May 1997.

29. G. Navarro. A guided tour to approximate string matching. *ACM Computing Surveys*, 33(1):31–88, 2001.

30. J. Neville, M. Adler, and D. Jensen. Clustering relational data using attribute and link information. In Proceedings of the Text Mining and Link Analysis Workshop, Eighteenth International Joint Conference on Artificial Intelligence, 2003.

31. H. B. Newcombe, J. M. Kennedy, S. J. Axford, and A. P. James. Automatic linkage of vital records. *Science*, 130:954–959, 1959.

32. Parag and P. Domingos. Multi-relational record linkage. In Proceedings of 3rd Workshop on Multi-Relational Data Mining at ACM SI GKDD, Seattle, WA, August 2004.

33. H. Pasula, B. Marthi, B. Milch, S. Russell, and I. Shpitser. Identity uncertainty and citation matching. In *Advances in Neural Information Processing Systems 15*. MIT Press, Cambridge, MA, 2003.

34. V. Raman and J. M. Hellerstein. Potter's wheel: An interactive data cleaning system. In Proceedings VLDB, 2001.

35. P. Ravikumar and W. W. Cohen. A hierarchical graphical model for record linkage. In UAI 2004, Banff, CA, July 2004.

36. E. Ristad and P. Yianilos. Learning string edit distance. *IEEE Transactions on PAMI*, 20(5):522–532, 1998.

37. S. Sarawagi and A. Bhamidipaty. Interactive deduplication using active learning. In Proceedings of the Eighth ACM SIGKDD International Conference on Knowledge Discovery and Data Mining (KDD-2002). Edmonton, Alberta, 2002.

38. J. Shi and J. Malik. Normalized cuts and image segmentation. *IEEE Transactions on Pattern Analysis and Machine Intelligence*, 22(8):888–905, 2000.

39. S. Tejada, C. A. Knoblock, and S. Minton. Learning object identification rules for information integration. *Information Systems Journal*, 26(8):635–656, 2001.

40. W. E. Winkler. The state of record linkage and current research problems. Technical Report, Statistical Research Division, U.S. Census Bureau, Washington, DC, 1999.

41. W. E. Winkler. Methods for record linkage and Bayesian networks. Technical Report, Statistical Research Division, U.S. Census Bureau, Washington, DC, 2002.

Part III

APPLICATIONS

14

MINING FROM CHEMICAL GRAPHS

TAKASHI OKADA

Department of Informatics, School of Science & Engineering,
Kwansei Gakuin University, Sanda, Japan

14.1 INTRODUCTION AND REPRESENTATION OF MOLECULES

This chapter provides an overview of developments in chemical graph mining, with a focus centered somewhere between computer science and chemistry, reflecting the author's background. After a brief introduction to the representation of chemical structure, we discuss mining issues and developments in this field. Those who require further details about chemical information should consult a chemoinformatics textbook.

14.1.1 Molecular Structure and Structure Diagrams

Chemistry is fundamentally the science of molecules. Its main subjects are the elucidation of molecular structures, the establishment of relationships between structures and various properties, and the conversion of chemical structures (chemical reactions). Molecules consist of atoms and the bonds linking them. For example, a carbon atom has four valences, which can be used to form various stable bonds with other atoms, such as carbon, oxygen, and nitrogen. There are infinite combinations of atoms and bonds, and a huge number of chemical compounds have already been synthesized. In fact, more than 40 million compounds are currently registered

Mining Graph Data, Edited by Diane J. Cook and Lawrence B. Holder
Copyright © 2007 John Wiley & Sons, Inc.

with the Chemical Abstracts Service. Since each molecule must be identified separately, chemists have expended a great deal of time and energy expressing and naming unique structures since the nineteenth century.

Before beginning a discussion of detailed structure expression, it is important to point out that there are many levels to their representation, including structure diagrams, three-dimensional steric conformations, and electronic and vibrational structures. Our observations of molecules involve mixtures of various conformations, and the lifetime of a conformation is very short. The excited states of electronic and vibrational structures also have short lifetimes, and we cannot isolate them. Therefore, molecular structure diagrams have played a central role in chemistry, as the descriptive level of the diagrams matches the concept of a stable molecule. This means of expression has been used since the nineteenth century, and it is really the international language of chemistry, akin to the mathematical formulas in other sciences.

The chemical formula is another means of expressing a molecule; it describes the composition of elements. For example, two compounds with the chemical formula C_2H_6O can exist: dimethyl ether and ethanol; Figure 14.1 shows their structure diagrams. These two molecules are called *isomers*; their conversion does not occur at room temperature unless they are placed under certain reactive conditions.

In these diagrams, an atom is denoted by its element symbol. Each element has a unique number of valences (C, 4; H, 1; O, 2), which are used to constitute bonds between atoms. All the bonds in this figure are single bonds in which the atoms at each end use one valence. However, many molecules contain bonds using two or three valences, as shown in Figure 14.2. Here, acetic acid has a double bond using two valences each from C and O, and acetylene has a triple bond between two carbons. Sometimes, an atom has a positive or negative charge, as seen in the lower oxygen atom of the acetic acid anion. The charge of an atom usually affects the number of valences.

It is easy to understand that structure diagrams are undirected, labeled graphs in which atoms are nodes and bonds are the edges of the graph. Sometimes we find arrows in a structure diagram, which denote shifting of an electron pair from one atom to its neighbor. However, such arrows should be interpreted as a kind of edge label and should not be treated as a directed graph. Examples appear in many

$$C_2H_6O$$

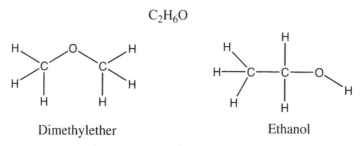

Dimethylether   Ethanol

Figure 14.1.   Isomers with the chemical formula C_2H_6O.

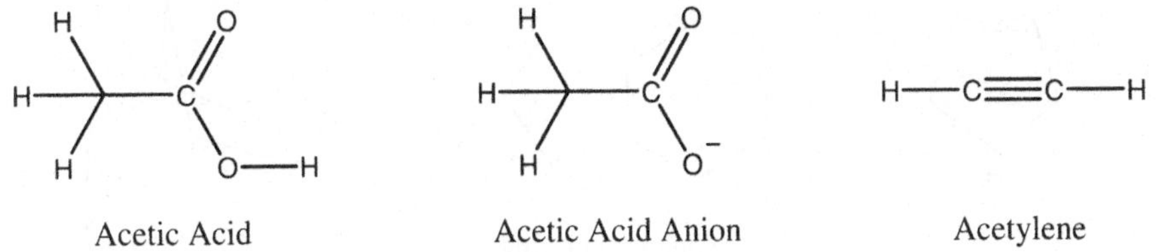

Acetic Acid   Acetic Acid Anion   Acetylene

Figure 14.2. Various structure diagrams.

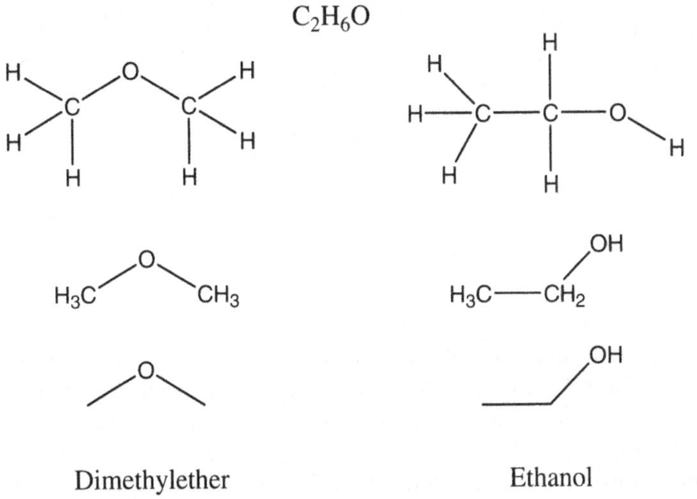

C_2H_6O

Dimethylether   Ethanol

Figure 14.3. Abbreviation of hydrogen and carbon atoms.

inorganic and organometallic compounds, and their expression as a graph is not yet well established, so we confine ourselves to using structure diagrams of organic molecules.

Hydrogen atoms are often considered attachments to fill open atom valences, and they are aggregated in CH_3 and CH_2 groups in the second row of Figure 14.3. Carbon atoms, which are so common, are also abbreviated in the third row. These three structure diagrams have the same meaning for chemists.

The diagrams of most structures contain rings, as shown in Figure 14.4. The bonds in the nonaromatic rings are single or double. By contrast, the single and double benzene bonds can be described in two different ways, as shown in the lower left graphs. These bonds are delocalized and exhibit a very different character from single and double bonds; they are called "aromatic" bonds, and chemists sometimes emphasize their nature by drawing a circle in the ring. This aromatic nature is also found in many rings, for example, pyridine and furan. When a structure diagram is given, an algorithm can determine whether a ring is aromatic. However, a furan ring can exhibit dual characteristics of aromatic bonds or single/double bonds, depending on the context under discussion.

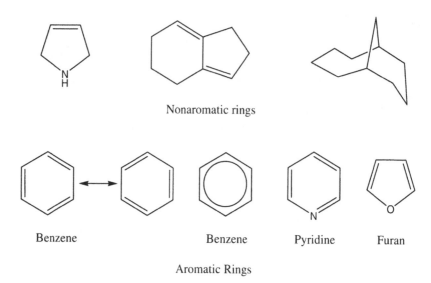

Nonaromatic rings

Benzene   Benzene   Pyridine   Furan

Aromatic Rings

Figure 14.4. Molecules with nonaromatic and aromatic rings.

Combinations of atoms and bonds with a variety of labels allow potentially infinite numbers of molecules, all of which can be expressed using a labeled graph. There are naming rules for these structure diagrams, which chemists have used in many documents, but currently the compound name has a limited role, and chemists' ways of thinking depend solely on structure diagrams. The compelling need for storage and retrieval of structure diagrams was the mother of chemoinformatics, and the connection table described in the next section is the basic tool for exchanging information on structure diagrams.

14.1.2 Connection Table

Figure 14.5 depicts a typical connection table for acetic acid. Here, the first line contains the numbers of atoms ($= 4$) and bonds ($= 3$). The following four lines (a1–a4) show atom coordinates and element symbols, using an arbitrary numbering of atoms. For example, the first atom at line a1 is carbon, located at position: $x = -0.665$, $y = -0.0248$, $z = 0.0$. These coordinates are only used to illustrate the structure diagram and have no physical reality. The last three lines (b1–b3) are used to denote bonds among atoms; line b3 indicates that there is a single bond (1) between atoms 2 and 4.

Hydrogen atoms are usually considered attachments to heavy atoms and are not included as a separate line in connection tables, which helps to decrease the order of a graph. However, chemists always consider veiled hydrogen atoms because they play important roles in all chemical phenomena.

An atom line sometimes has modifiers displaying information about charges and isotopes, as well as the number of attached hydrogen atoms. We can also attach flags to bond entries, denoting whether the bond is aromatic, and whether it is in

	4	3		
a1	-0.6650	-0.0248	0.0000	C
a2	0.1600	-0.0248	0.0000	C
a3	0.5935	0.6771	0.0000	O
a4	0.6650	-0.6771	0.0000	O
b1	1	2	1	
b2	2	3	2	
b3	2	4	1	

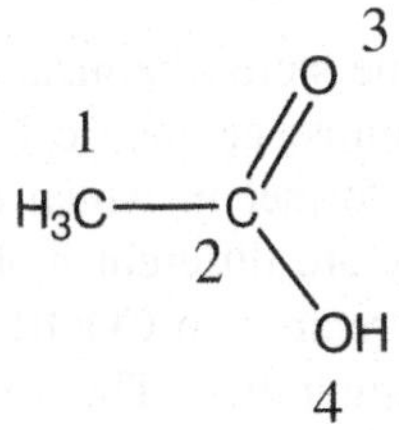

Figure 14.5. Sample connection table for acetic acid.

a ring. Usual datasets do not contain this information, however, and software must infer it from a simple connection table and the number of atom valences.

Most chemists consider a nitrogen atom to have two kinds of valence, depending on the molecular environment. For example, a nitrogen atom has three valences in methylamine and five in nitromethane, as shown in Figure 14.6. However, nitromethane can also be written as a resonance hybrid of two ionized structures in which the cationic nitrogen atom can be interpreted as having three valences (the valences in methylamine) plus one (modification by a positive charge). The latter assumes that neutral nitrogen always has three valences, and this scheme is usually used in connection tables because of its computational simplicity. The interpretation that the real molecule is a resonance hybrid of two ionized structures, like the acetic acid anion in Figure 14.6, is left to the minds of chemists.

Depending on the commercial software used, there are a variety of exchange formats for connection tables. MDL molfile [24] is the most frequently used format, and most commercial software can import/export files in this format.

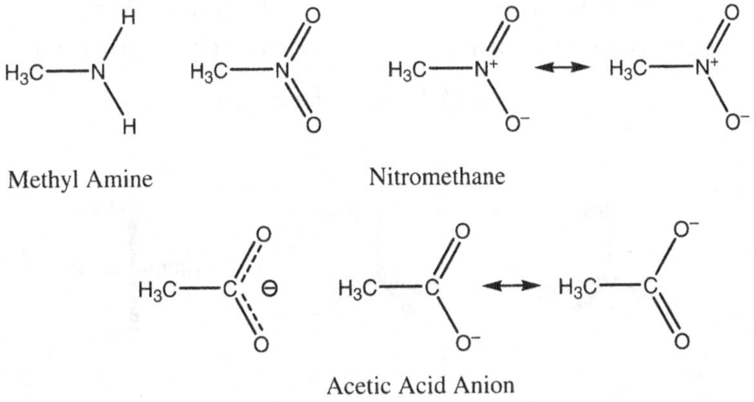

Methyl Amine   Nitromethane

Acetic Acid Anion

Figure 14.6. Multiple valence numbers and resonance.

14.1.3 Stereochemistry

Some stereochemical features cannot be described as a graph, including two major features: geometric isomers and chiral centers. Figure 14.7 shows an example of the former in which the same graph expresses maleic acid and fumaric acid, but they are different molecules, because the double bond at the center cannot rotate. There are two COOH groups: one has a *cis* configuration, and the other has a *trans* configuration. These geometric isomers have very different physical and chemical properties, between which we need to discriminate.

Figure 14.8 shows an example of a chiral center, another major feature. Here, wedge-shaped and dashed bonds are used to denote the relative positions of the attached groups. In these molecules, there are four different groups attached to the central carbon atom, and we cannot superimpose Figure 14.8(a) and 14.8(b). The central carbon atom in these molecules is called a chiral center, and it appears when there is no mirror symmetry in the structure of a molecule. In physical reality, wedge-shaped/dashed bonds are a single bond, and these expressions are useful only when a chiral center is specified. Therefore, diagrams 14.8(b) and 14.8(c) express the same molecule. Many biological molecules are chiral, which greatly affects their interactions with other biological molecules. Therefore, the expression of chiral molecules is an important subject in chemoinformatics.

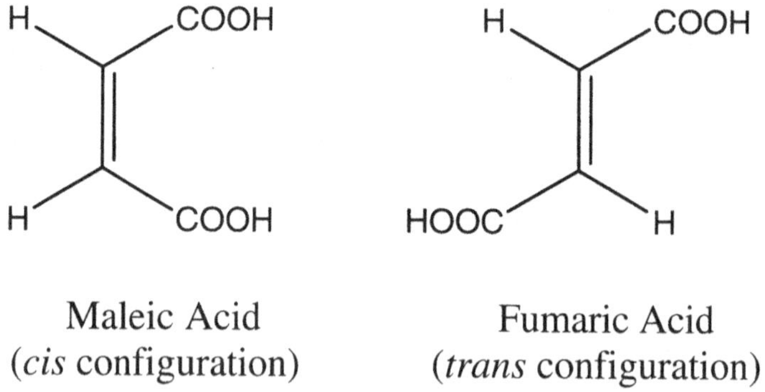

Maleic Acid
(*cis* configuration)

Fumaric Acid
(*trans* configuration)

Figure 14.7. Geometric isomers.

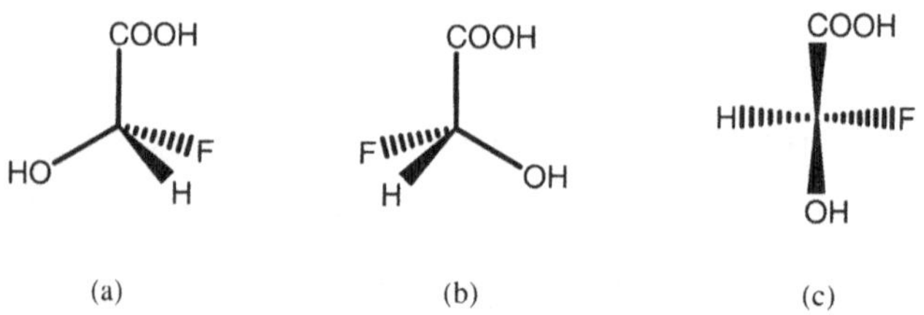

(a)   (b)   (c)

Figure 14.8. Chiral isomers.

One way to code stereochemical information is to use three-dimensional coordinates in a connection table, which allows us to detect *cis–trans* isomers using a written structure diagram. To express chirality, we usually annotate a wedge-shaped/dashed property in the bond entry, where the order of the constituent atoms has a meaning, as in a directed graph. These expressions are sometimes called 2.5-dimensional structure expressions, as compared to two-dimensional (pure graph) and full three-dimensional expressions. However, these stereochemical features create many difficulties for the storage and retrieval of molecular structures, which will be discussed later.

Three-dimensional molecular coordinates are also important in all aspects of chemistry. For example, we can estimate the docking ability of a molecule with proteins by using their coordinates to evaluate the energy of the interaction between them. However, the experimental determination of stable geometry requires significant effort. Even if we could determine the geometry, real compounds exist as thermodynamic mixtures of various conformations, and interactions with water and other molecules might change these conformations. Therefore, data mining research using three-dimensional coordinates requires consideration of the coordinates themselves and is best left to the hands of chemists.

14.1.4 Structure and Substructure Search

Chemists often need to retrieve a variety of molecular properties and synthetic methods, and the structure diagram is a key to their retrieval. They may also wish to retrieve information about molecules with similar structures sharing some common skeleton, which also makes a substructure search indispensable. Since structure diagrams are basically stored as graphs, computations for graph isomorphism and subisomorphism are necessary.

The largest molecular database found at the Chemical Abstracts Service currently includes 40 million chemical compounds that have appeared in articles or patents, along with links to abstracts. The service has a monopoly on the market and provides access to chemists worldwide, so they have made a big effort toward the efficient retrieval of molecular structures.

The Morgan algorithm is the core of the molecular registry system; it assigns canonical numbers to the atoms in a molecule [26]. In this algorithm, atoms of the same element in a molecule are differentiated by considering neighboring atoms. Once the atom numbering has been determined, it is easy to create a canonical name string that can be used as a hash key during retrieval. The system provides a filter for substructure searches, which involve thousands of small fragment structures. Prescreening using this filter decreases the number of candidate molecules that must be processed by the subgraph isomorphism algorithm.

The SEMA algorithm [43] has accomplished a similar naming of stereochemical information; the search for a substructure with stereochemical features is more difficult and is still under investigation (for an example of a recent development, see [39]).

14.1.5 SMILES and Markush Structures

Connection tables play a central role in structure diagrams, but chemists have made serious efforts to develop linear string notation that describes structure diagrams accurately. If chemists can read/write string notation easily, it could provide a convenient input method and a simple output method for structures in working text.

Of the variety of string notations so far proposed, some chemists currently prefer to use SMILES [6], which is also used as the core expression in a commercial chemical database management system, and has affected some fragment notations designed for attributes to be used in data mining.

SMILES uses simple compact notation; for example, dimethyl ether and ethanol (Fig. 14.1) are written as COC and CCO (or OCC), respectively. Here, hydrogen atoms and default single bonds are omitted and parentheses are used to indicate branching, so acetic acid (Fig. 14.5) is written as CC(=O)O. Branches may be stacked and acetic acid can also be coded as C(C)(O)=O. Bonds can also be represented using pairs of matching digits, for example, to denote ring closures. Therefore, the three-ring structures from the upper row of Figure 14.4 are denoted by N1CC=CC1, C1C=C2C(=CCC2)CC1, and C1CC2CCCC(C2)CC1. We can write benzene as C1=CC=CC=C1 or c1ccccc1, if we use the option of showing aromatic atoms using lowercase letters.

SMILES even allows stereochemistry to be described by introducing special characters such as "/" and "\" (for *cis* and *trans* configurations, respectively) and "@" (for chiral centers). We will not explain these characters here; there is a good SMILES tutorial on the Web, which also provides a function to generate structure diagrams from SMILES strings interactively [6].

Markush structures are another important way to express chemical structures. They appear mainly in patents when chemists make a claim to protect compounds related to an invention. Figure 14.9 shows an example. The structure includes three

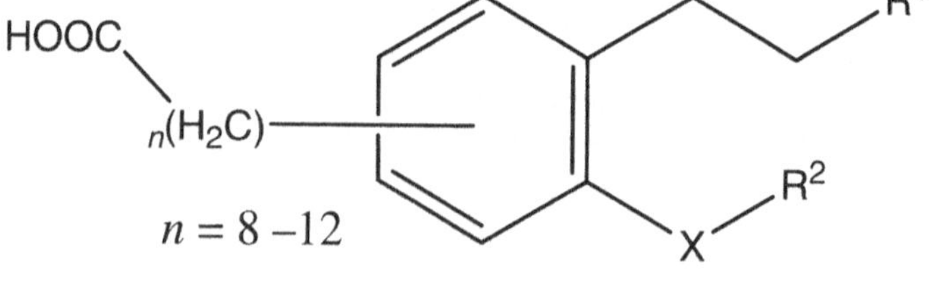

$$R^1 = \text{H or small alkyl, halogen, OH}$$
$$R^2 = \text{H, CH}_3$$
$$X = \text{O, NH}$$

Figure 14.9. Sample Markush structure.

variable substituents and contains many specific structures. The substituent at the left has a variable attachment position, and even a vague term like *small alkyl* is contained in R^1. We can use more complex expressions using conditional clauses or substituents defined by physical properties, so this type of expression is essentially a natural language. Retrievals from patent databases suffer from significant noise and incomplete answers, but this type of expression would be valuable if a mining system could return a Markush structure to summarize a cluster of molecular structures.

14.2 ISSUES FOR MINING

There are two ways to investigate relationships between a chemical structure and various molecular properties, including bioactivity and chemical reactions. One way is based on physical chemistry and uses computational methods to estimate properties. When we can get an accurate estimate, the resulting properties are used directly. Sometimes calculated results are not accurate enough, in which case the results can be used to develop a model to elucidate theoretical relationships between structures and properties. However, these theory-based methods encounter problems when examining the molecules found in a biological system.

The other way is based on data. The collection of experimental results is at the core of chemistry, which is why academic databases were first developed in this field. Experimental chemists survey data and create their own lines of investigation, so data mining has been essential for them. However, advances in experimental techniques and database development have dramatically increased the amount of available data, and data mining technology is expected to contribute to the work of all chemists.

With regard to relevant structures, molecular properties can be classified into three categories: local properties arising from a specific local substructure, additive properties to which all components of a molecule contribute, and complex properties affected by various factors. The next sections describe typical properties and mining issues related to these categories.

14.2.1 Local Properties

Typical examples of local properties include peak positions on various spectra [infrared (IR), nuclear magnetic resonance (NMR), and electron spin resonance (ESR)] and the acidity of a molecule. Chemists understand the fundamental theories behind these phenomena well, but the surrounding molecular environment affects these properties. For example, a carbonyl group (C=O) has an IR absorption band between $1650-1800$ cm^{-1}, and a carboxylic group (COOH) has a medium level of acidity, but atoms connected to these groups can change the peak position and acidity.

Quantitative estimation continues to be a problem, and techniques such as regression or artificial neural networks can be useful if we limit molecules to those with a common skeleton.

14.2.2   Additive Properties

All of the components of a molecule contribute to its additive properties. Molecular weight is the simplest example; it is simply the sum of atomic weights. Most thermodynamic properties fall into this group, and interactions among constituents produce changes in this property. Chemically important properties in this category include hydrophobicity (log P: octanol/water partitioning coefficient) and solubility in water.

The group contribution method usually allows the quantitative estimation of these properties; a least-square technique can be used to determine estimates from a component. However, when a property is a result of strong interactions among molecular constituents, only an estimate that is based on physicochemical theories can be made. A typical example in this class is the electronic energy of the ground/excited states of a molecule.

14.2.3   Complex Properties

Many important properties are affected by multiple factors; most of the biological activities of chemical compounds fall into this category. For example, a chemical compound must meet several conditions in order to be used as a drug clinically. First, it must not decompose in the stomach; it must be absorbed in the small bowel, move to the appropriate body site, and finally bind adequately to the target protein. Even when we limit our study to protein–drug interactions, protein structures are often unknown, and interactions between proteins and chemical compounds can change the conformations of both. Many physicochemical properties essential to functional materials have a similarly complex nature, and in this complex domain we need to have not only quantitative estimates of a property but also an understanding of related molecular structural characteristics.

The above discussions highlight the important issues of data mining: the evaluation of local properties and an understanding of the structural characteristics of a complex property. The following sections discuss examples of acidity and bioactivity (for more information on some of the bioactivity databases used in these examples, see [5, Database]).

14.3   CASE: A PROTOTYPE MINING SYSTEM IN CHEMISTRY

Klopman's CASE system was one of the first examples of chemical graph mining research [19]. While this system was intended to analyze bioactivity, for the sake of simplicity we introduce it by applying it to acidity analysis.

CASE automatically generates its attributes (called *descriptors* in the study of structure–activity relationships) from a learning set composed of active and inactive molecules. Attributes are normally linear fragments, although they include the effects of some branches. Table 14.1 shows fragments of lengths 3 and 4 generated from aspartic acid (its structure diagram is shown in the table).

TABLE 14.1   List of Fragments (Length = 3, 4) using CASE[a]

$$\text{HO}\text{---}\underset{\displaystyle\overset{\displaystyle\|}{\text{O}}}{\text{C}}\text{---}\text{CH}_2\text{---}\underset{\displaystyle\overset{\displaystyle\text{NH}_2}{\overset{\displaystyle|}{\,}}}{\underset{H}{\text{C}}}\text{---}\underset{\displaystyle\overset{\displaystyle\|}{\text{O}}}{\text{C}}\text{---}\text{OH}$$

3	4
C″–CH₂–CH–	C″–CH₂–CH–C≡
C″–AH–CH₂–	C″–AH–CH₂–C≡
NH₂–CH–CH₂–	NH₂–CH–CH₂–C≡
NH₂–CH–C≡	OH–C″–AH–CH₂–
O=C–OH	C≡C–AH–CH₂–
OH–K″–CH–	OH–K″–CH–CH₂–
OH–K″–CH₂–	OH–K″–CH₂–CH–
O=K–CH–	O=K–CH–CH₂–
O=K–CH₂–	O=K–CH₂–CH–
	OH–K″–CH–NH₂
	O=K–CH–NH₂

[a] A and K indicate carbon atoms to which an amine and a side oxygen are attached, and a double prime indicates that the atom is unsaturated.

Once all the molecules are entered, the fragment distribution is analyzed statistically, assuming a binomial distribution. Any considerable deviation of a fragment from a random distribution among active and inactive classes indicates potentially significant activity (the F partial statistic is used at the 95% confidence level). Fragments are characterized as activating (biophore) and inactivating (biophobe).

Table 14.2 shows sample fragments from an acidity analysis using a small database of 100 miscellaneous organic compounds, of which 40 were acids. As expected, COOH (a carboxyl acid group) appeared as the strongest candidate as the component responsible for acidity. In this case, deactivating fragments (biophobe) were also detected for two groups containing NH_2; these are basic and prevent potential acidity through formation of a zwitterion in the molecule.

Klopman applied the CASE system to a carcinogenicity analysis of 38 unsubstituted polycyclic aromatic hydrocarbons; he recognized the two biophores and one biophobe shown in Figure 14.10. The biophores explained the famous bay region substructure established in carcinogenicity research.

Klopman used a naïve Bayes technique to estimate the activity of new compounds, where fragments were selected using an F statistic at a predetermined threshold (e.g., 99 or 85%). Later, a multiple regression capability was applied to allow quantitative estimation of the activity, where the numbers of biophores and biophobes were used as explanatory variables.

This CASE method can be considered the starting point of various mining methods that handle structure diagrams directly. Here, a relatively small number of linear

TABLE 14.2   Sample Fragments from Acidity Analysis[a]

Fragment	Inactives	Actives	
$CH_3-CH_2-CH_2-$	16	9	
$CH_2-CH_2-CH_2-$	35	28	
$CH=CH-CH_2-$	7	4	
$C''-CH_2-CH-$	7	7	
$O=C-CH_2-$	25	30	+
$O=C-OH$	4	38	+++
$CH=C-OH$	2	6	+
$CH_2-CH-Cl$	6	4	
$NH_2-CH-CH_2-$	6	1	−
$NH_2-CH-C=$	3	0	−

[a]Total number of molecules = 100. Number of actives = 40.

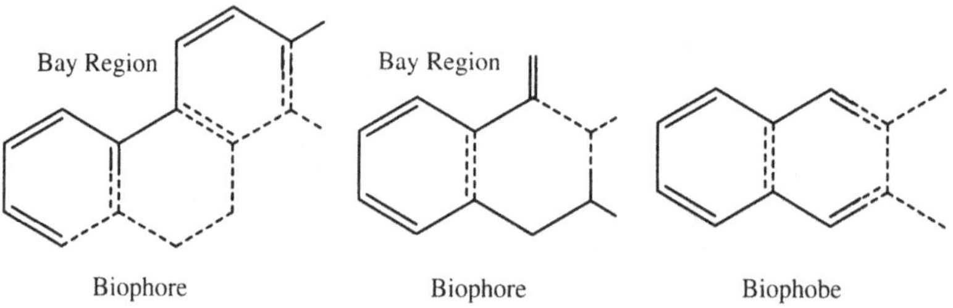

Figure 14.10.   Fragments from polycyclic aromatic hydrocarbons using CASE.

fragments explains the activity, and no conjunctive conditions are introduced. Later developments might include more quantitative estimates, extensions of biophore expressions using a general graph, the introduction of conjunctive combinations of various features, and a system that uses a more user-friendly human–computer interaction.

14.4   QUANTITATIVE ESTIMATION USING GRAPH MINING

Acidity estimation is not so important in chemistry, but it can be used as a model of local properties. When we study molecules that share a common skeleton, classical regression studies provide good estimates using substituent parameters as explanatory variables. However, they cannot handle a database with a variety of structures. Another well-used method depends on quantum chemical computations applicable to any group of structures, but these are outside the scope of this book.

In a previous work, we proposed a discrimination net of molecular graphs [36]. The objective was to provide a general framework for quantitative analogical reasoning methods for various properties. We subsequently applied this system to acidity estimation [28]. The oxygen atom in an OH group is the key element determining acidity and is called the "anchor atom" of the molecule. We introduced a new restricted graph subisomorphism, named *anchored subisomorphism*, in which the anchor atom, O, must be matched first. In this formulation, C–O is a subgraph of C–C–O, but C–C is not. Here, graphs do not include hydrogen atoms.

We used this relationship as the basis for partial ordering, and we could arrange molecular graphs for several oxy acids in a lattice, as shown in Figure 14.11, where the anchor OH group is underlined. Following this scheme, a molecule with two OH groups has two anchors, and the molecule appears at two positions in the lattice,

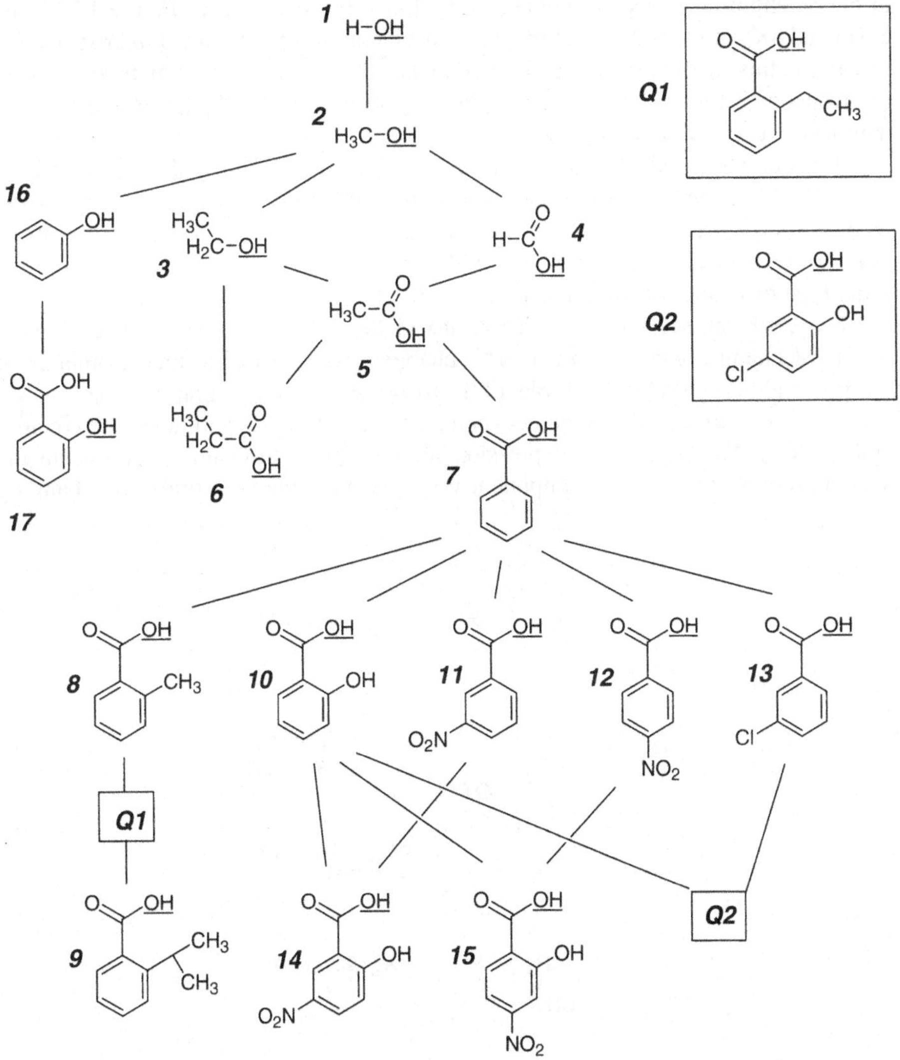

Figure 14.11. Sample discrimination net for oxy acids.

such as 10 and 17 in this figure. Such organization is reasonable since this molecule has two pK_a values (dissociation constants of an acid).

An algorithm was developed to locate a new molecule at the appropriate location in the net. Figure 14.11 shows two query molecules and their final location in the net. Therefore, we could use this net as a database of chemical graphs. The connecting link retained mapping of atoms between molecules, and molecule storage and retrieval using the discrimination net was fast.

When pK_a values are stored along with structures in the net, we can use them to infer an unknown pK_a value for a query molecule. We proposed two inference methods: The first was called the *direct method* and simply used the average pK_a of the neighboring molecules connected by links. For example, in Figure 14.12, the estimated pK_a of Q1 is 3.772. If there are sufficient molecules with similar structures in the net, this method provides good estimates. Nevertheless, this method can lead to erroneous estimates since acetic acid and ethanol are connected in the net, but they have very different pK_a values.

The other method was named the *parallelogram method* and is illustrated in Figure 14.13. Here, Q2 is the query molecule, and we select molecule 13 as the reference structure. In this case, the structural change from 13 to Q2 is the substitution of OH for H at the position neighboring the anchor COOH. We can find the same type of change at three positions: 7–10, 11–14, and 12–15.

Now, we can see three parallelograms in the net. If we assume that identical structural changes cause identical pK_a changes, we can make three estimates of the pK_a value, as shown in Table 14.3. To detect a parallelogram in the net, we must express a structural change between a pair of molecules and judge the equality among them. We used a tree expression for the structural change; paths from the anchor atom to the changed component were placed at the tree trunk, and changing

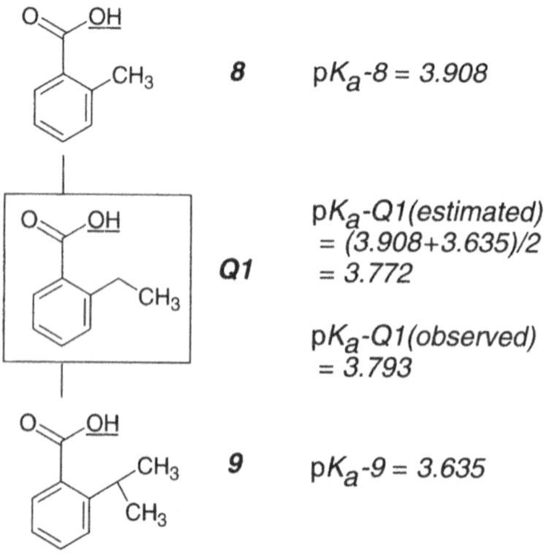

Figure 14.12. Direct method of estimation. The property value is estimated to be the median of all the property values on the adjacent nodes.

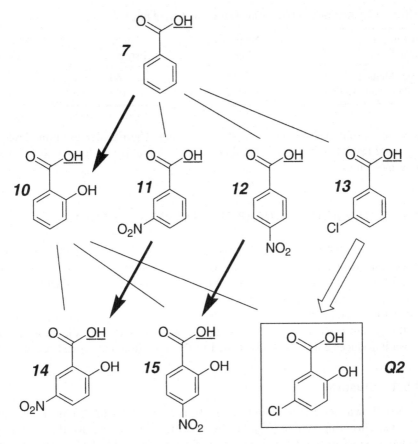

Figure 14.13. Parallelogram method of estimation.

TABLE 14.3 Estimated Values Using the Parallelogram Method

pK_a(mol_ID)	➡	pK_a(mol_ID)	pK_a-13	⟹	Estimated pK_a-Q2	Error[a]
4.200 (7)		2.754 (10)	3.821		2.375	−0.254
3.493 (11)		2.121 (14)	3.821		2.449	−0.180
3.425 (12)		2.231 (15)	3.821		2.627	−0.002

[a]The observed pK_a-Q2 is 2.629.

components were placed at leaves. When there were rings in the structure, we approximated them using linear structures formed by cutting the rings at the farthest position. The search range of parallelograms in the net was limited to a small number of intervening links.

We examined the accuracy of the results provided by these two inference methods using acidity data for 160 oxyacids listed in a chemistry handbook. Table 14.4

TABLE 14.4   Accuracy of Leave-One-Out Estimation of pK_a Values

Inference Method	Number of Molecules	Number of Estimations	Median of Errors
Direct Method	160	160	0.63
Parallelogram Method	37	153	0.12

shows the results of leave-one-out pK_a estimation. The parallelogram method gave very good results, although the number of parallelograms was limited to 153 for 37 query molecules.

14.5   EXTENSION OF LINEAR FRAGMENTS TO GRAPHS

One of the straightforward developments from the CASE method is the use of graphs instead of linear fragments as attributes. Originally, the maximal common subgraph from a group of active compounds was thought to give chemists a good substructure suggesting a biophore. However, when the data contained various types of skeleton, attempts often resulted in trivial graphs. Therefore, researchers learned that the system needed to generate multiple candidate graphs for the biophore. This section introduces some methods related to this direction and typical results.

14.5.1   Clusmol

CLUSMOL was an early example of such trials; it was a hierarchical conceptual molecular clustering system [37]. Given a set of molecular graphs, the system constructed a set of trees, in which given graphs were placed at leaf nodes, and other nodes expressed subgraphs shared by some of the starting graphs.

The example shown in Figure 14.14 provides a brief overview of the system. It shows the results of applying CLUSMOL to penicillin molecules in which nine molecules are clustered in two trees. Here, the nine leaf nodes in the two trees show individual penicillin molecules. They share the pc substructure shown at the right, and its substituent R varies in each molecule. The diagrams located at the upper nodes illustrate characteristic substructures shared by descendent molecules.

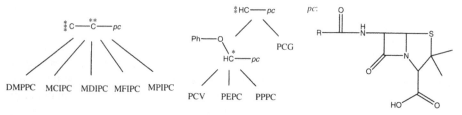

Figure 14.14.   Conceptual clustering of penicillin. An asterisk shows a free valence on the atom, and pc denotes the skeleton on the right.

The computation of a maximal common subgraph plays an essential role in tree construction. If we compute a maximal common subgraph from a pair of graphs, then we can postulate a new tree. Its root node expresses the maximal common subgraph obtained and this is linked to two leaf nodes that contain the original graphs. We can interpret this new tree as a kind of Markush structure, the length of which can be used as a measure to judge the quality of the expression using techniques such as the minimal description length.

In the example shown in Figure 14.14, the system first accepted nine molecular graphs and constructed nine trivial trees with only one node in each tree. Then, the system computed all the candidate trees constructed from all pairs of starting molecules. These candidate trees were sorted by their scores, and a kind of beam search was used to construct the best tree. Here, the maximal common subgraph on the root node of the new tree replaced those of the old smaller trees. The recursive construction of a larger tree continued until there were no further improvements in the tree set score.

Another scoring scheme was added so that the clustering of molecules in the same class is favored over clustering that mixes together. Using this scheme, five penicillinase-resistant forms of penicillin G were placed in the left tree in Figure 14.14, while basic forms of penicillin G were placed in the right tree. Inspection of the substructures appearing on the root nodes of these trees showed that a hydrogen atom does not appear at the adjacent position to pc in the left tree. The absence of this hydrogen atom was actually the cause of the penicillinase-resistant feature.

The CLUSMOL system succeeded in extracting important information discriminating activity classes. However, although the system reused previously computed results, the method still suffered from a long CPU (central processing unit) time when computing the many maximal common subgraphs.

14.5.2 Frequent Subgraph Approach

Finding subgraphs that appear in a set of molecules frequently is a naïve idea that is still attractive to computer scientists and chemists. In fact, some pharmaceutical companies have recently used ClassPharmer, which identifies frequently appearing skeletons and gives hints to chemists [3]. This is a commercial product and its detailed algorithm is therefore not public knowledge, but its results are favorites with chemists. The association rule mining proposed by Agrawal stimulated various extensions, including the study of graph mining [1]. These modern approaches are introduced here from a chemist's viewpoint.

Greedy Approach: Subdue and GBI. Cook and Holder developed the Subdue system, which extracts molecular substructures (see [4] and Chapter 7 of this book). They applied the system to the repetitive substructures found in rubber, cortisone, and DNA (deoxyribonucleic acid) databases, and succeeded in finding subgraphs such as the benzene ring and isoprene, although these concepts were already well established in chemistry.

This approach has two shortcomings: The search space is limited since it uses a greedy search, and it can produce small, trivial subgraphs. The system grows a subgraph starting from an atom, and an inadequate search direction tends to result in meaningless subgraphs.

Motoda et al. attempted a similar approach by developing the GBI system (see [12, 44] and Chapter 9 of this book), which expands a subgraph by chunking neighboring atoms that appear frequently. They also added a function to favor the chunking with the greatest capability to discriminate class labels of molecules. However, chemical properties are often specified by atoms separated by several bonds, and this strategy does not work well in many problems. Therefore, the researchers shifted their focus to the use of A priori-based graph mining (AGM), which is described below.

Molfea. De Raedt and Kramer proposed Molfea for efficient mining of molecular fragments [8]. Although their work did not extend linear fragments to general subgraphs, their methodology was a pioneer in this domain and can be easily extended to more general graphs. It analyzed a relatively large human immunodeficiency virus (HIV) dataset [22], a capability that is necessary in current chemistry mining tasks. Their method is also closely connected to methods discussed in the following subsection.

Their algorithm uses a levelwise version space search of frequent fragments, which constitutes nodes in the lattice. Fragment expression uses both SMILES coding and an extension using SMARTS language. This language can also describe rings and apply generic elements to atoms and bonds, although they did not use these functions.

The core idea was to use various constraints for fragments, that is, if fragment $f1$ is more specific than fragment $f2$, the support for $f1$ should be lower than that for $f2$. Furthermore, the database was divided into active, moderately active, and inactive compounds, and the researchers set a minimum support for the active compounds of 13 and a maximum support for the inactive compounds of 515.

The resulting lattice contained many fragments, and they proposed eight significant fragments, including N–C–c:c:c:o (21 active, 25 inactive), and N=N=N–C–C–C–n:c:c:c=O (51 active, 11 inactive), which were selected using statistical significance and accuracy. Although they do not provide detailed inspections of significant fragments from a chemist's viewpoint, they can identify substructures related to the famous azidothymidine group.

Methods of Constructing a Subgraph Lattice. Several attempts at graph mining have extended the association rule to graphs. Since this type of methodology can recognize all subgraphs with more than minimum support, it is likely that this direction will aid in discovering substructures responsible for complex bioactivities.

A priori-based graph mining (AGM) was the first attempt [14, 15]. It attempted to generate general subgraphs, including unconnected ones, but suffered from the problem of combinatorial explosion in lattice size. Later, Kuramochi proposed the frequent subgraph (FSG) algorithm, which used the same levelwise search, but

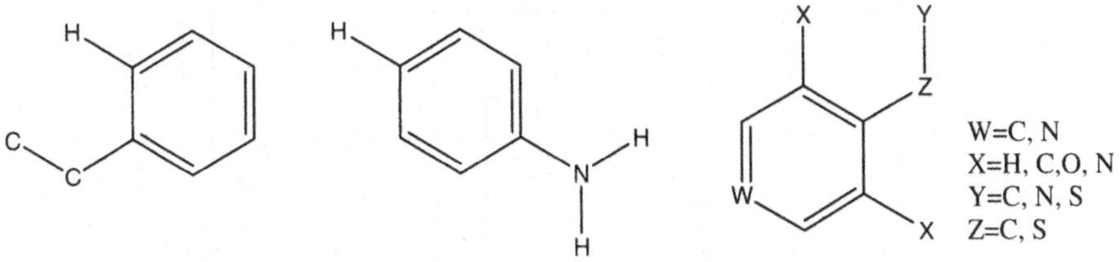

Support=7.3%	Support=6.2%	Active= 8/182
pos=6, neg=19	pos=17, neg=4	Inactive= 38/155
(a)	(b)	(c)

W=C, N
X=H, C,O, N
Y=C, N, S
Z=C, S

Figure 14.15. Results of Inokuchi's analysis on the PTC dataset.

limited the target to connected subgraphs (see [23] and Chapter 6 of this book). Borgelt proposed a different kind of method that used a depth-first search [2], unfortunately the search space was incomplete. They also attempted to detect a subgraph containing a wild card atom. This type of graph could be considered a kind of Markush structure that gives a concise representation of many sorts of chemical graphs, and it is likely to improve the ability to discriminate between active and inactive compounds. Inokuchi later improved his system along similar lines and identified a chemically essential substructure from a group of active compounds [16]. The following paragraphs describe a few results.

Inokuchi applied his method to the Predictive Toxicology Challenge (PTC) and to HIV datasets. Figures 14.15(a) and 14.15(b) show the resulting subgraph obtained from the PTC data, but these results are not particularly interesting to chemists. However, Inokuchi was able to obtain a generic structure [Fig. 14.15(c)], which might still be too general to be useful to chemists but is nonetheless more informative. This result shows that incorporating the Markush structure can be very useful when summarizing a group of structures.

By contrast, Figure 14.16 shows that with the HIV dataset, Inokuchi was able to obtain a pair of subgraphs showing active and inactive substructures, where the

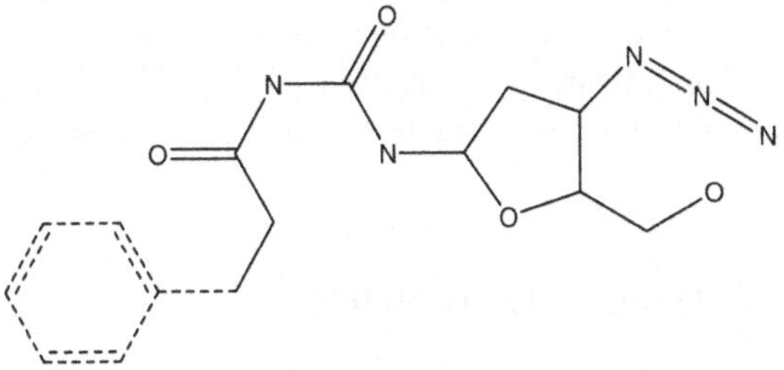

With Phenyl	Without Phenyl
Active = 0 / 422	Active = 64 / 422
Inactive = 3 / 41,884	Inactive = 16/ 41884

Figure 14.16. Results of Inokuchi's analysis on the HIV dataset.

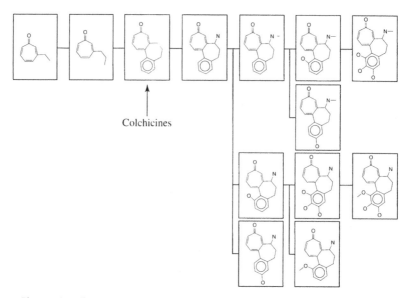

Figure 14.17. Expression of frequent subgraphs in the analysis by Borgelt.

steric hindrance caused by a phenyl ring (depicted by the dashed substructure) likely prevents activity.

The density of similar molecules seems to cause the contrast between these two datasets. Since the HIV dataset contains many compounds with the azidothymidine skeleton (shown in Fig. 14.16), the system was able to detect a meaningful difference between active and inactive substructures. Conversely, the structures in the PTC dataset have a variety of skeletons, and there are too few compounds with a specific skeleton to clearly distinguish between active and inactive molecules.

Borgelt also found an interesting result [2]. After discriminative fragments were discovered, they not only identified the most discriminative subgraph but also a group of related subgraphs. Figure 14.17 shows how the fragments that they found were organized into a treelike format. This format is very helpful to chemists because it aids in the selection of the most meaningful fragments. In the example shown in Figure 14.17, chemists select colchicines but may also draw valuable information from variants that are also displayed. If the distributions of active/inactive compounds are shown along with structures, this might be more helpful.

14.6 COMBINATION OF CONDITIONS

The next group of chemical graph mining systems uses a conjunction of structural descriptors to express a target concept. Descriptor complexity ranges from a simple atom and bond to linear fragments. This section introduces three groups of systems in this category.

14.6.1 MultiCASE

This system is the successor to CASE [20]. MultiCASE is now a commercial product, and it is currently one of the best mining systems for practical use. The MCASE/MC4PC system, developed in cooperation with the U.S. Food and Drug Administration, is used in routine day-to-day drug evaluation [42].

The CASE system generated linear fragments and selected biophores and biophobes. It treated all the selected fragments equally and was not able to differentiate between the substructures responsible for an activity and those that influenced the activity. For example, the F–CH–F fragment shown in Figure 14.18 dramatically increases the acidity of the COOH fragment but does not itself have acidic functionality. CASE used this fragment as a biophore, if there were such pairs of fragments in the database. This type of modulation is also common in various bioactivities.

A major improvement made in MultiCASE was the construction of a hierarchical model to solve this problem. As an example, let us examine the MultiCASE procedure for oxyacids data. Starting with a small database of 121 miscellaneous organic compounds, of which 42 are acidic and 79 are not, MultiCASE generates several candidate biophores. The system first identifies the strongest biophore fragment and performs a CASE analysis to detect modulators.

The first biophore identified in the oxyacid database was COOH, which appeared in 45 molecules. Table 14.5 lists the modulator fragments detected in this subset and the regression coefficients for these modulators. Compounds containing the first biophore were eliminated from the database, and the remaining compounds were used in a new analysis until all molecules were eliminated or no statistically significant fragments were found. Table 14.6 shows the six resulting biophore candidates. Therefore, we can summarize the MultiCASE system as a combination of a decision list and statistical analysis using linear fragments.

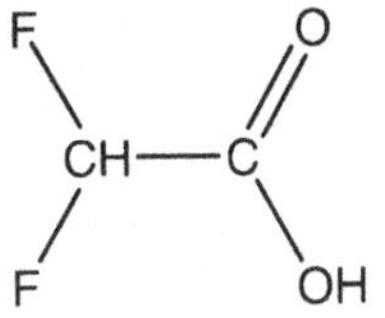

Acidic Functionality: COOH

Moduation of Acidity: F-CH-F

Figure 14.18.   Two fragments from difluoroacetic acid.

TABLE 14.5   List of Modulators Found in Molecules Containing COOH[a]

	Fragment	No	In	Ma	Ac	QSAR
1	NH_2–CH–	10	7	0	3	2.97
2	F–CH–F	1	0	0	1	−3.34
3	CO–C–F	1	0	0	1	−2.47
4	CO–C–Cl	1	0	0	1	−2.13
5	CO–CH–Cl	2	0	0	2	−1.70
6	$CO–CH_2–NH_2$	1	1	0	0	22.70

[a] In, inactive; Ma, marginally active; Ac, active; QSAR, regression coefficient.

TABLE 14.6   List of MultiCASE Biophores for Acidity

	Fragment	No	In	Ma	Ac	Av.	pK_a
1	CO–OH–	45	8	0	37	4.5	+++
2	NO_2–CH_2–	2	0	0	2	2.9	
3	NO_2–CH–	1	0	0	1	0.0	
4	CO–CH–CO–	1	0	0	1	6.0	
5	SO_2–CH–SO_2–	1	0	0	1	0.0	
6	N SC–CH–C SN	1	0	0	1	0.0	

MultiCASE also incorporates some new expressions for fragments. It can describe the *cis–trans* configuration of a fragment as well as the measure of congestion. It also introduces a sophisticated algorithm that uses biophore fragments that are chemically similar to the known biophore but do not have enough support in the database.

Klopman used MultiCASE to propose characteristic substructures relevant to many bioactivities. Figure 14.19 shows the results of a dopamine D_2 agonist study [21]. Six biophores (illustrated by the bold lines in the upper part of the figure)

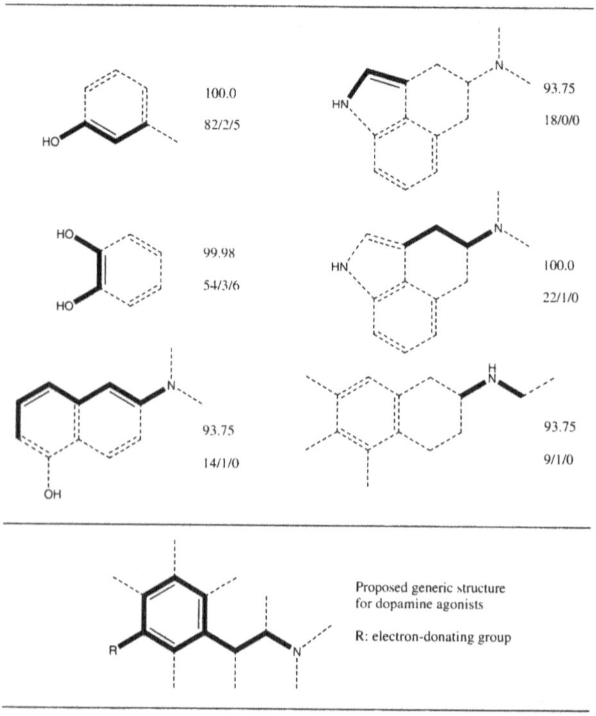

Figure 14.19.   Characteristic substructures that Klopman found in D_2 agonists. Numbers show the calculated probability of relevancy, as well as the numbers of compounds with active/inactive/marginal compounds.

were detected. After consulting the supporting molecular structures, representative skeletons (depicted by dotted bonds) containing these biophores were found. Finally, the generic substructure shown in Figure 14.19 was proposed as a characteristic substructure to explain this activity. This result seems reasonable, although the proposed structure does not seem to cover all important biophores.

14.6.2 Inductive Logic Programming

Inductive logic programming (ILP) methodology is well known for its flexibility, as it can incorporate any kind of background knowledge in the learning process. Researchers consider its main drawback to be the huge search space, which might limit its application. Mining from chemical graphs was originally considered intractable, if ILP begins its learning with elementary atoms and bonds.

In 1966, King et al. used Progol to find the characteristic substructures responsible for mutagenicity [18]. This was surprising because they predicted several characteristic substructures, and their accuracy was comparable to results obtained using conventional methods. Their results were published in one of the top-rated natural science journals. This success further stimulated structure–activity relationship (SAR) research using ILP.

King et al. [18] used a mutagenicity dataset of 230 aromatic nitro compounds collected by Debnath et al. [7]. The molecules contained a variety of structures, although when compared to the entire range of organic molecules, the domain was restricted to a relatively narrow region. Figure 14.20 shows the resulting characteristic substructures.

The resulting substructures contained relatively large skeletons and a distribution of partial charges estimated using Quanta software. The success and weakness of this result seem to be a result of the partial charges. Chemists often perceive partial charges, and they use terms like electron-donating and -withdrawing groups. However, the charges in the example shown in Figure 14.20 are too detailed to

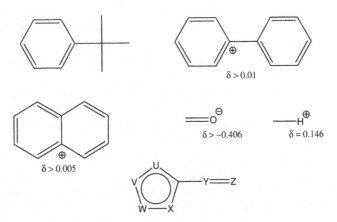

Figure 14.20. Characteristic substructures found in a mutagenicity dataset.

discriminate, and it is difficult for a chemist to imagine molecules supporting these substructures. By contrast, the introduction of a partially charged atom should restrict the search space to a tractable size. This shows the potential of ILP since the method can utilize any kind of knowledge, including that provided by chemists.

Researchers have devised various extensions of ILP, such as three-dimensional pharmacophore recognition [10], the inclusion of found substructures as variables in a regression analysis [17], and the construction of rules that occasionally use regression using knowledge of the regression method itself [40]. However, most chemists have hesitated to use these systems because they are difficult to use and the process of analysis is very sophisticated in comparison with conventional methods.

Lastly, a recent study incorporated the existing top-down ILP system (FOIL) and applied a multiple-instance-based measure to find common characteristics among parts of positive examples [27]. In addition to the usual atom and bond predicates, the researchers added a link predicate to describe the relationship between two atoms that are a specified distance apart within molecules. This relatively simple search method is accurate for mutagenesis and dopamine data. Furthermore, a rule derived from the dopamine dataset suggested an important feature: An atom has one link to oxygen (length ≥ 3.7 Å) and another link to nitrogen (length ≥ 3.3 Å). Finding this feature depends on knowing the three-dimensional molecular structure, so it is not always applicable, but combined with background knowledge provided by domain experts, it is a good way to utilize the flexibility of the ILP systems.

14.6.3 Cascade Model

We developed the cascade model, an extension of association rule mining, and applied it to SAR analysis. We used linear fragments extracted from structural formulas as attributes and used the conjunction of their presence/absence as rule conditions. It was first applied to a mutagenicity dataset [31], and to analyze carcinogenicity in the Predictive Toxicology Challenge [34]. The early applications provided some characteristic substructures and received high marks at the PTC challenge [13].

Recently, effort has focused on the analysis of dopamine agonists and antagonists. Unfortunately, our original method did not provide results sufficient to explain the pharmacophore of these drugs. After trying to improve the resulting rule set in various ways, we finally succeeded in extracting valuable knowledge about the pharmacophore.

This section shows the latest results for characteristic substructures obtained for D_1 agonist activity. In our opinion, these represent state-of-the-art chemical structure mining results, and we introduce a detailed mining framework and interpretation process.

We used the MDDR database [25], which contains more than 100,000 pharmaceuticals, and extracted 369 records with dopamine (D_1, D_2, and D_{auto}) agonist activities. First, we extracted 4626 linear fragments with lengths under 10. We considered fragments appearing in 3–97% of molecules as attributes and applied a correlation-based scheme to select 306 fragments [35]. Finally, hand selection by chemists provided 345 fragments for analysis using the cascade model. The model expresses linear fragments in a similar way to CASE and Molfea, but we omit atom symbols except for the two terminal elements on each end. We also attach an atom flag that shows whether a hydrogen atom is connected [38].

The cascade model can be considered an extension of association rule mining. It creates an itemset lattice in which an [attribute: value] pair is used as an item to constitute itemsets. Links in the lattice are selected and interpreted as rules [29, 30]; that is, we observe the distribution of activity (y/n) along all links, and if a distinct change in distribution appears along some link, then we focus on the two terminal nodes of the link. Suppose that the itemset at the upper end of a link is [A: y], and item [B: y] is added along the link. If a marked activity change occurs along this link, we can write the following rule:

```
Cases: 200 ==> 50,      BSS=12.5
IF [B: y] added on [A: y]
THEN [Activity]:               (.80 .20)   ==> (.30 .70)    (y n)
THEN [C]:                      (.50 .50)   ==> (.94 .06)    (y n)
Ridge: Pre outside [A: n]: (.70 .30)/100 ==> (.70 .30)/50 (y n)
```

where the added item [B: y] is the main condition of the rule, and the items at the upper end of the link ([A: y]) are preconditions. The main condition changes the ratio of active compounds from 0.8 to 0.3, while the number of supporting instances decreases from 200 to 50. BSS refers to the between-groups sum of squares, which is derived from the decomposition of the sum of squares for a categorical variable. Its value can be used to measure the strength of a rule. The second THEN clause indicates that there is also a sharp change in the distribution of attribute [C] values when the main condition is applied. This description is called *collateral correlation*.

Usually, too many links with large BSS values are detected to be interpreted by human analysts, so we introduced two schemes to decrease the number of rules [32]. First, a rule candidate link found in the lattice is optimized greedily in order to produce the rule with the local maximum BSS value. Since many rules converge on the same rule expression, this optimization is useful for decreasing the number of resulting rules. The second step involves the facility to organize rules into principal and relative rules. A group of rules that share a relatively large number of supporting instances is expressed as one principal rule (with the largest BSS value) and its relative rules. This function can be used to decrease the number of principal rules that must be inspected and to indicate relationships among rules.

We also added ridge information, shown on the last line of the previous example [33]. This describes [A: n] as the ridge region detected at the outside of the current precondition. A change of "activity" distribution is denoted in this ridge region. Compared to the large change in activity distribution for instances with [A: y], the distribution on this ridge does not change. This means that the BSS value decreases sharply if we expand the rule region to include this ridge region. This ridge information is expected to guide the *datascape* survey.

Using these schemes, we inspected 2 principal and 14 relative final rules; Table 14.7 shows the four most important. Rules not cited in the table were interpreted as paraphrases of rule R1.

TABLE 14.7 Rules Suggesting Characteristics of D₁ Agonist Activity

Rule ID	Number of Compounds and Conditions of a Rule		Distribution Changes in D1Ag and Collateral Correlations		
			Descriptor	Before	After
R1	Number of compounds	369 → 52	D1Ag:	17% →	96%
			DAuAg:	50% →	0%
			C4H-C4H-:::c3-O2H:	19% →	92%
	Main condition	[O2H-c3:c3-O2H: y]	C4H-C4H-::c3-O2H:	18% →	98%
			N3H-C4H-:::c3-O2H:	14% →	67%
			N3H-C4H-::c3-O2H:	10% →	69%
	Preconditions	None	N3H-C4H-C4H-c3:	18% →	81%
			C4H-N3H-C4H-c3	15% →	62%
	Ridge 1: new inside		[N3H-C4H-:::c3-O2H: y]	66% in 53 →	97% in 35
R1- UL9	Number of compounds	288 → 16	D1Ag:	19% →	100%
			DAuAg	51% →	0%
			D2Ag:	36% →	100%
	Main condition	[C4H-N3—-:::c3-O2H: y]	C4H-N3—-:::c3-O2H:	5% →	87%
			C4H-N3—C4H-c3:	11% →	100%
	Preconditions	[N3H: y]	O2H-c3:c3-O2H:	17% →	100%

R1- UL12			
Number of compounds	170 → 12		
Main condition	[N3H-C4H—::c3-O2H: y]		
		D1Ag:	14% → 100%
		DAuAg	40% → 0%
		D2Ag:	50% → 0%
		C4H-N3H—:::c3-O2H:	8% → 100%
		C4H-N3H—C4H-c3:	7% → 100%
		C4H-N3H-C4H-c3:	8% → 100%
		O2H-c3:c3-O2H	12% → 100%
Preconditions	[N3H-C4H-:c3H:c3H: n]	O2H-c3:::-C4H-c3:	7% → 100%
	[O2-c3:c3: n]	O2H-c3::-C4H-c3:	7% → 100%

R14			
Number of compounds	72 → 11	D1Ag:	15% → 9%
Main condition	[C4H-C4H-O2-c3: n]	DAuAg:	72% → 0%
	[LUMO: 1–3]	CO2-O2-c3:c3H:	31% → 64%
	[C4H-N3—c3:c3: n]	O2-c3:c3-O2:	22% → 64%
Preconditions	[O2-c3:c3:c3H: y]	N3H-C4H—:c3H:c3H:	7% → 45%

Rule R1 is the strongest rule derived from the D_1 agonist study. There do not appear to be any preconditions, and the activity ratio (D_1Ag) increases from 17% in 369 compounds to 96% in 52 compounds by including a catechol structure (O2H-c3:c3-O2H). Use of this main condition depresses D_{auto} activity (DAuAg) to 0%. Other collateral correlations suggest that N3H-C4H-C4H-c3, and OH groups likely exist at the positions *meta* and *para* to this ethylamine substituent. However, this ethylamine group is not an indispensable substructure, as the presence of N3H-C4H-C4H-c3 is limited to 81% of the compounds. This observation is also supported by the ridge information. That is, selection of a region inside the rule on adding a new condition [N3H-C4H-:::c3-O2H: y] results in 53 (35 actives, 18 inactives) and 35 (34 actives, 1 inactive) compounds before and after the main condition is applied, respectively. This means that when the catechol structure is absent, there are 17 inactive and 1 active compounds, and N3H-C4H-:::c3-O2H cannot show D_1Ag activity without the catechol substructure. Therefore, we can construct a hypothesis for D_1Ag activity: catechol is the active site and the appearance of activity is supported by an ethylamine substituent at the *meta* and *para* positions as the binding site.

A catechol supported by an ethylamine substituent is the structure of dopamine molecule I in Figure 14.21, and it covers 50 out of 63 D_1 agonists. Therefore, this hypothesis seems rational.

At first glance, the third entry in Table 14.7, R1-UL12, seems to provide new substructures. Its collateral correlations indicate an N3H group at the atoms 1, 2, and 3 positions away from an aromatic ring, as well as a diphenylmethane substructure (c3-C4H-c3). However, visual inspection of the supporting structures found skeleton II in these molecules. Therefore, we do not need to change our hypothesis since it contains dopamine structure I.

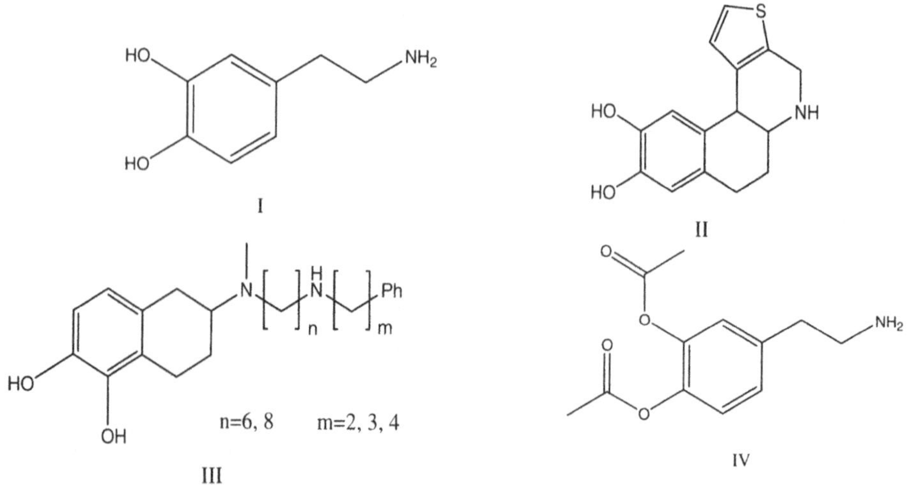

Figure 14.21. Characteristic substructures of dopamine D_1 agonists.

The only exceptional relative rule was R1-UL9, the second entry in Table 14.7, which could not be explained by structure I. Interesting aspects of this rule include the 100% co-appearance of D_2 agonist activity, as well as the *tert*-amine structure (N3) in the main condition. These points markedly contrast with those found in R1, where *prim*- and *sec*-amines (N3H) aided the appearance of D_1Ag activity, and D_2Ag activity was found in 38% of 52 compounds. The precondition and collateral correlations also suggest the importance of *prim*- or *sec*-amines, an ethylamine substituent, and a catechol structure. Inspection of the supporting structures showed that this rule was derived from compounds with skeleton III. In some compounds, we found a dopamine structure around the phenyl ring at the right, but this could not explain the D_1Ag activity in all the supporting compounds. Therefore, we proposed a new hypothesis: The active site is the catechol (O2H-c3:c3-O2H) in the left ring, but the binding site is the *sec*-amine in the middle of the long chain. This *sec*-amine can locate itself close to the catechol ring by folding the $(CH_2)_n$ ($n = 6, 8$) chain. The characteristic substructure proposed by MultiCASE, shown in Figure 14.19, seems to have identified this group of compounds.

Rule R14 is the second and last principal rule leading to D_1Ag activity. Unlike the R1 group rules, the substitution of OHs on an aromatic ring, which plays an essential role in the above hypothesis, does not appear. It was difficult to interpret this rule as the main condition and the second precondition are designated by the absence of ether and *tert*-amine substructures. However, we found that 6 out of 11 compounds had skeleton IV, where catechol OHs were transformed to esters. These esters likely decompose into OHs after absorption to cells and act as prodrugs.

The analyses of other activities were much more difficult and depended heavily on the insights of chemists. However, their efforts have led to the discovery of many important skeletons from many compounds with a variety of structures. Without the combination of cascade model analysis and the insights of chemists, these discoveries would have been impossible.

14.7 CONCLUDING REMARKS

This section begins with a description of various chemical graph treatments that are not mentioned in the preceding sections. The chemical reaction is an important class of information that has not yet been mentioned in this chapter. Essentially, a reaction can be described by chemical graphs of reactants and products, along with additional information about solvents, temperatures, catalysts, yields, and selectivity. There are a few commercial databases of chemical reactions used by many organic chemists. However, mining from these databases has not been successful. The sparseness of reaction data may explain this; a database usually contains a prototype of successful chemical reactions. When a reaction condition changes, the reaction itself may often be useless because its yield or selectivity decreases. Unfortunately, these negative data are not described in the database, which may cause the sparseness of the database.

It is worth noting an interesting new expression for chemical reactions that gives a new type of chemical graph. An atom in a molecule after a chemical reaction corresponds to some atom before the reaction. Consequently, we can superimpose the chemical graphs from before and after the reaction to obtain a new graph in which an edge between two atoms has a label specified by a pair of bond types (before/after the reaction), for example, double $\rightarrow$ single, single $\rightarrow$ none, and none $\rightarrow$ single. This expression, named an *imaginary transition structure*, does not add new information but can be useful in ordering various types of chemical reaction [11].

Researchers have used regression analysis in a variety of fields in chemometrics and chemoinformatics. The method can be very useful when molecules share a common skeleton, and researchers can use some substituent properties as explanatory variables during a regression. The success of the regression method has led to many trials in which researchers have mapped a chemical graph onto some numerical variables expressing the topological characteristics of the graph; these numerical variables are called *topological indices*. In early attempts, researchers succeeded in correlating a physical property (e.g., the boiling point of hydrocarbons) to a topological index. Many indices were invented, and some were widely used in quantitative structure–activity relationship studies. However, these indices have a well-known shortcoming: It is difficult for chemists to understand the meaning of regression formulas including these indices (for details, see [9]).

Takahashi et al. [41] proposed topological fragment spectra (TFS), which are sets of numerical attributes generated from the fragments contained in a molecule. They generated every fragment smaller than a specified size and generated some characteristic numbers (such as mass or the number of valences) from the constituting atoms. When the number of fragments is counted along with the characteristic number, the resulting plot, the TFS, can be considered a spectrum that shows molecular characteristics, and spectrum heights can be used as attributes for mining. Combining the support vector machine and TFS, they were able to predict dopamine-related activities with high accuracy [41]. However, each peak in a spectrum corresponds to multiple fragments. Even if some peak height determined the activity, they could not identify fragments important to the activity directly. Nevertheless, they did develop a fragment inspection system for each TFS peak and succeeded in finding a fragment important to dopamine D_4 agonist activity.

Medicinal chemists have long awaited a mining system that grasps characteristic substructures to active compounds. Recently, they have adopted a decision tree method (called recursive partitioning) in day-to-day work, but they are generally not satisfied with this system and are looking forward to a more useful and convenient mining system.

As we have discussed, various mining systems have succeeded in finding common skeletons. However, chemists already know the major skeletons responsible for important activities, although these are seldom published in research articles. Chemists' visions for these systems are unlimited, and they are not satisfied by the rediscovery of known skeletons. They wish to find exceptional molecules, which might be the basis for a new drug. The hypotheses proposed by a system must cover the most active compounds and must also show exceptions. Although data contain substantial noise, quantitative treatments of activities are also necessary.

Chemical mining systems also need to handle a vast number of molecules. High-throughput screening technology can provide more than 100,000 of SAR data daily. A chemical genomics project that began recently is collecting data on a variety of genome expressions for one million compounds. Computer scientists will need to provide very fast, user-friendly data mining systems in the future.

REFERENCES

1. R. Agrawal and R. Srikant. Fast algorithms for mining association rules in large databases, Proc. VLDB 1994, pp. 487–499, 1994.
2. C. Borgelt, H. Hofer, and M. Berthold. Finding discriminative molecular fragments. Workshop Information Mining—Navigating Large Heterogeneous Spaces of Multimedia Information, German Conference on Artificial Intelligence, Hamburg, Germany, 2003.
3. Bioreason: ClassPharmer. http://www.bioreason.com/.
4. D. J. Cook and L. B. Holder. Substructure discovery using minimum description length and background knowledge. *Journal of Artificial Intelligence Research*, 1:231–255, 1994.
5. Database: (a) Mutagenicity data: http://www.clab.kwansei.ac.jp/mining/datasets/PAKDD2000/okd.htm. (b) Predictive Toxicology Challenge (PTC): http://www.predictive-toxicology.org/ptc/. (c) Anti-HIV activity: http://dtp.nci.nih.gov/docs/aids/aids_data.html. (d) MDDR: http://www.mdl.com/products/knowledge/drug_data_report/index.jsp. MDDR is a commercial product.
6. Daylight: http://www.daylight.com/smiles/f_smiles.html.
7. A. K. Debnath, R. L. Lopez de Compadre, G. Debnath, A. J. Shusterman and C. Hansch. Structure-activity relationship of mutagenic aromatic and heteroaromatic nitro compounds. *J. Med. Chem.* 34:786–797, 1991.
8. L. De Raedt and S. Kramer. The levelwise version space algorithm and its application to molecular fragment finding. Proc. IJCAI 2001, pp. 853–862, 2001.
9. J. Devillers and A. T. Balaban, eds. *Topological Indices and Related Descriptors in QSAR and QSPR*. Gordon and Breach, Amsterdam, 1999.
10. P. Finn, S. Muggleton, D. Page, and A. Srinivasan. Pharmacophore discovery using the inductive logic programming. *Machine Learning*, 30:241–270, 1998.
11. S. Fujita. *Computer-Oriented Representation of Organic Reactions*. Yoshioka Shoten, Kyoto, Japan, 2001.
12. W. Geamsakul, T. Matsuda, T. Yoshida, H. Motoda, and T. Washio. *Classifier Construction by Graph-Based Induction for Graph-Structured Data, Advances in Knowledge Discovery and Data Mining: Proc. PAKDD 2003*, pp. 52–62, LNCS 2637, Springer, Berlin, 2003.
13. C. Helma, R. D. King, S. Kramer, and A. Srinivasan. The Predictive Toxicology Challenge (PTC) for 2000–2001: http://www.predictive-toxicology.org/ptc/.
14. A. Inokuchi, T. Washio, and H. Motoda. *An Apriori-Based Algorithm for Mining Frequent Substructures from Graph Data. Proc. PKDD 2000*, pp. 13–23, LNAI 1910, Springer, Berlin, 2000.
15. A. Inokuchi, T. Washio, and H. Motoda. Complete mining of frequent patterns from graphs: Mining graph data, *Machine Learning*, 50(3):321–354, 2003.
16. A. Inokuchi, T. Washio, and H. Motoda. A general framework for mining frequent subgraphs from labeled graphs. *Fundamenta Informaticae*, 66:53–82, 2005.

17. R. D. King and A. Srinivasan. The discovery of indicator variables for QSAR using inductive logic programming. *Journal of Computer-Aided Mol. Design*, 11:571–580, 1997.

18. R. D. King, S. H. Muggleton, A. Srinivasan, and M. J. Sternberg. Structure-activity relationships derived by machine learning. *Proc. Natl. Acad. Sci. USA*, 93:438–442, 1996.

19. G. Klopman. Artificial intelligence approach to structure-activity studies. *Journal of American Chemical Society*, 106:7315–7321, 1984.

20. G. Klopman. MULTICASE 1. A hierarchical computer automated structure evaluation program. *Quant. Struct.-Act. Relat.*, 11:176–184, 1992.

21. G. Klopman and A. Sedykh. An MCASE approach to the search of a cure for Parkinson's disease. *BMC Pharmacology*, 2:8, 2002.

22. S. Kramer, L. De Raedt, and C. Helma. Molecular feature mining in HIV data, *Proc. KDD '01*, pp. 136–143, ACM Press, Addison-Wesley, Boston, 2001.

23. M. Kuramochi and G. Karypis. Frequent subgraph discovery, Proc. 2001 IEEE International Conference on Data Mining (ICDM01), pp. 313–320, 2001.

24. CTfile formats: http://www.mdl.com/solutions/white_papers/ctfile_formats.jsp.

25. Drug Data Report: http://www.mdl.com/products/knowledge/drug_data_report/index.jsp.

26. H. L. Morgan. The generation of a unique machine description for chemical structures: A technique developed at Chemical Abstracts Service. *J. Chem. Doc.*, 5:107–113, 1965.

27. C. Nattee, S. Sinthupinyo, M. Numao, and T. Okada. Learning first-order rules from data with multiple parts: Applications on mining chemical compound data. *Proc. ICML 2004*, No. 77, ACM Press, Addison-Wesley, Boston, 2004.

28. T. Okada. Similarity and analogy based on discrimination net. In W. A. Warr, ed. *Chemical Structures 2*, pp. 389–398, Springer, Berlin, 1993.

29. T. Okada. Rule induction in cascade model based on sum of squares decomposition. *Proc. PKDD 1999*, pp. 468–474, LNAI 1704, Springer, Berlin, 1999.

30. T. Okada. Efficient detection of local interactions in the cascade model. *Proc. PAKDD 2000*, pp. 193–203, LNAI 1805, Springer, Berlin, 2000.

31. T. Okada. Discovery of structure activity relationships using the cascade model: The mutagenicity of aromatic nitro compounds. *Journal of Computer Aided Chemistry*, 2:79–86, 2002.

32. T. Okada. Datascape survey using the cascade model. *Discovery Science 2002*, LNCS 2534, pp. 233–246, Springer, Berlin, 2002.

33. T. Okada. Topographical expression of a rule for active mining. In H. Motoda, eds. *Active Mining*, pp. 247–257, IOS Press, Amsterdam, 2002.

34. T. Okada. Characteristic substructures and properties in chemical carcinogens studied by the cascade model, *Bioinformatics*, 19:1208–1215, 2003.

35. T. Okada. Attribute selection in chemical graph mining using correlations among linear fragments. *Proc. Workshop on Mining Graphs, Trees and Sequences (MGTS'04)*. http://hms.liacs.nl/mgts2004/.

36. T. Okada and T. Kawai. Analogical reasoning in chemistry. *Tetrahedron Computer Methodology*, 2(part 1):327–336; (part 2):337–347, 1989.

37. T. Okada and W. T. Wipke. CLUSMOL: A system for the conceptual clustering of molecules, *Tetrahedron Computer Methodology*, 2:249–264, 1989.

38. T. Okada, M. Yamakawa, and H. Niitsuma. Spiral mining using attributes from 3D molecular structures. *Active Mining: Second International Workshop, AM 2003*, pp. 287–302, LNCS 3430, Springer, Berlin, 2005.

39. H. Satoh, H. Koshino, and T. Nakata. Extended CAST coding method for exact search of stereochemical structures. *Journal of Computer Aided Chemistry*, 3:48–55, 2002.

40. A. Srinivasan and R. D. King. Using inductive logic programming to construct structure-activity relationships. AAAI Spring Symposium, Predictive Toxicology of Chemicals: Experiences and Impact of AI Tools, pp. 64–73, AAAI SS-99-01, 1999.

41. Y. Takahashi, S. Fujishima, K. Nishikoori, H. Kato, and T. Okada. Identification of dopamine D1 receptor agonists and antagonists under existing noise compounds by TFS-based ANN and SVM, *J.Comput. Chem. Jpn.*, 4:43–48, 2005.

42. U.S. FDA, Center for Drug Evaluation and Research: http://www.fda.gov/cder/Offices/OPS_IO/ICSAS.htm.

43. W. T. Wipke and T. M. Dyott. Stereochemically unique naming algorithm. *Journal of American Chemical Society*, 96:4834–4842, 1974.

44. K. Yoshida and H. Motoda. CLIP: Concept learning from inference patterns, *Artificial Intelligence*, 75:63–92, 1995.

15

UNIFIED APPROACH TO ROOTED TREE MINING: ALGORITHMS AND APPLICATIONS

MOHAMMED ZAKI

Department of Computer Science, Rensselaer Polytechnic Institute,
Troy, New York

15.1 INTRODUCTION

Tree patterns typically arise in applications such as bioinformatics, Web mining, mining semistructured documents, and so on. For example, given a database of XML documents, one might like to mine the commonly occurring "structural" patterns, that is, subtrees, that appear in the collection. As another example, given several phylogenies (i.e., evolutionary trees) from the Tree of Life [15], indicating evolutionary history of several organisms, one might be interested in discovering if there are common subtree patterns.

Recently, there has been tremendous interest in mining increasingly complex pattern types such as trees [1, 2, 4–6, 18, 26, 30] and graphs [12, 14, 27]. For example, several algorithms for tree mining have been proposed recently, which include TreeMiner [30], which mines embedded, ordered trees; SLEUTH [31], which mines embedded, unordered trees; FreqT [1], which mines induced ordered trees, FreeTreeMiner [4], which mines induced, unordered, free trees (i.e., there is no distinct root); TreeFinder [22], which mines embedded, unordered trees (but it may miss some patterns; it is not complete); and PathJoin [26], uFreqt [18], uNot [2], CMTreeMiner [6], and HybridTreeMiner [5], which mine induced, unordered trees.

In this chapter we extend SLEUTH[1] to obtain an efficient, unified algorithm for the problem of mining frequent subtrees. The key contributions of our work are as follows: (1) We present a unified approach to tree mining that can handle both ordered and unordered and both induced and embedded trees. (2) We propose a new self-contained equivalence class extension scheme to generate all candidate trees. Only potentially frequent extensions are considered, but some redundancy is allowed in the candidate generation to make each class self-contained. We study the trade-off between nonredundant versus potentially frequent candidate generation. (3) We extend the notion of scope-list joins (first proposed in [30]) for fast frequency computation for unordered/induced trees. (4) We also propose a method to count only distinct tree occurrences, instead of all mappings. (5) Finally, we conduct performance evaluation on several synthetic datasets and a real Web log dataset to show that SLEUTH is an efficient algorithm for different types of tree patterns. We present applications of tree mining in bioinformatics, such as mining frequent RNA (ribonucleic acid) structures and common phylogenetic tree patterns.

15.2 PRELIMINARIES

A *rooted tree* $T = (V, E)$ is a directed, acyclic, graph with vertex set $V = \{0, 1, \cdots, n\}$, edge set $E = \{(x, y)|x, y \in V\}$, and with one distinguished vertex $r \in V$ called the *root* such that for all $x \in V$, there is a *unique* path from r to x. In a *labeled tree*, $l : V \to L$ is a labeling function mapping vertices to a set of *labels* $L = \{\ell_1, \ell_2, \ldots\}$. In an *ordered tree* the children of each vertex are ordered (i.e., first child, second child, etc.), otherwise, the tree is *unordered*.

If $x, y \in V$ and there is a path from x to y, then x is called an *ancestor* of y (and y a *descendant* of x), denoted as $x \leq_p y$, where p is the length of the path from x to y. If $x \leq_1 y$ (i.e., x is an immediate ancestor), then x is called the *parent* of y, and y the *child* of x. If x and y have the same parent, x and y are called *siblings*, and if they have a common ancestor, they are called *cousins*.

We also assume that vertex $x \in V$ is synonymous with (or numbered according to) its position in the depth-first (preorder) traversal of the tree T (e.g., the root r is vertex 0). Let $T(x)$ denote the subtree rooted at x, and let y be the rightmost leaf (or highest numbered descendant) under x. Then the *scope* of x is given as $s(x) = [x, y]$. Intuitively, $s(x)$ demarcates the range of vertices under x.

As suggested in [30], we represent a tree T by its *string encoding*, denoted $\mathcal{T}$, generated as follows: Add vertex labels to T in a depth-first preorder traversal of T, and add a unique symbol $\$ \notin L$ whenever we backtrack from a child to its parent. For example, for T shown in Figure 15.1, its string encoding is $A\ B\ A\ C\ \$\ B\ \$\ \$\ C\ \$\ \$\ C\ \$$. We use the notation $\mathcal{T}[i]$ to denote the element at position i in $\mathcal{T}$, where $i \in [1, |\mathcal{T}|]$, and $|\mathcal{T}|$ is the length of the string $\mathcal{T}$.

Given a tree $S = (V_s, E_s)$ and tree $T = (V_t, E_t)$, we say that S is an *isomorphic subtree* of T *iff* there exists a one-to-one mapping $\varphi : V_s \to V_t$, such that $(x, y) \in E_s$

---

[1]SLEUTH is an anagram of the bold letters in the phrase: Listing "**H**idden" or **E**mbedded/induced (**U**n)ordered Sub**T**rees).

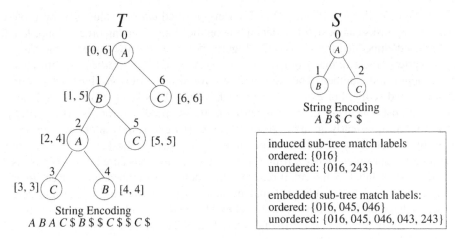

Figure 15.1. An example: Tree and subtree.

iff $(\varphi(x), \varphi(y)) \in E_t$. If φ is onto, then S and T are called *isomorphic*. S is called an *induced subtree* of $T = (V_t, E_t)$, denoted $S \preceq_i T$, iff S is an isomorphic subtree of T, and φ preserves labels, that is, $l(x) = l(\varphi(x))$, $\forall x \in V_s$. That is, for induced subtrees φ preserves the parent–child relationships, as well as vertex labels. The induced subtree obtained by deleting the rightmost leaf in T is called an *immediate prefix* of T. The induced tree obtained from T by a series of rightmost node deletions is called a *prefix* of T. In the sequel we use prefix to mean an immediate prefix, unless we indicate otherwise.

$S = (V_s, E_s)$ is called an *embedded subtree* of $T = (V_t, E_t)$, denoted as $S \preceq_e T$ iff there exists a 1-to-1 mapping $\varphi : V_s \to V_t$ that satisfies: (i) $(x, y) \in E_s$ iff $\varphi(x) \leq_p \varphi(y)$, and (ii) $l(x) = l(\varphi(x))$. That is, for embedded subtrees φ preserves ancestor–descendant relationships and labels. A (sub)tree of size k is also called a k-(sub)tree. If $S \preceq_e T$, we also say that T *contains* S or S *occurs* in T. Note that each occurrence of S in T can be identified by its unique *match label*, given by the sequence $\varphi(x_0)\varphi(x_1) \cdots \varphi(x_{|S|})$, where $x_i \in V_s$. That is, a match label of S is given as the set of matching positions in T.

Let $\delta_T(S)$ denote the number of occurrences (induced or embedded, depending on context) of the subtree S in a tree T. Let d_T be an indicator variable, with $d_T(S) = 1$ if $\delta_T(S) > 0$ and $d_T(S) = 0$ if $\delta_T(S) = 0$. Let D denote a database (a *forest*) of trees. The *support* of a subtree S in the database is defined as $\sigma(S) = \sum_{T \in D} d_T(S)$, that is, the number of trees in D that contain at least one occurrence of S. The *weighted support* of S is defined as $\sigma_w(S) = \sum_{T \in D} \delta_T(S)$, that is, total number of occurrences of S over all trees in D. Typically, support is given as a percentage of the total number of trees in D. A subtree S is *frequent* if its support is more than or equal to a user-specified *minimum support (minsup)* value. We denote by F_k the set of all frequent subtrees of size k. In some domains one might be interested in using weighted support, instead of support. Both of them are allowed using our mining approach, but we focus mainly on support.

Given a collection of trees D and a user-specified *minsup* value, several rooted tree mining tasks can be defined, depending on the choices among ordered/unordered or induced/embedded trees. Consider Figure 15.1, which shows an example tree T with vertex labels drawn from the set $L = \{A, B, C\}$, and vertices identified by their depth-first number. The figure shows for each vertex, its label, depth-first number, and scope. For example, the root is vertex 0, its label $l(0) = A$, and since the rightmost leaf under the root is vertex 6, the scope of the root is $s(0) = [0, 6]$. Consider S; it is clearly an induced subtree of T. If we look only at ordered subtrees, then the match label of S in T is given as: $012 \rightarrow \varphi(0)\varphi(1)\varphi(2) = 016$ (we omit set notation for convenience). If unordered subtrees are considered, then 243 is also a valid match label. S has additional match labels as an embedded subtree. In the ordered case, we have additional match labels 045 and 046, and in the unordered case, we have on top of these two, the label 043. Thus the induced weighted support of S is 1 for ordered and 2 for the unordered case. The embedded weighted support of S is 3, if ordered, and 5, if unordered. The support of S is 1 in all cases.

15.3 RELATED WORK

Recently, tree mining has attracted a lot of attention. Zaki proposed TreeMiner [30] to mine labeled, embedded, and ordered subtrees. The notions of scope lists and rightmost extension were introduced in that work. TreeMiner was also used in building a structural classifier for XML data [32]. XSpanner [24] is a pattern-growth-based method for mining embedded ordered subtrees. Asai et al. [1] presented FreqT, an Apriori-like algorithm for mining labeled ordered trees; they independently proposed the rightmost candidate generation scheme. Wang and Liu [25] developed an algorithm to mine frequently occurring subtrees in XML documents. Their algorithm is also reminiscent of the levelwise Apriori approach, and they mine induced subtrees only. There are several other recent algorithms that mine different types of tree patterns, which include FreeTreeMiner [4], which mines induced, unordered, free trees (i.e., there is no distinct root); SingleTreeMining [21], which mines rooted, unordered, trees, with application to phylogenetic tree pattern mining; and PathJoin [26], uFreqt [18], uNot [2], and HybridTreeMiner [5], which mine induced, unordered trees. CMTreeMiner [6] mines maximal and closed induced, unordered trees. TreeFinder [22] uses an inductive logic programming (ILP) approach to mine unordered, embedded subtrees, but it is not a complete method, that is, it can miss many frequent subtrees, especially as support is lowered or when the different trees in the database have common node labels. Dryade [23] is a recent method to mine embedded unordered subtrees, with the restriction that no two siblings have the same labels. In recent work, Zaki [31] proposed SLEUTH, the first complete algorithm to mine embedded unordered trees. Our focus here is on an efficient, unified approach to mine the complete set of frequent, induced/embedded, ordered/unordered trees.

Frequent tree mining is also related to tree isomorphism [20] and tree pattern matching [7]. The tree inclusion problem was studied in [13], that is, given labeled trees P and T, can P be obtained from T by deleting nodes? This problem is equivalent to checking if P is embedded in T. Here we are interested in enumerating all common subtrees in a collection of trees.

There has also been recent work in mining frequent graph patterns. The AGM algorithm (see [12] and Chapter 9 of this book) discovers induced (possibly disconnected) subgraphs. The FSG algorithm (see [14] and Chapter 6 of this book) improves upon AGM, and mines only the connected subgraphs. Both methods follow an Apriori-style levelwise approach. Recent methods to mine graphs using a depth-first tree-based extension have been proposed in [27, 28] (see also Chapter 5 of this book). Another method uses a candidate generation approach based on Canonical Adjacency Matrices [11]. GASTON [17] adopts an interesting stepwise approach using a combination of path, free tree, and finally graph mining to discover all frequent subgraphs. The Subdue system (see [8] and Chapter 7 of this book) also discovers graph patterns using the minimum description length principle (MDL). An approach termed graph-based induction (GBI) was proposed in [29], which uses beam search for mining subgraphs. In contrast to these approaches, we are interested in developing efficient, complete algorithms for tree patterns.

15.4   GENERATING CANDIDATE SUBTREES

There are two main steps for enumerating frequent subtrees in D. First, we need a systematic way of generating *candidate* subtrees whose frequency is to be computed. The candidate set should be nonredundant to the extent possible; ideally, each subtree should be generated at most once. Second, we need efficient ways of counting the number of occurrences of each candidate tree in the database D, and to determine which candidates pass the *minsup* threshold. The latter step is data structure dependent and will be treated in Section 15.5. We begin with the problem of candidate generation in this section.

An *automorphism* of a tree is an isomorphism with itself. Let Aut(T) denote the *automorphism group*, that is, the set of all label preserving automorphisms, of T. Henceforth, by automorphism, we mean label preserving automorphisms. The goal of candidate generation is to enumerate only one *canonical* representative from Aut(T). For an unordered tree T, there can be many automorphisms. For example, Figure 15.2 shows some of the automorphisms of the same tree. On the other hand, ordered trees have either only one automorphism (the trivial one that maps T to

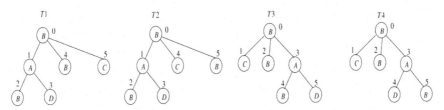

Figure 15.2.   Some automorphisms of the same tree.

itself) or, if there are several of them, they are indistinguishable (e.g., when some node has at least two identical subtrees). Whether a tree is induced or embedded does not impact its automorphism group.

Let there be a linear order $\leq$ defined on the elements of the label set L. Given any two trees X and Y, we can define a linear order $\leq$, called *tree order* between them, recursively as follows: Let r_x and r_y denote the roots of X and Y, and let $c_1^{r_x}, \ldots, c_m^{r_x}$ and $c_1^{r_y}, \ldots, c_n^{r_y}$ denote the ordered list of children of r_x and r_y, respectively. Also let $T(c_i^{r_x})$ denote the subtree of X rooted at vertex $c_i^{r_x}$. Then $X \leq Y$ [alternatively, $T(r_x) \leq T(r_y)$] iff either:

1. $l(r_x) < l(r_y)$, or

2. $l(r_x) = l(r_y)$, and either (a) $n \leq m$ and $T(c_i^{r_x}) = T(c_i^{r_y})$ for all $1 \leq i \leq n$, that is, Y is a prefix (not necessarily immediate prefix) of or equal to X, or (b) there exists $j \in [1, \min(m, n)]$, such that $T(c_i^{r_x}) = T(c_i^{r_y})$ for all $i < j$, and $T(c_j^{r_x}) < T(c_j^{r_y})$.

This tree ordering is essentially the same as that in [18], although their tree coding is different.

We can also define a *code order* on the tree encodings directly as follows: Assume that the special backtrack symbol $\$ > \ell$ for all $\ell \in L$. Given two string encodings $\mathcal{X}$ and $\mathcal{Y}$. We say that $\mathcal{X} \leq \mathcal{Y}$ iff either:

(i) $|\mathcal{Y}| \leq |\mathcal{X}|$ and $\mathcal{X}[k] = \mathcal{Y}[k]$ for all $1 \leq k \leq |\mathcal{Y}|$, or

(ii) There exists $k \in [1, \min(|\mathcal{X}|, |\mathcal{Y}|)]$, such that for all $1 \leq i < k$, $\mathcal{X}[i] = \mathcal{Y}[i]$ and $\mathcal{X}[k] < \mathcal{Y}[k]$.

Incidentally, a similar tree code ordering was independently proposed in CMTree-Miner [6].

LEMMA 15.1 $X \leq Y$ iff $\mathcal{X} \leq \mathcal{Y}$.

Proof. Condition (i) in code order holds if $\mathcal{X}$ and $\mathcal{Y}$ are identical for the entire length of $\mathcal{Y}$, but this is true iff Y is a prefix of (or equal to) X.

Condition (ii) holds if and only if $\mathcal{X}$ and $\mathcal{Y}$ are identical up to position $k - 1$, that is, $\mathcal{X}[1, \ldots, k - 1] = \mathcal{Y}[1, \ldots, k - 1]$. This is true iff both X and Y share a common prefix tree P with encoding $\mathcal{P} = \mathcal{X}[1, \ldots, k - 1]$). Let v_X^i (and v_Y^j) refer to the node in tree X (and Y), which corresponds to position $\mathcal{X}[i] \neq \$$ (and $\mathcal{Y}[j] \neq \$$).

If $k = 1$, then P is an empty tree with encoding $\mathcal{P} = \emptyset$. It is clear that $l(r_x) < l(r_y)$ iff $\mathcal{X}[1] < \mathcal{Y}[1]$. If $k > 1$, then $\mathcal{X}[k] < \mathcal{Y}[k]$, iff one of the following cases is true: (a) $\mathcal{X}[k] \neq \$$ and $\mathcal{Y}[k] \neq \$$: We immediately have $\mathcal{X}[k] < \mathcal{Y}[k]$ iff $T(v_X^k) < T(v_Y^k)$ iff $X < Y$. (b) $\mathcal{X}[k] \neq \$$ and $\mathcal{Y}[k] = \$$: let v_X^j be parent of node v_X^k ($j < k$), and let v_Y^j be the corresponding node in Y (which refers to $\mathcal{Y}[j] \neq \$$). We then immediately have that $T(v_Y^j)$ is a prefix of $T(v_X^j)$, since $\mathcal{X}[j, \ldots, k - 1] = \mathcal{Y}[j, \ldots, k - 1]$, and v_X^j has an extra child v_X^k, whereas v_Y^j does not.   □

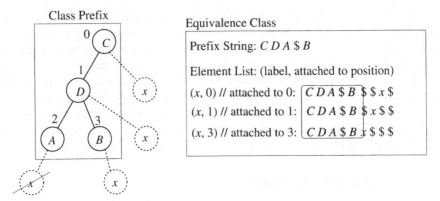

Figure 15.3. Prefix extension and equivalence class.

Given $Aut(T)$ the canonical representative $T_c \in Aut(T)$ is the tree, such that $T_c \leq X$ for all $X \in Aut(T)$. For any $P \in Aut(T)$ we say that P is in *canonical form* if $P = T_c$. For example, $T_c = T_1$ for the automorphism group $Aut(T_1)$, four of whose members are shown in Figure 15.2. We can see that the string encoding $T_1 = BAB\$D\$\$B\$C\$$ is smaller than $T_2 = BAB\$D\$\$C\$B\$$ and also smaller than other members.

LEMMA 15.2   A tree T is in canonical form iff for all vertices $v \in T$, $T(c_i^v) \leq T(c_{i+1}^v)$ for all $i \in [1, k]$, where $c_1^v, c_2^v, \ldots, c_k^v$ is the list of ordered children of v.

Proof. T is in canonical form implies that $T \leq X$ for all $X \in Aut(T)$. Assume that there exist some vertex $v \in T$ such that $T(c_i) > T(c_{i+1})$ for some $i \in [1, k]$, where $c_1, c_2, \ldots, c_k$ are the ordered children of v. But then, we can obtain tree T' by simply swapping the subtrees $T(c_i)$ and $T(c_{i+1})$ under node v. However, by doing so, we make $T' < T$, which contradicts the assumption that T is canonical.   □

Let $R(P) = v_1 v_2, \ldots, v_m$ denote the rightmost path in tree P, that is, the path from root P_r to the rightmost leaf in P. Given a seed frequent tree P, we can generate new candidates P_x^i obtained by adding a new leaf with label x to any vertex v_i on the rightmost path $R(P)$. We call this process as *prefix-based* extension since each such candidate has P as its prefix tree.

It has been shown that prefix-based extension can correctly enumerate all ordered trees [1, 30]. SLEUTH follows the same strategy for ordered, embedded trees. For unordered trees, we only have to do a further check to see if the new extension is the canonical form for its automorphism group, and if so, it is a valid extension. For example, Figure 15.3 shows the seed tree P, with encoding $\mathcal{P} = CDA\$B$ (omitting trailing \$'s). To preserve the prefix tree, only rightmost branch extensions are allowed. Since the rightmost path is $R(P) = 013$, we can extend P by adding a new vertex with label x any of these vertices, to obtain a new tree P_x^i ($i \in \{0, 1, 3\}$). Note, how adding x to node 2 gives a different prefix tree encoding $CDAx$, and is thus disallowed, as shown in Figure 15.3.

In [18] it was shown that for any tree in canonical form its prefix is also in canonical form. Thus starting from vertices with distinct labels, using prefix extensions, and retaining only canonical forms for each automorphism group, we can enumerate all unordered trees nonredundantly. For each candidate, we can count the number of occurrences in database D to determine which are frequent. Thus the main challenges in tree extension are to: (i) efficiently determine whether an extension yields a canonical tree, and (ii) determine extensions that will potentially be frequent. The former step considers only valid candidates, whereas the latter step minimizes the number of frequency computations against the database.

15.4.1 Canonical Extension

To check if a tree is in canonical form, we need to make sure that for each vertex $v \in T$, $T(c_i) \le T(c_{i+1})$ for all $i \in [1, k]$, where $c_1, c_2, \ldots, c_k$ is the list of ordered children of v. However, since we extend only canonical trees, for a new candidate, its prefix is in canonical form, and we can do better.

LEMMA 15.3   Let P be a tree in canonical form, and let $R(P)$ be the rightmost path in P. Let P_x^k be the tree extension of P when adding a vertex with label x to some vertex v_k in $R(P)$. For any $v_i \in R(P_x^k)$, let $c_{l-1}^{v_i}$ and $c_l^{v_i}$ denote the last two children of v_i.[2] Then P_x^k is in canonical form iff for all $v_i \in R(P_x^k)$, $T(c_{l-1}^{v_i}) \le T(c_l^{v_i})$.

Proof. Let $R(P) = v_1 v_2 \cdots v_k v_{k+1} \cdots v_m$ be the rightmost path in P. By Lemma 15.2, P is in canonical form implies that for every node $v_i \in R(P)$, we have $T(c_{l-1}^{v_i}) \le T(c_l^{v_i})$.

When we extend P to P_x^k, we obtain a new rightmost path $R(P_x^k) = v_1 v_2, \ldots, v_k v_n$, where v_n is the new last child of v_k (with label x). Thus both $R(P)$ and $R(P_x^k)$ share the vertices $v_1 v_2, \ldots, v_k$ in common. Note that for any $i > k$, $v_i \in R(P)$ is unaffected by the addition of vertex v_n. On the other hand, for all $i < k$, the last child $c_l^{v_i}$ of $v_i \in R(P)$ [i.e., $v_i \in R(P_x^k)$] is affected by v_n, whereas $c_{l-1}^{v_i}$ remains unchanged. Also for $i = k$, the last two children of v_k change in tree P_x^k; we have $c_{l-1}^{v_k} = v_{k+1}$ and $c_l^{v_k} = v_n$.

Since P is in canonical form, we immediately have that for all $v_i \in \{v_1, v_2, \ldots, v_k\}$, $T(c_j^{v_i}) \le T(c_{l-1}^{v_i})$ for all $j < l - 1$. Thus we only have to compare the new subtree $T(c_l^{v_i})$ with $T(c_{l-1}^{v_i})$. If $T(c_{l-1}^{v_i}) \le T(c_l^{v_i})$ for all $v_i \in R(P_x^k)$, then by Lemma 15.2, we immediately have that P_x^k is in canonical form. On the other hand if $T(c_{l-1}^{v_i}) > T(c_l^{v_i})$ for some $v_i \in R(P_x^k)$, then P_x^k cannot be in canonical form. $\square$

According to Lemma 15.3 we can check if a tree P_x^k is in canonical form by starting from the rightmost leaf in $R(P_x^k)$ and checking if the subtrees under the last two children for each node on the rightmost path are ordered according to $\le$. By Lemma 15.1 it is sufficient to check if their string encodings are ordered by $\le$.

---

[2]If v_i is a leaf, then both children are empty, and if v_i has only one child, then $c_{l-1}^{v_i}$ is empty.

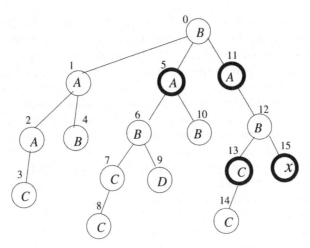

Figure 15.4. Check for canonical form.

For example, given the candidate tree P_x^{12} shown in Figure 15.4 which has a new vertex 15 with label x attached to node 12 on the rightmost path, we first compare 15 with its previous sibling 13. For $T(13) \leq T(15)$, we require that $x \geq C$. After skipping node 11 (with empty previous sibling), we reach node 0, where we compare $T(5)$ and $T(11)$. For $T(5) \leq T(11)$ we require that $x \geq D$, otherwise P_x^{12} is not canonical. Thus for any $x < D$ the tree is not canonical. It is possible to speed-up the canonicality checking by adopting a different tree coding [18], but here we will continue to use the string encoding of a tree. The corresponding checks for canonicality based on Lemma 15.1 among the subtree encodings are shown below:

T(13) vs. T(15): BAAC$$B$$ABCC $$D $$B $$AB<u>CC</u> $$<u>x</u>

T(5) vs. T(11): BAAC$$B$<u>$ABCC $$D $$B</u> $$<u>ABCC $$x</u>

Based on the check for canonical form, we can determine which labels are possible for each rightmost path extension. Given a tree P and the set of frequent edges F_2 (or the frequent labels F_1), we can then try to extend P with each edge from F_2 (or each item in F_1) that leads to a canonical extension. Even though all of these candidates are nonredundant (i.e., there are no isomorphic duplicates), this extension process may still produce too many candidate trees, whose frequencies have to be counted in the database D, and many of the candidates may not be frequent. To reduce the number of such trees, we try to extend P with a vertex that is more likely to result in a frequent tree, using the idea of a prefix equivalence class.

15.4.2 Equivalence Class-Based Extension

We say that two k-subtrees X, Y are in the same *prefix equivalence class* iff they share the same prefix tree. Thus any two members of a prefix class differ only in the last vertex. For example, Figure 15.3 shows the class template for subtrees with the same prefix subtree P with string encoding $\mathcal{P} = C\ D\ A\ \$\ B$. The figure shows the actual format we use to store an equivalence class; it consists of the class prefix string and a list of elements. Each element is given as a (x, i) pair, where x is the *label* of the last vertex, and i specifies the vertex in P to which x is attached. For example $(x, 1)$ refers to the case where x is attached to vertex 1. The figure shows the encoding of the subtrees corresponding to each class element. Note how each of them shares the same prefix up to the $(k - 1)$th vertex. These subtrees are shown only for illustration purposes; we only store the element list in a class.

Let P be a prefix subtree of size $k - 1$; we use the notation $[P]$ to refer to its class (we will use P and its string encoding $\mathcal{P}$ interchangeably). If (x, i) is an element of the class, we write it as $(x, i) \in [P]$. Each (x, i) pair corresponds to a subtree of size k, sharing P as the prefix, with the last vertex labeled x, attached to vertex i in P. We use the notation P_x^i to refer to the new prefix subtree formed by adding (x, i) to P. Let P be a $(k - 1)$ subtree, and let $[P] = \{(x_i) | P_x^i$ is frequent$\}$ be the set of all possible frequent extensions of prefix tree P. Then the set of potentially frequent candidate trees for the class $[P_x^i]$ [obtained by adding an element (x, i) to P], can be obtained by prefix extensions of P_x^i with each element $(y, j) \in [P]$, given as follows: (i) *cousin extension*: If $j \le i$ and $|P| = k - 1 \ge 1$, then $(y, j) \in [P_x^i]$, and in addition (ii) *descendant extension*: If $j = i$ then $(y, k - 1) \in [P_x^i]$.

Consider Figure 15.5, showing the prefix class $\mathcal{P} = AB$, which contains 2 elements, $(C, 1)$ and $(D, 0)$. Let us consider the extensions of first element, that is, of $[P_C^1] = [ABC]$. First we must consider element $(C, 1)$ itself. As descendant extension, we add $(C, 2)$ (tree C_1), and as cousin extension, we add $(C, 1)$ (tree C_2).

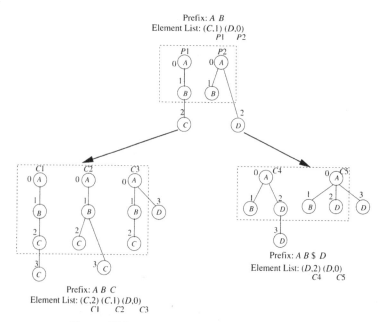

Prefix: *A B*
Element List: (C,1) (D,0)
              P1   P2

Prefix: *A B $ D*
Element List: (D,2) (D,0)
              C4   C5

Prefix: *A B C*
Element List: (C,2) (C,1) (D,0)
              C1   C2   C3

Figure 15.5. Equivalence-class-based extension.

Extending with $(D, 0)$, since $0 < 1$, we only add cousin extension $(D, 0)$ (tree C_3) to $[ABC]$. When considering extensions of $[P_D^0] = [AB\$D]$, we consider $(C, 1)$ first. But since C is attached to vertex 1, it cannot preserve the prefix tree P_D. Considering $(D, 0)$, we add $(D, 0)$ as a cousin extension and $(D, 2)$ as a descendant extension, corresponding to trees C_4 and C_5.

15.4.3 Discussion

Note that for ordered and embedded tree mining, we can guarantee that each equivalence class is complete, that is, all potential embedded, ordered $(k + 1)$ subtrees can be obtained by joining two embedded, ordered k subtrees in the class. For unordered subtrees, we have to allow noncanonical subtrees to preserve completeness. Thus any embedded, unordered $(k + 1)$ subtree can be obtained by joining two embedded, unordered k subtrees in the class, where the tree being extended is canonical. For induced subtrees also the equivalence class is not complete, if we only keep induced subtrees as class members. For example, in Figure 15.1, there are two induced 2 subtrees with B as a root, namely, $BA\$$ and $BC\$$. Thus $[B] = \{(A, 0), (C, 0)\}$. It is clear that we cannot obtain the pattern $BAB\$\$$ by joining only two elements within the class $[B]$. To guarantee completeness, we must allow embedded subtrees to be class members. For example, if we add the embedded pattern $BB\$$ to $[B]$, then we will be able to obtain $BAB\$\$$ by joining only two elements within the class $[B]$. In general, we can obtain any induced, (un)ordered $(k + 1)$ subtree, from two k subtrees within a class, provided the tree being extended is both canonical and induced. Thus the main observation behind equivalence class extension is the $F_k \times F_k$ candidate generation process, where only known frequent k elements from the same class are used for extending P_x^i. Furthermore, we only extend P_x^i, if it is in canonical form and satisfies the given tree properties. However, to guarantee that all possible extensions are members of $[P]$, we have to relax the canonicality or induced requirements.

As opposed to equivalence class extensions, for pure canonical extensions, an equivalence class contains only canonical members, and only those members that satisfy the tree properties (embedded or induced). To guarantee that all possible extensions will be tried, we extend $[P_x^i]$ by considering all members of the form $(x, y) \in F_2$, which itself stores only canonical or embedded/induced elements as the case may require. Canonical extension thus corresponds to an $F_k \times F_2$ (or $F_k \times F_1$) candidate generation process. In essence canonical and equivalence class extensions represent a trade-off between the number of redundant (isomorphic) candidates generated and the number of potentially frequent candidates to count. Canonical extensions generate nonredundant candidates, but many of which may turn out not to be frequent. On the other hand, equivalence class extension generates redundant candidates, but considers a smaller number of (potentially frequent) extensions. In our experiments we found equivalence class extensions to be more efficient (see Section 15.8). One consequence of using equivalence class extensions is that SLEUTH does not depend on any particular canonical form; it can work with any systematic way of choosing a representative from an automorphism group. Provided only one representative is extended, its class contains all information about the extensions that can be potentially frequent. This can provide a lot of flexibility on how tree enumeration is performed.

15.5 FREQUENCY COMPUTATION

The candidate generation step allow us to enumerate potentially frequent ordered/ unordered subtrees in a systematic manner. The goal of the frequency counting step is to quickly find the support of a candidate. We first look at the task of finding the frequency of embedded subtrees, and then extend the method to compute the support of induced subtrees.

15.5.1 Embedded Subtrees

In SLEUTH, we represent the database in the vertical format [30] in which for every distinct label we store its scope-list, which is a list of tree IDS and vertex scopes where that label occurs. For label ℓ, we denote its scope list as $\mathcal{L}(\ell)$; each entry in the scope list is a pair (t, s), where t is a tree ID (TID) in which ℓ occurs, and s is the scope of a vertex with label ℓ in TID t. Figure 15.6 shows a database of three trees, and the scope lists for each label. Consider label A; since it occurs at vertex 0 with scope $[0, 3]$ in tree T_0, we add $(0, [0, 3])$ to its scope list. A also occurs in T_1 with scope $[1, 3]$, and in T_2 with scopes $[0, 7]$ and $[4, 7]$. Thus we add $(1, [1, 3])$, $(2, [0, 7])$ and $(2, [4, 7])$ to $\mathcal{L}(A)$. In a similar manner, the scope lists for other labels are created.

We also use the scope lists to represent the list of occurrences in the database, for any k-subtree S. Let x be the label of the rightmost leaf in S. The scope list of S consists of triples (t, m, s), where t is a TID where S occurs, s is the scope of vertex with label x in TID t, and m is a match label for the prefix subtree of S. Thus the vertical database is in fact the set of scope lists for all 1 subtrees (and since they have no prefix, there is no match label).

SLEUTH uses scope-list joins for fast frequency computation for a new embedded extension. We assume that each element (x, i) in a prefix class $[P]$ has a scope list that stores all occurrences of the tree P_x^i [obtained by extending P with (x, i)]. The vertical database contains the initial scope lists $\mathcal{L}(\ell)$ for each distinct label ℓ. To compute the scope lists for members of $[P_x^i]$, we need to join the scope lists of (x, i) with every other element $(y, j) \in [P]$. If the resulting tree is frequent, we insert the element in $[P_x^i]$.

Let $s_x = [l_x, u_x]$ be a scope for vertex x, and $s_y = [l_y, u_y]$ a scope for y. We say that s_x is *strictly less* than s_y, denoted $s_x < s_y$, if and only if $u_x < l_y$, that is, the interval s_x has no overlap with s_y, and it occurs before s_y. We say that s_x *contains* s_y, denoted $s_x \supset s_y$, if and only if $l_x < l_y$ and $u_x \geq u_y$, that is, the interval s_y is a proper subset of s_x.

Recall from the equivalence class extension that when we extend element $[P_x^i]$ there can be at most two possible outcomes, that is, descendant extension or cousin extension. The use of scopes allows us to compute in constant time whether y is

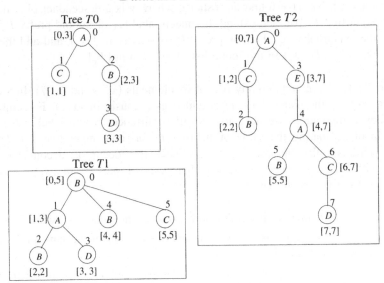

Figure 15.6. Scope lists.

a descendant of x or y is a cousin of x. We describe below how to compute the embedded support for (un)ordered extensions, using the descendant and cousin tests.

Descendant Test. Given $[P]$ and any two of its elements (x, i) and (y, j). In a descendant extension of P_x^i the element (y, j) is added as a child of (x, i). For embedded frequency computation, we have to find all occurrences where label y occurs as a descendant of x, sharing the same prefix tree P_x^i in some $T \in D$, with TID t. This is called the *descendant test*. To check if this subtree occurs in an input tree T with TID t, we search if there exists triples $(t_y, m_y, s_y) \in \mathcal{L}(y)$ and $(t_x, m_x, s_x) \in \mathcal{L}(x)$, such that:

1. $t_y = t_x = t$, that is, the triples both occur in the same tree, with TID t.
2. $m_y = m_x = m$, that is, x and y are both extensions of the same prefix occurrence, with match label m.

3. $s_y \subset s_x$, that is, y lies within the scope of x. If the three conditions are satisfied, we have found an instance where y is a descendant of x in some input tree T. We then extend the match label m_y of the old prefix P, to get the match label for the new prefix P_x^i (given as $m_y \cup l_x$), and add the triple $(t_y, \{m_y \cup l_x\}, s_y)$ to the scope list of $(y, |P|)$ in $[P_x^i]$.

Cousin Test. Given $[P]$ and any two of its elements (x, i) and (y, j). In a cousin extension of P_x^i the element (y, j) is added as a cousin of (x, i). For embedded frequency computation, we have to find all occurrences where label y occurs as a cousin of x, sharing the same prefix tree P_x^i in some input tree $T \in D$, with TID t. This is called the *cousin test*. To check if y occurs as a cousin in some tree T with TID t, we need to check if there exists triples $(t_y, m_y, s_y) \in \mathcal{L}(y)$ and $(t_x, m_x, s_x) \in \mathcal{L}(x)$, such that:

1. $t_y = t_x = t$, that is, the triples both occur in the same tree, with TID t.
2. $m_y = m_x = m$, that is, x and y are both extensions of the same prefix occurrence, with match label m.
3. $s_x < s_y$ or $s_x > s_y$, that is, either x comes before y or y comes before x in depth-first ordering, and their scopes do not overlap. This allows us to find the unordered frequency and is one of the crucial differences compared to ordered tree mining, as in TreeMiner [30], which only checks if $s_x < s_y$. If these conditions are satisfied, we add the triple $(t_y, \{m_y \cup l_x\}, s_y)$ to the scope list of (y, j) in $[P_x^i]$.

Figure 15.7 shows an example of how scope-list joins work, using the database D from Figure 15.6. The initial class with empty prefix consists of four frequent labels (A, B, C, and D), with their scope lists. All pairs (not necessarily distinct) of elements are considered for extension. Two of the frequent trees in class $[A]$ are shown, namely $AB\$$ and $AC\$$. $AB\$$ is obtained by joining the scope lists of A and B and performing descendant tests, since we want to find those occurrences of B

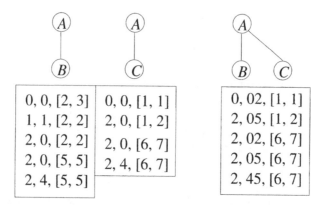

Figure 15.7. Scope-list joins: embedded.

that are within some scope of A (i.e., under a subtree rooted at A). Let s_x denote a scope for label x. For tree T_0 we find that $s_B = [2, 3] \subset s_A = [0, 3]$. Thus we add the triple $(0, 0, [2, 3])$ to the new scope list. Similarly, we test the other occurrences of B under A in trees T_1 and T_2. If a new scope list occurs in at least *minsup* TIDS, the pattern is considered frequent. The next candidate shows an example of testing frequency of a cousin extension, namely, how to compute the scope list of $AB\$C$ by joining $\mathcal{L}(AB)$ and $\mathcal{L}(AC)$. For finding all unordered embedded occurrences, we need to test for disjoint scopes, with $s_B < s_C$ or $s_C < s_B$, which have the same match label. For example, in T_0, we find that $s_B = [2, 3]$ and $s_C = [1, 1]$ satisfy these condition. Thus we add the triple $(0, 02, [1, 1])$ to $\mathcal{L}(AB\$C)$. Notice that the new prefix match label (02) is obtained by adding to the old prefix match label (0), the position where B occurs (i.e., 2). The other occurrences are noted in the final scope list.

15.5.2 Induced Subtrees

For counting the support of only induced trees, SLEUTH extends the scope list to be a five-tuple of the form (t, m, s, d, i), where in addition to the tid t, prefix match-label m, and last node scope s, we keep the last node's depth d (i.e., the number of edges on the path from the root to the given node), and a Boolean flag i indicating whether this tuple contributes to the induced support of the candidate. Initially, for single items, d is the actual depth of the item in tree t, but for k subtrees ($k \geq 2$), d denotes the depth of the node in the candidate subtree. Figure 15.8 shows the single item scope lists for induced mining; only the triples (t, s, d) are shown for single items, since the match-label $m = \emptyset$, and the induced flag $i = 1$ for all elements. For example, in tree T_0, A occurs at vertex 0 with scope $[0, 3]$ and at depth 0, in tree T_0, we add $(0, [0, 3], 0)$ to its scope list. In tree T_1, A occurs at node 1 with scope $[1, 3]$ and at depth 1, so we add $(1, [1, 3], 1)$ to its scope list, and so on.

Instead of cousin and descendant tests, for induced mining, we have to consider only sibling and child tests. Let $[P]$ be an equivalence class, let $(x, i) \in [P]$. For canonical extensions, let $(y, j) \in F_2$, and for equivalence class extensions, let $(y, j) \in [P]$.

Child Test. In a child extension of P_x^i the element (y, j) is added as a child of (x, i). For induced frequency computation, we first find all occurrences where label y occurs as a descendant of x, but we increment the support only for those tuples where y is a direct child of x. Note that we keep all embedded occurrences to preserve the equivalence-class completeness property. Thus for induced support counting, like in the embedded case, we begin by searching if there exists tuples $(t_y, m_y, s_y, d_y, i_y) \in \mathcal{L}(y)$ and $(t_x, m_x, s_x, d_x, i_x) \in \mathcal{L}(x)$, such that the TIDS ($t_y = t_x = t$) and match labels ($m_y = m_x = m$) are equal, and y lies within the scope of x ($s_y \subset s_x$). In addition, we compute the difference in the depth of nodes y and x, $\delta = d_y - d_x$. If $i_x = 1$, then x represents an induced subtree, and therefore, if $\delta = 1$, then y must also be an induced extension of x. In this case we add the new tuple $(t_y, \{m_y \cup l_x\}, s_y, d, 1)$ to the scope list of $(y, |P|)$ in $[P_x^i]$, where $d = \delta$ when

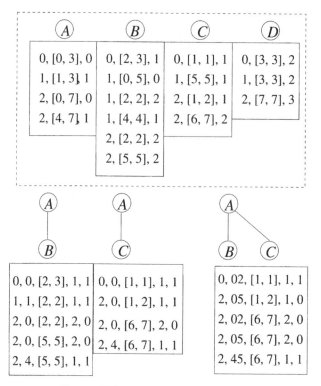

Figure 15.8.   Scope-list joins: induced.

transitioning from absolute depth of y in t to relative depth of y in the candidate, that is, for 2 subtrees, otherwise $d = d_y$ (for $k > 2$). If $i_x \neq 1$, then y cannot be an induced extension of x, but rather is an embedded extension. In this case we add the new tuple $(t_y, \{m_y \cup l_x\}, s_y, d, 0)$ to the scope list, but only if we are using equivalence-class extensions.

Cousin Test. In a cousin extension of P_x^i the element (y, j) is added as a cousin of (x, i). For induced support counting, we require that both y and x are induced extensions of the same parent node. That is we look for tuples $(t_y, m_y, s_y, d_y, i_y) \in \mathcal{L}(y)$ and $(t_x, m_x, s_x, d_x, i_x) \in \mathcal{L}(x)$, such that $t_y = t_x = t$, $m_y = m_x = m$, and $s_x < s_y$ for ordered trees, and in addition if $s_x > s_y$ for unordered trees. Further, if $i_x = i_y = 1$, then it is an induced extension, and we add the tuple $(t_y, \{m_y \cup l_x\}, s_y, d_y, 1)$ to the scope list of (y, j) in $[P_x^i]$. Otherwise, if we are using equivalence-class extensions, we add $(t_y, \{m_y \cup l_x\}, s_y, d_y, 0)$ to the scope list. Note that since a cousin test can only be applied to k subtrees, with $k \geq 3$, the depth d_y is already relative to the root of the candidate.

Figure 15.8 shows an example of how induced scope-list joins work, using the database D from Figure 15.6. It shows the initial scope lists for the four frequent items. Consider the candidate $AB\$$ obtained by joining the scope lists of A and B

and performing child tests. Consider $(0, [2, 3], 1) \in \mathcal{L}(B)$ and $(0, [0, 3], 0) \in \mathcal{L}(A)$. We find that $s_B = [2, 3] \subset s_A = [0, 3]$, and $\delta = d_y - d_x = 1 - 0 = 1$, which means it is an induced occurrence of the pattern, so we add the tuple $(0, 0, [2, 3], 1, 1)$ to the new scope list $\mathcal{L}(AB)$. As another example, for T_2 we have $(2, [2, 2], 2) \in \mathcal{L}(B)$ and $(2, [0, 7], 0) \in \mathcal{L}(A)$, with $s_B = [2, 2] \subset s_A = [0, 7]$, and $\delta = d_y - d_x = 2 - 0 = 2$, which means it is only an embedded extension, so we add $(2, 0, [2, 2], 2, 0)$ to the new scope list. In a similar manner we compute other elements of $\mathcal{L}(AB)$ and of $\mathcal{L}(AC)$. For induced support we only count those tuples with the induced flag $i = 1$. Thus the induced support of $AB\$$ is 3 (since there is at least one element with $i = 1$ for each t), and the induced support of $AC\$$ is 2. As an example of cousin testing, consider the scope list of $AB\$C$, obtained by joining $\mathcal{L}(AB)$ and $\mathcal{L}(AC)$. For finding all (un)ordered induced occurrences, we need to test for disjoint scopes, with $s_B < s_C$ or $s_C < s_B$, which have the same match label, and where both tuples are induced. For example, for T_0, we find that the tuple $(0, 0, [2, 3], 1, 1)$ and $(0, 0, [1, 1], 1, 1)$, in $\mathcal{L}(AB)$ and $\mathcal{L}(AC)$, respectively, satisfy these conditions. Thus we add the tuple $(0, 02, [1, 1], 1, 1)$ to $\mathcal{L}(AB\$C)$. If we were interested only in ordered subtrees, this tuple would not be valid, since $[1, 1] < [2, 3]$. Note also that tuples $(2, 0, [5, 5], 2, 0)$ and $(2, 0, [1, 2], 1, 1)$, in $\mathcal{L}(AB)$ and $\mathcal{L}(AC)$, respectively, represent a sibling, rather than a cousin extension. So we add the tuple $(2, 05, [1, 2], 1, 0)$ to the new list, The other tuples are obtained similarly, and the induced support of $AB\$C$ is 2. In this example, we showed the scope lists assuming equivalence class extensions; for pure canonical extension, all those tuples with $i = 0$ will not be added to the scope lists.

15.6   COUNTING DISTINCT OCCURRENCES

SLEUTH is inherently a very efficient method for weighted support computation since it counts all embeddings of a frequent pattern within each database tree, using the scope-list joins. Many applications, however, may require only the support, that is, instead of finding all embeddings of a subtree in the entire database, we may simply want to know the number of database trees that contain at least one embedding of a subtree. If there are relatively few embeddings per tree, SLEUTH continues to be very effective for support counting. On the other hand, if there are many duplicate labels, and if the tree is highly branched, the number of embeddings can get large, resulting in long scope lists and increased running time. If the application calls for the use of weighted support, the increased cost is acceptable, but if we want only support, it is possible to optimize SLEUTH to count only distinct occurrences of each pattern.

To count only distinct occurrences SLEUTH uses a different scope-list representation for computing pattern frequency. It does not maintain the match labels, which keep track of all embeddings. Instead, it stores the scopes for all nodes on the rightmost path within a tree; we call the new scope lists as *scope-vector lists* or *SV lists* for short. Thus each element of the new list is a pair of the form $(t, \mathbf{s})$, where t is a tree ID and $\mathbf{s} = \{s_1, s_2, \dots, s_m\}$ is the scope vector of matching node scopes s_i on the rightmost path. Furthermore, $\mathbf{s}$ represents a *minimal* occurrence of the pattern within a database tree, that is, there does not exist another scope-vector $\mathbf{s}'$ strictly contained in $\mathbf{s}$, [3] such that the pattern also occurs at nodes with scopes given by $\mathbf{s}'$. Note that for induced mining, we can extend the tuple to be of the form $(t, \mathbf{s}, d, i)$, where d is the depth information, and i is an induced flag, as previously described in Section 15.5.2. For simplicity, we only illustrate the embedded case below; it is straightforward to extend it to the induced case.

15.6.1 SV-List Joins

Given two trees (x, i) and (y, j) within the same equivalence class $[P]$, we perform SV list as follows: Let $(t_x, \mathbf{s}_x)$ and $(t_y, \mathbf{s}_y)$ be any SV-list elements for nodes x and y, respectively. Let $\mathbf{s}_x = \{s_x^1, s_x^2, \ldots, s_x^m\}$ and $\mathbf{s}_y = \{s_y^1, s_y^2, \ldots, s_y^n\}$.

Descendant Test. For the descendant test we first make sure that $t_x = t_y$, that is, both nodes x and y occur in the same database tree. Next, we look at the last node scope of scope vectors $\mathbf{s}_x$ and $\mathbf{s}_y$, namely s_x^m and s_y^n. If $s_y^n \subset s_x^m$, and there does not exist another last node scope, say $s_x^{m'}$, in another element of x's SV list, such that $s_y^n \subset s_x^{m'} \subset s_x^m$ (i.e., this is a minimal occurrence of the pattern), then we add the pair $(t_x, \mathbf{s}' = \{s_x^1, \ldots, s_x^k, \ldots, s_y^n\})$ new SV list (where $\mathbf{s}'$ represents the scope vector for only those nodes on the rightmost path of the extended pattern).

Cousin Test. After checking $t_x = t_y$, we make sure that $s_x^m < s_y^n$ or $s_x^m > s_y^n$, that is, the last nodes of each element are disjoint. Note that when extending P_x^i with (y, j) we obtain a new tree with prefix P_x^i, and which has y as the label of the rightmost node, attached to node j in the prefix. The next step in the cousin test is to compare the scopes at position j in both x and y (i.e., s_x^j and s_y^j) and s_y^n. There are two cases to consider: (a) $s_y^n \subset s_x^j$ and either $s_y^n > s_x^{j+1}$ or $s_y^n < s_x^{j+1}$ [i.e., the last node of y is contained within the jth node of x (say, with label z), but it is not contained within the $(j+1)$th node's scope], or (b) either $s_y^n > s_x^j$ or $s_y^n < s_x^j$, and $s_x^j \subset s_y^j$, that is, the last node of y is before or after the jth node of x and the jth node of x is contained in the jth node of y. If (a) is true, then we add the pair $(t_x, \{s_x^1, \ldots, s_x^j, s_y^n\})$ to the SV list of the new candidate, or if (b) is true, we add $(t_x, \{s_y^1, \ldots, s_y^j, s_y^n\})$. To maintain minimality, we store the pair only for the nearest jth node to y in a database tree.

Figure 15.9 shows an example of how SV-list joins work, using the database D from Figure 15.6. The initial SV lists are the same as the item scope lists in Figure 15.6. While computing the new SV lists for the subtrees $AB\$$ and $AB\$$, we have to perform only descendant tests. The key is to keep only minimal occurrences. For example, in tree T_2, the node scopes $[0, 7]$ and $[4, 7]$ for label A both contain the scope $[5, 5]$ for label B. In this case, the SV list for $AB\$$ contains only the pair $(2, [4, 7], [5, 5])$. In a similar manner the complete SV lists for both patterns

---

[3]We say that a scope-vector $\mathbf{s}' = \{s_1', s_2', \ldots, s_n'\}$ is contained within another scope-vector $\mathbf{s} = \{s_1, s_2, \ldots, s_m\}$ if $(s_1 < s_1' \wedge s_n' \leq s_m)$ or $(s_1 \leq s_1' \wedge s_n' < s_m)$.

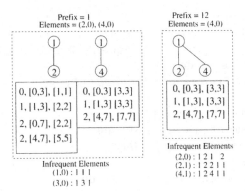

Figure 15.9. SV-list joins.

are obtained, as shown in Figure 15.9 These two lists are joined to compute the frequency of $AB\$D\$$, using the cousin test. In our example, all tree IDS belong to case (a) of the cousin test. For example, for T_2, node label D has scope $[7, 7]$, whereas node label B has occurrences at scopes $[2, 2]$ and $[5, 5]$. Here $j = 0$ and thus D's scope $[7, 7]$ is contained in B's jth node's scope $[0, 7]$, and also it is after B's $(j + 1)$th node's scope $[2, 2]$. The cousin test is true, but it is not minimal since the test is also satisfied for B's scope $[4, 7]$, and thus we add $(2, [4, 7], [7, 7])$ to the new candidate's SV list.

15.7 THE SLEUTH ALGORITHM

Figure 15.10 shows the high-level structure of SLEUTH. The main steps include the computation of the frequent labels (1 subtrees) and 2 subtrees, and the enumeration of all other frequent subtrees via recursive (depth-first) equivalence-class extensions of each class $[P]_1 \in F_2$. We will now describe each step in some more detail.

Computing F_1 and F_2. SLEUTH assumes that the initial database is in the horizontal string encoded format. To compute F_1 (line 1), for each label $i \in T$ (the string encoding of tree T), we increment i's count in a count array. This step also computes other database statistics such as the number of trees, maximum number of labels, and so on. All labels in F_1 belong to the class with empty prefix, given as $[P]_0 = [\emptyset] = \{(i, \_), i \in F_1\}$, and the position $\_$ indicates that i is not attached to any vertex. Total time for this step is $O(n)$ per tree, where $n = |T|$.

For efficient F_2 counting (line 2) we compute the supports of all candidates by using a two-dimensional integer array of size $F_1 \times F_1$, where $cnt[i][j]$ gives the count of the candidate (embedded) subtree with encoding $(i \ j \ \$)$. Total time for this step is $O(n^2)$ per tree. While computing F_2 we also create the vertical scope-list representation for each frequent item $i \in F_1$, and before each call of Enumerate-FrequentSubtrees $([P]_1 \in E)$ (line 4) we also compute the scope lists of all frequent elements (2 subtrees) in the class.

```
SLEUTH (D, minsup):
1. F₁ = { frequent 1-subtrees };
2. F₂ = { classes [P]₁ of frequent 2-subtrees };
3. for all [P]₁ ∈ F₂ do
4.   Enumerate-Frequent-Subtrees([P]₁);

ENUMERATE-FREQUENT-SUBTREES([P]):
5. for each element (x, i) ∈ [P] do
6.   if check-canonical(Pₓⁱ) then
7.     [Pₓⁱ] = ∅;
8.     for each element (y, j) ∈ [P] do
9.       if do-descendant-extension then
               ℒ_d = descendant-scope-list-join((x, i), (y, j));
10.          if do-cousin-extension then
               ℒ_c = cousin-scope-list-join((x, i), (y, j));
11.          if descendant or cousin extension is frequent then
12.             Add (y, j) and/or (y, k−1) & their scope-lists
                   to equivalence class [Pₓⁱ];
13.          Enumerate-Frequent-Subtrees([Pₓⁱ]);
```

Figure 15.10. SLEUTH algorithm.

Computing $F_k(k \geq 3)$. Figure 15.10 shows the pseudocode for the recursive (depth-first) search for frequent subtrees (Enumerate-Frequent-Subtrees). The input to the procedure is a set of elements of a class $[P]$, along with their scope lists (or SV lists). Frequent subtrees are generated by joining the scope lists (SV lists) of all pairs of elements.

Before extending the class $[P_x^i]$ we first make sure that P_x^i is the canonical representative of its automorphism group (line 6). If not, the pattern will not be extended. If yes, we try to extend P_x^i with every element $(y, j) \in [P]$. We try both descendant and cousin extensions, and perform descendant or cousin tests during scope-list join or SV-list join (lines 9–10). If any candidate is frequent, it is added to the new class $[P_x^i]$. This way, the subtrees found to be frequent at the current level form the elements of classes for the next level. This recursive process is repeated until all frequent subtrees have been enumerated. If $[P]$ has n elements, the total cost is given as $O(qn^2)$, where q is the cost of a scope-list join. The cost of scope-list join is $O(me^2)$, where m is the average number of distinct TIDs in the scope list of the two elements, and e is the average number of embeddings of the pattern per tid. The total cost of generating a new class is therefore $O(m(en)^2)$.

In terms of memory management, we need memory to store classes along a path in Depth First Search (DFS) search. In fact we need to store intermediate scope lists for two classes at a time, that is, the current class $[P]$, and a new candidate class $[P_x^i]$. Thus the memory footprint of SLEUTH is not much, unless the scope lists become too big, which can happen if the number of embeddings of a pattern is large. If the lists are too large to fit in memory, we can do joins in stages. That is, we can bring in portions of the scope lists for the two elements to be joined, perform descendant or cousin tests, and write out portions of the new scope list.

LEMMA 15.4 SLEUTH correctly generates all possible induced/embedded, ordered/unordered, frequent subtrees.

15.7.1 Equivalence Class versus. Canonical Extensions

As described above, SLEUTH uses equivalence class extensions to enumerate the frequent trees. For comparison we also implemented the pure-canonical extension in a method called SLEUTH-FκF2. The main idea is to extend a canonical and frequent (induced/embedded) subtree, with a known frequent (induced/embedded) subtree from F_2. The main difference is that Enumerate-Frequent-Subtrees takes as input a class $[P]$, all of whose elements are known to be both *frequent and canonical*. Each member (x, i) of $[P]$ is either extended with another element of $[P]$ or with elements in $[x]$, where $[x] \in F_2$ denotes all possible frequent 2 subtrees of the form $xy\$$; to guarantee correctness we have to extend $[P_x^i]$ with all $y \in [x]$. Note that elements of both $[P]$ and $[x]$ represent canonical subtrees, and if the descendant or cousin extension is canonical, we perform descendant and cousin joins, and add the new subtree to $[P_x^i]$ if is is frequent. This way, each class only contains elements that are both canonical and frequent and is induced/embedded as the case requires.

As we mentioned earlier pure canonical and equivalence-class extensions denote a trade-off between the number of redundant candidates generated and the number of potentially frequent candidates to count. Canonical extensions generate nonredundant candidates, but many of which may turn out not to be frequent (since, in essence, we join F_k with F_2 to obtain F_{k+1}). On the other hand, equivalence-class extension generates redundant candidates but considers a smaller number of (potentially frequent) extensions (since, in essence, we join F_k with F_k to obtain F_{k+1}). In Section 15.8 we compare these two methods experimentally; we found SLEUTH, which uses equivalence-class extensions to be more efficient, than SLEUTH-FκF2, which uses only canonical extensions.

15.8 EXPERIMENTAL RESULTS

All experiments were performed on a 3.2-GHz Pentium 4 processor with 1-GB main memory, and with a 200-GB, 7200-rpm disk, running RedHat Linux 9. Timings are based on total wall-clock time and include all preprocessing costs (such as creating scope lists).

Synthetic Datasets. We used the synthetic data generation program to create a database of artificial website browsing behavior [30]. The program constructs a master website browsing tree W based on parameters supplied by the user. These parameters include the maximum fanout F of a node, the maximum depth D of the tree, the total number of nodes M in the tree, and the number of node labels N. For each node in master tree W, the generator assigns probabilities of following its children nodes, including the option of backtracking to its parent, such that sum of all the probabilities is 1. Using the master tree, one can generate a subtree $T_i \preceq W$ by randomly picking a subtree of W as the root of T_i and then recursively picking children of the current node according to the probability of following that link.

We used the following default values for the parameters: the number of labels $N = 100$, the number of vertices in the master tree $M = 10,000$, the maximum depth $D = 10$, the maximum fanout $F = 10$, and total number of subtrees $T = 100,000$. We use three synthetic datasets: $D10$ dataset had all default values, $F5$ had all values set to default, except for fanout $F = 5$, and for $T1M$ we set $T = 1,000,000$, with remaining default values.

CSLOGS Dataset. Consists of Web log files collected over 1 month at the CS department [30]. The logs touched 13,361 unique Web pages within the department's website. After processing the raw logs 59,691 user browsing subtrees of the CS department website were obtained. The average string encoding length for a user subtree was 23.3.

15.8.1 Performance Evaluation

We first compare four options for SLEUTH for different types of tree mining tasks, namely SLEUTH-EU (embedded, unordered), SLEUTH-EO (embedded, ordered), SLEUTH-IU (induced, unordered), and SLEUTH-IO (induced, ordered). Note that SLEUTH-EO is essentially the same as the TreeMiner algorithm [30] (which also uses vertical scope lists to mine embedded, ordered trees). Figure 15.11 shows the length distribution of the different types of frequent trees for the highest minsup value. Figure 15.12 shows their performance on different datasets for different values of minimum support.

Let us consider the $F5$ dataset. We find that for embedded trees, there is a gap between unordered and ordered pattern mining. Ordered pattern mining (SLEUTH-EO) is slower, even though unordered pattern mining (SLEUTH-EU) needs to check for canonical forms, whereas SLEUTH-EO does not. The reason this happens is because there are more ordered rather than unordered frequent patterns for this dataset (as shown in Fig. 15.11). Looking at the length distribution, we find it to be mainly symmetric across the support values, and also, generally speaking more ordered trees are found as compared to unordered ones, especially as minimum support is lowered. Similar trends are obtained for the $D10$ and $T1M$ datasets.

Comparing the induced, unordered (SLEUTH-IU) and ordered (SLEUTH-IO) mining methods, we see once again that SLEUTH-IU is slightly faster then SLEUTH-IO; the difference is not large since there is little difference in the length distributions of these pattern types. Similar trends are obtained for the $D10$ and $T1M$ datasets.

Comparing embedded versus induced trees, we find a very big difference in the running times (embedded mining can be four to five times slower than induced mining). One look at the length distributions explains why this is the case. We see

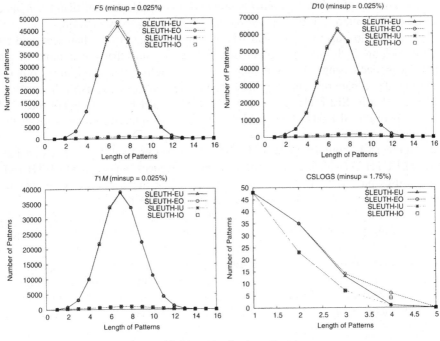

Figure 15.11. Distribution of patterns.

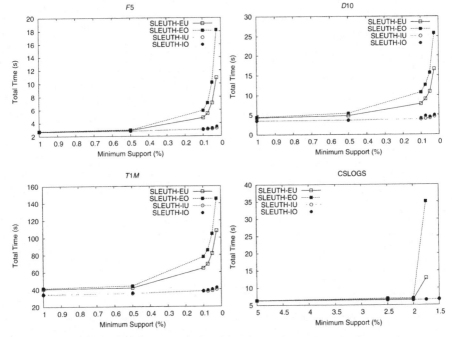

Figure 15.12. SLEUTH-EU vs. SLEUTH-EO vs. SLEUTH-IU vs. SLEUTH-IO.

that the number of induced patterns is orders of magnitude smaller than the number of embedded patterns. The shape of the distribution is also symmetric.

The Web log dataset *CSLOGS* has different characteristics than the synthetic ones. Looking at the pattern length distribution, we find that the number of patterns keep decreasing as length increases. Like before there are more ordered, than unordered patterns, and the timing trends remain the same as in the synthetic datasets. That is, SLEUTH-EO is slower than SLEUTH-EU, there is not much difference between SLEUTH-IO and SLEUTH-IU, and induced mining is faster than embedded mining.

Figure 15.13 compares SLEUTH-EO (which uses equivalence class extensions) with SLEUTH-D (which mines only distinct occurrences) and with SLEUTH-F$_K$F2 (which uses pure canonical extensions). We evaluate the case only for embedded, ordered trees since the results are similar for other pattern types.

Comparing SLEUTH-EO with SLEUTH-F$_K$F2, for all the datasets, we find that there is a big performance loss for the pure canonical extensions due to the joins of F_k with F_2, which result in many infrequent candidates; SLEUTH-F$_K$F2 can be five times slower than SLEUTH-EO. This shows clearly that the equivalence-class-based strategy of generating some redundant candidates, but extending only canonical prefix classes, is superior to generating many infrequent but purely canonical candidates. Comparing SLEUTH-EO with SLEUTH-D, we find that for the

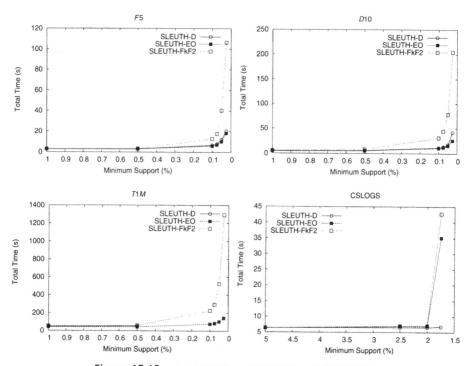

Figure 15.13. SLEUTH-EO vs. SLEUTH-D vs. SLEUTH-F$_K$F2.

synthetic datasets, SLEUTH-D is slightly slower. This happens because, for these datasets, the number of possible embeddings is small, and SLEUTH-EO requires potentially smaller memory (since an SV-list element can be twice the size of a scope-list element). On the other hand, *CSLOGS* has highly branched trees, and consequently, there are many embeddings, as support is lowered. In this case, we find that SLEUTH-D can be about seven times faster than SLEUTH-EO, confirming that counting distinct occurrences is clearly beneficial when the number of mappings increases rapidly.

Summarizing from the results over synthetic and real datasets, we can conclude that SLEUTH is an efficient, complete, unified algorithm for mining ordered/unordered, induced/embedded trees. Furthermore, it has optimizations to mine only distinct occurrences, and its equivalence-class extension scheme is more effective than a pure canonical extension process.

15.9   TREE MINING APPLICATIONS IN BIOINFORMATICS

In this section we look at two applications of tree mining within the domain of bioinformatics: RNA structure and phylogenetic tree analysis.

15.9.1   RNA Structure

RNA molecules perform a variety of important biochemical functions, including translation; RNA splicing and editing, and cellular localization. Predicting RNA structure is thus an important problem; if a significant match to an RNA molecule of known structure and function is found, then a query molecule may have a similar role. Here we are interested in finding common motifs in a database of RNA structures [10].

Whereas RNA has a three-dimensional (3D) shape, it can be viewed in terms of its secondary structure, which is composed mainly of double-stranded regions formed by folding the single-stranded RNA molecule back on itself. To produce these double-stranded regions a subsequence of bases (made up of four letters: A,C,G,U) must be complementary to another subsequence so that base pairing can occur (G-C and A-U). It is these pairings that contribute to the energetic stability of the RNA molecule. Moreover, bulges may also form, for example, when the middle portion of a complementary subsequence does not participate in the base pairing. Thus there are different RNA secondary structures that are possible, such as single-stranded RNA, double-stranded RNA helix, stem and loop or hairpin loop, bulge loop, interior loop, junction and multiloops, and so on [16]. In addition there may be tertiary interactions between RNA secondary structures, for example, pseudoknots, kissing hairpins, hairpin-bulge contacts, and so on. Figure 15.14 shows a two-dimensional (2D) representation of a (transfer) RNA secondary structure. There are 5 loops (as numbered in the center); loop 1 is a bulge loop, 3, 4, and 5 are hairpin loops, and 2 is a multijunction loop.

To mine common RNA motifs or patterns, we use a tree representation of RNA secondary structure obtained from the RNA Matrix method used in the RAG

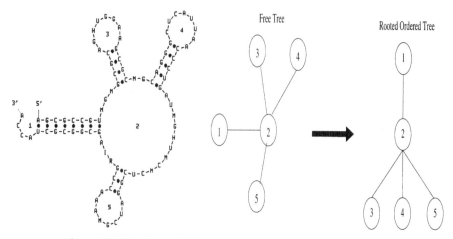

Figure 15.14.   Example RNA structure and its tree representation.

(RNA-as-graph) database [9]. In the RNA tree, a nucleotide bulge, hairpin loop, or internal loop is considered a vertex if there is more than one unmatched nucleotide or noncomplementary base pair. The 3' and 5' ends of a helical stem are considered vertices and so is a junction. An RNA stem with complementary base pairs (more than 1) is considered an edge. The resulting free tree captures the topological aspects of RNA structure. To turn the free tree into a rooted labeled tree, we label each vertex from 1 to n, numbered sequentially from the 5' to the 3' end of the RNA strand. We choose the root to be vertex 1, and children of a node are ordered by their label number. For example, Figure 15.14 shows the RNA free tree representing the RNA secondary structure and its rooted version.

We took 34 Eukarya RNA structures from the Ribonuclease P (Rnase P) database [3]. Rnase P is the ribonucleoprotein endonuclease that cleaves transfer (and other) RNA precursors. Rnase P is generally made up of two subunits, an RNA and a protein, and it is the RNA subunit that acts as the catalytic unit of the enzyme. The RNase P database is a compilation of currently available RNase P sequences and structures. For a given RNase P RNA subunit, we obtained a free tree using the RNA Matrix program [4] and then converted it into a rooted ordered tree. The resulting RNA tree dataset has 34 trees, with the smallest having 2 vertices and the largest having 12 vertices. We then ran SLEUTH on this RNA tree dataset. Figure 15.15 shows the total time taken to mine the dataset and the number of patterns found at different values of minimum support. We observe that mining at minimum support of one occurrence took less than 0.1 s, and found 5593 total patterns. An example of a common topological RNA pattern is also shown (rightmost figure); this pattern appears in at least 10 of the 34 Eukarya RNA. By applying tree mining, it is thus possible to analyze RNA structures to enumerate all the frequent topologies. Such information can be a useful step in characterizing existing RNA structures and may help in structure prediction tasks [10]. Enumerating

---

[4]http://monod.biomath.nyu.edu/rna/analysis/rna_matrix.php.

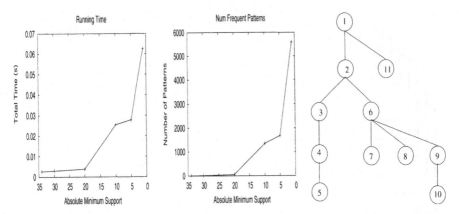

Figure 15.15.   RNase P database: (a) time, (b) number patterns, and (c) example pattern.

frequent RNA trees also helps in cataloging the kinds of RNA structures seen in nature [9].

15.9.2 Phylogenetic Trees

Given several phylogenies (i.e., evolutionary trees) from the Tree of Life [15], indicating evolutionary history of several organisms, one might be interested in discovering if there are common subtree patterns. This is an important task, since there are many algorithms for inferring phylogenies, and biologists are often interested in finding consensus subtrees (those shared by many trees) [19]. Tree mining can also be used to mine cousin pairs in phylogenetic trees [21]. A cousin pair is essentially a pair of siblings, and mining pairs that share common ancestors give important clues about the evolutionary divergence between two organisms or species.

TreeBASE is a relational database designed to manage and explore information on phylogenetic relationships.[5] It stores phylogenetic trees and data matrices used to generate them from published research papers. It includes bibliographic information on phylogenetic studies, as well as details on taxa, methods, and analyses performed; it contains all types of phylogenetic data (e.g., trees of species, trees of populations, trees of genes) representing all biotic taxa. The database is ideally suited to allow retrieval and recombination of trees and data from different studies; it thus provides a means of assessing and synthesizing phylogenetic knowledge.

Figure 15.16 shows part of the evolutionary relationship between organisms of the plylum Nematoda taken from the TreeBase site. This tree was produced using a parsimony-based phylogenetic tree construction method [16]; using different algorithms may produce several variants of the evolutionary relationships. Tree mining can help infer the parts of the phylogeny that are common among many alternate evolutionary trees.

[5]http://www.treebase.org/.

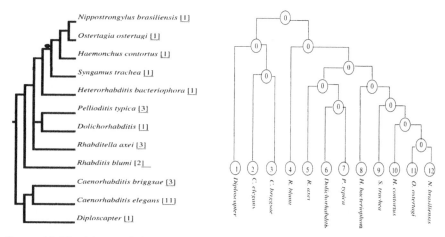

Figure 15.16. (a) Part of phylogenetic tree of phylum nematoda and (b) tree for mining.

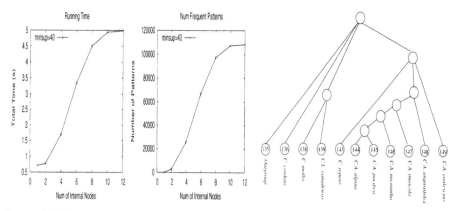

Figure 15.17. Phylogenetic data: (a) runtime and number patterns and (b) example pattern.

We took 1974 trees from the TreeBase dataset and converted them into a format suitable for mining. We give each organism a unique label (e.g., *C. elegans* has label 2), and we give each internal node the same label (e.g., 0). Given the resulting database of 1974 trees, we mine for frequent patterns. To prevent combinatorial effects, we also impose a constrain on the number of internal nodes allowed in the mined patterns; this constrain is incorporated during mining for efficiency reasons (as opposed to postprocessing). Figure 15.17 shows the running time and number of patterns found for an absolute support value of 40, as the number of internal nodes increase from 1 to 12. As we allow more internal nodes, more patterns are found. An example of a mined frequent pattern (with frequency 42) is also shown; this pattern shows the evolutionary relationship between members of the Circaea plant family. Notice how the most closely related organisms, for example, *Circaea alpina* (C.A.), group together (right branch under the root).

15.10 CONCLUSIONS

In this chapter we presented SLEUTH, a unified algorithm to mine induced/embedded, ordered/unordered subtrees and the procedure for systematic candidate subtree generation using self-contained equivalence prefix classes. All frequent patterns are enumerated by scope-list joins via the descendant and cousin tests. Our experiments show that SLEUTH is highly effective in mining various kinds of tree patterns. We studied two applications of tree mining: finding common RNA structures and mining common phylogenetic subtrees.

For future work we plan to extend our tree mining framework to incorporate user-specified constraints. Given that tree mining, though able to extract informative patterns, is an expensive task, performing general unconstrained mining can be too expensive and is also likely to produce many patterns that may not be relevant to a given user. Incorporating constraints is one way to focus the search and to allow interactivity. We also plan to develop efficient algorithms to mine maximal frequent subtrees from dense datasets that may have very large subtrees. Finally, we plan to apply our tree mining techniques to other compelling applications, such as the extraction of structure from XML documents and their use in classification, clustering, and so on.

ACKNOWLEDGMENTS

This work was supported in part by NSF Career Award IIS-0092978, DOE Career Award DE-FG02-02ER25538, and NSF grants EIA-0103708 and EMT-0432098.

REFERENCES

1. T. Asai, K. Abe, S. Kawasoe, H. Arimura, H. Satamoto, and S. Arikawa. Efficient substructure discovery from large semi-structured data. In *2nd SIAM Int'l Conference on Data Mining*, April 2002.

2. T. Asai, H. Arimura, T. Uno, and S. Nakano. Discovering frequent substructures in large unordered trees. In *6th Int'l Conf. on Discovery Science*, October 2003.

3. J. W. Brown. The ribonuclease p database. *Nucleic Acids Research*, 27(1):314–315, 1999.

4. Y. Chi, Y. Yang, and R. R. Muntz. Indexing and mining free trees. In *3rd IEEE International Conference on Data Mining*, 2003.

5. Y. Chi, Y. Yang, and R. R. Muntz. Hybridtreeminer: An efficient algorihtm for mining frequent rooted trees and free trees using canonical forms. In *16th International Conference on Scientific and Statistical Database Management*, 2004.

6. Y. Chi, Y. Yang, Y. Xia, and R. R. Muntz. Cmtreeminer: Mining both closed and maximal frequent subtrees. In *8th Pacific-Asia Conference on Knowledge Discovery and Data Mining*, 2004.

7. R. Cole, R. Hariharan, and P. Indyk. Tree pattern matching and subset matching in deterministic $o(n \log^3 n)$-time. In *10th Symposium on Discrete Algorithms*, 1999.

8. D. Cook and L. Holder. Substructure discovery using minimal description length and background knowledge. *Journal of Artificial Intelligence Research*, 1:231–255, 1994.

9. H. H. Gan, D. Fera, J. Zorn, N. Shiffeldrim, M. Tang, U. Laserson, N. Kim, and T. Schlick. RAG: RNA-As-Graphs database–concepts, analysis, and features. *Bioinformatics*, 20(8):1285–91, 2004.

10. H. H. Gan, S. Pasquali, and T. Schlick. Exploring the repertoire of rna secondary motifs using graph theory with implications for rna design. *Nucleic Acids Res.*, 31:2926–2943, 2003.

11. Jun Huan, Wei Wang, and Jan Prins. Efficient mining of frequent subgraphs in the presence of isomorphism. In *IEEE Int'l Conf. on Data Mining*, 2003.

12. A. Inokuchi, T. Washio, and H. Motoda. An apriori-based algorithm for mining frequent substructures from graph data. In *4th European Conference on Principles of Knowledge Discovery and Data Mining*, September 2000.

13. P. Kilpelainen and H. Mannila. Ordered and unordered tree inclusion. *SIAM J. of Computing*, 24(2):340–356, 1995.

14. M. Kuramochi and G. Karypis. Frequent subgraph discovery. In *1st IEEE Int'l Conf. on Data Mining*, November 2001.

15. V. Morell. Web-crawling up the tree of life. *Science*, 273(5275):568–570, aug 1996.

16. D. W. Mount. *Bioinformatics: Sequence and Genome Analysis*. Cold Spring Harbor Press, 2001.

17. S. Nijssen and J. N. Kok. A quickstart in frequent structure mining can make a difference. In *ACM SIGKDD Int'l Conf. on KDD*, 2004.

18. Siegfried Nijssen and Joost N. Kok. Efficient discovery of frequent unordered trees. In *1st Int'l Workshop on Mining Graphs, Trees and Sequences*, 2003.

19. R. D. Page and E. C. Holmes. *Molecular Evolution: A Phylogenetic Approach*. Blackwell Science, 1998.

20. R. Shamir and D. Tsur. Faster subtree isomorphism. *Journal of Algorithms*, 33:267–280, 1999.

21. D. Shasha, J. Wang, and S. Zhang. Unordered tree mining with applications to phylogeny. In *International Conference on Data Engineering*, 2004.

22. A. Termier, M-C. Rousset, and M. Sebag. Treefinder: a first step towards xml data mining. In *IEEE Int'l Conf. on Data Mining*, 2002.

23. A. Termier, M-C. Rousset, and M. Sebag. Dryade: a new approach for discovering closed frequent trees in heterogeneous tree databases. In *IEEE Int'l Conf. on Data Mining*, 2004.

24. C. Wang, M. Hong, J. Pei, H. Zhou, W. Wang, and B. Shi. Efficient pattern-growth methods for frequent tree pattern mining. In *Pacific-Asia Conference on KDD*, 2004.

25. K. Wang and H. Liu. Discovering typical structures of documents: A road map approach. In *ACM SIGIR Conference on Information Retrieval*, 1998.

26. Y. Xiao, J.-F. Yao, Z. Li, and M. H. Dunham. Efficient data mining for maximal frequent subtrees. In *International Conference on Data Mining*, 2003.

27. X. Yan and J. Han. gspan: Graph-based substructure pattern mining. In *IEEE Int'l Conf. on Data Mining*, 2002.

28. X. Yan and J. Han. Closegraph: Mining closed frequent graph patterns. In *ACM SIGKDD Int. Conf. on Knowledge Discovery and Data Mining*, August 2003.

29. K. Yoshida and H. Motoda. CLIP: Concept learning from inference patterns. *Artificial Intelligence*, 75(1):63–92, 1995.

30. M. J. Zaki. Efficiently mining frequent trees in a forest. In *8th ACM SIGKDD Int'l Conf. Knowledge Discovery and Data Mining*, July 2002.

31. M. J. Zaki. Efficiently mining frequent embedded unordered trees. *Fundamenta Informaticae*, 66(1-2):33–52, 2005.

32. M. J. Zaki and C. C. Aggarwal. Xrules: An effective structural classifier for xml data. In *9th ACM SIGKDD Int'l Conf. Knowledge Discovery and Data Mining*, August 2003.

16

DENSE SUBGRAPH EXTRACTION

DAVID GIBSON,* RAVI KUMAR,†
KEVIN S. MCCURLEY,‡ AND ANDREW TOMKINS†

*IBM Almaden Research Center, 650 Harry Road, San Jose, CA 95120.
Email: davgib@us.ibm.com
†Yahoo! Research, 701 First Avenue, Sunnyvale, CA 94089. Part of this work
was done while the author was at the IBM Almaden Research Center.
Email: {ravikumar,atomkins}@yahoo-inc.com
‡Google, Inc. 1600 Amphitheatre Parkway, Mountain View, CA 94043.
Part of this work was done while the author was at the
IBM Almaden Research Center, mccurley@digicrime.com

Oh, what a tangled web we weave, when first we practice to weave!

—Mignon McLaughlin

16.1 INTRODUCTION

A *clique* in a graph is a collection of vertices such that each pair of vertices in the collection has an edge between them. A clique is the extreme form of a dense subgraph: Every possible edge within the subgraph exists. In this chapter, we will consider weakening this notion by searching for collections of vertices such that many, but not necessarily all, pairs within the collection have an edge between them. We will call such a collection of vertices and the edges between them a *dense subgraph*. The particular definition we use will depend on the situation; we consider several variants.

Dense subgraphs arise frequently in real-world analysis of graphs and represent clusters within the graph. In social networks, they represent collections of interlinked

Mining Graph Data, Edited by Diane J. Cook and Lawrence B. Holder
Copyright © 2007 John Wiley & Sons, Inc.

individuals such as groups of friends or co-workers. In citation graphs, they represent collections of authors who co-cite one another's work. In hyperlink graphs such as the World Wide Web, they represent communities of Web pages or websites that link together in a dense manner. In other graphs they may have more specific meanings; for example, in a graph whose vertices are products and whose edges indicate that some individual bought both products, the dense subgraphs are taken to be groups of products that are commonly purchased together, often because they fall into the same category—dense subgraphs within books might represent genres like "science fiction" or "romance."

Algorithmically, the discovery of dense subgraphs in large graphs is notoriously difficult. We begin with a discussion of efficiency in the context of large disk-resident graphs.

16.1.1   Efficient Algorithms for Massive Graphs

Consider a small crawl of the Web containing 1 billion vertices and 10 billion edges, and an algorithm that simply tries pairs of vertices to see whether they have a significant number of neighbors in common. Such an algorithm will explore $10^{18}/2$ pairs, each of which will require two random accesses to extract the neighbor set. On a modern computer, such an operation will take approximately 95 million years to complete. Thus, the traditional view that algorithms are efficient if they operate in polynomial time is not appropriate in the world of massive graphs—the above example shows that quadratic-time operations are problematic. Indeed, even linear-time operations that must access the vertices in an unknown order must perform a single disk seek per vertex; this operation will require about 1 month on a modern computer with a single disk.

On the other hand, the graph described above can be streamed from disk in a relatively snappy 15 min or so, due to the disparity between sequential and random input/output (I/O) speeds. Similarly, a merge sort may be implemented using a small number of sequential scans, so sorting is also an operation we may consider. Finally, the concern around performing a random access for each vertex is quite valid, but in special cases the overhead can be reduced: If we are able to request many such random accesses at a time, say 100 accesses simultaneously per disk as a rule of thumb, then we may apply multiple threads or asynchronous I/O to send all these requests to the disk at once. The disk contains optimizations to reorder these requests for more efficient access, returning us again to the realm of sequential access speeds. Thus, the operations that we may realistically allow ourselves for massive disk-resident graphs are the following:

- Sequential scans through the vertices
- Sort operations, since these can be implemented in terms of sequential scans
- Linear numbers of random access operations, but only if we are able to request many such operations at once, allowing the disk to reorder the requests for more efficient processing

16.1.2 Preliminaries

A graph $G = (V, E)$ consists of a set V of vertices, and a set $E \subseteq V \times V$ of edges. If $(u, v) \in E$ whenever $(v, u) \in E$, we called the graph *undirected*; otherwise, we call it *directed*. For directed graphs, the *out-neighbors* of a vertex v are $out(v) = \{u \in V \mid (u, v) \in E\}$. The *in-neighbors* $in(v)$ are defined analogously. The *in-degree* and *out-degree* of a vertex are the number of in- and out-neighbors, respectively: $d_{in}(v) = |in(v)|$, and $d_{out}(v) = |out(v)|$. A path of length k from u to v is a sequence of $k + 1$ vertices $u = w_1, w_2, \ldots, w_{k+1} = v$ such that $(w_i, w_{i+1}) \in E$. Vertex u is said to be *connected* to vertex v if there is a path of any length between them; for directed graphs, this relation need not be symmetric, so a path from u to v does not imply a path from v to u. A *connected component* $C \subseteq V$ is a set of vertices such that u is connected to v for every $(u, v) \in C \times C$.

We identify a few common graphs by name. A K_j is a clique of size j; that is, a graph with j vertices and an edge from every vertex of the clique to every other vertex. A $K_{i,j}$ is a bipartite clique consisting of $i + j$ vertices such that each of the first i vertices contains a link to every one of the j remaining vertices, and no other edges exist.

To simplify our presentation, whenever we specify that a graph is undirected, we will use $nbr(v)$ to refer to the neighbors of vertex v, and $deg(v)$ to refer to the degree: $deg(v) = |nbr(v)|$.

16.1.3 Overview of Results

We begin with a detailed discussion of a commonly used measure in the theoretical algorithms literature and cover exact and provably approximate algorithms to find dense subgraphs under this measure. From there, we discuss three algorithms for dense subgraph extraction. The first algorithm is *trawling*, a procedure for automatically identifying fine-grained communities within a graph, using a measure based on local footprints [36, 37]. The procedure is motivated by the World Wide Web graph, in which *hub* Web pages exist, providing links to interesting content of a particular flavor. Observe that two hub pages linking to the same j pages of content will represent a $K_{2,j}$ in the graph. Based on this observation, the trawling algorithm identifies small complete bipartite graphs as the "signature" of communities, and then expands those signatures to find the full community.

From this automated approach to determining small-scale communities, we move to *graph shingling*, an approach to finding larger-scale dense subgraphs [25]. This approach is based on the observation that the vertices of a large dense subgraph tend to have significant overlap in their out-neighbors. By finding small patterns of common out-neighbors, it is possible to characterize each vertex of a large graph in terms of the out-neighbor patterns it contains. This representation induces a new type of graph in which dense subgraphs have become denser, while sparse subgraphs have become sparser. Repeated application of this procedure results in the discovery of dense subgraphs.

These two approaches are batch algorithms in the sense that they process the entire graph in order to produce a comprehensive list of dense subgraphs. The remaining topic we discuss covers a technique that may be applied in real time to produce a dense subgraph with certain properties in response to a user query. The algorithm is designed in the context of social network graphs in which vertices are people and edges are relationships denoting, for example, that one person is the friend of another. Given a query consisting of two vertices, the algorithm must produce a *connection subgraph*, or a dense subgraph that best captures the connections between the two vertices (see [19, 20] and Chapter 5 of this book).

16.2   RELATED WORK

16.2.1   Data Mining Applied to Graphs

Traditional data mining research, for instance [1], focuses largely on algorithms for inferring association rules and other statistical correlation measures in a given dataset. These notions may be applied to graphical data, but the algorithms are not by default tuned for Web-like graphs, and hence are not practical in those domains, even with efficient methods such as Apriori [1] or Query Flocks [50]. Unlike market baskets, where there are at most about a million distinct items, there are between two and three orders of magnitude more "items," that is, Web pages, in our case.

The relationship we will exploit, namely co-citation, is effectively the join of the Web "points to" relation and its transposed version, the Web "pointed to by" relation. This approach suggests that a database-style query language might be able to express certain types of subgraph extraction, and in fact a view of the Web as a semistructured database has been advanced by many authors. In particular, LORE [42] and WebSQL [43] use graph-theoretic and relational views of the Web. These views support a structured query interface to the Web (Lorel and WebSQL, respectively) which is evocative of and similar to SQL. An advantage of this approach is that many interesting queries, including methods such as HITS [32], can be expressed as simple expressions in the very powerful SQL syntax. The associated disadvantage is that the generality comes with a computational cost that is prohibitive in our context. Indeed, LORE and WebSQL are but two examples of projects in this space. Some other examples are W3QL [33], WebQuery [11], Weblog [38], and ParaSite [48]. For a survey of database techniques on the Web and the relationships between them, see [22].

Algorithmic and memory bottleneck issues in graph computations are addressed in [3, 28]. The focus in this work is in developing a notion of data streams, a model of computation under which data in secondary memory can be streamed through main memory in a relatively static order. They show a relationship between these problems and the theoretical study of communication complexity. They use this relationship mainly to derive impossibility results.

16.2.2   Web-Based Community Finding

Flake et al. [21] define a density-based notion of communities. Given an undirected graph G, a community is a subset C of vertices such that for all vertices $v \in C$, v has at least many edges connecting to vertices in C as it does to vertices in $V \setminus C$.

They show that a community can be identified by calculating the $s - t$ minimum cut of G and identifying vertices that are reachable from s to be the community. While the ideal method requires random access to in-links and out-links for the underlying graph, they approximate this by directing a focused crawler along link paths that are highly relevant to the community. They also augment the crawler by using a version of the Expectation-Maximization (EM) algorithm to reseed the crawler with relevant sites.

Kumar et al. [34] take a more combinatorial approach to finding communities. Their motivation was to discover storylines from search results, where storylines are windows that offer glimpses into interesting themes that are latent among the top search results for a query. To do this, they work with a term–document relation. The goal is then to find as many pairs $(D_1, T_1), \ldots, (D_k, T_k)$ as possible, where the D_i's and T_i's are fairly large subsets, respectively, of documents and terms, such that most terms in T_i occur in most documents in D_i, and very few terms in T_i occur in the documents in D_j, for $j \neq i$. Each (D, T) pair in the collection will represent a storyline. This combinatorial formulation is a generalization of the maximum induced bipartite matching problem [49] and is hence NP-hard. Nevertheless, they show that simple algorithms based on local search and dynamic programming work well in practice.

Gibson et al. [24] describe experiments on the Web in which they use spectral methods to extract information about "communities" on the Web. They use the nonprincipal eigenvectors of matrices arising in Kleinberg's HITS algorithm to define their communities. They give evidence that the nonprincipal eigenvectors of the co-citation matrix reveal interesting information about the fine structure of a Web community. While eigenvectors seem to provide useful information in the contexts of search and clustering in purely text corpora (see also Deerwester et al. [16]), they can be relatively computationally expensive on the scale of the Web. In addition they need not be complete, that is, interesting structures could be left undiscovered. The second issue is not necessarily a show-stopper, as long as not too many communities are missed. The complementary issue, "false positives", can be problematic. In contrast, the techniques described in Section 16.4 almost never find "coincidental" false positives.

Kumar et al. [35] give an algorithm for dense subgraph extraction in the context of Web logs, with an eye to watching how these communities of blogs evolve over time. Their algorithms are similar to (and in fact motivated by) the trawling algorithms described below, but do not make use of the bipartite structure characteristic of Web pages, as observed by Kleinberg [32], in which good resources pages, or "hubs," tend to link to good authoritative pages.

Communities in graphs may be viewed as clusters, which gives rise to a broad set of techniques from graph clustering, partitioning, and matrix reordering [7, 17, 30, 51, 52]. We do not describe this work in detail but merely observe that it remains a rich source of ideas and approaches for bringing together densely interlinked vertices.

16.2.3 Searching the Web Graph

The information foraging paradigm was proposed in [47]. In their study, the authors argue that Web pages fall into a number of types characterized by their role in helping an information forager find and satisfy his or her information need. The thesis is that recognizing and annotating pages with their type provides a significant "value add" to the browsing or foraging experience. The categories they propose are much finer than the hub and authority view taken in [32] and in the Clever project [13]. Their algorithms, however, appear unlikely to scale to sizes that are interesting to us.

A number of search engines and retrieval projects have used links, and in particular densely linked regions, to provide additional information regarding the quality and reliability of search results. See, for instance [5, 8, 13, 32, 45]. Link analysis has become popular as a search tool, and a number of algorithmic extensions have it possible to apply the approach in an increasingly efficient and personalized manner [27, 29]. However, our focus in this article is in using these dense regions for a different purpose: to understand and mine community structure on the Web.

16.2.4 Communities and Connections in Social Networks

The approaches to community finding that we discuss in Section 16.6 make use of a notion of distance between two vertices in the graph in order to estimate the relationship between two people. Many such measures have been proposed; see, for example, the work of Kleinberg and Liben-Nowell [40] and Palmer and Faloutsos [46].

Similarly, there is a deep theory connecting walks on undirected graphs to electrical circuits, and this work suggests that community relationship between vertices may be estimated by computing the expected numbers of steps to travel from one vertex to the other and back following a random walk; if this number is small, the vertices are taken to belong to the same community. See [14, 18, 46] for a review of this literature.

Finally, simpler combinatorial notions of connectedness, such as betweenness indices [6], may also be employed to study the role of a particular vertex in a community of vertices.

16.3 FINDING THE DENSEST SUBGRAPH

In this section we address the problem of finding the densest subgraph in a given graph. We first define the notion of density in an undirected graph and then describe two algorithms to find the densest subgraph. The first is an exact algorithm but is not very efficient to implement in practice. The second algorithm is highly efficient, and while it might not output the densest subgraph, it is still guaranteed to output a subgraph whose density is not far off from the best. Finally, we briefly describe the notion of density in directed graphs.

Let $G = (V, E)$ be an undirected graph and for $S, T \subseteq V$, let $e(S, T) = \{(u, v) \in E \mid u \in S, v \in T\}$, that is, the set of edges with one endpoint in S and another in T. For $S \subseteq V$, let $\bar{S} = V \setminus S$, and for a vertex $v \in V$, let $\deg_S(v)$ denote the degree of v in the subgraph of G induced by $S \cup \{v\}$. For a subset $S \subseteq V$, the *edge density* $\rho(S)$

is given by $\rho(S) = |e(S, S)|/|S|$, that is, the ratio of the number of edges inside S and the size of S. Clearly, the edge density of K_k, the k clique, is $k - 1/2$. The *densest subgraph problem* is to find a subset S of vertices in a graph that maximizes $\rho(S)$, and the *densest k subgraph problem* is the densest subgraph problem with the additional requirement that $|S| = k$.

The densest k-subgraph problem is NP-hard since it generalizes the maximum clique problem. Recently, Khot [31] showed that, under reasonable complexity-theoretic assumptions, the densest-k subgraph problem does not even admit a poly-nomial-time approximation scheme. On the other hand, the densest subgraph problem is solvable exactly in polynomial time. We illustrate this algorithm below (cf. [23, 39]).

The algorithm proceeds by checking if there exists a nonempty subset S of vertices such that $\rho(S) > c$; the densest subgraph can then be found by a binary search on c. The former can be checked in the following manner. First, note that $2|e(S, S)| = \left[\sum_{u \in S} \deg(u)\right] - |e(S, \overline{S})|$. Then, the condition $\rho(S) > c$ is equivalent to $\left[\sum_{u \in S} \deg(u)\right] - |e(S, \overline{S})| - 2c|S| > 0$. Adding and subtracting $\sum_{u \in \overline{S}} \deg(u)$ to this expression and noting that $\sum_{u \in V} \deg(u) = 2|E|$, it follows that $\rho(S) > c$ if and only if

$$\sum_{u \in \overline{S}} \deg(u) + 2c|S| + |e(S, \overline{S})| < 2|E| \qquad (16.1)$$

The existence of a nonempty S satisfying the left hand side of (16.1) can be ascertained by a weighted minimum cut computation in the following way. Con-struct a graph $G' = (V', E')$, where $V' = V \cup \{s, t\}$ and E' consists of all the edges in E each with a cost of 1, an edge from s to each vertex in V with a cost of $\deg(u)$, and an edge from each vertex $u \in V$ to t with a cost of $2c(u)$. The (s, t)-minimum cut in G', which can be computed in polynomial time, is of the form $(\{s\} \cup S, \overline{S} \cup \{t\})$ with cost equal to the left-hand side of (16.1). Thus, $\rho(S) > c$ if and only if $S \neq \emptyset$ satisfies (16.1).

The above algorithm, though it runs in polynomial time, is expensive because of the minimum cut subroutine. Charikar [15] obtained a very efficient greedy algorithm for the densest subgraph problem that computes a solution $\tilde{S}$ such that $\rho(\tilde{S}) \geq \left(\frac{1}{2}\right) \rho(S^*)$, where S^* is the optimal solution—in other words, a 2-approximation. His iterative algorithm maintains a subset S_i of vertices at the ith step. Initially, $S_1 = V$. At the i step of the iteration, a vertex with the least degree in the subgraph induced by S_i is removed from S_i to yield S_{i+1}. The algorithm terminates when $S_i = \emptyset$. The final output of the algorithm is $\tilde{S} = \arg \max_i \rho(S_i)$, that is, S_i that maximizes $\rho(S_i)$ for $i = 1, 2, \ldots$.

Using simple data structures, the above algorithm can be implemented in linear time, that is, $O(|V| + |E|)$. To do this, maintain lists of vertices with the same degree; this can be constructed in linear time. In each iteration, the minimum degree vertex is removed and the degrees of its neighbors are reduced by 1; the total time to do this is $O(|E|)$. Since the minimum degree drops by at most 1, if the minimum degree in the current iteration is d, the minimum degree for the next iteration is either $d - 1$, d, or $d + 1$. The time taken to do this is $O(|V|)$.

To analyze the performance of this algorithm, first note that the degree of each vertex in S^* has to be large, otherwise we could always remove this vertex and obtain a solution that is better than S^*. Formalizing this, for any vertex $v \in S^*$,

$$\frac{|e(S^*)|}{|S|} = \rho(S^*) \geq \rho(S^* \setminus \{v\}) = \frac{|e(S^* \setminus \{v\})| - \deg_{S^*}(v)}{|S^*| - 1}$$

which yields $\deg_{S^*}(v) \geq |e(S^*)|/|S^*|$. Let i be the first step in the iteration when some vertex v in the optimal solution S^* is deleted; there is always such an i since the algorithm iterates until $S_i = \emptyset$. By the choice of i, $S_i \supset S^*$ and so for every $u \in S_i \cap S^*$, $\deg_{S_i}(u) \geq \deg_{S^*}(u)$. Since the algorithm chose v with the least degree and $v \in S_i \cap S^*$, for all vertices $u \in S_i$, we have $\deg_{S_i}(u) \geq \deg_{S_i}(v) \geq |e(S^*)|/|S^*|$. Now,

$$\begin{aligned}
\rho(S_i) = \frac{|e(S_i)|}{|S_i|} &= \frac{\sum_{u \in S^*} \deg_{S_i}(u) + \sum_{u \in S_i \setminus S^*} \deg_{S_i}(u)}{2|S_i|} \\
&\geq \frac{\sum_{u \in S^*} \deg_{S^*}(u) + \sum_{u \in S_i \setminus S^*} \deg_{S_i}(u)}{2|S_i|} \\
&\geq \frac{2e(S^*) + |S_i \setminus S^*| \cdot |e(S^*)|/|S^*|}{2|S_i|} \\
&\geq \frac{|e(S^*)|}{|S^*|} \cdot \frac{2|S^*| + |S_i \setminus S^*|}{2|S_i|} \\
&\geq \frac{|e(S^*)|}{2|S^*|} = \frac{\rho(S^*)}{2}
\end{aligned}$$

Charikar also considers the densest subgraph problem in directed graphs. The definition of densest subgraph is modified as follows. For $S, T \subseteq V$, let $\rho(S, T) = |e(S, S)|/\sqrt{|S||T|}$ and the densest directed subgraph problem is to choose S, T to maximize $\rho(S, T)$. He obtains an exact algorithm for this problem; this algorithm is based on Linear programming (LP). He also obtains a greedy algorithm (similar in spirit as above) that obtains a 2-approximation to the optimum solution.

16.4 TRAWLING

In this section we describe heuristic approaches to automatically enumerate Web communities. Our main focus is the trawling algorithm proposed in [37] to do such an enumeration.

The graph under consideration is the Web graph where Web pages correspond to vertices and directed edges correspond to hyperlinks from one page to another. The main observation is the following: websites that should be part of the same community frequently do not reference one another. This could be because of competition, or because the sites do not share a common point of view, or simply because the sites are not aware of each other. In these situations, the footprint of a Web community can be recognized by co-citation, where pages that are related are frequently referenced together. Kumar et al. [37] postulate that co-citation is a characteristic of well-developed and explicitly known communities. Furthermore, they argue that co-citation is an early indicator of emerging communities.

Mathematically posed, this is a characterization of Web communities by dense directed bipartite subgraphs. Recall that a bipartite graph $G = (U, V, E)$ is a graph whose vertex set can be partitioned into two sets denoted U and V and every directed edge in E is directed from a vertex $u \in U$ to a vertex $v \in V$. A bipartite graph is *dense* if many of the possible edges between U and V are present; a bipartite graph is *complete* if $E = U \times V$. As mentioned earlier, we refer to a complete bipartite graph with $|U| = i$ and $|V| = j$ as a K_{ij}. An (i, j) core is a K_{ij} that optionally contains additional edges beyond those from U to V. Formally, a community contains an (i, j) core if it contains an edge-induced subgraph with $i + j$ vertices that is isomorphic to a K_{ij}.

The trawling algorithm is built on the following pair of assumptions:

1. Any sufficiently strong online community will almost surely contain an (i, j) core.
2. Almost all (i, j) cores appear due to the presence of a community rather than as a spurious event.

The first assumption states that finding all the cores is sufficient to find almost all the communities. The second assumption states that finding all the cores is necessary to find all the communities. The main idea behind trawling is to find a community by finding its core, and then to expand the core to find the rest of the community. This latter step can be done, for instance, by using an algorithm derived from the HITS/Clever algorithm [13, 32].

The following simple fact about random bipartite graphs lends support to the first assumption above. Let $G = (U, V, E)$ be a random bipartite graph where E is a set of uniformly chosen random edges. Then, there exist i, j that are functions of $|U|, |V|, |E|$ such that with high probability, G contains an (i, j) core. For instance, it is easy to show that if $|U| = |V| = 10$ and $|E| = 50$, then with probability more than 0.99, i and j will be at least 5. That is, for a community of 10 "hub-like" pages each of which links at random to 5 out of 10 candidate "authority-like" pages, there will almost certainly be a $(5, 5)$ core. While the Web cannot be modeled as a random graph, the above fact informally suggests that every community almost surely contains a large core. Thus, even though one cannot argue about the distribution of links in a Web community, the following hypothesis is plausible: A random large enough and dense enough bipartite directed subgraph of the Web almost surely has a core.

Conversely, we may ask whether cores form spuriously as people who are not members of communities simply happen to link to the same content. A quick thought experiment shows that this is unlikely. Consider a graph without communities in which each vertex links to 10 other vertices chosen uniformly at random from the entire vertex set. We ask whether such a random graph is likely to contain an (i, j) core, and for concreteness, we consider a $(2, 3)$ core; clearly, any larger core will be increasingly unlikely. It is easy to show that as the size of the graph grows to infinity, the expected number of $(2, 3)$ cores drops to zero. This suggests that the presence of a core is not due to random chance—there must be some reason that the vertices of U chose to perform the unlikely event of individually linking to the vertices of V.

Together, these observations motivate the trawling approach and give some intuition behind the two assumptions above.

Note that the cores are directed—there is a set of i pages all of which hyperlink to a set of j pages, while no assumption is made of links out of the latter set. Intuitively, the former are pages created by members of the community, focusing on what they believe are the most valuable pages for that community (of which the core contains j). For this reason the i pages that contain the links are called *fans* and the j pages that are referenced are called *centers*.

For the rest of this section, we will focus on how to efficiently find the cores from the Web graph using the *trawling* algorithm. The main idea is to do pruning on the Web graph in a way that eliminates as many vertices as possible, while simultaneously ensuring that pruned vertices or edges cannot be part of any yet-unidentified core. Each step of the algorithm is implementable in the data stream model augmented with sorting.

There are three main steps in the trawling algorithm.

- Iterative pruning: The idea here is that when looking for (i, j) cores, potential fans with out-degree less than j and potential centers with in-degree less than i can be deleted along with their edges from the graph. This process can be iteratively applied since deleting a vertex will affect the in- and out-degrees of its neighbors. Iterative pruning thus cuts down the size of the graph without outputting any cores.

  This pruning can be implemented in the data stream model with sorting, as follows. Suppose the graph is represented as a set of directed edges. First, the edges are sorted according to source and fans with low degree are eliminated. Next, the edges are sorted according to destination and centers with low degree are eliminated. This process can be iteratively applied. In fact, the number of sort passes can be reduced by holding a small in-memory index of the pruned vertices.

- Inclusion–exclusion pruning: The idea is that at each step, we either eliminate a page from contention or we output a new (i, j) core. The fans that are chosen in this step are the ones with out-degrees equal to i. Let u be a fan with out-degree exactly j and let $\Gamma(u)$ be the centers to which it points. Then, u is in an (i, j) core if and only if there are $i - 1$ other fans all pointing to each center in $\Gamma(u)$. If i and j are small, this condition can be checked quite easily. An analogous condition can be applied to choose potential centers in this step. Thus, inclusion–exclusion pruning also reduces the size of the graph while outputting cores.

Again, by careful engineering, these steps can be implemented in the data stream model with sorting.

- Exhaustive enumeration: After running a series of the above pruning steps until no further progress can be made, the size of the graph has shrunk dramatically and remaining cores can be enumerated using naive exhaustive techniques. Fix j and consider all $(1, j)$ cores; this is the set of vertices with out-degree at least j. Now, construct $(2, j)$ cores by checking every fan that also links to any center in a $(1, j)$ core. Proceeding in a similar manner, all the $(3, j)$ cores can be enumerated and so on.

Their experiments suggest that trawling a copy of the Web might result in the discovery of many more communities that will become explicitly recognized in the future. For more details, see [37].

16.5   GRAPH SHINGLING

The trawling and dense subgraph algorithms discussed above are intended for graphs of arbitrary size, but they typically discover relatively small structures. When the goal is to discover subgraphs that themselves are large, we need different techniques. The algorithm described here relies on the central idea of *shingling* [10] and achieves its scalability by following the data stream paradigm.

As a motivating example for this form of subgraph extraction, consider the problem of identifying collusion on the Web. Many search engines rely on link-based relevance calculation schemes, such as PageRank [45] and HITS [32]. These schemes begin with the notion that pages should be ranked highly if many pages link to them, and refine this notion by attempting to determine the quality of the originating pages. Many individuals and institutions, then, attempt to improve their search engine ranking by manipulating the link graph to their advantage. Typically this form of "spamming" involves creating a large number of links to their target sites. Search engines usually disregard nepotistic links, which are links within the pages on a single site, and so link spamming will be visible in the Web graph as a large set of originating sites, which behave as if they were colluding: Each links to many or all of a large set of target sites. Of course, this collusion may not actually be collusion between individuals, but rather that all the originating sites are controlled by the same entity. The goal is to identify these large colluding structures, so as to reduce or possibly eliminate their contributions to search engine ranking.

These large bipartite structures may also arise if an institution, wittingly or not, creates a large number of sites with very similar content. These deliberate or accidental mirrors then need to be treated as copies, so that they do not artificially inflate the ranking of the sites to which they link. On a much smaller scale than either spamming or mirroring, we also find genuine community structures, where a large community of individuals share many links.

The input to the algorithm, then, is a large directed graph $G = (V, E)$. The output is a set of dense subgraphs $C = \{S \subseteq V \mid S \text{ is dense}\}$. The density of the subgraphs found depends upon the parameters to the algorithm and the nature of the input graph.

16.5.1 Shingling

Shingling was introduced in [10] for identifying near duplicates in a corpus of text. They have since seen wide usage to estimate the similarity of Web pages, and they were also used to detect mirror sites (see Chakrabarti [12] and Bharat et al. [4]). We will employ the shingling technique to solve the following problem: Given subsets of a universe U of elements, generate a constant-size fingerprint such that two subsets A and B may be compared by simply comparing their fingerprints.

The name shingling refers to the way that housing shingles overlap, as illustrated by the following application to detecting near-duplicate text documents. Each document is first converted to a set of elements corresponding to overlapping sequences of words in the document, as follows:

```
Element 1:  'overlapping subsequences of'
Element 2:          'subsequences of words'
Element 3:                    'of words in'
Element 4:                      'words in the'
Element 5:                          'in the document'
```

Using this representation, some of the information about word sequence is retained, even though the set itself is simply an unordered bag of sequences. Two documents are declared near duplicate if they have many elements in common.

A simple and natural measure of the similarity of two sets is the *Jaccard coefficient*, defined as the size of the intersection of the sets divided by the size of their union: $J(A, B) = |A \cap B|/|A \cup B|$. Formally, if π is a random permutation of the elements in the ordered universe U from which A and B are drawn, then it can be shown that

$$\Pr[\pi^{-1}(\min_{a \in A}\{\pi(a)\}) = \pi^{-1}(\min_{b \in B}\{\pi(b)\})] = \frac{|A \cap B|}{|A \cup B|} = J(A, B)$$

That is, the probability that the smallest element of A and B is the same, where smallest is defined by the permutation π, is exactly the similarity of the two sets according to the Jaccard coefficient $J(A, B)$. Using this observation, we compute the fingerprint of A by fixing a constant number c of permutations $\pi_1, \ldots, \pi_c$ of U, and producing a vector whose ith element is $\min_{a \in A} \pi_i(a)$. The similarity of two sets is then estimated to be the number of positions of their respective fingerprint vectors that agree.

This formulation is not yet sufficient for our needs and we need a generalization. The formulation as given may be viewed as follows: Consider every one-element set contained entirely in A or B and measure agreement by the fraction of these one-element subsets that appear in both sets. Generalizing, by analogy to subsequences of words, we consider every s-element set contained entirely within either set and measure similarity by the fraction of these s-element subsets that appear in both. This is identical to measuring the similarity of A and B by computing the Jaccard coefficient of two sets A_s and B_s, where $A_s = \{\{a_1, a_2, a_3\} \mid a_i \in A\}$ and B_s is defined likewise. The same fingerprinting scheme applies unchanged to A_s and B_s. We will refer to each of the s-element subsets as a *shingle* and to the algorithm that produces the set as an (s, c) shingling algorithm. Thus, we imagine creating a fingerprint of a set by applying an (s, c) shingling in which each individual shingle is computed as a hash of s elements of the set, and c such shingles are kept as the fingerprint. These two parameters give a way to adjust the shingle computation so that overlapping sets generate a reasonable number of identical shingles. When s is small, more shingles will be identical. By increasing c, the opportunities for shingles to match is increased. In particular, if $J(A, B) = p$, then the function

$$h(s, c, p) = 1 - (1 - p^s)^c$$

captures the probability that at least one shingle is shared when using (s, c) shingling. Figure 16.1 illustrates the behavior of this function. We can see, for example, that if two sets are 20% similar, then $(3, 90)$ shingling will result in a shared shingle with probability about 0.5.

Fortunately, it is not necessary to consider all possible permutations π in the choice of c such permutations; a more succinct family of *min-wise independent permutations* will accomplish the task. For more details, see [9, 10]. In practice, two universal hash functions have been shown to work well.

In the first step of the algorithm, each set will represent the out-links of a particular vertex, and two vertices will be considered similar if they share many

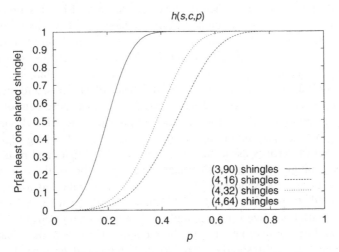

Figure 16.1.   Effect of s, c on the correctness of shingles.

Algorithm Shingle (X, s, c)

```
Let H be a hash function from strings to integers
Let p be a sufficiently large random prime (e.g., near 2^32)
Let a_1, b_1, ..., a_c, b_c be random integers in [1...p]

For j = 1...c do
  Let S_j = {(a_j · H(x) + b_j) mod p | x ∈ X}

  Let z_j = H(min(s, S_j))    Hash the concatenation of s minimal elements in S_j

Return z_1, ..., z_c
```

Figure 16.2.   An efficient shingler.

out-links. In later stages of the algorithm, exactly the same formulation is employed recursively to determine the similarity between two shingles in terms of the set of vertices that reference each shingle.

A time and memory efficient (s, c) shingling algorithm for a set X is given in Figure 16.2. To avoid the computation of a hash function for every s-element subset, we apply a weakly distributed integer residue function to each element, which suffices for choosing minimal elements uniformly, and then apply a well-distributed hash function to the resulting set of minimal elements.

16.5.2   High-Level Description

The algorithm seeks clusters of vertices that tend to link to the same destinations and proceeds as follows. First, for each vertex in the graph, it applies an (s, c) shingling algorithm to the set of destinations linked to from that vertex, resulting in c shingles per vertex. For each distinct shingle created by this process, it produces a list of all vertices referencing that shingle. We have now succeeded in bringing together the vertices that share a particular shingle (and therefore a particular set of s out-links). To begin clustering vertices together, the goal is find sets of vertices that jointly share a sufficiently large number of shingles. In the next phase, we would like to perform one further level of analysis, grouping together shingles that tend to occur on the same vertices, so that these sets of commonly co-occurring shingles may be used as the defining patterns of a dense subgraph. Such an analysis is yet another application of the (s, c) shingling algorithm, this time to the set of vertices associated with a particular shingle, and results in bringing together shingles that have significant overlap in the vertices on which they occur. If necessary, the sequence of operations may be repeated for as many levels as required. In practice, two levels of this algorithm suffice to convert dense subgraphs of arbitrary size (i.e., hundreds or hundreds of millions of vertices) into constant-size fingerprints that can then be recognized in a straightforward manner. The particular density threshold captured by the scheme is determined by parameters s and c.

Algorithm UnionFind (V, W, E)

```
Let  C = {{w} | w ∈ W}
For  v ∈ V  do
  Merge the sets {Find(w) | w ∈ E[v]}
Return  C
```

Figure 16.3. Performing a UnionFind.

16.5.3 Algorithm Details

We need two subroutines: the shingling algorithm given above in Figure 16.2 and the classic UnionFind algorithm in Figure 16.3. The Find operation returns the element of C, which contains the given w. This can be implemented in the usual in-memory fashion where the Find operation traces pointers to a canonical element of the set, simultaneously linking all elements thus traversed to the canonical element, while Merge links all subsequent canonical elements to the canonical element of the first set. However, the multiple levels of recursive shingling are able to reduce the problem size and avoid the need for out-of-core implementations.

The main algorithm is shown in Figure 16.4. The algorithm recurses up to a depth given by the value ℓ_{max}; in practice, ℓ_{max} will be typically 2 or 3. For each level ℓ we can control the shingler parameters s_ℓ and c_ℓ.

Unwinding the recursion, we see that RecursiveShingle proceeds in three phases:

- Produce shingles up to level ℓ_{max}
- Cluster the level $(\ell_{max} - 1)$ shingles according to their shared level ℓ_{max} shingles.
- Map these clusters back to the original source vertices.

16.5.4 An Example

Figure 16.5 illustrates the data maintained by the algorithm as it progresses. It shows the recursion up to $\ell_{max} = 3$. As input we have the source vertices V, the set of destination vertices W, and the links relation E. The set W is transformed using the shingler with parameters (s_1, c_1). For the next step of the recursion, this becomes V'. The new E relation is found by transposing the shingles relation. The recursion continues until $\ell = \ell_{max}$, at which point the clusters are determined by UnionFind.

As the recursion unwinds, the clusters of level $(\ell_{max} - 1)$ shingles are mapped back using the successive E relations until we have clusters of source vertices.

As an illustration of the behavior of the algorithm, consider its performance on a graph containing a complete subgraph (clique) of size n. The data needed to represent this subgraph is illustrated in Figure 16.6.

Algorithm RecursiveShingle (ℓ, V, W, E)

```
Fix ℓₘₐₓ, s₁, c₁, ..., sℓₘₐₓ, cℓₘₐₓ
  If ℓ = ℓₘₐₓ then
      Return UnionFind(V, W, E)    Partition W by links from V
  else
      For v ∈ V do
        S[v] = Shingle(E[v], sₗ₊₁, cₗ₊₁)
      V′ = ∪ᵥₑᵥS[v]

      For v′ ∈ V′ do                Transpose S : V → V′ into E′
        E′[v′] = {v | S[v] v′}
      clusters = RecursiveShingle(ℓ+1, V′, V, E′)
      If ℓ = 0 then                 Map back, as far as source vertices
        Return clusters
      else
        Return {∪ᵥₑꞔE[v] | C ∈ clusters}
```

Figure 16.4. RecursiveShingle algorithm. To be invoked as RecursiveShingle (0, V, V, E), where V is the vertex set and E[v] lists the out-links of v. Returns a collection of sets C of vertices of V, where (C, $\cup_{c \in C}$ E[c]) is a dense bipartite graph.

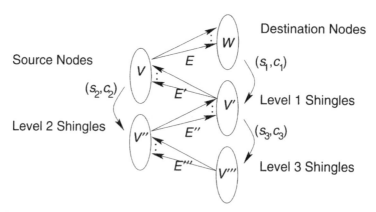

Figure 16.5. Data structures maintained by the recursive shingling algorithm.

Initially, the subgraph consists of n vertices $v_1, \ldots, v_n$, each of which contains n out-links $w_1, \ldots, w_n$. At this point, the representation of the subgraph, in the E relation, requires quadratic space. We apply our (s, c) shingling algorithm to the out-links of each vertex, creating c shingles of multiplicity s for each vertex. Since all the vertices of the clique will occur in each set of out-links, the same c shingles will occur for each vertex. The storage required in the E' relation is now $O(c \cdot n)$.

In the particular case of a clique, the algorithm could stop at $\ell_{max} = 1$, or even 0, since UnionFind will already bring together all the clique vertices. However,

Original Graph Matrix	After First-Level Shingling	After Second-Level Shingling
E is size $n \times n$	E' is $c \times n$	E'' is $c \times c$
$v_1: w_1, w_2, \cdots w_n$	$v'_1: v_1, v_2, \cdots v_n$	$v''_1: v'_1, v'_2, \cdots v'_c$
$v_2: w_1, w_2, \cdots w_n$	$v'_2: v_1, v_2, \cdots v_n$	$v''_2: v'_1, v'_2, \cdots v'_c$
$\cdots \cdots \cdots \cdots \cdots$	$\cdots \cdots \cdots \cdots \cdots$	$\cdots \cdots \cdots \cdots \cdots$
$v_n: w_1, w_2, \cdots w_n$	$v'_c: v_1, v_2, \cdots v_n$	$v''_c: v'_1, v'_2, \cdots v'_c$

Figure 16.6. Reduction of data by recursive shingling.

recursing further will reduce the size of the data set. In E' every first-level shingle v' refers to the same set of source vertices v, and so the same set of c third-level shingles v'' will be created. The original $n \times n$ representation now requires $O(c^2)$ storage.

Thus, for any graph, any clique of any size will be converted into a constant-size set of shingles. The process behaves similarly for subgraphs above a certain density (dependent on the count and size of shingles being produced at each level), giving us small representations of large dense subgraphs in the original graph.

16.5.5 Implementation

The stages of the RecursiveShingle algorithm are easy to implement efficiently in the data stream paradigm. As Figure 16.5 illustrates, the algorithm requires multiple passes over successive generations of the E relation. All the computations can be performed directly on relations, which can be stored in text files and handled using standard line-oriented sorting utilities. Storing these relations using simple compression requires about 10 MB per million edges. The sets V and W need not be maintained explicitly. All the steps in the algorithm are linear operations, with the exception of the transposition, which requires a sort.

16.5.6 Study of the Web Host Graph

To illustrate of the behavior of RecursiveShingle in practice, it was run on a large graph from a Web crawl. IBM's WebFountain project [26] crawls the Web and performs many forms of analysis and builds various databases. In particular, it builds a site database that includes a graph consisting of the links between websites. (In this context, a website is defined to be all the pages that are retrieved from a particular host name. While some sites may span multiple or only partial hosts, this heuristic works correctly in most cases.)

The dataset from September 2004 was used, representing 2.1 billion pages and 50 million sites. It contains 11 billion edges, implying a mean out-degree of 220.

Table 16.1 reports the sizes of the dataset processed by the successive phases of the algorithm, over two runs. In the first run, the (4, 16) shingling parameters were used throughout. UnionFind produced a total of 2.8 million components, each containing an average of 24 hosts.

TABLE 16.1   Two Runs of RecursiveShingle on Host Graph[a]

		$(s_1, c_1) = (3, 90),$	
$(s_1, c_1) = (s_2, c_2) = (4, 16)$		$(s_2, c_2) = (s_3, c_3) = (4, 16)$	
$\|V\| = 50$	$\|E\| = 11000$	$\|V\| = 50$	$\|E\| = 11000$
$\|V'\| = 275$	$\|E'\| = 420$	$\|V'\| = 957$	$\|E'\| = 2500$
$\|V''\| = 60$	$\|E''\| = 98$	$\|V''\| = 1000$	$\|E''\| = 1200$
		$\|V'''\| = 700$	$\|E'''\| = 750$

[a]All sizes are in millions.

In the second run, the first-level shingling was performed using $(3, 90)$ shingling. Running UnionFind at the second level produced a giant component of 1.8 million first-level shingles, indicating that the reduced s parameter caused too many shingles to be generated. However, recursing to depth 3 succeeded in removing the giant component.

Figure 16.7 shows how the out-degree of E is reduced by successive recursions. At the first level there are some enormous shingles, some occurring in excess of 100K hosts. But the out-degree reduction as discussed in Section 16.5.4 comes into play at higher levels. The second-level shingles all have out-degree at most 100, although this graph still has a giant component, and the third level has out-degree at most 10.

A strikingly large fraction of the dense subgraphs found turn out to be link spam. Random clusters were sampled, biased by cluster size, and the destination vertex with the highest in-degree, but which was not shared by other clusters, was examined. Of these vertices 88% were found to be link spam.

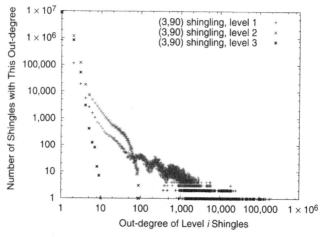

Figure 16.7.   Recursion reduces the out-degree of E.

16.6 CONNECTION SUBGRAPHS

We have considered algorithms for finding small-scale bipartite cores in huge graphs, and for finding larger dense subgraphs in huge graphs using shingling. Both of these techniques are applied in a batch processing mode: The graph is sorted and streamed through memory some number of times, and the overall running time may be quite significant. We turn now to an algorithm that may be applied in response to a user query in order to return a result in real time over huge graphs. The goal of the previous algorithms is to enumerate dense subgraphs: that is, to create a more or less exhaustive list of dense subgraphs in the graph. In the current section, we explore instead the idea of producing a single carefully chosen dense subgraph based on the particular query sent by the user.

We will discuss this algorithm in the context of social network graphs in which the vertices correspond to people, and the edges correspond to relationships such as friendship; a survey by Newman [44] gives an introduction to the field of analysis of such graphs. Social network graph analysis is an important tool for understanding social relationships between entities and has been employed with great success on very small graphs that are edited by hand. At this scale, visualization is a very useful tool for analysis, but it is very difficult to gain much from visualization of graphs of more than a few dozen vertices. Thus, when dealing with a huge social network graph such as a telephone call graph or a global email graph, we are faced with the task of extracting a subgraph that best captures the social relationships between a small number of vertices.

We will treat the social network as a graph $G = (V, E)$ with a weight function w that assigns a weight to each edge $e \in E$ corresponding to the strength of the relationship. There is a great deal of ambiguity in the problem of modeling the strength of social relationships in small subgroups of a larger population, but if we leave that aside for a moment, an abstract problem can be formulated as follows:

Connection Subgraph Problem. Given a weighted undirected connected graph G, a pair of distinguished vertices s and t, and an integer budget b, find a b-vertex connected subgraph H of G that contains both s and t and optimizes a "goodness" function $g(H)$.

In this section we discuss several aspects of this problem:

- We identify a candidate goodness function $g(\cdot)$ that we believe provides a good model of strength of connections in social networks.
- We describe a heuristic algorithm for computing the optimal subgraph that satisfies the optimality constraint and analyze its behavior.
- We further describe some heuristic optimizations for the algorithm that make the technique usable in an interactive situation.

16.6.1 Measure of Connection Strength in Social Networks

The definition of an "important relationship" in a connection subgraph is an inherently subjective notion, and one might imagine several such definitions. We begin by describing two well-known notions of "good connections" and explain why they fail to capture the notion of strength in social networks. By convention we consider the edge weights w_{uv} in the graph to indicate a strength of connection between the endpoints u, v of the edge (e.g., it might represent the volume of communication between the two, or the number of shared interests, etc).

The first such candidate notion for strength in social networks is that of simple path length. It seems intuitively obvious that short paths are more significant than long paths, and certainly direct connections are more important than third-party connections. In an edge-weighted graph in which the edge weights indicate strength of connection, we might replace path length by the sum of the reciprocal edge weights. Unfortunately we believe that such measures are insufficiently sensitive to a crucial feature of social networks, namely the fact that gregarious people tend to form many weak connections rather than a few very strong connections. Consider the example of Figure 16.8 in which the edge weights are equal to 1. In this case both paths $s \rightarrow 3 \rightarrow t$ and $s \rightarrow 4 \rightarrow t$ are of length 2, but the path through vertex 4 is intuitively less significant than the path through 3, since s and t are one of many neighbors of vertex 4, but vertex 3 has *only* the neighbors s and t, indicating that these connections are probably more important to vertex 3 than the connections are to vertex 4. Since both paths are the same length, the shortest path metric somehow fails to discriminate between the relative strength of social connections.

Another notion that one might consider for measuring the strength of connections in social networks is that of network flow. In this paradigm, the edge weights indicate a capacity of an edge, and the flow along a path is the minimum capacity of an edge on this path. Once again we see that the example shows how this fails to accurately capture the notion of connection strength in social networks, since the paths $s \rightarrow 1 \rightarrow 2 \rightarrow t$ and $s \rightarrow 3 \rightarrow t$ both have flow equal to 1, and yet somehow our intuition tells us that the path through 3 is more significant than the path through 1 and 2 since the former is of length 2 rather than length 3.

16.6.1.1 *Intuition from Random Walks.*
From these examples, it is apparent that a measure of strength of a connection in social networks needs to incorporate both the degrees of the intervening vertices as well as the length of the path. While it is possible to design ad hoc measures that mix these two features, we claim that

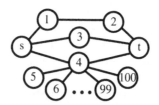

Figure 16.8. Example illustrating why both shortest path and network flow measures are inadequate to measure strength in social networks.

there is a natural measure of the strength of connection in a subgraph that arises from thinking of the likelihood that information would flow from s to t in the entire graph without leaving the subgraph, or alternatively that a message that follows a random walk from s to t would remain within the subgraph. Here we think of the edge weights as inducing a probability distribution for each vertex, in which the probability of leaving a vertex along a given edge emanating from the vertex is proportional to the weight of the edge.

Unfortunately, the paradigm of random walks suffers from the same defect as that of network flow, namely that short paths are not favored over longer paths of the same edge weight. In the case of random walks there is a very simple modification to the basic paradigm that compensates for this. Specifically, at each stage of the random walk from s to t we admit a small probability that the random walk will fail. This favors short paths over long paths since each step in the path has a nonnegligible chance of starting over again.

For specificity, we can define a random walk as follows. We start at s and proceed along the edges of the graph to t. We introduce a universal sink vertex z to the graph and make both z and t absorbing states, in the sense that once the walk encounters these states it never leaves. We then introduce a parameter α, and for every vertex $u \neq z, s$ we add edges (u, z) of weight $\alpha \sum_v w(u, v)$, so that the edge from u to z carries a fixed fraction of the weight of all edges emanating from u. In our experiments, we used $\alpha = 1$, but other choices can be used to vary the bias toward short paths. We now define a random walk along the graph as follows. We start at vertex s, and at each step of the way we follow an edge with probability proportional to its weight.

For the random walk defined in this way, we could define several measures of an optimal subgraph in the network, including the paradigm of maximizing the probability of remaining with the subgraph while walking from s to t. In the remainder of the discussion we will develop a slightly different alternative that follows the same kind of motivation.

16.6.1.2 *Electrical Networks.*

The theory of random walks on graphs is well studied, and there is an elegant correspondence between the theory of random walks on undirected graphs and that of current flow in electrical networks, in which edges have conductance specified by the edge weights (or equivalently, a resistance equal to the reciprocal of the weight). For surveys on the relationship between electrical current flow and random walks on graphs, see [2, 18, 41]. Assume that we apply a voltage of 1 V to vertex s, and ground vertices z and t (voltage = 0), and let $V(u)$ denote the induced voltage at vertex u. The voltages are known to satisfy Ohm's law, which states that the current flow along an edge is equal to the difference in potentials times the conductance of the edge. Let $I(u, v)$ be the current flow from u to v. Then Ohm's law states that

$$\forall u, v : I(u, v) = w_{uv}(V(u) - V(v)) \qquad (16.2)$$

Moreover, we must also satisfy Kirchoff's law, which states that the current flow on edges incident to a vertex must sum to zero, or

$$\forall v \neq s, t, z : \sum_u I(u, v) = 0 \tag{16.3}$$

Together these conditions imply that V is a *harmonic function* with poles s, z, t. By this we mean that if $\Gamma(u)$ denotes the neighbors of u, and $w_v = \sum_{u \in \Gamma(v)} w_{uv}$, then

$$V(u) = \sum_{v \in \Gamma(u)} \frac{w_{uv} V(v)}{w_v} \tag{16.4}$$

for every $u \neq z, s, t$. The function $V(u)$ also has a natural interpretation in this case (see [2, Section 3.2]), namely that it represents the probability that a walk that starts at v will reach u before it reaches either of the absorbing states z or t.

16.6.1.3 *Algorithms for Voltages.*

From Eq. (16.4) we see that the calculation of V reduces to an eigenvector calculation of the form

$$\begin{bmatrix} A & B \\ 0 & I \end{bmatrix} V = V \tag{16.5}$$

where B represents the boundary conditions connecting vertices to s, z, and t; I is a 3×3 identity matrix, and $a_{ij} = w_{ij}/w_j$. The simplest algorithm allows us to solve this set of linear equations in time $O(n^3)$ for a graph with n vertices, but in practice it can be solved approximately in $O(|E|)$ time, where $|E|$ denotes the number of edges in the graph. The reason is that we only need to calculate the principal eigenvector and under suitable nondegeneracy conditions, the power method can be employed, with a running time that is dependent on the desired precision and the gap between the second largest and the largest eigenvalues.

16.6.1.4 *Connection Subgraphs.*

We are now prepared to define our measure of connection strength for connection subgraphs. We phrase it in terms of the electrical paradigm, but the previous section illustrates that it could as easily be phrased in terms of random walks. We begin with several definitions:

Definition 16.1   *Vertex v is downhill from u ($u \rightarrow_d v$), if $I(u, v) > 0$.*

We can then define $I_{out}(u)$, the total flow leaving vertex u:

Definition 16.2   *Total out-flow from vertex u: $I_{out}(u) = \sum_{\{v \mid u \rightarrow_d v\}} I(u, v)$.*

Definition 16.3 (Prefix Path)   A prefix path *is any downhill path $\mathcal{P}$ that starts from the source s, that is, $\mathcal{P} = (s = u_1, \ldots, u_i)$ where $u_j \rightarrow_d u_{j+1}$.*

Obviously, a prefix path has no loops because of the downhill requirement.

Definition 16.4 (Delivered Current)   *The* delivered current $\hat{I}(\mathcal{P})$ *over a prefix-path* $\mathcal{P} = (s = u_1, \ldots, u_i)$ *is the volume of electrons that arrive at u_i from s, strictly through $\mathcal{P}$. Formally, we define $\hat{I}()$ inductively as follows, beginning with a single edge as base case:*

$$\hat{I}(s, u) = I(s, u)$$

$$\hat{I}(s = u_1, \ldots, u_i) = \hat{I}(s = u_1, \ldots, u_{i-1}) \frac{I(u_{i-1}, u_i)}{I_{out}(u_{i-1})}$$

In words, to estimate the delivered current to vertex u_i through path $\mathcal{P}$, we are pro-rating the delivered current to vertex u_{i-1} proportionately to the outgoing current $I(u_{i-1}, u_i)$. We are now ready to define the current delivered by a subgraph; notice that this definition is intentionally quite different from the current delivered by applying voltages and computing current flows on the subgraph alone.

Definition 16.5 (Captured Flow)   We *say the* captured flow $CF(\mathcal{H})$ *of a subgraph $\mathcal{H}$ of $\mathcal{G}$ is the total delivered current, summed over all source-sink prefix paths that belong to $\mathcal{H}$.*

$$CF(\mathcal{H}) \equiv g(\mathcal{H}) = \sum_{\mathcal{P}=(s,\ldots,z)\in\mathcal{H}} \hat{I}(\mathcal{P}) \tag{16.6}$$

Note that any individual positron may traverse only a single prefix path, so this measure does not double-count any flow.

EXAMPLE 16.1   Consider the graph shown in Figure 16.9. For simplicity of exposition, and without loss of generality, we do not have a universal sink z (i.e., we set $\alpha = 0$). After the voltages of the source and sink have been fixed to 1 and 0, respectively, the resulting voltages are shown for each other vertex. These voltages induce currents along each edge as shown. There are five *downhill* source-to-sink paths in the graph. These paths, with their delivered current are shown in Table 16.2. The path that delivers the most current (and the most current per vertex) is $s \to b \to z$. We can compute the 2/5 A delivered by this path by observing that, of the 0.5 A that arrive at vertex b on the $s \to b$ edge, 1/5 departs toward vertex c, while 4/5 depart toward vertex z. Thus, $4/5 \times 0.5$A gives the 2/5 A we seek.

Consider the $\{s, b, c, z\}$ subgraph. We can compute its captured flow by adding the delivered current of all paths that travel exclusively through the subgraph; namely, $s \to b \to c \to z$ and $s \to b \to z$; these paths together capture $2/5 + 1/10 = 0.5$ A of total current. We observe that this is one of two optimal 4-vertex subgraphs that could be produced.

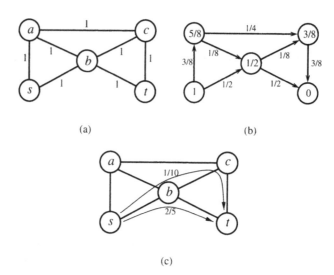

(a)　　　　　　　　　　　　　　(b)

(c)

Figure 16.9. (a) Sample network, showing (b) voltages, current, and (c) paths with delivered current.

TABLE 16.2　Current Flow
along Paths in Figure 16.9

$s \to b \to z$	2/5
$s \to a \to c \to z$	1/4
$s \to b \to c \to z$	1/10
$s \to a \to b \to z$	1/10
$s \to a \to b \to c \to z$	1/40

16.6.2　Algorithms for Connection Subgraphs

Our optimization problem is now to find a subgraph that maximizes the captured flow over all subgraphs of its size. For this we apply a greedy heuristic, as follows. First, it initializes an output graph to be empty. Next, it iteratively adds end-to-end paths (i.e., from source s to sink z) to the output graph. Since the output graph is growing, a new path may include vertices that are already present in the output graph; the algorithm will favor such paths. Formally, at each step the algorithm adds the path with the highest marginal flow per capita. That is, it chooses the path $\mathcal{P}$ that maximizes the ratio of flow along the path, divided by the number of new vertices that must be added to the output graph.

Notice that the inductive definition of delivered current given above could easily be computed using dynamic programming. We will modify this computation in order to compute the path that maximizes our measure.

We begin with a definition of entries in our dynamic programming table $D_{v,k}$ (for "delivery matrix"), to be interpreted in the context of a partially built output graph.

Definition 16.6 $D_{v,k}$ *is the current delivered from s to v along the prefix path* $\mathcal{P} = (s = u_1, \ldots, u_\ell = v)$ *such that:*

1. $\mathcal{P}$ *has exactly k vertices not in the present output graph.*
2. $\mathcal{P}$ *delivers the highest current to v among all such paths that end at v.*

To compute D we exploit the fact that the electric current flows $I(*, *)$ form an acyclic graph. Formally, we arrange the vertices into a sequence $u_1 = s, u_2, u_3, \ldots,$ $z = u_n$ such that if vertex u_j is downhill from u_i ($u_i \to_d u_j$), then u_j follows u_i in our ordering ($i < j$). That is, the vertices are sorted in descending order of voltage, and so electric current always flows from left to right in the ordering. We will fill in the table D in the order given by the topological sort above, guaranteeing that when we compute $D_{v,k}$, we will already have computed $D_{u,*}$ for all $u \to_d v$. The entries of D are computed in Algorithm 6.1.

Algorithm **16.1** *DisplayGraphGeneration*

```
Initialize output graph Gdisp = ∅
Let P be the maximum allowable path length
While output graph is not big enough do
  For i = 1 to |G| do
    Let v = ui
    For k = 2 to P do
      If v is already in the output graph then k' = k else k' = k − 1
      Let Dv,k = maxu | u→dv (Du,k' I(u, v)/Iout(u))
  Add the path maximizing Dt,k/k, k ≠ 0
```

Intuitively, $I(u, v)/I_{out}(u)$ represents the fraction of flow arriving at u that continues to v. Multiplying this by $D_{u,k'}$ gives the total flow that can be delivered to v through a simple path. The path maximizing our measure is then the path that maximizes $D_{z,k}/k$ over all $k \neq 0$; it can be computed by tracing back the maximal value of D from z to s.

16.6.3 Optimization Heuristics for Interactive Environments

The primary motivation for the connection subgraph problem is one of cognitive understanding of relationships by humans, and the primary use of such techniques is in interactive environments. Unfortunately, in the case of global social structures, we can easily imagine graphs with millions of vertices, which presents a challenge for rapid calculations on such graphs. Thus, while the algorithm described in the previous section is polynomial time in the size of the input, it is still too expensive to compute for extremely large graphs. Hence there is a need for algorithms that work with extremely low latency on extremely large graphs. Our approach is to employ a heuristic that quickly extracts a subgraph of the original graph that contains the most important paths from s to t, upon which we run the algorithm of Section 16.6.2. We call this subgraph the *candidate graph*.

Formally, the candidate generation process takes vertices s and t in the original graph $\mathcal{G}$ and produces a much smaller graph ($\mathcal{G}_{cand}$) by carefully growing neighborhoods around s and t. Because our intent is to apply this to graphs that are extremely large, we assume that vertices are discovered incrementally during the execution of the algorithm. In the framework, candidate generation algorithms strategically expand the neighborhoods of s and t until there is a significant overlap. We start by knowing only s and t, and we explore vertices further by retrieving their neighbor set. Thus a vertex in our $\mathcal{G}_{cand}$ graph may exist in one of two states: We may know the neighbors of the vertex, or we may only know the vertex as the neighbor of a vertex whose neighbors we already know. The operation of expanding a vertex consists of retrieving its neighbor set, and we think of this as a relatively expensive operation (since it will likely have to go to disk for a very large graph).

Let $D(s)$ be the set of vertices first discovered through a series of expansions beginning at s; we say that s is the *root* of all vertices in $D(s)$. We define $E(s)$ as the set of *expanded* vertices within $D(s)$, that is, those for which we have already retrieved their list of neighbors. Likewise, let $P(s)$ be the set of *pending* vertices within $D(s)$ that have not yet been expanded. Similarly, define $D(t)$, $E(t)$, and $P(t)$. Note that $D(s)$ is disjoint from $D(t)$ since each vertex is discovered only once, by expanding a vertex whose root is either s or t. Recall that for weighted graphs, we use w_{uv} as the weight of the edge from u to v. We define $\deg(u)$ to be the degree (number of neighbors) of u. Algorithm 16.2 shows CandidateGeneration.

Algorithm 16.2 *CandidateGeneration*

```
Set P(s) = {s} and P(t) = {t}
While not stoppingCondition() do
  v = pickHeuristic()
  Let r be the root of v
  Expand v, moving it from P(r) to E(r)
  Add all neighbors of v that are not already in E(r) to P(r)
```

Thus, the details of the algorithm lie in the process of deciding which vertex to expand next and when to terminate expansion. Our algorithm repeatedly expands carefully selected, unexpanded vertices, chosen by the pickHeuristic(), until a stopping condition stoppingCondition() is reached. These are the two major routines, and there is a lot of opportunity for experimentation within this framework. In effect, pickHeuristic() strives to suggest a vertex for expansion, estimating how much delivered current this vertex will carry. Thus, the heuristic favors vertices that are (a) close to the source s or the sink t (b) with strong connections (high conductance) and (c) low degree, to avoid the "famous-vertex" effect (recall vertex 4 of Figure 16.8.)

One appealing heuristic for pickHeuristic() is to run a real random walk on $D(s)$ and $D(t)$ and keep track of which vertices in $P(s)$ and $P(t)$ are hit most often. If we select a vertex u from $P(s)$ or $P(t)$ as the next vertex to be explored, and u had previously been visited k times in random walks, then before choosing the next vertex we should run k random walks from s or t to vertices in $P(s)$ and $P(t)$, so that we keep a constant number of walks distributed among the vertices that are yet to be expanded. Alternative approaches to selecting a vertex that is in some sense closest to the root are described in [19], but we omit details on these techniques.

The stoppingCondition() puts limits on the size of the output graph $\mathcal{G}_{cand}$ (count of expansions, count of distinct vertices discovered, etc.). In our experiments, we defined three thresholds for termination; the algorithm will stop as soon as any threshold is exceeded. First, we adopt a threshold on total expansions to limit the total number of disk accesses. Second, we adopt a larger threshold on discovered vertices, even if those vertices have not yet been expanded, to limit memory usage. And finally, we adopt a threshold on number of cut edges [edges between $D(s)$ and $D(t)$], as a measure of the connectedness of the set of vertices with the src as a root with the set of vertices with the sink as a root.

16.6.4 Experiments

The optimization techniques described in Section 16.6.3 turn out to be remarkably effective in practice. We performed our experiments on a graph that we call the *Web names graph*. This graph was constructed from a crawl of the World Wide Web and represents a social network graph between names of people. The first step in this is to run an entity extraction tool on the pages of the World Wide Web in order to recognize whenever a person's name is recognized. Techniques for devising such tools are fairly well developed and can extract names with high probability. The names that are discovered in this way become the vertices in our graph, and we create an edge between two names of weight w if in our crawl we find w distinct pages in which the two names are mentioned together in close proximity to each other. In our experiments, we used a window of 50 text tokens, declaring an edge if the two names appear within this many tokens of each other. From our crawl, we were able to recognize nearly 78 million distinct names, with approximately 1.4 billion edges between them.

It should be noted that the vertices in our graph are not people but rather the names of people. Thus when we find a connection graph between two names, we may be confused by names that represent two different people from relatively disconnected social groups. This might be troubling at first, but in practice this is *always* the case with real-world data since in nearly every application area there is difficulty in distinguishing people from each other based on only information stored in a database. This is particularly true in the case of fraud or other criminal investigations in which individuals strive to deliberately conceal or confuse their identity.

Even with the problem of confusing common names for different people, the techniques we have developed are remarkably effective, even for finding connections between people who are apparently disconnected from each other socially. An example subgraph from our dataset is shown in Figure 16.10 in which we show

Figure 16.10. Connection subgraph example from a social network graph extracted from the World Wide Web. In this example we see connections between two people who might be expected to have nothing to do with each other, namely the scientist Alan Turing and the actress Sharon Stone. In fact the algorithm was able to identify that Kate Winslett is the actress who is probably most closely related to Alan Turing.

a connection from the actress Sharon Stone to the computer scientist Alan Turing. At first glance these people might be thought to have nothing to do with each other, but in fact our system discovered some very interesting connections that were not immediately apparent. From the nearly 78 million vertices in our database, it selected the two names Kate Winslett and Gillian Anderson. These are actresses that might appear to have nothing to do with Alan Turing, but in fact these are the actresses that are probably *best* connected to Alan Turing. In the case of Kate Winslett, she starred in a movie titled *Enigma* whose title refers to the famous German cipher machine of World War II. It turns out that Alan Turing was one of the mathematicians who was most instrumental in breaking this cipher system, which changed the course of the war. In the case of Gillian Anderson, she stars in a science fiction television show that appeals strongly to a scientific audience that is likely to know who Alan Turing was.

16.7 CONCLUSIONS

Discovering dense subgraphs is an important problem with a wide variety of applications. While this problem is not easy, there are several heuristics that work very well in practice, even on massive graphs. Many of these heuristics are tailored to a particular application/dataset. It will be interesting to develop more efficient and effective heuristics, perhaps even with some provable guarantees, assuming an underlying model for graph generation. It will also be interesting to study the effectiveness of spectral methods for finding dense subgraphs.

REFERENCES

1. R. Agrawal and R. Srikant. Fast algorithms for mining association rules in large databases. In Proceedings of 20th International Conference on Very Large Data Bases, Santiago, Chile, pp. 487–499, 1994.
2. D. Aldous and J. A. Fill. Reversible markov chains and random walks on graphs, 2001, unpublished.
3. N. Alon, Y. Matias, and M. Szegedy. The space complexity of approximating the frequency moments. *Journal of Computer and System Sciences*, 58(1):137–147, 1999.

4. K. Bharat, A. Broder, J. Dean, and M. R. Henzinger. A comparison of techniques to find mirrored hosts on the WWW. *Journal of the American Society for Information Science*, 51(12):1114–1122, 2000.
5. K. Bharat and M. Henzinger. Improved algorithms for topic distillation in hyperlinked environments. In Proceedings of 21st International ACM SIGIR Conference on Research and Development in Information Retrieval, pp. 104–111, 1998.
6. U. Brandes. A faster algorithm for betweenness centrality. *Journal of Mathematical Sociology*, 25:163–177, 2001.
7. U. Brandes, M. Gaertler, and D. Wagner. Experiments on graph clustering algorithms. In Proceedings of 11th European Symposium on Algorithms, pp. 568–579, 2003.
8. S. Brin and L. Page. The anatomy of a large scale hypertextual web search engine. *Computer Networks*, 30(1–7):107–117, 1998.
9. A. Broder, M. Charikar, A. Frieze, and M. Mitzenmacher. Min-wise independent permutations. *Journal of Computer and System Sciences*, 60:630–659, 2000.
10. A. Broder, S. C. Glassman, M. S. Manasse, and G. Zweig. Syntactic clustering of the web. *WWW6/Computer Networks*, 29(8–13):1157–1166, 1997.
11. J. Carriere and R. Kazman. Webquery: Searching and visualizing the web through connectivity. *Computer Networks*, 29(8–13):1257–1267, 1997.
12. S. Chakrabarti. *Mining the Web: Discovering Knowledge from Hypertext Data*. Morgan Kaufmann, San Francisco, CA, 2002.
13. S. Chakrabarti, B. Dom, D. Gibson, R. Kumar, P. Raghavan, S. Rajagopalan, and A. Tomkins. Experiments in topic distillation. In SIGIR Workshop on Hypertext Information Retrieval on the Web, 1998.
14. A. K. Chandra, P. Raghavan, W. L. Ruzzo, R. Smolensky, and P. Tiwari. The electrical resistance of a graph captures its commute and cover times. *Computational Complexity*, 6(4):312–340, 1997.
15. M. Charikar. Greedy approximation algorithms for finding dense components in a graph. In Proceedings of 3rd International Workshop on Approximation Algorithms for Combinatorial Optimization (APPROX), pp. 84–95, 2000.
16. S. Deerwester, S. T. Dumais, G. W. Furnas, T. K. Landauer and R. Harshman. Indexing by Latent Semantic Analysis. *Journal of the American Society for Information Science*, 41(6): 391–407.
17. I. S. Dhillon, S. Mallela, and D. S. Modha. Information-theoretic co-clustering. In Proceedings of 9th ACM SIGKDD International Conference on Knowledge Discovery and Data Mining, pp. 89–98, 2003.
18. P. G. Doyle and J. L. Snell. *Random Walks and Electric Networks*. Carus Mathematical Monographs. Mathematical Association of America, Washington, DC, 1984.
19. C. Faloutsos, K. McCurley, and A. Tomkins. Fast discovery of *Connection Subgraphs*. In Proceedings of 10th ACM SIGKDD Conference on Knowledge Discovery and Data Mining, pp. 118–127, 2004.
20. C. Faloutsos, K. S. McCurley, and A. Tomkins. Connection subgraphs in social networks. In Workshop on Link Analysis, Counterterrorism, and Privacy. SIAM International Conference on Data Mining, 2004.
21. G. Flake, S. Lawrence, and C. Lee Giles. Efficient identification of web communities. In Proceedings of 6th ACM SIGKDD International Conference on Knowledge Discovery and Data Mining, pp. 150–160, 2000.
22. D. Florescu, A. Y. Levy, and A. O. Mendelzon. Database techniques for the world-wide web: A survey. *SIGMOD Record*, 27(3):59–74, 1998.

23. G. Gallo, M. D. Grigoriadis, and R. Tarjan. A fast parametric maximum flow algorithm and applications. *SIAM Journal on Computing*, 18:30–55, 1989.

24. D. Gibson, J. Kleinberg, and P. Raghavan. Inferring web communities from link topology. In Proceedings of 9th ACM Conference on Hypertext and Hypermedia, pp. 225–234, 1998.

25. D. Gibson, R. Kumar, and A. Tomkins. Extracting large dense subgraphs in massive graphs. In Proceedings of 31st International Conference on Very Large Data Bases, 2005.

26. D. Gruhl, L. Chavet, D. Gibson, J. Meyer, P. Pattanayak, A. Tomkins, and J. Zien. How to build a WebFountain: An architecture for very large-scale text analytics. *IBM Systems Journal*, 43(1):64–77, 2004.

27. T. H. Haveliwala. Topic-sensitive pagerank. Proceedings of 11th International World Wide Web Conference, pp. 517–526, 2002.

28. M. Henzinger, P. Raghavan, and S. Rajagopalan. Computing on data streams. In *DIMACS series in Discrete Mathematics and Theoretical Computer Science*, 50:107–118, 1999.

29. G. Jeh and J. Widom. Scaling personalized web search. Proceedings of 12th International World Wide Web Conference, pp. 271–279, 2003.

30. G. Karypis and V. Kumar. Parallel multilevel k-way partitioning for irregular graphs. *SIAM Review*, 41(2):278–300, 1999.

31. S. Khot. Ruling out PTAS for graph min-bisection, densest subgraph, and bipartite clique. In Proceedings of 45th IEEE Annual Symposium on Foundations of Computer Science, pp. 136–145, 2004.

32. J. Kleinberg. Authoritative sources in a hyperlinked environment. *Journal of the ACM*, 46(5):604–632, 2000.

33. D. Konopnicki and O. Shmueli. Information gathering in the world-wide web: The W2QL query language and the W3QS system. *ACM Transactions on Database Systems*, 23(4):369–410, 1998.

34. R. Kumar, U. Mahadevan, and D. Sivakumar. A graph-theoretic approach to extract storylines from search results. In Proceedings of 10th ACM SIGKDD International Conference on Knowledge Discovery and Data Mining, pp. 216–225, 2004.

35. R. Kumar, J. Novak, P. Raghavan, and A. Tomkins. Structure and evolution of blogspace. *Communications of the ACM*, 47(12):35–39, 2004.

36. R. Kumar, P. Raghavan, S. Rajagopalan, and A. Tomkins. Extracting large scale knowledge bases from the web. In Proceedings of 25th International Conference on Very Large Databases, pp. 639–650, 1999.

37. R. Kumar, P. Raghavan, S. Rajagopalan, and A. Tomkins. Trawling the web for emerging cyber-communities. *WWW/Computer Networks*, 31(11–16):1481–1493, 1999.

38. L. V. S. Lakshmanan, F. Sadri, and S. N. Subramanian. SchemeSQL: An extension of SQL for multidatabase interoperability. *ACM Transactions on Database Systems*, 26(4):476–519, 2001.

39. E. Lawler. *Combinatorial Optimization: Networks and Matroids*. Holt, Rinehart, and Winston, New York, 1976.

40. D. Liben-Nowell and J. Kleinberg. The link prediction problem for social networks. In Proceedings of 12th ACM International Conference on Information and Knowledge Management, pp. 556–559, 2003.

41. L. Lovász. *Random Walks on Graphs: A Survey*, pp. 353–398. Janos Bolyai Mathematical Society, Budapest, 1996.
42. J. McHugh, S. Abiteboul, R. Goldman, D. Quass, and J. Widom. LORE: A database management system for semistructured data. *SIGMOD Record*, 26(3):54–66, 1997.
43. A. O. Mendelzon, G. A. Mihaila, and T. Milo. Querying the world wide web. *International Journal on Digital Libraries*, 1(1):54–67, 1997.
44. M. E. J. Newman. Fast algorithm for detecting community structure in networks. *Physical Review E*, 69:066133, 2003.
45. L. Page, S. Brin, R. Motwani, and T. Winograd. The PageRank citation ranking: Bringing order to the web. Technical Report, Stanford Digital Library Technologies Project, Palo Alto, CA, 1998.
46. C. R. Palmer and C. Faloutsos. Electricity based external similarity of categorical attributes. In *Proceedings of 7th* Pacific-Asia Conference on Advances in Knowledge Discovery and Data Mining, pp. 486–500, 2003.
47. P. Pirolli, J. Pitkow, and R. Rao. Silk from a sow's ear: Extracting usable structures from the web. In Proceedings of ACM Conference on Human Factors in Computing Systems, pp. 118–125, 1996.
48. E. Spertus. Parasite: Mining structural information on the web. *Computer Networks*, 29(8–13):1205–1215, 1997.
49. L. Stockmeyer and V. Vazirani. NP-completeness of some generalizations of the maximum matching problem. *Information Processing Letters*, 15(1):14–19, 1982.
50. D. Tsur, J. D. Ullman, S. Abiteboul, C. Clifton, R. Motwani, S. Nestorov, and A. Rosenthal. Query Flocks: A generalization of association-rule mining. In Proceedings of 17th ACM SIGMOD International Conference on Management of Data, pp. 1–12, 1998.
51. S. van Dongen. *Graph Clustering by Flow Simulation*. Ph.D. thesis, University of Utrecht, 2000.
52. S. Virtanen. Clustering the Chilean web. In Proceedings of 1st Latin American Web Conference, pp. 229–233, 2003.

17

SOCIAL NETWORK ANALYSIS

SHERRY E. MARCUS, MELANIE MOY, AND THAYNE COFFMAN

21st Century Technologies, Inc. Austin, TX

17.1 INTRODUCTION

In this chapter, we would like to provide a brief introduction to social network analysis (SNA) as a graph mining tool. In Section 17.2, we introduce the basic concepts of social network analysis and the descriptions of a number of useful SNA metrics. In the following section, we discuss group detection as an application of SNA, including the introduction of a novel group detection algorithm implemented within the Terrorist Modus Operandi Detection System (TMODS) [5, 6, 11, 13] platform. We continue with a discussion of the TMODS platform, including graph matching and SNA capabilities. Section 17.5 presents the results of applying group detection to a small selection of representative datasets and, finally, we provide our conclusions.

17.2 SOCIAL NETWORK ANALYSIS

Social network analysis is an approach to the study of human social interactions. SNA can be used to investigate kinship patterns, community structure, or the organization of other formal and informal social networks [16, 19]. These "social"

Mining Graph Data, Edited by Diane J. Cook and Lawrence B. Holder
Copyright © 2007 John Wiley & Sons, Inc.

networks may be associated with corporations, family groups, filial groups, command and control structures, or covert organizations.

Two important problems in social network analysis are determining the functional roles of individuals in a social network and diagnosing network-wide conditions or states. Different individuals in a social network often fulfill different roles. Examples of intuitive roles include leaders, followers, regulators, "popular people," and early adopters. SNA researchers have also identified other roles that are less intuitive but are still common in normal and abnormal networks, like bridges, hubs, gatekeepers, and pulse takers. We would like to be able to automatically determine the role(s) that a particular actor fills in a social network.

Social network theorists have also identified characteristics that can be used to differentiate healthy organizations from those with underlying organizational problems, or to differentiate normal social network structures from those typically used for illicit activity, and we would like to classify networks into various "healthy" and "unhealthy" categories. Some of these characteristics are quantifiable as SNA metrics, while others lack simple indicators.

Social network analysts often represent the social network as a graph, and we follow this approach. In its simplest form, a social network graph (sometimes called an "activity graph") contains nodes representing actors (generally people or organizations) and edges representing relationships or communications between the actors. Analysts then reason about the individual actors and the network as a whole through graph-theoretic approaches, including (of particular interest to this study) SNA metrics that take different values at different nodes in the graph.

17.2.1 Bestiary of SNA Metrics[1]

In this subsection, we define a number of SNA metrics useful for the analysis of datasets. These metrics are:

- Betweenness centrality
- Closeness centrality
- Average path length
- Average cycle length
- Characteristic path length
- Global efficiency
- Clustering coefficient
- Cliqueishness
- Homogeneity
- E-to-I ratio
- Degree
- Density

---

[1]**Bestiary:** a descriptive or anecdotal treatise on various real or mythical kinds of animals, especially a medieval work with a moralizing tone.

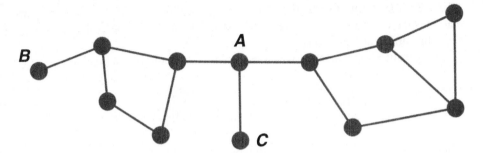

Figure 17.1. Betweenness centrality and average path length.

- Girth
- Diameter
- Radius
- Radiality and integration
- Circumference

All of these metrics (among others) are implemented within the TMODS platform, as described below.[2]

17.2.1.1 Brandes Betweenness Centrality (BR).

Brandes Betweenness centrality (BR) is a measure of the amount to which a given node lies on the shortest paths between other nodes in the graph. Figure 17.1 shows some examples of varying betweenness—node A has high betweenness because the shortest paths (called geodesics) between many pairs of other nodes in the graph pass through node A. Nodes B and C have low betweenness because most of the geodesics in the graph do not pass through them.

Formally, global betweenness centrality is defined, for a graph G with nodes $\{s, t, v\} \in N(G)$, as

$$\Omega(s, t) = \{\text{Number of distinct geodesics from } s \text{ to } t\}$$

$$\Omega_v(s, t) = \{\text{Number of geodesics from } s \text{ to } t \text{ that pass through } v\}$$

$$C_b(v) = \sum_{s, t \neq v} \frac{\Omega_v(s, t)}{\Omega(s, t)}$$

Typically, betweenness centrality is normalized to the range $[0, 1]$ within an individual graph. We use one of the more common (and efficient) known algorithms for computing betweenness centrality, developed by Ulrik Brandes [2].

---

[2] Some of these computations, such as finding cycles and cliques, are NP-complete.

17.2.1.2 Closeness Centrality. Closeness centrality is a category of algorithms that rates the centrality of a node by its closeness (distance) to other nodes. "Closeness" has different definitions.

This feature implements a measure of closeness centrality defined as

$$\text{CLC}(G, \text{keyNode}) = \frac{N_R^2}{N \sum_i \text{distance}(\text{keyNode}, n_i)}$$

where

$$
\begin{aligned}
N &= \text{number of nodes in } G \\
N_R &= \text{number of nodes in } G \text{ reachable from} \\
&\quad \text{keyNode (with the maximum possible} \\
&\quad N_R = N - 1) \\
\text{distance}(\text{keyNode}, n_i) &= N \text{ for all nodes not reachable from keyNode.}
\end{aligned}
$$

As fewer nodes are reachable from keyNode, CLC goes down. Also, as the distances between keyNode and nodes that are reachable increases, CLC goes down.

It is possible that this is a standard SNA metric, but enough assumptions and "tweaks" have been made that this is unlikely to be exactly the same as what other people refer to as closeness centrality.

Distances (and reachability) may be calculated using either directed or undirected edges.

17.2.1.3 Average Path Length (APL). Average path length computes the average distance between the key node and any other node in the neighborhood. Thus

$$\text{APL}(G, \text{keyNode}) = \text{average}_i(\text{distance}(\text{keyNode}, N_i))$$

for all nodes N_i in the neighborhood. This gets a little more complicated when not all nodes are reachable from the keyNode (because the graph is disconnected or because computation is being done in a directed fashion). There are three possible settings for how to handle unreachable nodes:

Ignore: The average is taken only over nodes reachable from keyNode.

Count as zero: The distance to unreachable nodes is considered to be zero.

Count as N: The distance to unreachable nodes is considered to be N, the size of the neighborhood.

In Figure 17.1, node B has a relatively high average path length, while nodes A and C have relatively low average path lengths.

17.2.1.4 Average Cycle Length (ACL). Average cycle length computes the average length of all the cycles in the graph. Only cycles of length ≤ 3 are considered. Average cycle length is always computed using directed edges.

17.2.1.5   Characteristic Path Length (CPL).

Characteristic path length is a measure of the average distance between any two nodes in the graph. Thus

$$CPL(G) = \text{average}_{i,j}(\text{distance}(N_i, N_j))$$

for all nodes N_i, N_j in the graph. The definition gets a bit more complicated when not all nodes are reachable from each other (because the graph is disconnected or because computation is being done in a directed fashion). There are three possible settings for how to handle unreachable nodes:

Ignore: The average is taken only over nodes reachable from each other.

Count as zero: The distance between unreachable nodes is considered to be zero.

Count as N: The distance between unreachable nodes is considered to be N, the size of the graph.

17.2.1.6   Global Efficiency (GE).

Global efficiency is a measure of the rate at which nodes can be reached from each other in the graph. It is approximately inversely related to characteristic path length. Global efficiency will always range $[0..1]$. Intuitively, high global efficiency means the distance between nodes is generally short ($GE = 1$ means every node can be reached from every other in one step), and low global efficiency means many steps must be taken between nodes:

$$GE(G) = \text{average}_{i,j}\left(\frac{1}{\text{distance}(N_i, N_j)}\right)$$

Unreachable nodes can be treated two ways: They can be ignored (equivalent to treating them as if they were at distance zero), or they can be treated as if they were at distance N (which imposes the maximum possible penalty on global efficiency for unreachable nodes), depending on configuration settings.

17.2.1.7   Clustering Coefficient (CC).

Clustering coefficient (CC) measures the density of the distance-one neighborhood around the keyNode (even if the search group is larger than this neighborhood). It is a measure of the likelihood of "If A knows B and B knows C, then A knows C" for fixed $B = $ keyNode. It is, equivalently, a measure of the percentage of closed triangles in the neighborhood immediately around keyNode. Thus you would have $CC(n) = 1.0$ for nodes in a clique, and $CC(n) = 0.0$ for the center of a star (or a hub in a hub-and-spoke network).

To compute CC, find the subgraph Glocal $= $ all nodes at distance exactly 1 from keyNode, and all edges between them. Note that this subgraph does not include the keyNode. Then compute the density (directed or undirected, depending on configuration) of that graph.

In this context, directed density is defined as

$$D_d(G) = \frac{E(G)}{N(G)(N(G) - 1)}$$

where $N(G)$ and $E(G)$ are the number of nodes and edges in the graph, respectively. Note that this allows the possibility that $D_d(G) > 1.0$ for multigraphs. The definition of undirected density is a bit more complex because we are operating on a directed multigraph representation. In this context, undirected density is defined as

$$D_u(G) = \frac{E(G)}{(N(G)(N(G) - 1)/2)}$$

where

$$E(G) = \frac{\sum_n (\text{Size}(\text{Neighbors}(n)))}{2}$$

Note that this is not the same definition of undirected density as used in the density feature.

17.2.1.8 Cliqueishness.
Cliqueishness is a measure of how much the node in question appears to belong to a clique. A clique is loosely defined as a group of nodes that is highly connected internally but loosely connected to nodes outside the clique. Determining whether a node is a member of an ideal (or pure) clique is relatively easy—the difficult part is quantifying a node's membership in groups that are "almost" cliques.

To compute cliqueishness on a node N, the metric takes the ratio of how similar the distribution sequence of N is to two archetypical distribution sequences—one for a node in an ideal clique and the other for a node in an ideal nonclique (a K-regular graph). The closer N's distribution sequence is to the ideal clique sequence, the larger the numerator is and the closer $\text{cliq}(N)$ is to $+\infty$. The closer N's distribution sequence is to the ideal nonclique, the larger the denominator is and the closer $\text{cliq}(N)$ is to 0.0.

The distribution sequence for node N, $\gamma(N)$, is a sequence that contains in each element the number of nodes at exactly that distance from N. For a node with 1 adjacent node, 2 nodes at distance 1, 5 nodes at distance 3, and 12 nodes at distance 4, the distribution sequence would be $\gamma(N) = [1\ 1\ 2\ 5\ 12]$ (because one node, N, is at distance zero).

The archetypical distribution sequences depend on the number of nodes in the graph. The archetypical clique distribution sequence is $[1\ k\ 0\ 0\ 0\ \ldots]$ for a k-member clique. This signifies a completely connected clique with no connections to any other nodes. The archetypical nonclique distribution sequence is formed by two factors. In an infinitely large graph, the sequence would grow exponentially, that is, $[1\ k\ k^2\ k^3\ k^4\ \ldots]$. However, in a finite graph the sequence growth is limited by the fact that you will run out of nodes, so we model it as a binomial distribution with parameter 0.45.

Similarity between two distribution sequences is measured by the correlation between those sequences.

17.2.1.9 Homogeneity. Homogeneity measures how many nodes in a neighborhood share the same attribute values on selected attributes. As a simple example, a group of 10 nodes that all have the value "Colombian" for their "Nationality" attribute would be considered very homogeneous. A group of 10 nodes each with a different value for the "Nationality" attribute would be considered very heterogeneous. A group of 10 nodes with 5 "Colombian" and 5 "Chinese" would be somewhere in between. Attribute values are considered either equal or unequal—there is no notion of "almost equal" or "similar" values.

The homogeneity metric builds a histogram of attribute values in a neighborhood, normalizes it to sum to 1, and measures the heterogeneity of that histogram. That heterogeneity result is normalized over the maximum possible heterogeneity (a completely flat histogram) and then subtracted from 1 to yield the homogeneity measure, which will always fall between 0 (maximally heterogeneous) and 1 (maximally homogeneous).

The metric used to measure histogram heterogeneity can be any of entropy impurity, Gini impurity, or misclassification impurity. Each of these are a little different, primarily in how sharply they peak around the extreme homogeneity value of 1. The sharper the peak, the less "forgiving" the metric is about histograms that are not quite the ideal homogeneous group (all the attribute values are the same). They are listed here in decreasing order of the sharpness of that peak.

Misclassification Impurity: $\text{Mis}(H) = 1 - \max_i[H(i)]$

Gini Impurity: $\text{Gini}(H) = \frac{1}{2}(1 - \sum_i[H(i)H(i)])$

Entropy Impurity: $\text{Ent}(H) = -\sum_i[H(i)\ \log_2(H(i))]$

17.2.1.10 E-to-I Ratio. External-to-internal (E-to-I) ratio measures the number of edges from a node in group A to a node not in A versus the number of edges from a node in A to a node in A. As a simple example, a graph representing a small community "Smallville" might have several nodes. An edge from node A to node B represents a communication from person A to person B. Within this graph, there exists a smaller group, "farmers," that consists of some number of nodes less than the size of the entire community. The E-to-I ratio measures the amount of communication that occurs between farmers and nonfarmers versus the communication that occurs between farmers.

- An E-to-I ratio of 1:1 indicates a group that communicates as often with people outside the group as within.
- An E-to-I ratio of 0:1 indicates a group that communicates ONLY with members inside the group.
- An E-to-I ratio of 1:0 indicates a group that communicates ONLY with members outside the group.

- An E-to-I ratio of 0:0 indicates that members of the group do not communicate with anyone in the community.

17.2.1.11 Degree. The degree of a node is the number of edges on the node. Depending on its configuration settings, degree can consider only edges starting at this node (outgoing degree), only edges ending at this node (incoming degree), or both (undirected degree). Also, depending on its configuration settings, degree can be computed either considering or ignoring the weights on the edges.

17.2.1.12 Density. Density measures the ratio of existing edges to potentially existing edges. Undirected density is defined as

$$\text{Den}_{\text{undirected}}(G) = \frac{E(G)}{(N(G)(N(G) - 1))}$$

Directed density is defined as

$$\text{Den}_{\text{directed}}(G) = \frac{E(G)}{2N(G)(N(G) - 1)}$$

Note that multigraphs may have densities greater than 1.0 or could result in a density of 1.0 on a not completely connected graph.

17.2.1.13 Girth. The girth of a graph is the length of the shortest cycle (of length ≥ 3) in the graph. Girth is always computed directed.

17.2.1.14 Diameter. Diameter is one measure of the "size" of a graph. The diameter is defined as

$$D(G) = \max_i (\max_j (\text{distance}(N_i, N_j))$$

Diameter measures the maximum number of steps ever necessary to get from any node in the graph to any other node in the graph. Starting at any node, you are guaranteed that you can reach any other reachable node in at most $D(G)$ steps.

Unreachable nodes can be treated two ways: They can be ignored or they can be treated as if they were at distance N, depending on configuration settings. Treating them as if they were at distance N ensures that diameter decreases monotonically as you add edges to the graph. It also has the effect that the diameter for any graph where one or more nodes are unreachable from each other will always be N from each other will always be N.

17.2.1.15   Radius. Radius is another measure of the "size" of a graph. The radius is defined as

$$R(G) = \min_i (\max_j (\text{distance}(N_i, N_j)))$$

The radius implicitly defines which nodes in the graph are "central" nodes. A node is central when it can reach any other node in the graph in the minimum number of steps, compared to other nodes in the graph. Stated another way, any noncentral node will require more steps to reach some other node in the graph than a central node requires to reach any other node in the graph.

Unreachable nodes can be treated two ways: They can be ignored or they can be treated as if they were at distance N, depending on configuration settings. Treating them as if they were at distance N ensures that radius is decreased monotonically as you add edges to the graph.

17.2.1.16   Radiality and Integration. Radiality is a centrality measure that characterizes the degree to which a node's outward nominations reach out into the network. Similarly, integration is a centrality measure that characterizes the degree to which a node's inward nomination integrates it into the network. The radiality of agent A increases when the agents connected to A are not connected together [7, 18]. A server connected to many clients would be very radial since most clients connect only with the server for information and not with each other. For a given node v, radiality computes [3]:

$$R(v) = \sum_{t \in V} \frac{\text{diameter}(G) + 1 + \delta(v, t)}{n - 1}$$

and integration computes [3]:

$$I(v) = \sum t \in V \frac{\text{diameter}(G) + 1 + \delta(t, v)}{n - 1}$$

where

$$
\begin{aligned}
G &= \text{graph containing } v \\
V &= \text{set of vertices in the graph} \\
t &= \text{node in } V \text{ that is not } v \\
\delta(v, t) &= \text{shortest distance from node } v \text{ to node } t \\
\text{diameter}(G) &= \text{diameter of graph } G \text{ where diameter is} \\
&\quad \text{another social analysis metric} \\
n &= \text{number of nodes in } V
\end{aligned}
$$

17.2.1.17   Circumference. The circumference of a graph is the length of the longest cycle (of length ≥ 3) in the graph. Circumference is always computed directed.

17.3   GROUP DETECTION

In its usual context, group detection refers to the discovery of underlying organizational structure that relates selected individuals with each other. In a broader context, it refers to the discovery of underlying structure relating instances of any type of entity among themselves. We could be looking for firms banding together to form a cartel, mixed drinks that can be grouped according to ingredients, or basketball teams that may be grouped into conferences.

17.3.1   Group Detection Using SNA

17.3.1.1   Best Friends Group Detection Algorithm. We can think of the Best Friends Group Detection algorithm as a computationally efficient approach to detecting groups based upon concepts similar to the E-to-I ratio. The Best Friends Group Detection algorithm uses user-defined parameters to form an initial group. A group begins with a "seed" node, and nodes with associations/ties to the seed node are used to form an initial group. The group is pruned according to user-defined parameters. For example, nodes that are not well-connected to the group may be removed. Finally, the group must meet the requirements specified in a user-defined fitness function. If the group meets these requirements, it is stored as a potential group.

   Instructions regarding how to run Best Friends Group Detection either interactively or through TMODS batch scripts may be found in [15].

17.4   TERRORIST MODUS OPERANDI DETECTION SYSTEM

The Terrorist Modus Operandi Detection System (TMODS) automates the tasks of searching for and analyzing instances of particular threatening activity patterns. With TMODS, the analyst can define an attributed relational graph (ARG) to represent the pattern of threatening activity he or she is looking for. TMODS then automates the search for that threat pattern through an input graph representing the large volume of observed data. TMODS pinpoints the subsets of data that match the threat pattern defined by the analyst, transforming an arduous manual search into an efficient automated tool. It is important to note that the activity graph analyzed by TMODS can be derived from multiple sources, allowing for an analysis of information across several domains without extra effort. TMODS is a mature, distributed Java software application that has been under development by 21st Century Technologies since October 2001. Before we describe TMODS capabilities in further detail, a brief overview of graph matching is provided below.

   Graph matching is often known as subgraph isomorphism [10, 12] and is a well-known problem in graph theory. Informally, graph matching finds subsets of a large input graph that are "equivalent to" a pattern graph (threat pattern). These sections of the input graph, called matches, are "equivalent" in the sense that their nodes and edges correspond one-to-one with those in the pattern graph. Figure 17.2 illustrates a sample pattern graph and an inexact match highlighted within one

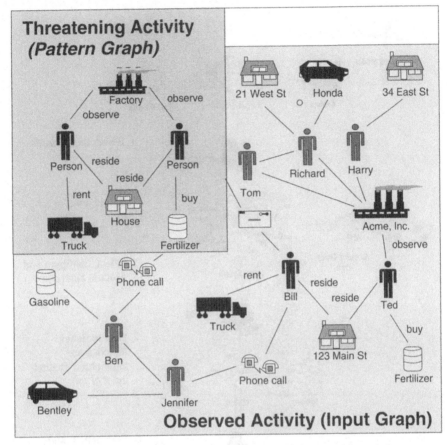

Figure 17.2. Inexact match to a pattern graph in an activity graph.

possible input graph. The match is inexact because there is a missing edge between Bill and Acme Inc.

As defined formally, subgraph isomorphism is an 'NP'-complete problem (no known polynomial-time solution exists) first identified in 1978. An adjacency matrix is one representation for an abstract graph. For a graph $G = (N, E)$ with nodes in set N and edges in set E, the adjacency matrix is a matrix where each element $G(i, j)$ equals 1 if there is an edge from node n_i to node n_j, and 0 otherwise. Inputs to the subgraph isomorphism problem are a pattern graph with adjacency matrix G_p, and an input graph with adjacency matrix G_i. The solutions to the problem (the matches) are expressed in the form of a permutation matrix, M, such that $G_p = MG_iM^T$. Because M is a permutation matrix, elements of M must be in the set $\{0, 1\}$ and M may have at most one nonzero element in each row and column. See Chapter 2 for an in-depth discussion of graph matching.

The TMODS turns its ability to solve the subgraph isomorphism problem into a powerful tool for intelligence analysts. Figure 17.3 shows a typical usage scenario

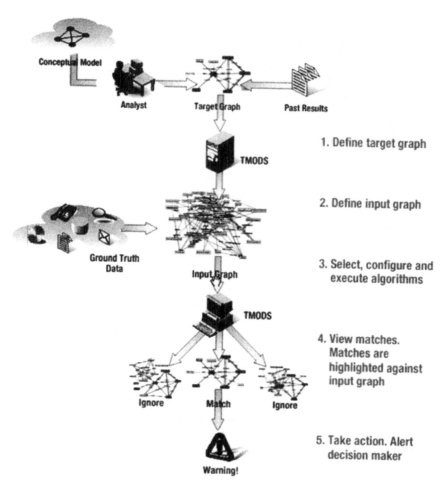

Figure 17.3.   Typical TMODS usage pattern.

for TMODS. First, the analyst defines a pattern graph to represent activities they consider threatening based on past results or an idea they have about a potential threat. The analyst also specifies a set of data sources from which TMODS imports and fuses information on observed activity to form the input graph. The analyst then selects, configures, and executes the search algorithms. TMODS highlights matches to the threat pattern against surrounding activity in the input graph. Finally, the analyst views the set of matches and decides (possibly based on further investigation) which need to be acted on.

In the asymmetric threat detection domain there is a core set of graph patterns that are of general interest to the counterterrorist (CT) analyst because they represent known methods of illicit operations. In many such transactions, there are a multitude of middlemen or "front organizations or persons" used to shield and protect the leaders involved. These graph patterns may be implemented as: (1) the movement of money or physical objects between several organizations and persons over a period of time, (2) the shipment of physical assets between multiple "front" countries to deflect suspicion of those assets, or (3) the churning of identifies for a specific individual. The key point behind the development of these patterns is that there are really only a handful of methods by which an organization or individual can operate covertly. These graph patterns are genuinely different than that of normal business activity. This is true not only for terrorist detection but for narcotics detection as well as transnational criminal activity. While TMODS will never detect all these patterns nor will the analyst be able to define all possible threat patterns in advance, the ability to detect unexpected variations of known threat patterns, within a reasonable period of time, is a significant achievement. Analysts have the ability to draw graph patterns that are stored within TMODS and as graph patterns are matched, alerts are sent to the analyst.

17.4.1 TMODS Extensions to Standard Graph Matching

The TMODS capabilities have been extended beyond approximations to the standard subgraph isomorphism problem to provide additional capabilities to the analyst.

17.4.1.1 Inexact Matching. Perhaps the most important extension to the basic subgraph isomorphism problem that TMODS supports is inexact matching. When analysts define a pattern they are looking for, TMODS will find and highlight activity that exactly matches that pattern. It will also, however, find and highlight activity that comes close to the pattern, but is not an exact match. The analyst can define a cutoff score that defines the quality of matches that he or she wishes to be informed about.

Inexact matching provides huge practical benefits in the world of intelligence analysis. First, analysts never have a perfect view of all activity. Inexact matching lets TMODS detect threats even when some activity is hidden. While not all the information about preparations for an attack may be present, often enough telltale signs are visible that an incomplete picture can still be constructed. Second, inexact matching lets the analyst find variations on earlier attack strategies—it prevents them from always fighting yesterday's war. Even if terrorist groups are varying their strategies (either intentionally or because of practical considerations), their overall plan will have to be at least somewhat similar to previous plans because the goals they are trying to achieve will be similar. Those similarities, combined with inexact matching, give TMODS enough of a match to alert analysts to variations of previous attacks. Finally, inexact matching insulates TMODS from analyst error, where some aspects of the pattern may simply be defined incorrectly. Inexact matching is an important capability for counterterrorism and related homeland security activities. Inexact matching finds sections of the graph that are "almost like" the target graph the analyst is searching for. Target graphs (or threat patterns) are representations of suspicious or illicit activity, potential terrorists, or criminal organizations. The input graph represents the real-world data where TMODS searches for patterns.

Graph matching can be compared to regular expression matching; where regular expressions search for patterns in text files, graph matching searches for patterns in abstract graphs. Inexact graph matching provides analogous capability to "wildcards" in regular expression matching and is vital for coping with real-world data.

We define a weighted graph edit distance [4] between a pattern and its returned match, which formally defines the similarity measure. Figure 17.4 shows an abstract representation of the progression from an exact match to inexact matches. Inexact matches may have missing edges or nodes, missing or incorrect attribute values, missing or incorrect edge labels, or any combination of these. Partial graph matches refer to a subset of the pattern graph identified within the input graph. Based on the weighted edit distance criteria, a 95% match in this case represents an inexact match on an edge attribute (e.g., in the target pattern we are looking for an "arrest" edge attribute and in the pattern returned by TMODS we find "killer" as the edge attribute).

However, the ranking is still high because we are able to find a connection between two entities. In the 90% match case, we can only indirectly (through an intermediate node) link a "weapon" to an individual, and thus the ranking is lower at 90%. TMODS has the capability to allow the analyst to assign ranking criteria for inexact matches.

17.4.1.2 *Multiple Choices and Abstractions.* TMODS patterns support variations on how the pattern may be instantiated. For example, a communication between two individuals might be achieved through email, a face-to-face meeting, or a phone call. For an analyst's particular purposes, the medium through which the communication occurred may not matter. TMODS lets the analyst define multiple alternative graphs for each pattern, called choices. Regardless of which choice graph was matched, the match is represented by its abstraction, which defines the salient aspects of the match (usually a subset of the full match information). In our previous example, for instance, the salient information is likely the identities of the two people involved. Regardless of how they communicated—be it email, meeting,

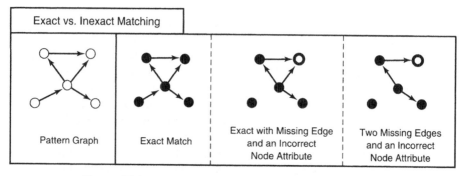

Figure 17.4. TMODS exact and inexact matching capability.

or phone call—the analyst can define the pattern such that their matches would be presented uniformly.

17.4.1.3 Hierarchical Patterns. TMODS allows analysts to define patterns that are built from other patterns. Rather than requiring analysts to describe their entire pattern in one graph, TMODS lets the analyst modularize their patterns.

Figure 17.5 shows an example from the counterterrorism domain, where the pattern actually represents a composition of multiple graph-based patterns. When combined with TMODS' capabilities for multiple choices and abstractions, the ability to match hierarchical patterns is extremely powerful. Take, for example, the problem of representing a "murder for hire" in a pattern. All such contract killings will have a customer, a killer, and a victim, but aside from that, there may be a wide variety of possible instantiations. The customer may plan the killing directly with the killer or the planning may be done through a middleman. Planning might entail a single communication or a number of communications, and each communication could be made through any number of media. The combination of multiple choices, abstractions, and the ability to define hierarchical patterns lets TMODS represent extremely complicated and varied activity patterns while controlling algorithmic complexity and the complexity presented to the user. In one specific model of contract killings, TMODS was able to represent 13 million different variations on how a contract kill could be executed. The top-level pattern had 11 subpatterns with a total of 31 choices, and each instantiation had an average of 75 total nodes. The use of hierarchical patterns changes an exponential complexity into an additive

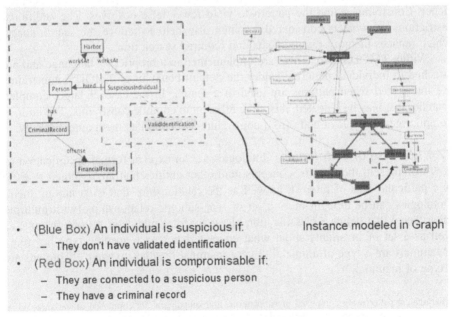

- (Blue Box) An individual is suspicious if:
  - They don't have validated identification
- (Red Box) An individual is compromisable if:
  - They are connected to a suspicious person
  - They have a criminal record

Instance modeled in Graph

Figure 17.5. TMODS exact and inexact matching capability.

complexity,[3] to great advantage. A discussion of specific algorithms used can be found in Section 17.4.2.

The TMODS is able to process approximately 900,000 graph elements per hour using an average desktop computer (i.e., 3-GHz Pentium 4). A previous benchmark exercise, in a specific problem domain, showed that TMODS can achieve 55% recall and 71.6% precision in search results at the stated processing rate; these results are without fusion algorithms. With fusion algorithms added, TMODS achieves 83.8% recall and 65% precision.

17.4.1.4 Constraints. TMODS lets the analyst define constraints between attribute values on nodes and edges. Constraints restrict the relationships between attributes of actors, events, or relationships. Using constraints, the analyst can restrict timing or relative ordering and represent stateful threat patterns. In a counterterrorism context, constraints can be used to specify restrictions like:

- The event must have occurred within 50 miles of Houston.
- The intelligence must have come from a reliable source in country X.
- The person observed must belong to the same organization as person Y.
- The money must come from an organization with at least $10M/year in income.
- The quantity must be larger than 10,000.

Support for constraints in TMODS patterns offer a number of advantages, all of which stem from the fact that constraints make the TMODS pattern representation richer. Constraints refine the pattern to yield fewer false positives. The additional restrictions on what is considered a match also help to reduce the search space, which reduces memory requirements and required search time.

Within TMODS, constraints are implemented in a proprietary language and are attached to individual patterns. Under the default behavior of TMODS, constraints are interpreted when patterns are used in a search. In the case of large, complex searches that may be repeated, it is more efficient to compile constraints. The built-in compiler generates Java code from constraints and compiles them automatically.

17.4.1.5 Use of Ontologies. Ontologies are an explicit formal specification of how to represent the objects, concepts, and other entities that are assumed to exist in a particular area of interest, as well as the relationships that exist among them. Ontologies can be thought of as a set of type-subtype relationships (with multiple inheritance) that define a classification hierarchy for a particular set of concepts, as well as a set of invariants about what characteristics each specific type has (e.g., "Mammals are a type of animal." "Every mammal breathes oxygen." "Humans are a type of mammal.").

---

[3]Hierarchical patterns are composed of subpatterns, and subpatterns are composed of variants. While the complexity is exponential in the number of nodes in the largest variant, it is only additive in the total number of variants.

The use of ontologies in TMODS allows for patterns that refer to general object or event types, instead of specific instances of those types. As a result, patterns can be written very generally, so that they cover many different types of threat or many variants of a threat. This lets the analyst define fewer patterns and makes those patterns more flexible. For performance purposes, 21st Century Technologies primarily uses domain-specific ontologies. Occasionally, a TMODS user can use an ontology previously defined for the domain of interest, but it is usually necessary to work with a domain expert to create an ontology that supports the application.

Figure 17.6 is an example from the counterterrorism domain. Instead of writing a pattern that only matches a specific combination of activities leading to compromised security (and having to write new patterns for each new combination), the analyst writes the single "generic compromised security" pattern. The criminal act and criminal record events in the pattern are tied to specific classes in a terrorism ontology. Any specialization of those classes will trigger a pattern match.

17.4.2 Algorithms

The field of graph isomorphism detection dates back to the early 1970s. The most relevant proposal for a general solution was that of Berztiss [1] in 1973; his algorithm performed favorably against Ullman's algorithm, but he only handled the case of full graph isomorphism. Ullman [17] published the seminal work in subgraph isomorphism in 1976. Ullman's algorithm dominated the field for two decades, and in fact most approaches since that time borrow heavily from Ullman, first in that they perform an exhaustive search that matches nodes one by one, and second in that they rely heavily on edge-based pruning to winnow out the search space. In a practical sense, because of the complexity of the problem, it is not currently expected that exhaustive algorithms such as those listed above will be able to solve the problem in reasonable time for particularly large graphs [8, 9, 14]. As a result, nonexhaustive, local search techniques are used in practical implementations in order to achieve results quickly. TMODS employs two major algorithms. The Merging Matches algorithm is an exhaustive search used for small input patterns. For larger patterns, TMODS uses a fast genetic search.

17.4.2.1 Merging Matches. Merging Matches is an algorithm developed by 21st Century Technologies to search for matches with small to medium size input patterns. It is a complete search; that is, the matches it finds are guaranteed to be the best matches possible. Any other potential matches must be worse.

Merging Matches operates by building up a list of potential matches. The initial entries to the list match one node from the input pattern to one node in the pattern graph. The algorithm then iteratively merges matches with each other until the best matches are found. At each step, any two entries that differ by exactly one node are combined if all the nodes they match are connected by an edge in the pattern graph. The resulting entry is guaranteed to be internally connected and to contain exactly one more node than either of its parents. These entries are then ranked according to their probability of matching the input graph. Entries that cannot

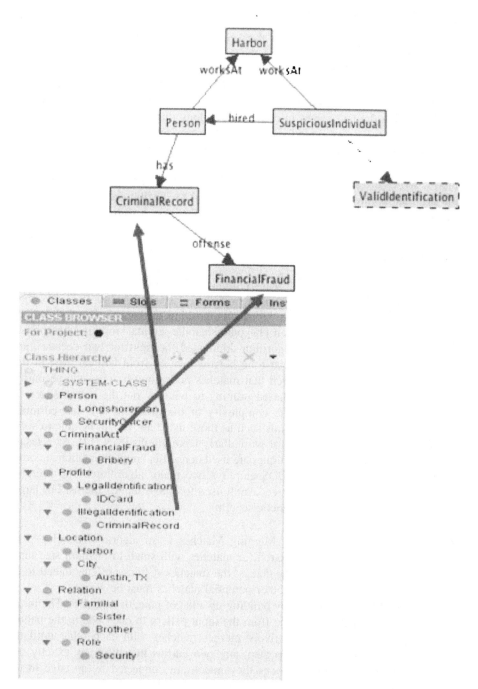

Figure 17.6.   Subset of a terrorist ontology.

constitute a reasonable match under the best circumstances are pruned, and the process of merging continues until the entire pattern is matched or no new candidate matches can be generated. At this point, if any of the entries are reasonably similar to the input pattern, the algorithm selects these entries and provides them to the analyst to examine.

As mentioned, Merging Matches is complete and executes fairly quickly for medium size input graphs. For large input graphs, however, the list of entries grows too rapidly for this approach to be practical, and other, less exhaustive methods are needed.

17.4.2.2 *Genetic Search.*

As mentioned, the Merging Matches algorithm is insufficient to handle searches involving large input patterns. Yet analysts may need to search for patterns that contain more than a dozen nodes in activity graphs with thousands of nodes and tens of thousands of edges. In such cases, TMODS leverages a Genetic Search algorithm to efficiently search for input patterns within the activity graph.

Genetic search algorithms are modeled on the Darwinian theory of evolution. This approach involves the creation of a population of representative members of the state space and the successive development of this population through processes modeling reproduction and natural selection. In TMODS, the initial population is generated by randomly selecting nodes in the pattern graph to create potential matches. These matches are then ranked by a fitness function to determine their closeness to the input pattern, and the members of the population with the highest scores reproduce with a randomly selected member of the general population. A new member is next added to the population by combining aspects of each parent. This new match may undergo stochastic variation, or mutation, in which a node in the match is altered so that it does not reflect the content of either of the parents. Finally, the algorithm ranks the new population, consisting of all the previous members plus newly generated children. The population size is maintained constant at this point by culling the lowest ranking members. The process of reproduction is then repeated through successive generations, until the average fitness of the population ceases to increase in a statistically significant manner. At this point, the best specimens are selected as the result of the search. While genetic algorithms are not complete and may not find every possible solution within the search space, they have proven effective at efficiently locating solutions to large problems within enormous search spaces.

The fitness measure used by TMODS calculates a probability of a match based on the weighted edit distance between a potential match and the search pattern. To compute the weighted edit distance, a cost is assigned to each node, edge, and attribute in the search pattern. The probability that a generated pattern matches the search input is a ratio between the total cost of all nodes, edges, and attributes and the cost of those elements that the potential match lacks. Some attributes in the search pattern may be designated as required, in which case the probability of a match is zero unless these necessary attributes are present.

17.4.2.3 Distributed Genetic Search. TMODS distributes its genetic search over several processes to increase the speed and completeness of the search. TMODS assigns a limited search domain to each process, potentially running on different computer systems. These limited domains require a particular node or nodes in the pattern graph to be included at a fixed position in any potential match. Each process then performs its own genetic search on the limited domain, and if any of its candidate matches are reasonably similar to the input pattern, the process returns its findings. Once all subordinate processes have reported their results, TMODS passes on all acceptable matches to the analyst.

This distributed approach performs several searches at once on multiple processors, delivering final results much faster. The efficiency of TMODS' search is further enhanced by using a reduced candidate set (the set of possible matches to a pattern), which allows the exclusion of many possible matches without an actual search, so that considerably less computation will have to be performed.

17.4.3 Other TMODS Capabilities

While the core technologies of TMODS are stable and mature, new facilities continue to be added to complement TMODS' ability to detect threatening patterns. Two major recent additions are statistical classification and analysis based on social network analysis (SNA) and pattern layering.

17.4.3.1 Social Network Analysis. Social network analysis (SNA) is a technique used by intelligence analysts to detect abnormal patterns of social interaction [3, 10, 12]. SNA arose out of the study of social structure within the fields of social psychology and social anthropology in the 1930s [15]. It can be used to investigate kinship patterns, community structure, or the organization of other formal and informal networks such as corporations, filial groups, or computer networks.

Social network analysts typically represent the social network as a graph. In its simplest form, a social network graph (sometimes called an "activity graph") contains nodes representing actors (generally people or organizations) and edges representing relationships or communications between the actors. Analysts then reason about the individual actors and the network as a whole through graph-theoretic approaches, including SNA metrics that take different values at different nodes in the graph. This approach is similar to and compatible with TMODS and other graph matching technology.

Central problems in social network analysis include determining the functional roles of individuals and organizations in a social network (e.g., gatekeepers, leaders, followers) and diagnosing network-wide conditions [19]. For example, studies of interlocking directorships among leading corporate enterprises in the 1960s have shown that banks were the most central enterprises in the network.

17.4.3.2 TMODS SNA Capabilities. TMODS has been developed to identify characteristics that differentiate normal social networks from those used for illicit activity. Many of these characteristics are quantifiable with multivariate SNA metrics. For example, in legitimate organizations, there are usually many redundant communication paths between individuals. If an individual wants to get information to A, he may contact her directly, or go through B or C. Covert organizations generally do not have many paths of communications flow between individuals. Redundant paths lead to increased risk of exposure or capture. Redundancy is one of many properties that TMODS can recognize, represent, and exploit.

The TMODS can compute the functional roles of individuals in a social network and diagnose network-wide conditions or states. Different individuals in a social network fulfill different roles. Examples of intuitive roles include leaders, followers, regulators, "popular people," and early adopters. Other roles that are less intuitive but are still common in both normal and covert/abnormal social networks include bridges, hubs, gatekeepers, and pulse takers.

Much of our work in SNA metrics focuses on distilling complex aspects of a communications graph into simple numerical measures, which can then be used with traditional statistical pattern classification techniques to categorize activity as threatening or nonthreatening. TMODS can then measure communications efficiency and redundancy, identify which participants are "central" to the communication structure, and single-out those groups of participants that form tightly knit cliques. SNA metrics are a unique approach to quantifying and analyzing human interpersonal (or online) communication that provide TMODS with novel detection abilities.

17.4.3.3   *Event Detection via Social Network Analysis.* The unifying goal of the various applications of SNA described above is to quickly differentiate between threatening and nonthreatening activity. These approaches use abnormal group structure, behavior, and communication patterns as the indicators of threatening activity. In some cases, static SNA is sufficient to characterize a group's structure as normal or abnormal. In other cases, dynamic SNA is required to differentiate between normal and abnormal group structure and behavior, based on a group's evolution over time. These approaches require user-specified a priori models (which are often but not always available) of both normal and abnormal behavior; equivalently, they require labeled training data. This places them in the general category of supervised learning algorithms. When we do not have a sufficient amount of labeled training data, we must use unsupervised learning techniques, which perform the task of event detection (also called anomaly detection).

The goal of SNA event detection is to flag any significant changes in a group's communication patterns, without requiring users to specify a priori models for normal or abnormal behavior. Where supervised learning techniques perform the task of "Tell me which groups behave in this particular way," SNA event detection performs the task of "Tell me if any group begins to significantly change its behavior." As described in the sections above and in many counterterrorism studies, the lead-up to major terrorist events is almost always preceded by significant changes in the threat group's communication behavior. SNA event detection finds these internal indicators and gives the intelligence analyst the chance to alert decision makers in time for preemptive action.

21st Century Technologies has developed an initial prototype of our SNA event detection capability, as well as many of the building blocks required to move that capability to the next level. The application of SNA event detection to predicting terrorist activity is very similar to the application of SNA event detection to detecting anomalous digital network behavior.

The SNA event detection begins by tracking the SNA metric values of various groups over time, just as it is done for dynamic SNA. The left half of Figure 17.7 shows two visualizations of SNA metric values gathered over time. The figure shows a time-series view and a scatterplot view.

As information is gathered over time, TMODS automatically builds its own models of normal behavior for the groups being observed.

The right half of Figure 17.7 shows observed SNA metric values. This is TMODS' model of behavior. The figure also shows five points (circled) that do not fit with the other observed behavior. These are the anomalies we are looking for—previously unmanifested communication patterns that may indicate internal threats. There are a number of approaches for deciding if an alert should be raised, given observations and a learned behavior model.

Supervised learning algorithms are used when we have labeled training data, and unsupervised learning algorithms are used when we have unlabeled training data. A third approach is called "learning with a critic." As TMODS raises detections via the unsupervised event detection algorithms, the analyst can make a case-by-case decision on whether the detection is a true or false positive. This feedback can be incorporated in the event detection algorithms to tune or train them to improve their results. Ultimately, this set of case-by-case decisions and corresponding evidence can constitute a labeled training dataset, which can be used by our supervised learning methods.

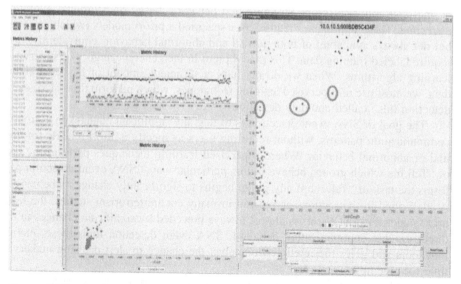

Figure 17.7. SNA event detection: (left) SNA metric history and (right) anomaly detection.

17.5 COMPUTATIONAL EXPERIMENTS

In this section, we will present a few illustrative examples of the use of Best Friends Group Detection.

The first example is determining conference membership of NCAA basketball teams based upon the schedule of games. Teams play more often within their own conferences than they do against teams outside the conference. This sort of problem is extremely well suited to an SNA-based algorithm because the connections between teams are explicitly related to the conference structure. In Figure 17.8, we can see that the subgraph made up of a conference's members is quite dense, but there are few links between conferences.

In Figure 17.9, we can see that the detected group of papers all are related to gravitation. However, the algorithm does not look at titles of papers or keywords. It strictly concentrates on the edge structure of the data. The edges involved are citations. When one paper cites another, an edge is present in the graph. Intuitively, it makes sense that many of the citations of a particular paper would point to papers with a closely related topic. In effect, citations are acting as a proxy for topic.

As with many structure-based methods, Best Friends Group Detection does not perform as well when observability is limited. In other words, if important information connecting entities is not included in the graph, then the algorithm cannot take advantage of that information. Data can be omitted due to data entry

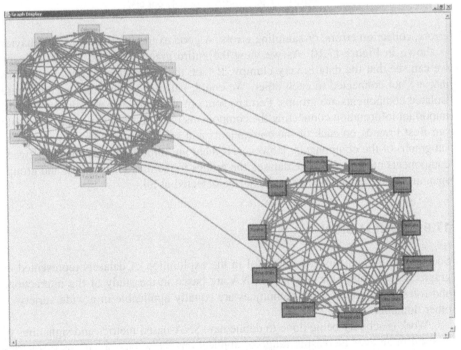

Figure 17.8. NCAA basketball conferences.

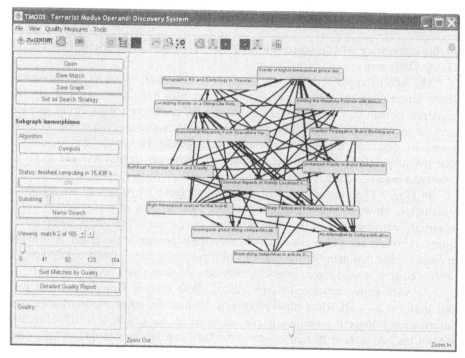

Figure 17.9. Group of gravity papers.

errors, collection errors, or sampling errors. A good example of this sort of behavior is shown in Figure 17.10. As we view the entire graph from the 30,000-ft level, we can see that the data is very clumpy. If fact, there are many graph components that are not connected to each other. We could, naively, assume that each of these isolated components are groups. Perhaps some of them are, but we could also have important information connecting the components missing from the graph. We could run Best Friends on each of the components and see if they contain any groups as subgraphs of the components. However, if there are important connections between components missing from the dataset, the groups found may be spurious, and groups spanning multiple components will not be detected at all.

17.6 CONCLUSION

Social network analysis can be useful in the exploration of datasets represented as graphs. Even though the origins of SNA are based in the study of the interactions and roles of individuals, the techniques are equally applicable in a wide variety of other domains.

Work is actively being done to define new SNA-based metrics and signatures to be used in an ever-increasing range of applications. New interpretations of metrics

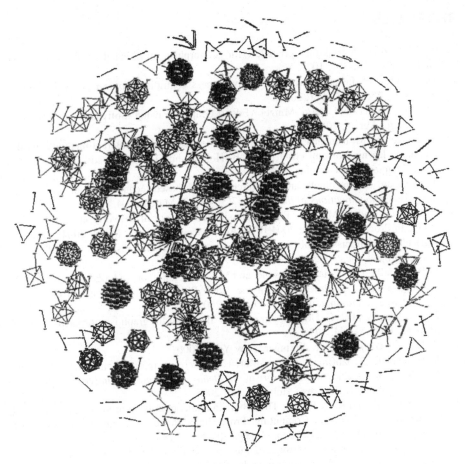

Figure 17.10. Zdata overview.

for increased understanding of the new domains are constantly being fleshed out. New representations of the underlying graphs are enabling the development of new algorithms, effectively addressing the requirements of larger and more complex datasets. This evolution will allow SNA techniques to remain relevant for a long time to come.

17.6.1 Acknowledgment

This research was managed by the Air Force Research Laboratory in Rome, New York. The views and conclusions contained in this document are those of the authors and should not be interpreted as necessarily representing the official policies, either expressed or implied, of Rome Laboratory, the United States Air Force, or the United States government.

REFERENCES

1. A. T. Berztiss. A backtrack procedure for isomorphism of directed graphs. *Journal of ACM*, 20(3):365–377, 1973.
2. U. Brandes. A faster algorithm for betweenness centrality. *Journal of Mathematical Sociology*, 25(2):163–177, 2001.
3. Michael Jünger and Petra Mutzel (Eds.): Graph Drawing Software, pp. 321–340. © Springer-Verlag, 2004.
4. H. Bunke and M. Neuhaus. Graph matching—exact and error-tolerant methods and the automatic learning of edit costs. In L. B. Holder and D. J. Cook, eds. *Mining Graph Data*, Chapter 2, this volume.
5. T. Coffman, S. Greenblatt, and S. Marcus. Graph-based technologies for intelligence analysis. *Communications of the ACM*, 47(3):45–47, March 2004.
6. T. Coffman and S. Marcus. Pattern classification in social network analysis: A case study. In Proceedings of the 2004 IEEE Aerospace Conference, March 2004.
7. Agriculture Commission of the European Communities and Fisheries. Final report: Images project.
8. L. P. Cordella, P. Foggia, C. Sansone, and M. Vento. An improved algorithm for matching large graphs. In Proceedings of the 3rd IAPR-TC-15 International Workshop on Graph-based Representations, pp. 149–159, 2001.
9. D. G. Corneil and C. C. Gotlieb. An efficient algorithm for graph isomorphism. *Journal of ACM*, 17(1):51–64, 1970.
10. R. Diestel. *Graph Theory*. Springer, New York, 2000.
11. S. Greenblatt, T. Coffman, and S. Marcus. Emerging information technologies and enabling policies for counter terrorism. In *Behavioral Network Analysis for Terrorist Detection*. Wiley-IEEE Press, Hoboken, NJ, 2005.
12. A. S. LaPaugh and R. R. Rivest. The subgraph homeomorphism problem. Annual ACM Symposium on Theory of Computing, Proceedings of the tenth annual symposium on Theory of Computing, San Diego, CA pgs. 40–50, 1978.
13. S. Marcus and T. Coffman. Terrorist modus operandi discovery system 1.0: Functionality, examples, and value. 21st Century Technologies Internal Publication, January, Austin, TX, 2002.
14. B. T. Messmer. Efficient graph matching algorithms for preprocessed model graphs. Ph.D. thesis, Institut fur Informatik und Angewandte Matheatik, Universitat Bern, Switzerland, 1995.
15. M. Moy. Using tmods to run the best friends group detection algorithm. 21st Century Technologies Internal Publication, May 2005.
16. J. Scott. *Social Network Analysis: A Handbook*, 2nd ed. SAGE, London, 2000.
17. J. R. Ullman. An algorithm for subgraph isomorphism. *Journal of the ACM*, 23(1):31–42, 1976.
18. T. W. Valente and R. K. Foreman. Integration and radiality: Measuring the extent of an individual's connectedness and reachability in a network. *Social Networks*, 20:89–105, 1998.
19. S. Wasserman and K. Faust. *Social Network Analysis: Methods and Applications*. Cambridge University Press, New York, NY, 1994.

INDEX

Mining Graph Data, Edited by Diane J. Cook and Lawrence B. Holder
Copyright © 2007 John Wiley & Sons, Inc.

CPSIA information can be obtained at www.ICGtesting.com
Printed in the USA
BVOW08*1436290714

360794BV00009B/159/P